Electricity

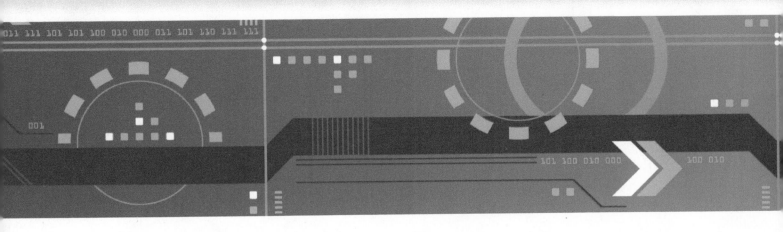

Eighth Edition

Electricity
Principles & Applications

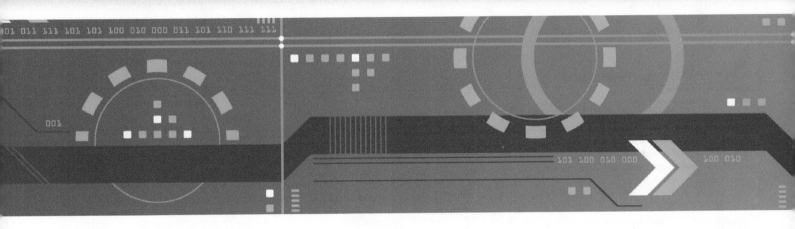

Richard J. Fowler

Connect
Learn
Succeed™

ELECTRICITY: PRINCIPLES & APPLICATIONS, EIGHTH EDITION

Published by McGraw-Hill, a business unit of The McGraw-Hill Companies, Inc., 1221 Avenue of the Americas, New York, NY, 10020. Copyright © 2013 by The McGraw-Hill Companies, Inc. All rights reserved. Printed in the United States of America. Previous editions © 1979, 1984, 1989, 1994, 1999, 2003, and 2008. No part of this publication may be reproduced or distributed in any form or by any means, or stored in a database or retrieval system, without the prior written consent of The McGraw-Hill Companies, Inc., including, but not limited to, in any network or other electronic storage or transmission, or broadcast for distance learning.

Some ancillaries, including electronic and print components, may not be available to customers outside the United States.

This book is printed on acid-free paper.

4 5 6 7 8 9 0 QVS/QVS 19 18 17 16 15

ISBN 978-0-07-337376-8
MHID 0-07-337376-1

Vice president/Director of marketing: *Alice Harra*
Publisher: *Scott Davidson*
Sponsoring editor: *Sarah Wood*
Director, digital products: *Crystal Szewczyk*
Development editor: *Vincent Bradshaw*
Marketing manager: *Kelly Curran*
Digital development editor: *Kevin White*
Director, Editing/Design/Production: *Jess Ann Kosic*
Project manager: *Jean R. Starr*
Buyer II: *Laura M. Fuller*
Senior designer: *Srdjan Savanovic*
Senior photo research coordinator: *Jeremy Cheshareck*
Photo researcher: *Jennifer Blankenship*
Manager, digital production: *Janean A. Utley*
Cover and interior design: *Jesi Lazar*
Typeface: *11/13 Times LT Std*
Compositor: *MPS Limited, a Macmillan Company*
Printer: *Quad/Graphics*
Cover credit: *©2008 Ian Grainger, Getty Images*

Credits: The credits section for this book begins on page 517 and is considered an extension of the copyright page.

Library of Congress Cataloging-in-Publication Data

Fowler, Richard J.
  Electricity : principles & applications / Richard J. Fowler. — 8th ed.
    p. cm.
  Includes index.
  ISBN-13: 978-0-07-337376-8 (alk. paper)
  ISBN-10: 0-07-337376-1 (alk. paper)
  1. Electricity. I. Title.
QC523.F75 2013
621.3—dc23

                             2011020111

The Internet addresses listed in the text were accurate at the time of publication. The inclusion of a Web site does not indicate an endorsement by the authors or McGraw-Hill, and McGraw-Hill does not guarantee the accuracy of the information presented at these sites.

www.mhhe.com

Contents

Editor's Foreword

The McGraw-Hill Career Education Trade and Technology list has been designed to provide entry-level competencies in a wide range of occupations in the electrical and electronic fields. It consists of coordinated instructional materials designed especially for the career-oriented student. A textbook, an experiments manual, and an instructor productivity center support each major subject area covered in the series. All of these focus on the theory, practices, applications, and experiences necessary for those preparing to enter technical careers.

There are two fundamental considerations in the preparation of a text like *Electricity: Principles and Applications*: the needs of the learner and needs of the employer. This text meets these needs in an expert fashion. The authors and editors have drawn upon their broad teaching and technical experiences to accurately interpret and meet the needs of the student. The needs of business and industry have been identified through personal interviews, industry publications, government occupational trend reports, and reports by industry associations.

The processes used to produce and refine the series have been ongoing. Technological change is rapid and the content has been revised to focus on current trends. Refinements in pedagogy have been defined and implemented based on classroom testing and feedback from students and instructors using the series. Every effort has been made to offer the best possible learning materials.

These include animated PowerPoint presentations, circuit files for simulation, a test generator with correlated test banks, dedicated websites for both students and instructors, and basic instrumentation labs. All of these are well coordinated and have been prepared by the author.

The widespread acceptance of *Electricity: Principles and Applications* and the positive responses from users confirm the basic soundness in content and design of all of the components as well as their effectiveness as teaching and learning tools. Instructors will find the texts and manuals in each of the subject areas logically structured, well-paced, and developed around a framework of modern objectives. Students will find the materials to be readable, lucidly illustrated, and interesting. They will also find a generous amount of self-study materials, review items, and examples to help them determine their own progress.

Both the initial and the ongoing success of this text and others within the McGraw-Hill Career Trade and Technology list are due in large part to the wisdom and vision of Gordon Rockmaker, who was a magical combination of editor, writer, teacher, electrical engineer, and friend. Gordon has retired but he is still our friend. The publisher and editor welcome comments and suggestions from instructors and students using this series.

Charles A. Schuler, Project Editor

Basic Skills in Electricity and Electronics

Charles A. Schuler, Project Editor

New Editions in This Series
Electricity: Principles and Applications, Eighth Edition, Richard J. Fowler
Electronics: Principles and Applications, Eighth Edition, Charles A. Schuler

Preface

The eighth edition of *Electricity: Principles and Applications* utilizes the same philosophy of learning and teaching as was used in the previous editions. It assumes that the student has very limited or no knowledge of electrical theory and principles. Mastery of the material in this text will aid the student desiring an entry-level job in an occupation requiring an understanding of electricity. Mastering this material will also provide the student with the knowledge and skills needed to pursue further education in electricity and electronics.

This text has been written so that students with limited math and reading skills can gain a clear understanding of electricity and electric devices. Concepts are explained and developed by using a number of short, simple sentences rather than one long, complicated, convoluted sentence. It is never assumed that something is intuitively obvious.

Any mathematics beyond simple arithmetic is carefully explained and illustrated with examples before it is used to solve electric circuit problems. Although simultaneous equations, matrixes, and determinants, are introduced in Chap. 6, they are defined and explained in some detail before actually being used. Similarly, the elements of trigonometry used with AC circuits are fully explained and illustrated with examples before being applied to AC circuit problems.

Chapters 1 through 6 of this text are devoted, in general, to the fundamentals of direct current, and Chaps. 8 through 13 focus on subjects usually associated with alternating current. This arrangement provides students with balanced coverage of basic concepts. The transition from direct current to alternating current through the study of magnetism and electromagnetism is distinct enough to allow use of the material in a traditional dc/ac sequence. However, all the material is structured to provide a unified introduction to the broad subject area called *electricity*.

Chapter-by-Chapter Changes to This Edition

Chapter 1

- Expanded the section on "efficiency" to include information on neon lightbulbs and LEDs.

- Added a new example that examines the efficiency of a complete system (i.e., a battery and light bulb circuit). The new Self-Test question 11, and the answer to question 11 at the end of the chapter, follow up on working with system efficiency.
- Added new Self-Test and Chapter Review questions and problems.

Chapter 2

- Changed the concluding paragraph of the "current in a vacuum" section to emphasize that devices that relied on thermionic emission and current in a vacuum have mostly been replaced by LCDs and solid-state devices.
- Added discussion of when to italicize a symbol (abbreviation) used in formulae.
- Added more examples that show how to use "powers of 10." Example 2-9 shows how to express a base 10 number of less than one in the power-of-ten format. Example 2-10 shows how to add, subtract, and divide numbers expressed in powers-of-ten. Also added Self-Test question 59 and its answer to give more practice on the procedures shown in the two added examples.
- Added new Self-Test and Chapter Review questions and problems.
- Added to Chapter Summary and Review.

Chapter 3

- Modified the section on circuit essentials.
- Clarified the discussion on calculating power.
- Added more Chapter Review questions and problems.

Chapter 4

- Added Self-Test and Chapter Review questions.
- Added a discussion on testing sealed lead-acid cells.
- Added a section on fuel cells.
- Added to the discussion of LEDs.
- Added an example for switch ratings.

Chapter 5

- Added new examples that:

  1. Require solving a series circuit containing a rheostat.
  2. Require determining the internal resistance of a battery.
  3. Require determining currents in a parallel circuit.
  4. Require determining the currents in a loaded zener-diode regulator circuit.

- Extended the discussion of "rules of thumb" for estimating current.
- Added material that illustrates the advantage of using the current divider formula.
- Added Chapter Review problems.
- Added questions and problems to the *Experiments Manual* Chapter 5 test.

Chapter 6

- Added a new example illustrating the use of simultaneous and loop equations.

Chapter 7

- Added a new figure showing flux pattern around a MIG-arc-welder cable carrying 60 A to 120 A.
- Replaced three figures with new pictures to better show flux patterns under a variety of conditions.
- Added an example showing how to calculate relative permeability.
- Added more Chapter Review problems.

Chapter 8

- Added new examples in the Three-Phase Alternating Current section.
- Added more Chapter Review problems.

Chapter 9

- Added more Chapter Review problems.

Chapter 10

- Expanded the introduction to the chapter to include use of large capacitors for PF correction.
- Added examples emphasizing C, V, and Q relationships and RC time constants.

Chapter 11

- Added an example on calculating the inductance of an inductor.
- Expanded the section on "Type of Inductors."

- Added the inductance formula to the Related Formulas.
- Added more Chapter Review Problems.

Chapter 12

- Added examples showing how to calculate coefficient of coupling, mutual inductance, and transformer ratings.
- Expanded discussion of transformer "Windings in Parallel."
- Added formulas to Related Formulas.
- Added more Chapter Review problems.

Chapter 13

- Added to the discussion of example 13-5 to show other ways to calculate the cos θ.
- Added an example dealing with Z and BW of a series resonant circuit.
- Added more Chapter Review problems.

Chapter 14

- Added new examples dealing with temperature rating of a motor and the R of a cable connecting to a motor.
- Expanded the discussion of reversing rotation of single-phase and three-phase motors.
- Added seven new problems to the Chapter Review problems.

Chapter 15

- Expanded the explanation of how a DMM measures current.
- Added an example dealing with ratings of meter movements.
- Expanded the discussion of sensitivity ratings of ac and dc voltmeters.
- Added more problems to the Chapter Review problems.

Chapter 16

- Re-emphasized the color and proper use of the neutral conductor.
- Added an example illustrating the current flow in each conductor when a short develops.
- Expanded the discussion on AFCI protection.

Appendix K

- A new appendix has been added that discusses renewable resources.

New Multisim-11 files have been created and added for the chapters. The files are located on the CD-ROMs for the textbook and the *Experiments Manual,* as well as on the Online Learning Center.

Additional Resources

An *Experiments Manual* designed specifically for this textbook is available. It contains a comprehensive test, a wide variety of lab exercises and experiments, and additional problems for each chapter in the textbook. Some experiments must be done with physical (real) components, some must be done with electronic-circuit simulation software, and some can be done with either. Multisim files are provided for those experiments that can be completed with simulation software. These files are located on the bound-in CD-ROM, along with a Multisim Primer tutorial (written by Patrick Hoppe of Gateway Technical College) explaining how to get up and running with the program.

The **Student CD-ROM,** also provided with the textbook, provides Multisim simulation files for many of the circuits in the textbook. They are arranged by chapter for easy reference.

The **Online Learning Center (OLC)** website (Student Center) provides extra review questions, links to industry sites, assignments, and tests. The student will also find the *Circuit Solver* program on the OLC. This program does not duplicate any of the typical simulation programs. It will help you select the value of a specific component needed to provide the results you desire from a circuit you are designing.

The Instructor Center of the **OLC** provides a wide selection of information for the instructor. It contains the following features:

- An Instructor's Manual that includes a list of the parts and equipment needed to perform lab experiments, Learning Outcomes for each chapter, answers to chapter review questions and problems, detailed instructions for seven projects, and more
- Twelve simulated instrumentation labs
- Instructor PowerPoint slide show for each chapter in the textbook. These shows use the student PowerPoint slides plus some instructional slides added to each chapter.
- Test generator software with a test bank for each chapter

About the Author

Richard J. Fowler has spent four years in the USAF testing, repairing, and maintaining radio and navigation equipment. He has taught electricity and electronics for 30 years—one year in a public high school and 29 years in three universities. He has an Ed.D. from Texas A&M University (1965) and has published two textbooks, two laboratory manuals, and one chapter in a professional yearbook.

Walkthrough

Electricity: Principles and Applications was written to be a concise and practical introduction to this subject. The textbook's easy-to-read style, color illustrations, clear explanations, and simple math problems make it ideal for students who want to learn the essentials of electricity and apply them to real job-related situations.

I think the images contained in the text really bring the document to life for the students. I think this motivation is most important . . . just to get the students to open the book and look at the content. Additionally, the content itself is most important. It is correct, well presented and written just for the grade level that I am most interested in. I choose a textbook based upon its content, its content being correct and well presented. This text provides all that.

—Kenneth P. De Lucca
Millersville University of Pennsylvania

Learning Outcomes

This chapter will help you to:

1-1 *Clearly differentiate* between work and energy.

1-2 *Use* base units for specifying and calculating energy and work.

1-3 *Understand* energy conversion.

1-4 *Understand* and calculate efficiency of energy conversions.

1-5 *Explain* the characteristics of the particles of an atom.

Each chapter starts with **Learning Outcomes** that give the reader an idea of what to expect in the following pages, and what he or she should be able to accomplish by the end of the chapter. These outcomes are now distinctly linked to the chapter subsections.

The text contains numerous worked-out **Examples** that emphasize the importance of using a systematic, step-by-step approach to problem solving. Students should make rough approximations of the answer to a problem before doing the detailed math necessary to find the exact answer. This can detect major mathematical errors that are sometimes overlooked.

EXAMPLE 1-1

If it requires a steady force of 150 newtons to pull a boat, how much work is required to pull a boat 8 meters?

Given:	Force = 150 newtons
	Distance = 8 meters
Find:	Work
Known:	Work (W) = force × distance
	1 newton-meter = 1 joule (J)
Solution:	Work = 150 newtons
	× 8 meters
	= 1200 newton-meters
	= 1200 joules
Answer:	Work = 1200 joules
	W = 1200 J

EXAMPLE 1-2

It requires 500 joules of energy and 100 newtons of force to move an object from point A to point B. What is the distance between point A and point B?

Given:	Energy = 500 joules
	Force = 100 newtons
Find:	Distance
Known:	W = force × distance, and by rearranging
	Distance = $\dfrac{W}{\text{force}}$
	1 newton-meter = 1 joule
Solution:	Distance = $\dfrac{500 \text{ newton-meters}}{100 \text{ newtons}}$
	= 5 meters
Answer:	Distance = 5 meters

Self-Test

Answer the following questions.

1. Define work and energy.
2. What is a base unit?
3. What is the base unit of energy?
4. How much energy does it take to push a car 120 meters with a steady force of 360 newtons?

Answers to Self-Tests

1. Work is a force moving through a distance; energy is the ability to do work.
2. A base unit is a term that is used to specify the amount of something.
3. joule (J)
4. **Given:**   Distance = 120 meters
                  Force = 360 newtons
   Find: Work
   Known: Work = force × distance
   Solution: Work = 360 newtons × 120 meters
   Answer: Work = 43,200 joules

Self-Tests serve as positive reinforcement for material that students already know, and they help students identify areas that need further study. Students may check their responses in the *Answers to Self-Tests* sections at the end of each chapter.

Key Terms are carefully defined and explained in the text and listed in the margins so students can easily find them.

Electron
Atoms
Proton
Neutron
Nucleus

Molecule

1-5   Structure of Matter

All matter is composed of *atoms*. Atoms are the basic building blocks of nature. Regardless of their physical characteristics, glass, chalk, rock, and wood are all made from atoms. Rock is different from wood because of the type of atoms of which it is composed.

The smallest particle of a substance that still has all of its characteristics is called a *molecule*. A molecule consists of two or more atoms. If a molecule of chalk is divided into smaller parts, it is no longer chalk.

There are more than 100 different types of atoms. Matter composed of a single type of

ABOUT ELECTRONICS

Exercise Caution with Batteries Although low-voltage, high-current sources such as lead-acid batteries cannot deliver an electric shock, they can cause severe burns when shorted by jewelry such as rings and watchbands.

History of Electronics and *About Electronics* add historical depth to the topics and highlight new and interesting technologies or facts.

 History of
Electronics

James Prescott Joule
The SI (*Système Internationale*) unit of measure for electric energy is the joule (J), named for James Prescott Joule (1 joule is equal to 1 volt-coulomb).

Internet Connection

Visit the website for the International Brotherhood of Electrical Workers (IBEW) for career and other information.

You May Recall sections connect the text on the page with concepts that were previously introduced in the textbook, while *Internet Connections* encourage students to do online research on certain topics.

Critical facts and principles are reviewed in the *Summary and Review* section at the end of each chapter.

Chapter 1 Summary and Review

Summary

1. Energy is the ability to do work.
2. The symbol for energy is W.
3. Electricity is a form of energy.
4. Work equals force times distance.
5. The joule is the base unit of energy.
6. The joule is the base unit of work.
7. The symbol for the joule is J.
8. Under ordinary conditions, energy is neither created nor destroyed.
9. All matter is made up of atoms.
10. Hydrogen has only one electron and one proton. All other atoms contain electrons, protons, and neutrons.
11. Protons and neutrons are found in the nucleus of the atom.
12. Electrons have a negative electric charge.
13. Protons have a positive electric charge.
14. Neutrons do not have an electric charge.
15. Atoms have an equal number of protons and electrons.

16. Valence electrons are found in the outermost shell of the atom.
17. Valence electrons can become free electrons when their energy level is raised.
18. Negative ions are atoms that have attracted and captured an extra (free) electron.
19. Positive ions are atoms that have given up an electron.
20. Static charges are the result of an object's possessing either more or fewer electrons than protons.
21. A positive charge means a deficiency of electrons.
22. A negative charge means an excess of electrons.
23. Like charges repel each other.
24. Unlike charges attract each other.
25. A charged object can attract an uncharged object.
26. When an uncharged object touches a charged object, it becomes charged.

The chapters (are) clearly and logically organized, building from one concept to another.

—James Fischer
San Joaquin Valley College

All of the important chapter formulas are summarized at the end of each chapter in **Related Formulas**. **Chapter Review Questions** also are found at the end of each chapter, as are **Chapter Review Problems** and **Critical Thinking Questions**.

Related Formulas

$$\text{Percent efficient} = \frac{W_{out}}{W_{in}} \times 100$$

Chapter Review Questions

For questions 1-1 to 1-7, determine whether each statement is true or false.

1-1. Opposite electric charges repel each other. (1-6)
1-2. Free electrons are at a higher energy level than valence electrons are. (1-8)
1-3. The same base unit is used for both work and energy. (1-2)
1-4. Atoms have twice as many protons as electrons. (1-6)
1-5. An object that has a static charge can possess either more or fewer electrons than protons. (1-10)

1-6. The second shell, or orbit, of an atom can contain any number of electrons. (1-5)
1-7. If an electric device is less than 100 percent efficient, part of the energy provided to it is destroyed. (1-3)

For questions 1-8 to 1-14, choose the letter that best completes each sentence.

1-8. The symbol for energy is (1-1)
   a. *D*
   b. *W*

Chapter Review Problems

1-1. What is the efficiency of a motor which requires 914 base units of electric energy to produce 585 joules of mechanical energy? (1-4)
1-2. How much work is done when an object is moved 3 meters by applying a force of 80 newtons? (1-2)
1-3. How much energy is involved in problem 1-2? (1-2)

1-4. How much energy is required by an electric lamp that is 18 percent efficient and provides 5463 J of light energy? (1-4)
1-5. A motor is 70 percent efficient and requires 1200 base units of energy. How much energy does it provide? (1-4)
1-6 How much heat energy is produced by a light bulb that requires 400 joules of electric energy to produce 125 joules of light energy? (1-3)

Critical Thinking Questions

1-1. Discuss the probable changes in your lifestyle if the electric service to your home were disconnected for one week.
1-2. In addition to decreasing costs, why is it important to increase the efficiency of electric devices as much as is practical?
1-3. large, continuously operated motors more efficient than small, inter-

1-4. Is the low efficiency of the lights in our homes a greater disadvantage during the winter months or the summer months? Why?
1-5. Many electric devices could be designed and constructed to operate more efficiently. Why aren't they?

Acknowledgments

Without the help, support, and encouragement of many people this book would never have materialized. Deep appreciation goes to Arla, my wife, who has proofread and typed everything from the original manuscript through the eighth edition. The author thanks the editors, project manager, and production staff of McGraw-Hill for their skillful and precise work in publishing the book. Finally, the author is grateful for the ideas, suggestions and criticisms provided by the following professionals who reviewed the manuscript for this edition:

James Rueckert
Redstone College

Edward S Abrasley, Jr
Chattahoochee Technical College

Ahmad Hemami
Iowa Lakes Community College

Richard Fornes
Johnson College

Jonathan White
Harding University

Kenneth Markowitz
New York City College of Technology

Kevin Maki
Fond du Lac Tribal and Community College

Steven Hintz
Wisconsin Indianhead Technical College

JC Morrow
Hopkinsville Community College

Marvin Moak
Hinds Community College

Kenneth P. De Lucca
Millersville University of Pennsylvania

Randy Owens
Henderson Community College

Erica Matthew
Florida Career College

Patrick J. Klette
Kankakee Community College

James Antonakos
Broome Community College

Fred M. Cope
Northeast State Technical Community College

James Fischer
San Joaquin Valley College

Gwen Oster
Northwest Technical College

Jeffrey A. Clade
York Technical College

Ed Margraff
Marion Technical College

Stuart Hilton
ITT

Chrys Panayiotou
Indian River State College

Wade Wittmus
Lakeshore Technical College

Bill Hessmiller
Technical Editor

Patrick Hoppe
Author, MultiSim Primer 11.0

Safety

Electric and electronic circuits can be dangerous. Safe practices are necessary to prevent electrical shock, fires, explosions, mechanical damage, and injuries resulting from the improper use of tools.

Perhaps the greatest hazard is electrical shock. A current through the human body in excess of 10 milliamperes can paralyze the victim and make it impossible to let go of a "live" conductor or component. Ten milliamperes is a rather small amount of current flow: it is only *ten one-thousandths* of an ampere. An ordinary flashlight can provide more than 40 times that amount of current!

Flashlight cells and batteries are safe to handle because the resistance of human skin is normally high enough to keep the current flow very small. For example, touching an ordinary 1.5-V cell produces a current flow in the microampere range (a microampere is one-millionth of an ampere). This amount of current is too small to be noticed.

High voltage, on the other hand, can force enough current through the skin to produce a shock. If the current approaches 100 milliamperes or more, the shock can be fatal. Thus, the danger of shock increases with voltage. Those who work with high voltage must be properly trained and equipped.

When human skin is moist or cut, its resistance to the flow of electricity can drop drastically. When this happens, even moderate voltages may cause a serious shock. Experienced technicians know this, and they also know that so-called low-voltage equipment may have a high-voltage section or two. In other words, they do not practice two methods of working with circuits: one for high voltage and one for low voltage. They follow safe procedures at all times. They do not assume protective devices are working. They do not assume a circuit is off even though the switch is in the OFF position. They know the switch could be defective.

Even a low-voltage, high-current-capacity system like an automotive electrical system can be quite dangerous. Short-circuiting such a system with a ring or metal watchband can cause very severe burns—especially when the ring or band welds to the points being shorted.

As your knowledge and experience grow, you will learn many specific safe procedures for dealing with electricity and electronics. In the meantime:

1. Always follow recommended safety procedures.
2. Use service manuals as often as possible. They often contain specific safety information. Read, and comply with, all appropriate material safety data sheets. Material Safety Data Sheets (MSDS) are provided with many products, and they discuss any hazards associated with the product as well as the safe use, storage, and disposal of the product.
3. Investigate before you act.
4. When in doubt, *do not act*. Ask your instructor or supervisor.

General Safety Rules for Electricity and Electronics

Safe practices will protect you and your fellow workers. Study the following rules. Discuss them with others, and ask your instructor about any you do not understand.

1. Do not work when you are tired or taking medicine that makes you drowsy.
2. Do not work in poor light.
3. Do not work in damp areas or with wet shoes or clothing.
4. Use approved tools, equipment, and protective devices.
5. Avoid wearing rings, bracelets, and similar metal items when working around exposed electric circuits.
6. Never assume that a circuit is off. Double-check it with an instrument that you are sure is operational.
7. Always use the lockout/tagout (LO/TO) procedure, which is locking power OFF and tagging. If not possible, then use the 'buddy system' to guarantee that power will not be turned on while a technician is still working on a circuit.
8. Never tamper with or try to override safety devices such as an interlock (a type of switch that automatically removes power when a door is opened or a panel removed).

9. Keep tools and test equipment clean and in good working condition. Replace insulated probes and leads at the first sign of deterioration.

10. Some devices, such as capacitors, can store a *lethal* charge. They may store this charge for long periods of time. You must be certain these devices are discharged before working around them.

11. Do not remove grounds and do not use adaptors that defeat the equipment ground.

12. Use only an approved fire extinguisher for electrical and electronic equipment. Water can conduct electricity and may severely damage equipment. Carbon dioxide (CO_2) or halogenated-type extinguishers are usually preferred. Foam-type extinguishers may also be desired in *some* cases. Commercial fire extinguishers are rated for the type of fires for which they are effective. Use only those rated for the proper working conditions.

13. Follow directions when using solvents and other chemicals. They may be toxic or flammable, or may damage certain materials such as plastics. Always read and follow the appropriate material safety data sheets.

14. A few materials used in electronic equipment are toxic. Examples include tantalum capacitors and beryllium oxide transistor cases. These devices should not be crushed or abraded, and you should wash your hands thoroughly after handling them. Other materials (such as heat shrink tubing) may produce irritating fumes if overheated. Always read and follow the appropriate material safety data sheets.

15. Certain circuit components affect the safe performance of equipment and systems. Use only exact or approved replacement parts.

16. Use protective clothing and safety glasses when handling high-vacuum devices such as picture tubes and cathode-ray tubes.

17. Don't work on equipment before you know proper procedures and are aware of any potential safety hazards. MSDS (material safety data sheets) are documents provided with many products that generally warn users of any hazard associated with the product.

18. Many accidents have been caused by people rushing and cutting corners. Take the time required to protect yourself and others. Running, horseplay, and practical jokes are strictly forbidden in shops and laboratories.

Circuits and equipment must be treated with respect. Learn how they work and the proper way of working on them. Always practice safety: your health and life depend on it.

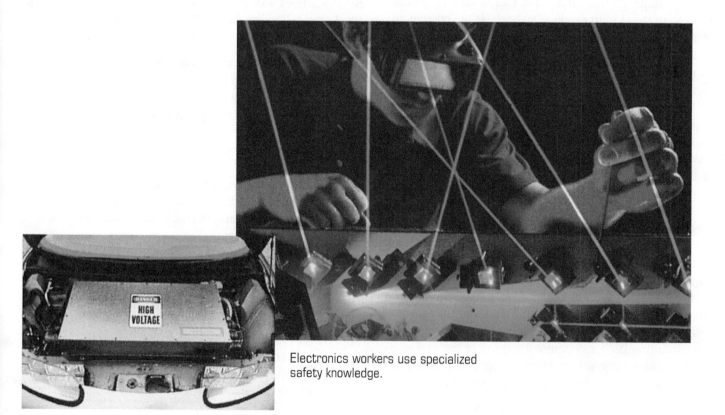

Electronics workers use specialized safety knowledge.

Basic Concepts

Learning Outcomes

This chapter will help you to:

1-1 *Clearly differentiate* between work and energy.

1-2 *Use* base units for specifying and calculating energy and work.

1-3 *Understand* energy conversion.

1-4 *Understand* and calculate efficiency of energy conversions.

1-5 *Explain* the characteristics of the particles of an atom.

1-6 *Explain* the nature of electric charge.

1-7 *Define* "Valence Electron."

1-8 *Define* "Free Electrons."

1-9 *Explain* why some ions are positive and other ions are negative.

1-10 *Understand* the cause and characteristics of static electricity.

1-11 *Explain* what causes static discharge.

1-12 *Explain* several industrial uses of static charges and static electricity.

Electricity is a form of energy. The study of electricity is concerned primarily with learning how to control electric energy. When properly controlled, electricity can do much of the work required to keep our society going. However, uncontrolled electric energy, such as lightning, can be very destructive.

Electric energy is so much a part of our daily lives that we tend to take it for granted. Yet without it our lives would be quite different, and much harder. Electric energy lights our homes and industries, operates our computers, radios, cell phones, and television sets, and turns the many motors used in clocks, washing machines, clothes dryers, vacuum cleaners, and so forth. It is hard to imagine a home that does not require electric energy.

Too much of the electric energy, and other forms of energy that our country relies on, is produced from oil and coal. Both are nonrenewable energy sources and both do serious damage to our environment when they are converted to other forms of energy. See Appendix K for information on some common forms of renewable energy sources and systems.

1-1 Work and Energy

Energy and *work* are very closely related terms. Although they are abbreviated (represented) with the same symbol and use the same units, they are not interchangeable terms. Each is defined and used in a different way.

Work is defined as a force moving through a distance. Energy is the ability, or capacity, to do work. In other words, it takes energy to do work. For example, it requires energy to pull a boat out of the water onto the beach, and work is done in pulling the boat out of the water.

The energy required to pull the boat from the water comes from the human body. A force is

Energy (W)

Work (W)

required to overcome the friction of the boat on the sand. A force is also required to overcome the gravitational pull on the boat as it is raised out of the water. The work then consists of the force required to move the boat some distance as it is pulled onto the shore.

The symbol (or abbreviation) for either work or energy is *W*. The same symbol is used for work and energy because the two terms are so closely related to each other. Yet the two terms must not be confused with each other. Energy is independent of work; it exists whether or not work is being done. Work, however, requires energy.

1-2   Unit of Energy

Base units

Joule (J)

Base units are the terms used to indicate the amount of something. The *joule* is the base unit of energy and work. The symbol for the joule is J. Specifying energy in joules is the same as specifying butter in pounds or money in dollars. All are base units used to specify amount. Base units are important because nearly all relationships in electricity are expressed in base units.

A joule of energy (or work) is very small compared with the amount of energy you use each day. For example, an electric toaster uses approximately 100,000 joules of energy to toast two slices of bread. It requires 360,000 joules to operate a small (100-watt) table lamp for 1 hour.

The work or energy involved in a mechanical system (such as pulling a boat) can be determined by the following relationship:

$$\text{Work} = \text{force} \times \text{distance}$$

Newton

Meter

In the metric system, the base unit for force is the *newton* (approximately 0.2248 pounds). The base unit for distance is the *meter* (approximately 39.4 inches), and the base unit for work (energy) is the joule. The joule is equal to the newton-meter, which is a convenient unit for mechanical energy.

EXAMPLE 1-1

If it requires a steady force of 150 newtons to pull a boat, how much work is required to pull a boat 8 meters?

Given: Force = 150 newtons
                Distance = 8 meters
Find: Work
Known: Work (*W*) = force × distance
                1 newton-meter = 1 joule (J)

Solution: Work = 150 newtons
                     × 8 meters
                   = 1200 newton-meters
                   = 1200 joules
Answer: Work = 1200 joules
                W = 1200 J

Note the procedure used in solving the problem given in the above example. First, the information (values) *given* in the problem is listed. Next, the information you are required to *find* is recorded. Finally, the relationship (formula) between the two is written. For simple problems this formal procedure may seem unnecessary, but solving more complex problems will be easier if you establish the habit of using this formal procedure.

In example 1-1, we found that the work done in moving the boat was 1200 joules (J). The amount of energy required to move the boat is also 1200 J. Work and energy have the same base unit. They are basically the same thing. Work is the use of energy to perform some task. For example, a car battery has energy stored in it. When the car engine is started, energy from the battery is used to do the work of cranking the engine. The *work done* and the *energy used* are two ways of saying the same thing.

EXAMPLE 1-2

It requires 500 joules of energy and 100 newtons of force to move an object from point A to point B. What is the distance between point A and point B?

Given: Energy = 500 joules
                Force = 100 newtons
Find: Distance
Known: *W* = force × distance, and by rearranging
                Distance = $\dfrac{W}{\text{force}}$
                1 newton-meter = 1 joule
Solution: Distance = $\dfrac{500 \text{ newton-meters}}{100 \text{ newtons}}$
                   = 5 meters
Answer: Distance = 5 meters

So far, the calculation of specific amounts of work or energy has been limited to mechanical examples. Once we learn some new terms, like *voltage, current,* and *power,* we will be able to solve problems involving electric energy.

Self-Test

Technician examining a fax machine.

1-3 Energy Conversion

One of the fundamental laws of classical physics states that, under ordinary conditions, energy can be neither created nor destroyed. The energy in the universe exists in various forms, such as heat energy, light energy, and electric energy. When we say we "use" electric energy, we do not mean that we have destroyed, or lost, the energy. We mean that we have *converted* that electric *energy* into a more useful form of energy. For example, when we operate an electric lamp, we are converting electric energy into light energy and heat energy. We have used the electric energy in the sense that it no longer exists as electric energy, but we have not used up the energy. It still exists as heat energy and light energy.

The study of electricity deals with the study of converting energy from one form to another form. Electric energy itself is obtained by converting other forms of energy to electric energy. Batteries convert chemical energy to electric energy, solar cells convert light energy

Converted energy

Lightning—an example of uncontrolled electric energy.

Electric energy is converted for use in the home.

to electric energy, and generators convert mechanical (rotational) energy to electric energy.

We seldom use energy directly in the form of electric energy. Yet the electrical form of energy is very desirable because it can be easily moved from one location to another. Electric energy produced at an electric power plant many miles from your home can be easily transferred from the plant to your home. Once the electric energy is delivered to your home, it can be converted to a more useful form.

We have already noted that a light bulb converts electric energy into light energy and heat energy. Another familiar object that converts energy is the electric stove, which converts electric energy to heat energy. Changing electric energy to mechanical (rotational) energy with an electric motor is also a common conversion.

Although the process is very involved and complex, radio receivers convert electric energy into sound energy. A very small amount of the electric energy comes from electric signals sent through the air. The rest of the electric energy comes from a battery or an electric outlet. The sound energy radiates from the speaker. In a similar fashion, television receivers convert electric energy into sound energy and light energy.

1-4 Efficiency

Efficiency

No conversion process is 100 percent efficient. That is, not all the energy put into a device or a system is converted into the form of energy we desire. When 1000 joules of electric energy is put into an incandescent light bulb, only about 200 joules of light energy is produced. The other 800 joules is converted into heat energy. We could say that the *efficiency* of an incandescent light bulb is low. Neon light bulbs are more than three times as efficient at converting electric energy to light energy than are incandescent light bulbs. Light-emitting diodes (LEDs) are even more efficient than are neon bulbs. They are about 10 times more efficient than are incandescent bulbs. With increased emphasis on reduced energy consumption, these devices are being recommended as replacements for incandescent light bulbs.

The efficiency of a system is usually expressed as a percentage. It is calculated by the formula

$$\text{Percent efficiency} = \frac{\text{useful energy out}}{\text{total energy in}} \times 100$$

Converting electric energy to light and heat energy.

By abbreviating percent efficiency to % eff. and using the symbol W for energy, we can write this formula as

$$\% \text{ eff.} = \frac{W_{out}}{W_{in}} \times 100$$

Let us determine the efficiency of the incandescent light bulb mentioned earlier.

EXAMPLE 1-3

What is the efficiency of a light bulb that uses 1000 joules of electric energy to produce 200 joules of light energy?

Given: Energy in = 1000 joules
Energy out = 200 joules

Find: Percent efficiency

Known: $\% \text{ eff.} = \dfrac{W_{out}}{W_{in}} \times 100$

Solution: $\% \text{ eff.} = \dfrac{200 \text{ joules}}{1000 \text{ joules}} \times 100$
$= 0.2 \times 100$
$= 20$

Answer: Efficiency = 20 percent

Notice that in the efficiency formula both the denominator and the numerator have base units of joules. The base units therefore cancel, and the answer is a pure number (it has no units). We could reword the answer to read, "The efficiency of the light bulb is 20 percent" or "The light bulb is 20 percent efficient."

EXAMPLE 1-4

How much energy is required to produce 460 joules of light energy from a light bulb that is 25 percent efficient?

Given: $\% \text{ eff.} = 25$
Energy out = 460 joules

Find: Energy in

Known: $\% \text{ eff.} = \dfrac{W_{out}}{W_{in}} \times 100$
by rearranging:
$W_{in} = \dfrac{W_{out}}{\% \text{ eff.}} \times 100$

Solution: $W_{in} = \dfrac{460 \text{ joules}}{25} \times 100$
$= 18.4 \text{ joules} \times 100$
$= 1840 \text{ joules}$

Answer: Energy in = 1840 joules

Not all electric devices have such a low efficiency as the incandescent light bulb. Electric motors like those used in washing machines, clothes dryers, and refrigerators have efficiencies of 50 to 75 percent. This means that 50 to 75 percent of the electric energy put into the motor is converted to mechanical (rotational) energy. The other 25 to 50 percent is converted into heat energy.

So far, we have illustrated the efficiency of converting electric energy into other desired forms of energy. Of course, we are just as interested in the efficiency of converting other forms of energy into electric energy.

EXAMPLE 1-5

What is the efficiency of an electric generator that produces 5000 joules of electric energy from the 7000 joules of mechanical energy used to rotate the generator?

Given: Energy in = 7000 joules
Energy out = 5000 joules

Find: Percent efficiency

Known: $\% \text{ eff.} = \dfrac{W_{out}}{W_{in}} \times 100$

Solution: $\% \text{ eff.} = \dfrac{5000 \text{ joules}}{7000 \text{ joules}} \times 100$
$= 0.714 \times 100$
$= 71.4$

Answer: Efficiency = 71.4 percent

Sometimes we want to know the efficiency of a system that has one device (like a battery) to convert some form of energy into electric energy and another device (like a light bulb) to convert the electric energy into other forms of energy.

EXAMPLE 1-6

What is the percent efficiency of a system in which a battery converts 820 joules of chemical energy to provide 150 joules of light energy from a light bulb?

Given: W_{in} (chemical energy converted by battery) = 820 joules
W_{out} (light energy from light bulb) = 150 joules

Find: Percent efficiency

Known: $\% \text{ eff.} = \dfrac{W_{out}}{W_{in}} \times 100$

Solution: $\% \text{ eff.} = \dfrac{150 \text{ joules}}{820 \text{ joules}} \times 100$
$= 18.3$

Answer: Efficiency = 18.3 percent

Answer the following questions.

5. List the forms of energy into which the electric energy from a car battery is converted.
6. What is the undesirable form of energy produced by both light bulbs and electric motors?
7. What happens to the temperature of an electric battery when the battery is discharging? Why?
8. An electric motor requires 1760 joules of electric energy to produce 1086 joules of mechanical energy. What is the efficiency of the motor?
9. A flashlight battery uses 110 joules of chemical energy to supply 100 joules of electric energy to the flashlight bulb. What is the efficiency of the battery?
10. How much mechanical energy will be provided by a motor that is 70 percent efficient and requires 1960 joules of electric energy?
11. In example 1-6, what is the efficiency of the battery if the light bulb is 20 percent efficient?

1-5 Structure of Matter

Electron

Atoms

Proton

Neutron

Nucleus

Molecule

Element

Compounds

All matter is composed of *atoms*. Atoms are the basic building blocks of nature. Regardless of their physical characteristics, glass, chalk, rock, and wood are all made from atoms. Rock is different from wood because of the type of atoms of which it is composed.

The smallest particle of a substance that still has all of its characteristics is called a *molecule*. A molecule consists of two or more atoms. If a molecule of chalk is divided into smaller parts, it is no longer chalk.

There are more than 100 different types of atoms. Matter composed of a single type of atom is called an *element*. Thus, there are as many elements as there are types of atoms. Some common elements are gold, silver, and copper.

There are thousands and thousands of different materials in the world. Obviously, most materials must be composed of more than one element. When different kinds of atoms combine chemically, they form materials called *compounds*. An example of a simple compound is water, which is composed of the elements oxygen and hydrogen. Many of the materials used in electronic circuits are composed of compounds.

To really understand electricity, we must "break the atom down" into still smaller particles. We need to be familiar with the three major particles of the atom. These are the *electron, proton,* and *neutron*. A pictorial representation of a helium atom showing its three major particles is illustrated in Fig. 1-1. The center of the atom is called the *nucleus*. Except for hydrogen atoms, the nucleus of atoms contains both protons and neutrons. The nucleus of the hydrogen atom consists of a single proton. The electrons revolve around the nucleus in elliptical paths. The electron is much larger (nearly 2000 times larger) than either the proton or the neutron. Although larger, the electron is much lighter than either the proton or the neutron (about $\frac{1}{2000}$ as heavy). Thus, the center (nucleus) of the atom contains most of the weight, and the electrons make up most of the volume. It should also be noted that the distance between the nucleus

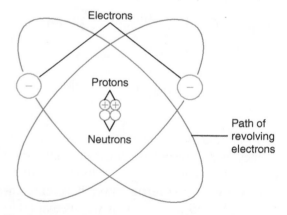

Fig. 1-1 Structure of an atom of helium.

and the electron is very great compared with the size of the electron. In fact, this distance is approximately 60,000 times greater than the diameter of the electron.

An analogy may help you to visualize the relative sizes of the atomic particles and the spaces between them. The simplest atom is the hydrogen atom, which contains one proton, one electron, and no neutrons. Let the nucleus of the hydrogen atom be represented by a common marble. The electron could then be represented by a 31-meter (100-foot) sphere located 1610 kilometers (1000 miles) from the marble. Although the distance between the nucleus and the electron is very great relative to the size of either, we must remember that these sizes and distances are submicroscopic. For example, the diameter of an electron is only 0.0000000000004 centimeter (4×10^{-13} cm). A cm is approximately 0.39 inch.

Electrons rotate, or orbit, around the nucleus of the atom in much the same manner as the earth rotates around the sun (Fig. 1-1). In atoms that contain more than one electron (that is, in all atoms except hydrogen atoms), each electron has its own orbit. With proper coordination of the orbiting electrons, it is possible for two or more atoms to share common space. Indeed, in many materials, neighboring atoms share electrons as well as space.

Figure 1-2 represents the aluminum atom in a two-dimensional form. Remember that each electron is actually orbiting around the nucleus in its own elliptical path. The two

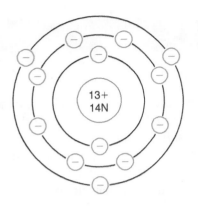

Fig. 1-2 Simplified presentation of an aluminum atom showing its 13 electrons (−), 13 protons (+), and 14 neutrons (N).

electrons closest to the nucleus do not actually follow the same orbital path. Their orbital paths are merely the same average distance from the nucleus. The two electrons closest to the nucleus are said to occupy the first *shell*, or *orbit*, of the atom. This first shell of the atom can accommodate only two electrons. Atoms that have more than two electrons, such as the aluminum atom, must have a second shell, or orbit.

Shell

Orbit

The second shell of the aluminum atom contains 8 electrons. This is the maximum number of electrons that the second shell of any atom can contain. The third shell can contain a maximum of 18 electrons, and the fourth shell a maximum of 32 electrons. Since the aluminum atom (Fig. 1-2) has only 13 electrons, its third shell has 3 electrons.

History of Electronics

James Prescott Joule
The SI (*Système Internationale*) unit of measure for electric energy is the joule (J), named for James Prescott Joule (1 joule is equal to 1 volt-coulomb).

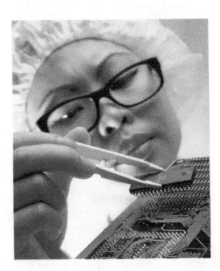

Technician working in a lab.

Answer the following questions.

12. What are the major particles of an atom?
13. True or false. The diameter of the electron is small compared with the distance between the proton and the electron.
14. True or false. The diameter of the proton is greater than the diameter of the electron.
15. True or false. The electron weighs less than the proton.
16. True or false. All electrons in the second shell of an atom follow the same orbital path.

Internet Connection

Visit the website for the International Brotherhood of Electrical Workers (IBEW) for career and other information.

Polarity

1-6 Electric Charge

Both electrons and protons possess electric charges, but these charges are of opposite *polarity.* Polarity refers to the type (negative or positive) of charge. The electron possesses a negative (−) charge, and the proton possesses a positive (+) charge. These electric charges create electric fields of force that behave much like magnetic fields of force. In Fig. 1-3, the lines with arrows on them represent the electric fields. Two positive charges (Fig. 1-3*b*) or two negative charges (Fig. 1-3*a*) repel each other. Two opposite electric charges attract each other (Fig. 1-3*c*). The force of attraction between the positive proton and the negative electron aids in keeping the electron in orbit around the nucleus. The neutron in the nucleus of the atom has no electric charge. The neutron can be ignored when considering the electric charge of the atom.

An atom in its natural state always has a net electric charge of zero; that is, it always has as many electrons as it has protons. For example, look at the aluminum atom illustrated (in simplified form) in Fig. 1-2. It has 13 electrons orbiting around the nucleus, and the nucleus contains 13 protons in addition to the 14 neutrons. We can say that the aluminum atom is electrically neutral, even though individual electrons and protons are electrically charged.

1-7 Valence Electrons

Valence electrons

The electrons in the outermost shell of an atom are called *valence electrons.* Valence electrons are atomic particles that are involved in chemical reactions and electric currents.

One of the forces that helps to hold electrons in orbit is the force of attraction between unlike

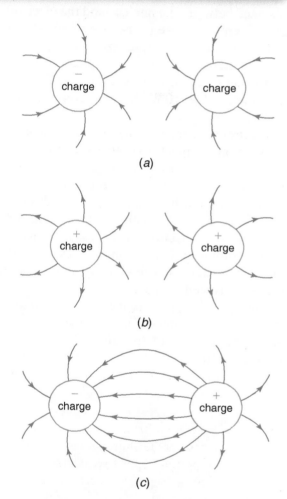

Fig. 1-3 Electric fields between charges. *[a]* and *[b]* Like charges repel each other. *[c]* Unlike charges attract each other.

charges. The closer together two particles of opposite electric charges are, the greater the electrical attraction between them. Therefore, the attraction between the proton of the nucleus and the electron decreases as the electron gets farther from the nucleus. Thus, the valence electrons are held to the nucleus with less attraction than the electrons in the inner shells are. The valence electrons can be more easily removed from the parent atom than the electrons in the inner shells can.

All electrons possess energy. They possess energy because they have weight and they are moving. Thus, they are capable of doing work. Valence electrons possess more energy than electrons in the inner shells. In general, the farther the electron is from the nucleus of the atom, the more energy it possesses.

1-8 Free Electrons

Free electrons are valence electrons that have been temporarily separated from an atom. They are free to wander about in the space around the atom. They are unattached to any particular atom. Only the valence electrons are capable of becoming free electrons. Electrons in the inner shells are very tightly held to the nucleus. They cannot be separated from the parent atom. A valence electron is freed from its atom when energy is added to the atom. The additional energy allows the valence electron to escape the force of attraction between the electron and the nucleus. As a free electron, the electron possesses more energy than it did as a valence electron. One way to provide the additional energy needed to free an electron is to heat the atom. Another way is to subject the atom to an electric field.

1-9 Ions

When a valence electron leaves an atom to become a free electron, it takes with it one negative electric charge. This absence of one negative electric charge from the parent atom leaves that parent atom with a net positive charge. In the case of the aluminum atom, there would be 13 protons (positive charges) but only 12 electrons (negative charges). Atoms which have more than or less than their normal complement of electrons are called *ions*. When an atom loses electrons, it becomes a *positive ion*. Conversely, atoms with an excess of electrons contain a net negative charge and become *negative ions*. The amount of energy required to free a valence electron and create an ion varies from element to element.

The energy required to create a free electron is related to the number of valence electrons

Free electrons

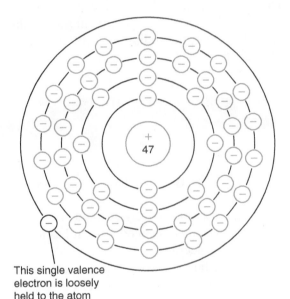

This single valence electron is loosely held to the atom

Fig. 1-4 Simplified silver atom.

contained in the atom. In general, the fewer the electrons in the valence shell, the smaller the amount of additional energy needed to free an electron. The silver atom, illustrated in Fig. 1-4, with its single valence electron requires relatively little energy to free the valence electron. Carbon, with four valence electrons, requires much more energy to free an electron. Elements with five or more electrons in the outer shell do not readily release their valence electrons.

Ions

Positive

Negative ions

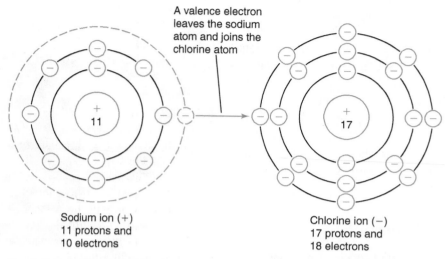

Sodium ion (+)
11 protons and
10 electrons

Chlorine ion (−)
17 protons and
18 electrons

Fig. 1-5 Creation of positive (sodium) ions and negative (chlorine) ions.

Negative ions are created when an atom accepts additional electrons. For example, in the compound sodium chloride (which is ordinary table salt), the sodium atoms share their lone valence electrons with the chlorine atoms to form salt crystals. When the sodium chloride is dissolved in water, the sodium and chlorine atoms separate from each other and the chlorine atom takes with it the sodium atom's valence electron. Thus the chlorine atom becomes a negative ion. At the same time, the sodium atom, which gave up an electron, becomes a positive ion (Fig. 1-5). The concept of ions is important in understanding electric circuits involving batteries and gas-filled devices.

Self-Test

Answer the following questions.

17. What polarity is the charge of an electron? A proton?
18. Is an atom electrically charged? Explain.
19. What is an atom called when it has lost a valence electron?
20. True or false. A free electron is at a higher energy level than a valence electron is.
21. True or false. An atom with seven valence electrons provides a free electron more readily than an atom with two valence electrons.
22. True or false. Ions can possess either a negative or a positive charge.

1-10 Static Charge and Static Electricity

Static electricity is a common phenomenon that all of us have observed. Probably the most dramatic example of static electricity is lightning. Static electricity is responsible for the shock you may receive when reaching for a metal doorknob after walking across a thick rug. It is also responsible for hair clinging to a comb and for the way some synthetic clothes cling to themselves.

All the above phenomena have one thing in common. They all involve the transfer of electrons from one object to another object or from one material to another material.

A *positive static charge* is created when a transfer of electrons leaves an object with

Positive static charge

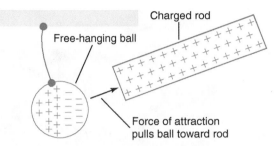

Fig. 1-6 Inducing a static charge. When the ball touches the rod, electrons are transferred to the rod. The ball is left with a positive charge.

a deficiency of electrons. A *negative static charge* results when an object is left with an excess of electrons. Static charges can be created by rubbing a glass rod with a piece of silk cloth. Some of the valence electrons from the glass rod become free electrons and are transferred to the cloth. The cloth takes on a negative charge, and the glass rod takes on a positive charge. The charges on the glass and the cloth tend to remain stationary, thus the name *static electricity*.

An object that possesses a static charge can attract objects that are not charged. This happens because the charged object, when placed near an uncharged object but not touching it, can induce a charge on the surface of the uncharged object (Fig. 1-6). The *induced charge,* because its polarity is the opposite of that of the charged object, is then attracted to the charged object. If the two objects—for example, the ball and the rod in Fig. 1-6—are allowed to touch, part of the positive charge of the rod is transferred to the ball. Then both objects have positive charges, and a force of repulsion results. The ball and the rod then move away from one another because their like charges repel each other.

Let us clarify how a charge is transferred from one object to the other, again using as an example the ball and the rod of Fig. 1-6. When the rod and the ball touch, protons *do not* travel from the rod to the ball. Remember, the electron is the only particle of the atom that can be easily detached and moved about. When the rod and the ball touch, electrons travel from the ball to the rod. This leaves the ball with a shortage of negative charge (electrons). The ball still

History of
Electronics

Niels Bohr

In 1913 Niels Bohr theorized that the atoms of all substances contain negatively charged particles, called *electrons,* in orbit about positively charged particles, called *nuclei.* Bohr attributed the inward force that keeps the electrons from flying off into space to an electrical attraction between the nucleus and electron, caused by opposite polarity. (*Encyclopedia of Electronics,* Gibilisco and Sclater, McGraw-Hill, 1990)

has its normal number of positive charges (protons). Since the ball now has more protons than electrons, it possesses a net positive charge.

1-11 Static Discharge

Static discharge occurs when the electric force field (Fig. 1-3) between a positive charge and a negative charge becomes too strong. Then electrons are pulled off the negatively charged object and travel through the air to the positively charged object. The observable spark is the result of the air path between the objects becoming ionized and deionized. When the air is ionized, electrons are raised to a higher energy level. When the air is deionized, the electrons return to a lower energy level. The difference in the two energy levels is given off as light energy when the electron returns to the lower energy level. Lightning is a column of ionized air caused by electrons traveling between a charged cloud and a portion of land with an opposite charge.

1-12 Uses of Static Electricity

Most of the useful applications of static electricity do not rely on discharges that ionize the air. Rather, they make use of the force of attraction between unlike charges or the force of repulsion between like charges. These forces are used to move charged particles to desired locations. For example, dust particles can be

Static discharge

Induced charge

Negative static charge

Electric power plant control room.

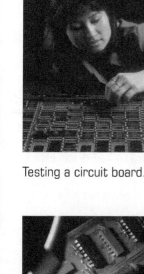

Testing a circuit board.

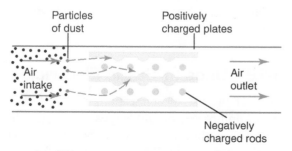

Modern circuit board.

removed from air by this method. In Fig. 1-7, the air is forced between negatively charged rods and positively charged plates. A negative charge is transferred from the rods to the dust particles. When the airstream goes by the positively charged plate, the negatively charged dust particles are pulled out. Electric devices of this type are often called *electrostatic precipitators.*

Static charges are also used in some spray-painting operations. The paint mist leaving the

spray gun nozzle is negatively charged, and the object to be painted is positively charged. This process tends to give a uniform coat of paint on an object, even one that has an irregular surface. As paint builds up on one part of the object, the charge on that part of the object is canceled. The force of attraction disappears. If excess paint builds up on part of the object, then that part becomes negatively charged and repels additional paint. The repelled paint is attracted to those parts of the object that are still positively charged.

Several properties of static electricity can be used to advantage in manufacturing abrasive paper (sandpaper). The backing paper is coated with an adhesive (glue) and given a static charge. The abrasive particles are given the opposite charge. As the paper is passed over the abrasive particles, the particles are attracted to the paper and adhere to it (Fig. 1-8). Once the abrasive particles are distributed on the adhesive backing paper, both the paper and the abrasive particles are given like charges. The

Particles of dust   Positively charged plates

Air intake

Air outlet

Negatively charged rods

Fig. 1-7 Principle of a dust precipitator. Dust particles are given a negative charge so that they are attracted to positively charged plates.

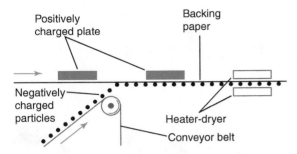

Fig. 1-8 Principles of making abrasive paper. Static charges position the abrasive particles so that their sharpest points are exposed.

like charges repel each other and try to push the abrasive particles from the paper. However, the adhesive is strong enough to hold the particles

to the paper. The abrasive particles "stand up" on the paper so that the sharpest point of each particle is exposed as the cutting surface. The adhesive material is then heated and hardened to hold the abrasive particles in place. The reason the abrasive particles stand up is that static charges concentrate on the sharpest point of an object. Since like charges repel each other, they push away from each other as far as possible. The sharpest point of the abrasive particle is the most highly charged part of the particle. Therefore, it is pushed farthest from the surface of the backing paper.

Self-Test

Answer the following questions.

23. What is a static charge?
24. List three industrial applications of static charges.
25. True or false. After a positively charged rod touches a neutral free-hanging ball, the ball will be repelled.
26. True or false. Static discharge causes electrons to transfer from one object to another object.
27. True or false. An object with a static charge can attract only other charged objects.

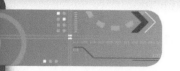

Chapter 1 Summary and Review

Summary

1. Energy is the ability to do work.
2. The symbol for energy is W.
3. Electricity is a form of energy.
4. Work equals force times distance.
5. The joule is the base unit of energy.
6. The joule is the base unit of work.
7. The symbol for the joule is J.
8. Under ordinary conditions, energy is neither created nor destroyed.
9. All matter is made up of atoms.
10. Hydrogen has only one electron and one proton. All other atoms contain electrons, protons, and neutrons.
11. Protons and neutrons are found in the nucleus of the atom.
12. Electrons have a negative electric charge.
13. Protons have a positive electric charge.
14. Neutrons do not have an electric charge.
15. Atoms have an equal number of protons and electrons.
16. Valence electrons are found in the outermost shell of the atom.
17. Valence electrons can become free electrons when their energy level is raised.
18. Negative ions are atoms that have attracted and captured an extra (free) electron.
19. Positive ions are atoms that have given up an electron.
20. Static charges are the result of an object's possessing either more or fewer electrons than protons.
21. A positive charge means a deficiency of electrons.
22. A negative charge means an excess of electrons.
23. Like charges repel each other.
24. Unlike charges attract each other.
25. A charged object can attract an uncharged object.
26. When an uncharged object touches a charged object, it becomes charged.

Related Formulas

$$\text{Percent efficient} = \frac{W_{out}}{W_{in}} \times 100$$

Chapter Review Questions

For questions 1-1 to 1-7, determine whether each statement is true or false.

1-1. Opposite electric charges repel each other. (1-6)
1-2. Free electrons are at a higher energy level than valence electrons are. (1-8)
1-3. The same base unit is used for both work and energy. (1-2)
1-4. Atoms have twice as many protons as electrons. (1-6)
1-5. An object that has a static charge can possess either more or fewer electrons than protons. (1-10)

1-6. The second shell, or orbit, of an atom can contain any number of electrons. (1-5)
1-7. If an electric device is less than 100 percent efficient, part of the energy provided to it is destroyed. (1-3)

For questions 1-8 to 1-14, choose the letter that best completes each sentence.

1-8. The symbol for energy is (1-1)
   a. *D*
   b. *W*

c. *N*

d. *j*

1-9. The base unit of energy is the (1-2)

a. Ion

b. Proton

c. Joule

d. Pound

1-10. The abbreviation for the base unit of energy is (1-2)

a. D

b. W

c. N

d. J

1-11. Static electricity can be used in (1-12)

a. Spray painting

b. Manufacturing abrasive paper

c. Removing dust from air

d. All of the above

1-12. The electrons in the outermost shell of atoms are called (1-7)

a. Free electrons

b. Valence electrons

c. Neutral electrons

d. Active electrons

1-13. A positive ion is an atom which has (1-9)

a. Captured one or more electrons

b. Given up one or more electrons

c. Captured one or more protons

d. Given up one or more protons

1-14. A negative charge can be created on an object by (1-10)

a. Removing protons from the object

b. Removing neutrons from the object

c. Adding electrons to the object

d. Adding neutrons to the object

Chapter Review Problems

1-1. What is the efficiency of a motor which requires 914 base units of electric energy to produce 585 joules of mechanical energy? (1-4)

1-2. How much work is done when an object is moved 3 meters by applying a force of 80 newtons? (1-2)

1-3. How much energy is involved in problem 1-2? (1-2)

1-4. How much energy is required by an electric lamp that is 18 percent efficient and provides 5463 J of light energy? (1-4)

1-5. A motor is 70 percent efficient and requires 1200 base units of energy. How much energy does it provide? (1-4)

1-6 How much heat energy is produced by a light bulb that requires 400 joules of electric energy to produce 125 joules of light energy? (1-3)

Critical Thinking Questions

1-1. Discuss the probable changes in your lifestyle if the electric service to your home were disconnected for one week.

1-2. In addition to decreasing costs, why is it important to increase the efficiency of electric devices as much as is practical?

1-3. Why are large, continuously operated motors designed to be more efficient than small, intermittently operated motors?

1-4. Is the low efficiency of the lights in our homes a greater disadvantage during the winter months or the summer months? Why?

1-5. Many electric devices could be designed and constructed to operate more efficiently. Why aren't they?

1. Work is a force moving through a distance; energy is the ability to do work.

2. A base unit is a term that is used to specify the amount of something.

3. joule (J)

4. **Given:**   Distance = 120 meters
   Force = 360 newtons
   Find: Work
   Known: Work = force × distance
   Solution: Work = 360 newtons × 120 meters
   Answer: Work = 43,200 joules

5. heat (due to inefficiencies of lights and motors), light, and mechanical energy.

6. heat energy

7. It increases, because inefficiency converts some of the battery's energy to heat.

8. **Given:** W_{out} = 1086 joules
   W_{in} = 1760 joules
   Find: Percent efficiency
   Known: % eff. $= \dfrac{W_{out}}{W_{in}} \times 100$

   Solution: % eff. $= \dfrac{1086 \text{ joules}}{1760 \text{ joules}} \times 100$

   $= 61.7$
   Answer: Efficiency = 61.7 percent

9. **Given:** W_{out} = 100 joules
   W_{in} = 110 joules
   Find: Percent efficiency
   Known: % eff. $= \dfrac{W_{out}}{W_{in}} \times 100$

   Solution: % eff. $= \dfrac{100 \text{ joules}}{110 \text{ joules}} \times 100$

   $= 90.9$
   Answer: Efficiency = 90.9 percent

10. **Given:** W_{in} = 1960 joules (J)
    % eff. = 70
    Find: W_{out}
    Known: % eff. $= \dfrac{W_{out}}{W_{in}}$

    by rearranging:
    $$W_{out} = \dfrac{W_{in} \times \% \text{ eff.}}{100}$$

Solution: $W_{out} = \dfrac{1960 \text{ J} \times 70}{100} = 1372 \text{ J}$

Answer: Mechanical energy out = 1372 J

11. **Given:**   Chemical energy in = 820 joules
    Light energy out = 150 joules
    Efficiency of light bulb = 20 percent
    Find: Efficiency of the battery
    Known: % eff. $= \dfrac{W_{out}}{W_{in}} \times 100$

    W_{out} of battery $= W_{in}$ of light bulb

    W_{in} of light bulb $= \dfrac{W_{out} \text{ of light bulb}}{\text{eff. of light bulb}}$

    % eff. of battery $= \dfrac{\text{electric energy out}}{\text{chemical energy in}}$
    $\times 100$

    Solution: W_{in} of light bulb $= \dfrac{150 \text{ joules}}{0.2}$

    $= 750 \text{ joules}$

    % eff. of battery $= \dfrac{750 \text{ joules}}{820 \text{ joules}} \times 100$
    $= 91.5$

    Answer: Efficiency of the battery = 91.5 percent

12. proton, electron, and neutron

13. T

14. F

15. T

16. F

17. Electron is negative, and proton is positive.

18. No, the negative charge of the electrons cancels the positive charge of the protons, leaving the atom neutral.

19. positive ion

20. T

21. F

22. T

23. A static charge is the excess or deficiency of electrons on an object.

24. removal of dust particles from air, spray painting, and manufacturing of abrasive paper

25. T

26. T

27. F

Electrical Quantities and Units

Learning Outcomes

This chapter will help you to:

2-1 *Compare* the polarity of the charge of a proton to that of an electron.

2-2 *Specify* the symbol for charge and its base unit as well as the symbol for the base unit of charge.

2-3 *Explain* when ions can be current carriers.

2-4 *Compare* the speed of travel of an individual electron to speed of travel of the effects of a current.

2-5 *List* the current carriers used to support current flow in a liquid.

2-6 *Explain* how current flow is supported in a vacuum.

2-7 *Define* the base unit of current in terms of charge and time.

2-8 *List* the two terms that can be used instead of voltage to refer to electric pressure that causes current to flow.

2-9 *Using* the symbols for voltage, energy and charge, express the relationship between them.

2-10 *Discuss* several ways in which "polarity" is used in describing electrical concepts and devices.

2-11 *List* two forms of energy often converted to electric energy. Also name the device used in each conversion process.

2-12 *Explain* how electric energy is converted to heat energy.

2-13 *Realize* that good conductors have very little resistance but very few have no resistance unless the temperature is close to, or at, absolute zero.

2-14 *Compare* the resistance of conductors and insulators and explain why good insulators are materials with covalent bonds.

2-15 *Understand* that some elements and compounds are semiconductors (neither good conductors or insulators). But, semiconductors are extensively used in electrical/electronic devices.

If you were describing a shopping trip to the grocery store, you would probably use many quantities and units in your description. You might start out by telling the distance (a quantity) in blocks or miles (units) to the store. In describing what you purchased, you could use many quantities and units, such as pounds of sugar, heads of lettuce, cans of soup, quarts of milk, dozens of eggs, or bars of soap.

In describing electric circuits, you also need to use quantities and units. This chapter tells you about many of the basic electrical quantities and units.

2-1 Charge

You May Recall

. . . that we previously defined an electric *charge* as the electrical property possessed by electrons and protons.

Charge (Q)

The proton has a positive charge and the electron a negative charge. However, we never did specify exact amounts of charge. That is, we never did define the base unit of charge so that we could specify exact amounts of charge.

2-2 Unit of Charge

The base unit of charge is the *coulomb*. A coulomb of charge is the amount of charge possessed by 6,250,000,000,000,000,000 electrons (6.25×10^{18} electrons). We do not use the charge on a single electron as the base unit because it is a very small charge—too small for most practical applications. The coulomb, the

Coulomb (C)

2-16 *Express* the amount of resistance in its base unit of the ohm by using the ohm symbol (Ω).

2-17 *Discuss* the fact that the resistance of a material can change with a change in temperature and the change is specified by the temperature coefficient of resistance of the material.

2-18 *Remember* that the resistance between any two opposite surfaces of a 1 cm cube of a material at 20°C is the resistivity or the specific resistance of the material.

2-19 *Realize* that resistors are physical devices manufactured to provide a specific amount of resistance.

2-20 *Understand* the difference (and relationship) between power and energy.

2-21 *Use* base units to express the relationship between energy, power, and time.

2-22 *Calculate* efficiency using the input power and the output power of a device or system.

2-23 *Utilize* powers of ten when working with large positive or negative numbers.

2-24 *Use* prefixes to express multiples and submultiples of base units.

2-25 *Convert* non-metric units to metric units and vice versa.

Electric current [/]

Current carrier

Speed of light

base unit of charge, is also used in defining the base units of other electrical quantities, such as current and voltage. The coulomb is named after Charles Augustin Coulomb, a French physicist.

History of Electronics

Charles Augustin Coulomb

French natural philosopher Charles Augustin Coulomb developed a method for measuring the force of attraction and repulsion between two electrically charged spheres. Coulomb established the law of inverse squares and defined the basic unit of charge quantity, the coulomb.

In electricity, we use many symbols (abbreviations) for electrical quantities and units. The symbol for charge is Q. The abbreviation for coulomb is C. Use of symbols in the study of electricity allows us to condense ideas and statements. For example, instead of writing "the charge is 5 coulombs," we can just write, "$Q = 5$ C."

2-3 Current and Current Carriers

Electric current is the movement of charged particles in a specified direction. The charged particle may be an electron, a positive ion, or a negative ion. The charged particle is often referred to as a *current carrier.* The movement of the charged particle may be through a solid, a gas, a liquid, or a vacuum. In a solid, such as copper wire, the charged particle (current carrier) is the electron. The ions in a copper wire, and in other solids, are rigidly held in place by the atomic (crystalline) structure of the material. Thus, ions cannot be current carriers in solid materials. However, in both liquids and gases, the ions are free to move about and become current carriers.

The symbol for current is I. The symbol I was chosen because early scientists talked about the intensity of the electricity in a wire.

2-4 Current in Solids

When thinking about current, keep two points in mind. First, the effect of current is almost instantaneous. Current in a wire travels at nearly the *speed of light:* 186,000 miles per second [3×10^8 meters per second (m/s)]. Second, an individual electron moves much more slowly than the effect of the current. It may take minutes for an individual electron to travel a few feet in the wire.

The ideas of the "effect of the current being instantaneous" and the "individual electron moving much more slowly" are illustrated in Fig. 2-1. Suppose you had a very long cardboard tube with a diameter just large enough to pass a tennis ball. You lay the tube on the floor and fill it full of tennis balls. When you push an extra ball into one end of the tube, another ball immediately comes out the other end of the tube (Fig. 2-1). If you did not know that the tube was full of tennis balls, you might think that

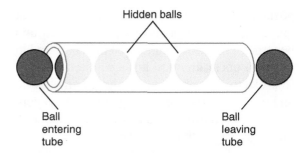

Fig. 2-1 Apparent speed illustrated. A ball exits the instant another ball enters the tube.

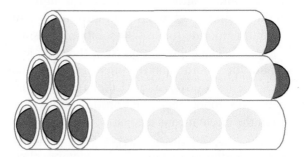

Fig. 2-2 Electron movement illustrated.

the tennis ball you pushed in one end of the tube traveled very quickly down the tube and out the other end. The effect (the one ball popping out the end) is very fast. Yet each tennis ball moves only a short distance.

Suppose you stacked up six tubes filled with balls (Fig. 2-2) and pushed balls first into one tube and then into another. Now you could have a steady stream of balls appearing at one end of the tubes, yet only the balls in one tube would be moving at any one time. Even within that tube each ball would move only a short distance. This is comparable to the way in which current carriers (electrons) move through a wire when current is flowing in the wire.

Assume you could look inside an aluminum wire and see the atoms and their particles (Fig. 2-3). The aluminum atom actually has 13 electrons and 13 protons, but, for simplicity, only 3 of each are shown. Now, suppose the ends of the wire are connected to a flashlight cell. The cell provides an electric field through the wire. The electric field frees some of the valence electrons of the aluminum atoms, as shown in Fig. 2-3, by giving them additional energy. At the moment that an individual electron is freed, it may be traveling in a direction opposite that of the main current. However, in the presence of the electric field, it soon changes its direction of movement. For every electron that is freed, a positive ion is created. This positive ion has an attraction for an electron. Eventually, one of the free electrons will migrate close to the positive ion. The electron will be captured by that positive ion, which then, of course, will become a neutral atom.

Notice that a free electron does not remain a free electron and travel the full length of the wire. Rather, it travels a short distance down the wire and is captured by one of the positive ions. At some later time this particular electron may again gain enough energy to free itself of its new parent atom. It then travels farther down the wire as a free electron. We can think of the electrons as hopping down the conductor from atom to atom to atom. As long as there is a new free electron created every time a free electron is captured, the net number of free electrons moving down the wire remains constant. Current continues to flow. It continues to flow in the same

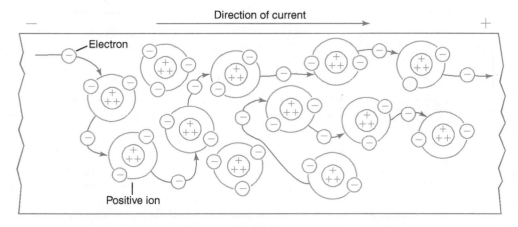

Fig. 2-3 Current in a solid. A free electron travels only a short distance before it combines with a positive ion.

direction through the conducting wire. Current that travels in the same direction all the time is called *direct current,* which is abbreviated dc. It is the type of current you get from cells and batteries such as those used in flashlights.

If you remember that the individual electron travels rather slowly from atom to atom, you will be able to understand alternating current. *Alternating current* (abbreviated ac) is the type of current you have in your home and your school. It is the type of current that periodically reverses the direction in which it is moving. The current in all the electric wires in your home reverses its direction every $\frac{1}{120}$ of a second. Some of the current in your television receiver reverses every $\frac{1}{67,000,000}$ of a second. Currents that reverse direction this often are easier to visualize if you think of the individual electrons as "swinging" back and forth between several atoms.

2-5 Current in Liquids and Gases

In gases, both positive ions and electrons are involved in current flow. When a gas is subjected to a strong electric field, the gas *ionizes.* Once ionized, the gas allows current to flow through it. Figure 2-4 illustrates current flow in ionized neon gas. The negative and positive signs indicate that the neon bulb is connected to a source of electric force, such as a battery. A neon atom has eight electrons in its outermost (valence) shell. When the atom is ionized, one electron is freed. The resulting positive neon ion travels toward the *negative plate* (Fig. 2-4). The resulting free electron travels toward the

positive plate. Once the positive ion arrives at the negative plate, it receives an electron and becomes a neutral atom. It then drifts around in the glass enclosure until it is again ionized. The free electron is received by the *positive plate* and travels out the connecting wire. The system of Fig. 2-4 requires that electrons be supplied to the negative plate and removed from the positive plate. That job is done by the power supply (battery), which also supplies the electric field.

Current flow in a liquid consists of both negative and positive ions moving through the liquid. A simplified diagram of current flow in a sodium chloride (table salt) solution is shown in Fig. 2-5. The abbreviation Na is the chemical symbol for sodium, and the plus sign (+) in the circle means a positive ion. Likewise, Cl is the chemical symbol for chlorine and the minus sign (−) in the circle means a negative ion. When an electric field is created between the plates, the positive sodium ions move to the negative plate and the negative chlorine ions move to the positive plate. Notice that the current in the liquid is composed entirely of ions. However, the current in the wires and plates connected to the liquid is composed of electrons in motion. The changeover from electron-charge carriers to ion-charge carriers occurs at the surface where the plates and liquid meet. This change of carrier is actually more complex than implied by Fig. 2-5. The change also involves some ions created by the water in which the salt is dissolved. However, whether or not we account for the water ions, the end results are the same. That is, negative ions give up electrons at the positive plate, and positive ions pick up electrons from the negative plate. A

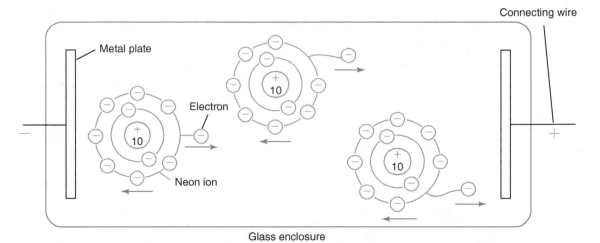

Fig. 2-4 Current in a gas. Both electrons and ions serve as current carriers.

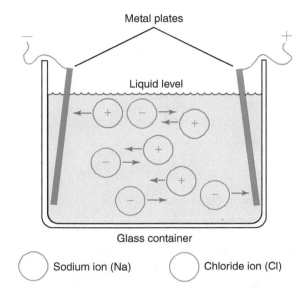

Fig. 2-5 Current in a liquid. Both positive and negative ions serve as current carriers.

André Marie Ampère
The unit of current, the ampere (A), was named for French physicist André Marie Ampère, who discovered that two parallel wires attract each other when currents flow through them in the same direction and repel each other when currents are made to flow in the opposite directions.

liquid solution which is capable of carrying current is called an *electrolyte*. A solution of seawater is an electrolyte; it is a solution containing ionized substances.

One industrial use of current flow in electrolytes is electroplating. *Electroplating* is a process by which a thin layer of one type of metal can be plated (surface-covered) over another material. The other material may be another metal or a piece of plastic coated with a conductive material. Figure 2-6 illustrates the electroplating of copper onto iron. The electrolyte is copper sulfate, which ionizes into a copper

ion (Cu^{++}) with two positive charges and a sulfate ion (SO_4^{--}) with two negative charges. The copper ions are attracted to the iron plate, where they pick up two electrons and adhere (stick to) to the iron plate as copper atoms. The sulfate ions move to the copper plate, where they chemically react with the copper to create more copper sulfate. The copper sulfate goes back into solution. The reaction that created the copper sulfate leaves two electrons on the copper plate. These two electrons move out through the wire connected to the copper plate. Notice that this electroplating system is always in balance. That is, for every two electrons that enter the negative iron plate, two more electrons leave the positive copper plate. This type of balance is present in all electric devices that carry current.

Electrolyte

Electroplating

2-6 Current in a Vacuum

Figure 2-7 shows how current flows in a vacuum. The cathode in Fig. 2-7 is a metal plate coated with a material that will emit electrons from its surface when heated to a temperature just below its melting point. The process of emitting electrons from a heated surface is called *thermionic emission*. Thermionic emission occurs when the valence electrons in the heated atoms gain sufficient energy to escape both the parent atoms and the surface barrier of the plate. When no positive or negative charges are connected to the plates, the surface of the cathode becomes positive as

Thermionic emission

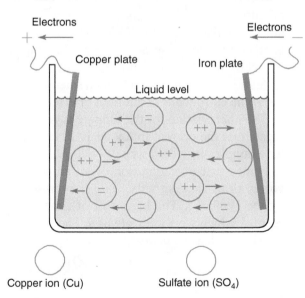

Fig. 2-6 Electroplating copper onto an iron plate. Current in the liquid consists of ions in motion.

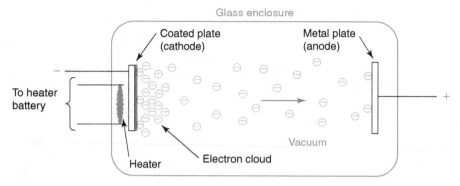

Fig. 2-7 Current in a vacuum. The electron cloud around the cathode provides a source of current carriers.

thermionic emission occurs. Thus, some of the emitted electrons are pulled back to the surface as they cool off and lose energy. After a cloud of electrons forms around the heated surface, an equilibrium exists in which as many electrons are returning as are being emitted. Now, when the plates are connected to a source of positive and negative charges, such as the terminals of a battery, electrons are pulled from the electron cloud to the anode (+ plate). Of course, the same number of electrons attracted to the anode are supplied to the cathode by the negative charge connected to the cathode. The current in the vacuum is continuous as long as the batteries continue to supply energy to the heater and charges to the plates.

Thermionic emission and current in a vacuum are at the heart of electronic vacuum tubes and cathode ray tubes (CRTs). These tubes were used extensively in electronic equipment such as oscilloscopes, radar, TV receivers, and computer monitors. However, these tubes have been largely replaced by liquid crystal displays (LCDs) and solid-state devices such as transistors, integrated circuits (ICs), and microprocessors.

2-7 Unit of Current — The Ampere

We have now developed a concept of electric current. Our next problem is to develop a logical unit for keeping track of the amount of current. The simplest method might be to keep track of the number of electrons or of the number of coulombs that move down the conductor. However, this method would leave much to be desired. It would not take into account the

amount of time required to move the charge. This would be somewhat comparable to tabulating automobile traffic without considering the time involved. For example, on a freeway, 1000 cars may pass a given point in 1 hour. On a two-lane country road it may take 20 hours for 1000 cars to pass a given point. Certainly the traffic is greater on the freeway. A more meaningful way to compare the traffic on these two roads would be to talk in terms of the number of cars per hour. Thus, the traffic on the freeway would be 1000 cars per hour whereas the traffic on the two-lane road would be 50 cars per hour.

In electricity, the amount of current is specified in terms of the charge and the time required to move the charge past a given point. The amount of electric current is therefore specified in *coulombs per second*. But, since coulombs per second is a rather long term, the base unit of current has been named the *ampere*. The ampere is the base unit of current. An ampere is equal to 1 coulomb per second. The ampere was chosen as the base unit of current in honor of André Marie Ampère, a French scientist who did some early work in the field of electricity.

The abbreviation for the ampere is A. To indicate that the current in a wire is 10 amperes, for example, we would write $I = 10$ A.

Notice that our definition of an ampere involves time. In electricity, *time* is represented by the symbol t. The base unit of time is the *second*, which we abbreviate as s. The relationship between time, charge, and current is

$$\text{Current } (I) = \frac{\text{charge } (Q)}{\text{time } (t)}$$

or

$$I = \frac{Q}{t}$$

Coulombs per second

Ampere (A)

Time (t)

Second (s)

$$I = \frac{Q}{t}$$

Expressed in the base units of the above quantities, the relationship is:

$$\text{Ampere} = \frac{\text{coulomb}}{\text{second}}$$

or, in abbreviated form:

$$A = \frac{C}{s}$$

Notice that the abbreviations (symbols) for electrical quantities are italicized (Q for charge, I for current, t for time, and W for energy and work) and the symbols for base units are not italicized (C for coulomb, A for ampere, s for second, and J for joule).

EXAMPLE 2-1

How much time is required for 12.5×10^{18} electrons to leave the negative terminal of a battery if the current provided by the battery is 0.5 A?

Given: $Q = 12.5 \times 10^{18}$ electrons
$I = 0.5$ A

Find: Time (t)

Known: 1 coulomb (C) $= 6.25 \times 10^{18}$ electrons

$I = \frac{Q}{t}$ and, rearranged,

$t = \frac{Q}{I}$

Solution: Convert from electrons to coulombs

$$Q = \frac{12.5 \times 10^{18} \text{ electrons}}{6.25 \times 10^{18} \text{ electrons/C}}$$

$$= 2 \text{ C}$$

Then calculate time

$$t = \frac{Q}{I} = \frac{2C}{0.5A} = 4 \text{ s}$$

Answer: Time = 4 seconds

$$A = \frac{C}{s}$$

Self-Test

Answer the following questions.

1. What is an electric charge?
2. The symbol for charge is _____.
3. A coulomb of charge is equal to the charge on _____ electrons.
4. The symbol for the base unit of charge is _____.
5. Either an _____ or an _____ can be a current carrier.
6. The current carrier in a vacuum is the_____.
7. True or false. An ionized gas has only one type of current carrier.
8. True or false. Random motion of a current carrier is considered to be an electric current.
9. True or false. The symbol for current is A.
10. Describe the way in which an electron travels through a copper wire.
11. How is alternating current different from direct current?
12. What is the base unit of current?
13. Define the ampere in terms of charge and time.
14. What is the abbreviation for the ampere?
15. Rewrite each of the statements below, using the correct symbols for the electrical quantities and units:
    a. The current is 8 amperes.
    b. The charge is 6 coulombs.
16. How much current is flowing when 16 coulombs pass a specified point in 4 seconds?

2-8 Voltage

Voltage is the electric pressure that causes current to flow. Voltage is also known as *electromotive force* (abbreviated emf), or *potential difference*. All these terms refer to the same thing, that is, the force that sets charges in motion.

Potential difference is the most descriptive term because a voltage is actually a potential energy difference that exists between two points. The symbol for voltage is V. To really understand voltage, one must first understand what is meant by potential energy and potential

Voltage (V)

Electromotive force

Potential difference

energy difference. We must, therefore, extend the discussion of energy that we started in Chapter 1.

Kinetic energy

All energy is either potential energy or kinetic energy. *Kinetic energy* refers to energy in motion, energy doing work, or energy being converted into another form. When you swing a baseball bat, the baseball bat has kinetic energy. It does work when it hits the ball; that is, it exerts a force on the ball that moves the ball through a distance. Anything that has mass (weight) and is in motion possesses kinetic energy.

Potential energy

Potential energy is energy at rest. It is energy that can be stored for long periods of time in its present form. It is capable of doing work when we provide the right conditions to convert it from its stored form into another form. Of course, when it is being converted, it is changed from potential to kinetic energy. Water stored in a lake behind the dam of a hydroelectric plant possesses potential energy due to gravitational forces. The potential energy of the water can be stored for long periods of time. When the energy is needed, it is converted to kinetic energy by letting the water flow through the hydroelectric plant.

An electric charge possesses potential energy. When you scuff around on the carpet, you collect an electric charge on your body. That charge (static electricity) is potential energy. Then, when you touch some object in the room, a spark occurs. The potential energy becomes kinetic energy as the electric charge is converted to the light and heat energy of the spark.

Electric fields

An object, such as this book, resting on a table also possesses potential energy. The book is capable of doing work when it moves from the table to the floor. Thus, the book has potential energy with respect to the floor. When the book falls from the table to the floor, its potential energy is converted into mechanical energy and heat energy. Potential energy is dependent on mass. If we replace the book with an object that weighs more, then the potential energy increases. The *potential energy difference* is independent of the amount of mass. It is a function of the distance between the two surfaces and of the gravitational force. Knowing the potential energy difference that exists between two points, we can easily figure the potential energy possessed by an object of any given weight.

Potential energy difference

EXAMPLE 2-2

What is the potential energy (with reference to the floor) of a 5.5-kilogram (kg) block resting on a table top if the potential energy difference between the tabletop and the floor is 8 joules per kilogram (J/kg)?

Given:	Weight = 5.5 kilograms, potential energy difference = 8 J/kg
Find:	Potential energy
Known:	Potential energy = potential energy difference × weight
Solution:	Potential energy = 8 J/kg × 5.5 kg
	= 44 J
Answer:	Potential energy = 44 joules

In the above discussions, we always started with the object on the tabletop and considered the energy it possessed with reference to the floor. We can just as well reverse the situation and consider the energy required to move the object from the floor to the tabletop. The potential energy difference is the same in either case. In the one case you remove energy from the system; in the other you must put energy into the system. That is, you do work in lifting the object from the floor to the tabletop.

So far in our discussion of potential energy, we have been using mechanical examples. In these examples, the potential energy of the object and the potential energy difference between the floor and the tabletop are due to weight and gravitational forces. In electricity, the potential energy and the potential energy differences are due to *electric fields* and electric charges.

Voltage is a potential energy difference similar to the potential energy difference in the mechanical case discussed above. Instead of being moved by the force of gravity, electric charges are moved by the force of an electric field. In Fig. 2-8, the electron loses energy as it moves from a negatively charged point to a positively charged point. (This is the same as the book's losing energy as it moves from the tabletop to the floor.) The lost energy of the electron could be converted to heat and light as the electron moved through a lamp. There is a voltage (potential

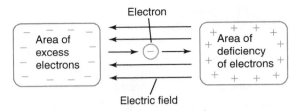

Fig. 2-8 Charge moving through an electric field. Part of the electron's energy is converted to another form of energy.

Fig. 2-9 Lead-acid battery.

energy difference) between the negative and positive areas shown in Fig. 2-8. These negative and positive areas could represent the terminals of an electric battery. A lead-acid storage battery, such as the one shown in Fig. 2-9, is a common source of voltage. A potential energy difference (voltage) exists between the negative and the positive terminals of a battery. This voltage is the result of an excess of electrons at the negative terminal and a deficiency of electrons at the positive terminal. When electrons move from the negative terminal to the positive terminal of the battery, work is done. Energy is taken from the battery and converted to another form of energy.

Like the mechanical system, the electric system can also be reversed. That is, the electron can gain energy when it is moved from the positively charged point to the negatively charged point. This is what happens when a battery is charged. The battery charger forces electrons back through the battery in the reverse direction.

2-9 Unit of Voltage—The Volt

We need a unit to indicate the potential energy difference (voltage) between two points, such as the terminals of a battery. This unit

must specify the energy available when a given charge is transported from a negative to a positive point. We already have the joule as the base unit of energy and the coulomb as the base unit of charge. Therefore, the logical unit of voltage is the *joule per coulomb*. The joule per coulomb is called the *volt*. The volt is the base unit of voltage. It is abbreviated V. A 12.6-V battery, like the ones used in automobiles, is shown in Fig. 2-9. In symbolic form we indicate the voltage of this battery as $V = 12.6$ V. A potential difference (voltage) of 12.6 V means that each coulomb of charge provides 12.6 J of energy. For example, 1 C flowing through a lamp converts 12.6 J of the battery's energy into heat and light energy.

The relationship between charge, energy, and voltage can be expressed as

$$\text{Voltage } (V) = \frac{\text{enery } (W)}{\text{charge } (Q)}$$

or, by rearranging,

$$W = V \times Q$$

This relationship can be used to determine electric energy in the same way as mechanical energy was found in example 2-2.

Joule per coulomb

Volt (V)

$W = V \times Q$

EXAMPLE 2-3

Determine the potential energy (W) of a 6-V battery that has 3000 C of charge (Q) stored in it.

Given:	$V = 6$ V
	$Q = 3000$ C
Find:	W
Known:	$W = VQ$
Solution:	$W = 6$ V $\times$ 3000 C
	$= 18{,}000$ J
Answer:	Potential energy $= 18{,}000$ joules

Notice in example 2-3 that multiplying volts by coulombs yields joules. This is because a volt is a joule per coulomb and the coulombs cancel:

$$\frac{\text{Joule}}{\text{Coulomb}} \times \frac{\text{coulomb}}{1} = \text{joule}$$

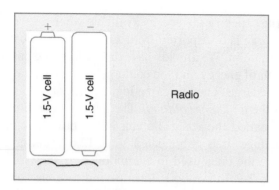

Fig. 2-10 When cells in a radio are installed, the correct polarity must be observed.

You May Recall

... that static electricity, which produces voltage, was discussed in a previous chapter. The mechanical energy required to rub a glass rod against a silk cloth is converted to electric energy. The silk cloth becomes the negative terminal, and the glass rod becomes the positive terminal of this voltage source.

Electric generator

Polarity

Electric chemical cell

Thermocouples

Crystals

Solar cells

Polarized

Seebeck effect

Piezoelectric effect

Straight polarity

Reverse polarity

2-10 Polarity

Polarity is a term that is used in several ways. We can say that the polarity of a charge is negative, or we can say that the polarity of a terminal is positive. We can also use the term to indicate how to connect the negative and positive terminals of electric devices. For example, when putting new cells in a transistor radio, we must install the cells correctly (Fig. 2-10). The positive terminal of one cell connects to the positive terminal of the radio, and the negative terminal of the other cell connects to the negative terminal of the radio.

Electric devices that have negative and positive identifications on their terminals are said to be *polarized*. When connecting such devices to a source of voltage (such as a battery), we must observe the polarity markings. Again, the negative terminal of the device is connected to the negative terminal of the source, and the positive to the positive. If polarity is not observed (that is, if a positive is connected to a negative), the device will not function and may be damaged or ruined.

In electric-arc welding, the welder can weld with either *straight polarity* or *reverse polarity*. The *straight* and *reverse* refer to whether the negative or the positive terminal of the voltage source is connected to the welding rod.

2-11 Sources of Voltage

Voltage can be created by a number of techniques. All involve the conversion of some other form of energy into electric energy. All of them create a voltage by producing an excess of electrons at one terminal and a deficiency of electrons at another terminal.

The most common way of producing a voltage is by an *electric generator*. Generators convert mechanical energy into electric energy. Large generators like the one shown in Fig. 2-11 produce the voltage (and energy) that is provided to our homes, schools, and industries. Most of these generators are turned by mechanical devices, such as steam turbines.

The *electric chemical cell* is the next most common source of voltage. This type of cell converts chemical energy into electric energy. Several cells can be connected together to form a battery. A wide variety of batteries and cells are manufactured. They range from dry cells that weigh a few grams to industrial batteries that weigh hundreds of kilograms.

Other devices that produce voltage are *thermocouples, crystals,* and *solar cells*.

The thermocouple converts heat energy into electric energy. This process is known as the *Seebeck effect*. Thermocouples are used extensively for measuring temperatures, especially high temperatures.

Crystals, like those in Fig. 2-12, produce voltage by the *piezoelectric effect*. A voltage is produced when a varying pressure is applied to

Fig. 2-11 Part of a large electric generator.

Fig. 2-12 Uncut quartz crystal. When stressed, quartz produces voltage.

the surface of the crystal. Crystals are used in such devices as phonograph pickup cartridges and microphones. In a microphone, the sound energy of the voice is first converted to mechanical energy by a diaphragm that applies pressure to the crystal. Thus, crystals convert mechanical energy into electric energy.

Solar cells are semiconductor devices. They convert light energy into electric energy. *Photovoltaic* is the term used to describe this conversion process. Solar cells can be used to provide the voltage needed to operate exposure meters in photography or a satellite communications system.

Photovoltaic

Self-Test

Answer the following questions.

17. A potential energy difference between two points in a circuit is called _____.
18. The base unit of voltage is the _____.
19. The symbol or abbreviation for voltage is _____.
20. _____ is abbreviated emf.
21. The symbol or abbreviation for the base unit of voltage is _____.
22. The electrolyte used to electroplate copper onto iron is _____.
23. Explain the difference between kinetic energy and potential energy.

24. Why is it incomplete to say, "The potential energy difference of the desk top is 9 joules per kilogram"?
25. Why is it incomplete to say, "The voltage at point *A* is 18 volts"?
26. Define the base unit of voltage in terms of energy and charge.
27. Define *polarized*.
28. List five devices that can produce a voltage. Also specify what type of energy they convert into electric energy.
29. What is the potential energy difference between two points if 100 J of energy is required to move 5 C of charge from one point to the other point?

2-12 Resistance

The opposition a material offers to current is called *resistance*. The symbol for resistance is R. All materials offer some resistance to current. However, there is extreme variation in the amount of resistance offered by various materials. It is harder to obtain free electrons (current carriers) from some materials than others. It requires more energy to free an electron in high-resistance materials than in low-resistance materials. Resistance converts electric energy into heat energy when current is forced through a material.

2-13 Conductors

Materials that offer very little resistance (opposition) to current are called *conductors*. Copper, aluminum, and silver are good conductors. They have very low resistance. In general, those elements that have three or fewer electrons in the valence shell can be classified as conductors. However, even within those elements classified as conductors, there is wide variation in the ability to conduct current. For example, iron has nearly six times the resistance of copper, though they are both considered conductors. Although silver is a slightly better conductor

Resistance (R)

Conductors

than copper, it is too expensive for common use. Aluminum is not as good a conductor as copper, but it is cheaper and lighter. Large aluminum conductors are used to bring electric energy into homes.

Superconductivity is the condition in which a material has no resistance. For many years, superconductivity could be demonstrated only at temperatures close to absolute zero, which is about −273°C (degrees Celsius) or −460°F (degrees Fahrenheit). Further research led to the development of new materials that exhibit superconductivity at temperatures well above absolute zero. The aim has always been to find materials that will superconduct at room temperature. Such materials would, obviously, greatly improve the efficiency of any electric system in which they could be used.

2-14 Insulators

Materials that offer a high resistance to current are called *insulators*. Even the best insulators do release an occasional free electron to serve as a current carrier. However, for most practical purposes we can consider an insulator to be a material that allows no current to flow through it. Common insulator materials used in electric devices are paper, wood, plastics, rubber, glass, and mica. Notice that common insulators are not pure elements. They are materials in which two or more elements are joined together to form a new substance. In the process of joining together, elements share their valence electrons. This sharing of valence electrons is called *covalent bonding*. It takes a lot of added energy to break an electron free of a covalent bond.

2-15 Semiconductors

Between the extremes of conductors and insulators are a group of elements known as *semiconductors*. Semiconductor elements have four valence electrons. Two of the best-known semiconductors are silicon and germanium. Semiconductors are neither good conductors nor good insulators. They allow some current to flow, yet they have a considerable amount of resistance. Semiconductors are extremely important industrial materials. They are the materials from which electronic devices such as transistors, integrated circuits (ICs), and solar cells are manufactured.

2-16 Unit of Resistance — The Ohm

So far we have discussed the amount of resistance in terms of *low resistance* and *high resistance*. In order to work with electric circuits, we must be able to state the amount of resistance more specifically. The unit used to specify the amount of resistance is the *ohm*. The ohm is the base unit of resistance. The symbol used as an abbreviation for ohm is Ω (the capital Greek letter *omega*). The ohm is named in honor of Georg Ohm, who worked out the relationship between current, voltage, and resistance. The ohm can be defined in several ways. First, it is the amount of resistance that allows 1 A of current to flow when the voltage is 1 V. Second, it is the amount of resistance of a column of mercury 106.3 centimeters (cm) in length, 1 millimeter square (mm²), and at a temperature of 0°C. From this second definition of the ohm, you can see that the amount of resistance of an object is determined by four factors: (1) the type of material from which the object is made, (2) the length of the object, (3) the cross-sectional area (height × width) of the object, and (4) the temperature of the object. The amount of resistance of an object is directly proportional to its length and inversely proportional to its cross-sectional area. For example, if the length of a piece of copper wire is doubled, then its resistance is also doubled (Fig. 2-13). If the cross-sectional area of the copper wire is made twice as great, then its resistance is one-half its former value. The shaded ends of the conductors in Fig. 2-13 are the cross-sectional areas of the conductors.

No simple relationships exist between resistance and temperature. The resistance of most materials increases as the temperature increases. However, with some materials, such as carbon, the resistance decreases as the temperature increases.

2-17 Temperature Coefficient

The change in resistance corresponding to a change in temperature is known as the *temperature coefficient of resistance*. Each material has its own temperature coefficient. Carbon has a negative temperature coefficient (the resistance decreases as the temperature increases), while most metals have a positive temperature

Superconductivity

Ohm (Ω)

Insulators

Covalent bonding

Semiconductors

Temperature coefficient of resistance

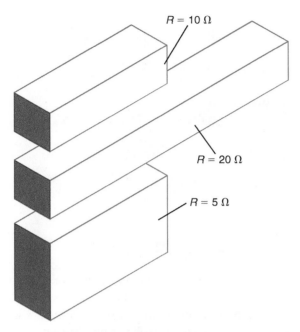

Fig. 2-13 Effects of length and cross-sectional area on resistance.

coefficient (the resistance increases as the temperature increases). A temperature coefficient is expressed as the number of ohms of change per million ohms per degree Celsius, which is abbreviated ppm/°C. For example, carbon has a negative temperature coefficient of 500 ppm/°C at 20°C. That is, a piece of carbon that has 1,000,000 Ω of resistance at 20°C has 1,000,500 Ω at 19°C, and at 18°C it has 1,001,000 Ω. In many electric and electronic devices, changes in resistance due to changes in temperature are so small that they can be ignored. In those devices in which small changes in resistance are important, such as electric meters, special low-temperature-coefficient materials are used. One of these materials is constantan (a mixture of copper and nickel). Constantan has a positive temperature coefficient of 18 ppm/°C at 20°C. Notice that the temperature coefficient is defined at a specific temperature. This means that the temperature coefficient itself changes with temperature. However, these changes are extremely small over the range of temperature in which most electric devices operate. A table of temperature coefficients of resistance for some common materials is in Appendix F.

2-18 Resistivity

The characteristic resistance of a material is given by its *resistivity* or its *specific resistance*.

These terms mean the same thing. The resistivity of a material is just the resistance (in ohms) of a specified-size cube of the material, usually a 1-cm, 1-m, or 1-ft cube. Resistivity ratings allow us to compare the abilities of various materials to conduct current. Figure 2-14 illustrates the way in which resistivity is determined in the base unit of ohm-centimeters ($\Omega \cdot cm$). Annealed copper at 20°C has a resistivity of 1.72×10^{-6} (0.00000172) $\Omega \cdot cm$. This means that a cube of copper 1 cm on each side has $1.72 \times 10^{-6}\ \Omega$ of resistance between any two opposite faces of the cube. A table of resistivity (in ohm-centimeters) is shown in Appendix E. The lower the resistivity of a material, the better a conductor it is.

The relationship of resistance to length, cross-sectional area, and resistivity is given by the formula

$$\text{Resistance } (R) = \frac{\text{resistivity } (K) \times \text{length } (L)}{\text{area } (A)}$$

or, using only symbols,

$$R = \frac{KL}{A}$$

$R = \dfrac{KL}{A}$

Very often the Greek letter rho (ρ) is used as the symbol for resistivity instead of the letter K. Resistance is in ohms if all other quantities are also in base units, that is, if resistivity is in ohm-centimeters, length in centimeters, and area in square centimeters.

A common unit used in electric wiring is the circular mil. In addition to listing wire size in AWG (American Wire Gage), many wire tables also specify the wire diameter in circular mils (CM). The resistance formula can be used with this unit if K is given in ohm-circular mils per foot, L in feet, and A in circular mils.

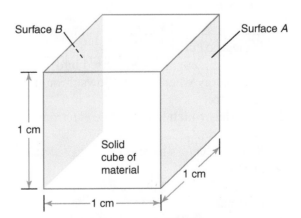

Fig. 2-14 Resistivity equals the resistance between surface A and surface B.

Resistivity

Specific resistance

Alessandro Volta

In 1796 Italian physicist Alessandro Volta developed the first chemical battery, which provided the first practical source of electricity. Named for Volta, the volt is the base unit of voltage.

Find: R

Known: $R = \dfrac{KL}{A}$

Solution: $R =$

$$\dfrac{0.00000172 \ \Omega \ \text{cm} \times 20{,}000 \ \text{cm}}{0.26 \ \text{cm}^2}$$

$$= 0.132 \ \Omega$$

Answer: Resistance = 0.132 ohm

2-19 Resistors

Many electronic devices, such as radio and television receivers, use large amounts of resistance to control the current. When we need resistance to control current, we use resistors. *Resistors* are physical devices that are manufactured in a wide variety of shapes and sizes. They are available in resistance values that range from less than 1Ω to many millions of ohms.

In most electric and electronic devices, the resistance of the conductors is so small compared with the resistance in the other parts of the device that the resistance of the conductors can be ignored. In most cases the resistance of the conductors is not calculated or figured into the design of the device.

Resistors

EXAMPLE 2-4

What is the resistance, at 20°C, of an electric motor winding that uses 200 m of copper conductor which is 0.26 centimeter square (cm²) (approximately 3⁄16 in. by 3⁄16 in.)? The resistivity of copper at 20°C is 0.00000172 $\Omega \cdot$ cm.

Given: $K = 0.00000172$ ohm-centimeter

$L = 20{,}000$ centimeters

$A = 0.26$ square centimeter

⎓ Self-Test

Answer the following questions.

30. _____ can be defined as opposition to electric current.
31. _____ energy is converted into heat energy when current flows through resistance.
32. Materials which have no free electrons are called _____.
33. Semiconductors have _____ valence electrons.
34. The _____ is the base unit of resistance.
35. _____ is the symbol for the base unit of resistance.

36. An ohm is equal to a (n) _____ per ampere.
37. List the four factors that determine the resistance of an object.
38. What is meant by the statement "This material has a negative temperature coefficient of 250 ppm/°C at 20°C"?
39. Classify the following materials as either insulators or conductors:
    a. Iron   e. Aluminum
    b. Rubber   f. Glass
    c. Paper   g. Copper
    d. Silver   h. Mica
40. Give an example of a material that has a negative temperature coefficient.

41. Define *resistivity*.
42. What is a common unit used to specify resistivity?
43. What is a resistor?

44. Aluminum has a resistivity of 17.0 ohm-circular mil per foot at 20°C. What is the resistance, at 20°C, of 200 ft of aluminum wire with a cross-sectional area of 6530 circular mils?

2-20   Power and Energy

Power refers to how rapidly energy is used or converted to another form of energy. Since energy is the ability to do work, we may also say power is concerned with how fast work is done. We will combine these two ideas in our definition of power. *Power* is the *rate of using energy* or doing work. The symbol for power is P.

The power needed to do a specified amount of work depends on how much time it takes to do the work. Suppose two workers are going to load 100 bricks onto a truck. They agree to each load one-half of the bricks. The first worker carries five bricks at a time, finishes half of the work in 10 trips to the truck, and is finished in 40 minutes (4 min per trip). The second worker carries two bricks at a time and finishes in 25 trips. This worker requires 100 minutes (4 min per trip). Both workers have done the same amount of work (loaded 50 bricks). However, carrying five bricks requires more power than carrying two bricks; so the first worker used more power than the second worker. The first worker has done more work per unit of time.

The companies from which we buy our electricity are very often named *power companies*. However, when we pay our electrical bill, we are paying for electric energy, not electric power. The electric power company is less concerned with the power we use (how rapidly we consume electric energy) than with the amount of energy we use.

2-21   Unit of Power

You May Recall

. . . that we use the joule as our base unit of energy and the second as our base unit of time.

Therefore, the logical unit for power is the *joule per second* (J/s). In honor of a Scottish scientist and inventor, James Watt, the joule per second has been named the *watt*. The base unit of power, therefore, is the watt. One watt is equal to 1 J/s. The abbreviation for the watt is W. As with other symbols in electricity, the letter W is used to denote two entirely different things. Don't confuse W, meaning energy, for W, meaning watts. In general, italicized letters (usually capital) represent electrical quantities like current (I) and voltage (V), and nonitalicized (roman) letters represent units like ampere (A) and volt (V).

The relationship between power, energy, and time is

$$\text{Power } (P) = \frac{\text{energy } (W)}{\text{time } (t)}$$

By rearranging terms, we can state the relationship as $W = Pt$. Let us use these relationships to solve some electrical problems.

Power (P)

Joule per second

Watt (W)

Rate of using energy

$W = Pt$

EXAMPLE 2-5

What is the power rating of an electric device that converts 940 J of energy in 10 s?

Given:	$W = 940$ J and $t = 10$ s
Find:	P
Known:	$P = \dfrac{W}{t}$
Solution:	$P = \dfrac{940 \text{ J}}{10 \text{ s}} = 94$ J/s
	$= 94$ W
Answer:	Power rating = 94 watts

EXAMPLE 2-6

How much energy would be required to operate a 60-W lamp for 30 min?

Given: $P = 60$ W and $t = 30$ min

Find: W

Known: $W = Pt$

Solution: Since there are 60 sec in 1 min, 30 min equals

$$30 \times 60 = 1800 \text{ s}$$

$$W = 60 \text{ W} \times 1800 \text{ s}$$

$$= 108{,}000 \text{ wattseconds}$$

$$= 108{,}000 \text{ J}$$

Answer: Energy required = 108,000 joules

Notice that in example 2-6 we first converted time from minutes to the base unit of seconds. By doing this, the answer came out in joules—the base unit of energy. When you are first learning electrical relationships, it is best if you work with all quantities in base units, whenever possible. Also notice in example 2-6 that watt × second = wattsecond. One wattsecond is equal to 1 J. This can be shown by substituting joule per second for watt:

$$\text{Wattsecond} = \frac{\text{joule}}{\text{second}} \times \frac{\text{second}}{1} = \text{joule}$$

Wattsecond

The term *wattsecond* is often used to express the amount of energy.

Since electric power companies charge their customers for energy use, electrical bills must be based on a unit of energy. The joule is much too small to be practical. Imagine getting a bill for several hundred million joules each month! Instead, power companies use the *kilowatthour*, which is equal to 3.6 million J.

2-22 Efficiency

You May Recall

. . . that we previously dealt with the concept of efficiency. We viewed efficiency in terms of useful energy out and total energy in.

Efficiency can also be viewed in terms of power. The formula is the same except that power is substituted for energy:

$$\% \text{ eff.} = \frac{P_{\text{out}}}{P_{\text{in}}} \times 100$$

EXAMPLE 2-7

What is the efficiency of a radio receiver that requires 4 W of electric power input to deliver 0.5 W of power output?

Given: $P_{\text{in}} = 4$ W

$P_{\text{out}} = 0.5$ W

Find: % eff.

Known: $\% \text{ eff.} = \dfrac{P_{\text{out}}}{P_{\text{in}}} \times 100$

Solution: $\% \text{ eff.} = \dfrac{0.5 \text{ W}}{4 \text{ W}} \times 100$

$$= 0.125 \times 100 = 12.5$$

Answer: Efficiency = 12.5 percent

The efficiency formula can also be arranged to allow us to determine the amount of input power.

$$P_{\text{in}} = \frac{P_{\text{out}}}{\% \text{ eff.}} \times 100$$

EXAMPLE 2-8

A stereo amplifier is producing 50 W of output power. How much power input is required if the amplifier is 30 percent efficient?

Given: $P_{\text{out}} = 50$ W

% eff. = 30

Find: P_{in}

Known: $P_{\text{in}} = \dfrac{P_{\text{out}}}{\% \text{ eff.}} \times 100$

Solution: $P_{\text{in}} = \dfrac{50}{30} \times 100$

$$= 1.7 \text{ W} \times 100$$

$$= 170 \text{ W}$$

Answer: Power input = 170 watts

Answer the following questions.

45. True or false. Power and work mean the same thing.
46. True or false. The base unit of power is the joule-second.
47. True or false. A joule per second equals one watt.
48. True or false. A wattsecond equals a joule.
49. True or false. Energy is equal to power divided by time.
50. True or false. A kilowatthour is equal to 3.6 million J.

51. An 850-W toaster requires 4 min to make a slice of toast. How much energy is required to make a slice of toast? Give your answer in joules.
52. A transistor receiver requires 6000 J to operate for 50 min. What is the power rating of the receiver?
53. What is the efficiency of an electric motor that requires 1200 W of electric power to deliver 800 W of mechanical power?

2-23 Powers of 10

By now you may have noticed that some very large numbers are used in electricity. These numbers are easier to manage if they are expressed in powers of 10. *Powers of 10* refer to numbers that are exponents of the base 10. *Base 10* refers to the number system we use in our everyday life. The *exponent* (or power) refers to the number of times that the digit 1 is multiplied by 10. For example, 10^2 means that 1 is multiplied by 10 twice:

$$1 \times 10 \times 10 = 100$$

Ten to the first power (10^1) is $1 \times 10 = 10$; $10^0 = 1$. Powers of 10 and their base-10 equivalents that are commonly used in electricity are listed in Table 2-1. Notice that the power (exponent) of 10 tells you how many places to move the decimal point. For example, 10^4 can be changed to a number without an exponent by writing a 1 and moving the decimal point four places to the right. Thus, 10^4 is equal to 10,000. Also, a number can be converted to a power of 10 by making the exponent (power) equal to the number of places you move the decimal point. Thus, 2100 can be expressed as 2.1×10^3, or 105,000 can be written 1.05×10^5. Notice that it is conventional to move the decimal so that there is only one digit to the left of it. The number multiplied by the power of 10 is called the *coefficient*. In the above examples, 1.05 and 2.1 are the coefficients.

Table 2-1	Powers of 10 and Base-10 Equivalents
Power of 10	Number (Base-10) Equivalent
10^{12}	1,000,000,000,000
10^{11}	100,000,000,000
10^{10}	10,000,000,000
10^9	1,000,000,000
10^8	100,000,000
10^7	10,000,000
10^6	1,000,000
10^5	100,000
10^4	10,000
10^3	1,000
10^2	100
10^1	10
10^0	1
10^{-1}	0.1
10^{-2}	0.01
10^{-3}	0.001
10^{-4}	0.0001
10^{-5}	0.00001
10^{-6}	0.000001
10^{-7}	0.0000001
10^{-8}	0.00000001
10^{-9}	0.000000001
10^{-10}	0.0000000001
10^{-11}	0.00000000001
10^{-12}	0.000000000001

Powers of 10

Base 10

Exponent

Coefficient

History of Electronics

Thomas Seebeck

In 1821 Thomas Seebeck discovered a thermoelectric phenomenon that became known as the Seebeck effect. He found that near a closed circuit composed of two linear conductors of different metals, a magnetic needle would be deflected if the two junctions were at different temperatures. Further, if the cooler junction were to become warmer, the direction of deflection would be reversed. (*McGraw-Hill Multimedia Encyclopedia of Science and Technology, Version 2.1, McGraw-Hill. 2000.*)

Internet Connection

Visit the website for the National Institute of Standards and Technology (NIST) for additional information.

Numbers smaller than 1 are expressed in powers of 10 with negative exponents. The negative exponent tells you how many times 1 is to be divided by 10. Thus 10^{-2} is the same as $1 \div 10 \div 10 = 0.01$. Notice again that the easy way to handle exponents is by moving the decimal point as many places as the exponent. When the exponent (power) is negative, move the decimal to the left. To convert 10^{-3}, just write a 1 and move the decimal three places to the left. This procedure yields 0.001, which is equal to 10^{-3}. To convert to a negative power of 10, just move the decimal to the right of the first digit larger than zero. The negative exponent equals the number of places you moved the decimal place. For example, 0.000000054 is equal to 5.4×10^{-8}, and 0.03816 is equal to 3.816×10^{-2}.

EXAMPLE 2-9

A conductor has a resistance (R) of 0.045 ohms (Ω). Express this resistance in the powers of 10 format.

Given: $R = 0.045\ \Omega$

Find: R in powers of 10

Known: The coefficient usually has only one digit greater than zero to the left of the decimal point.

When the decimal point is moved to the right, the exponent will be negative.

Solution: Move the decimal point two places to the right to provide a coefficient of 4.5.

The exponent will therefore be negative 2.

$R = 0.045\ \Omega = 4.5 \times 10^{-2}\ \Omega$

Answer: resistance = 4.5×10^{-2} ohms

The big advantage to expressing numbers in powers of 10 is that it simplifies arithmetic involving large numbers. To multiply two numbers expressed in powers of 10, just multiply the coefficients and then *add* the powers. The product is the new coefficient multiplied by the new power of 10. Some examples are

$$10^4 \times 10^2 = 10^6$$
$$10^{-2} \times 10^4 = 10^2$$
$$10^{-5} \times 10^3 = 10^{-2}$$
$$1.4 \times 10^2 \times 1.2 \times 10^6 = 1.4 \times 1.2 \times 10^2 \times 10^6$$
$$= 1.68 \times 10^8$$
$$6.3 \times 10^3 \times 8.4 \times 10^4 = 6.3 \times 8.4 \times 10^3 \times 10^4$$
$$= 52.92 \times 10^7$$
$$= 5.292 \times 10^8$$

To divide with powers of 10, first divide the coefficients. Then subtract the exponent in the divisor from the exponent in the dividend. This procedure is illustrated below:

$$(1 \times 10^4) \div (1 \times 10^2) = 1 \times 10^2$$
$$(4 \times 10^3) \div (2 \times 10^{-2}) = (4 \div 2)(10^3$$
$$\div 10^{-2})$$
$$= 2 \times 10^{3-(-2)}$$
$$= 2 \times 10^5$$
$$(6 \times 10^{-10}) \div (4 \times 10^{-8}) = (6 \div 4)$$
$$(10^{-10} \div 10^{-8})$$
$$= 1.5 \times 10^{-2}$$

If numbers expressed as powers of 10 are to be added or subtracted, both numbers must have the same exponent. The exponent then remains the same. For example:

$$(2.4 \times 10^6) + (3.5 \times 10^6) = 5.9 \times 10^6$$
$$(2.4 \times 10^4) + (3.5 \times 10^5) = (0.24 \times 10^5)$$
$$+ (3.5 \times 10^5)$$
$$= 3.74 \times 10^5$$
$$(3.8 \times 10^3) - (1.6 \times 10^3) = 2.2 \times 10^3$$

EXAMPLE 2-10

Solve this division problem and show your answer as a simple base-10 number.
$(2.1 \times 10^4 - 1.8 \times 10^3) \div (3.3 \times 10^2 + 1.5 \times 10^3)$

Given: Dividend $= (2.1 \times 10^4 - 1.8 \times 10^3)$
Divisor $= (3.3 \times 10^2 + 1.5 \times 10^3)$

Known: To either add or subtract numbers in powers of 10, both numbers must have the same exponent. (This may require that one of the coefficients has more than one digit to the left of the decimal point.) The answer will have the same exponent as the numbers being added or subtracted.

When adding, the coefficients of the numbers are added together to obtain the coefficient for the answer.

When subtracting, the coefficients of the numbers are subtracted to obtain the coefficient for the answer.

When dividing, the dividend coefficient is divided by the divisor coefficient and the divisor exponent is subtracted from the dividend exponent to get the coefficient and exponent of the answer.

Solution: $2.1 \times 10^4 - 1.8 \times 10^3 = 21 \times 10^3 - 1.8 \times 10^3 = 19.2 \times 10^3$
$3.3 \times 10^2 + 1.5 \times 10^3 = 3.3 \times 10^2 + 15 \times 10^2 = 18.3 \times 10^2$
Coefficient of the answer is
$19.2 \div 18.3 = 1.0492$
Exponent of the answer is
$10^3 - 10^2 = 10^1$

Answer: $1.0492 \times 10^1 = 10.492$

Self-Test

Answer the following questions.

54. Express the following numbers in powers of 10:
    a. 180
    b. 42,000
    c. 2,000,000

55. Convert the following powers of 10 to ordinary numbers:
    a. 3.1×10^3
    b. 10^4
    c. 2.46×10^3

56. Convert the following powers of 10 to ordinary numbers:
    a. 10^{-4}
    b. 2.81×10^{-3}
    c. 6.3×10^{-4}
    d. 6.3×10^2

57. Convert the following numbers to powers of 10:
    a. 0.0000001
    b. 0.028

c. 0.0072
d. 1000

58. Solve the following problems:
    a. $2 \times 10^8 \times 4 \times 10^3$
    b. $1.4 \times 10^2 \times 2.8 \times 10^{-3}$
    c. $\dfrac{6.6 \times 10^4}{3 \times 10^{-2}}$
    d. $\dfrac{4 \times 10^{-3}}{2 \times 10^2}$
    e. $(4 \times 10^3) + (6 \times 10^4)$

59. Solve this problem in these steps and show the results of each step in your answer:

    Step 1. Express the dividend with a single coefficient and exponent.

    Step 2. Express the divisor with a single coefficient and exponent.

    Step 3. Express the answer as a simple base-10 number.
    $$\dfrac{(3 \times 10^{-3}) + (4 \times 10^{-2})}{(2 \times 10^{-2}) - (1.2 \times 10^{-3})}$$

2-24 Multiple and Submultiple Units

For some applications of electricity, the base unit of a quantity may seem to be very large. For other applications, the same base unit may seem rather small. For example, in solid-state devices, we work with currents of less than 0.0000001 A. In an aluminum reduction plant, currents are greater than 110,000 A. Although these numbers could be shortened by expressing them in powers of 10, they would still be long expressions. Also, they would be long when spoken. For example, 1.1×10^5 A would be spoken as "one point one times 10 to the fifth amperes." To avoid such long expressions, scientists use *prefixes* to indicate units that are smaller and larger than the base unit.

The prefixes and their symbols commonly used in electricity are shown in Table 2-2.

Also shown in Table 2-2 are the relationships (both powers of 10 and base 10) of the prefixes to the base unit. Notice that adjacent prefixes are related by factors of 1000. Adjacent prefixes are either 1000 times larger or 1/1000 as large as their neighbor.

Multiple and submultiple units are designated by adding the appropriate prefix to the base unit. Now we can specify the 110,000 A used in an aluminum plant as

Prefixes

Multiple and submultiple units

Multiples of three

Table 2-2 Prefixes and Symbols

Prefix	Symbol	Number (Base 10)	Power of 10
Tera	T	1,000,000,000,000	10^{12}
Giga	G	1,000,000,000	10^{9}
Mega	M	1,000,000	10^{6}
Kilo	k	1000	10^{3}
Base unit		1	10^{0}
Milli	m	0.001	10^{-3}
Micro	μ	0.000001	10^{-6}
Nano	n	0.000000001	10^{-9}
Pico	p	0.000000000001	10^{-12}

0.11 megampere (MA) or 110 kiloamperes (kA). The 0.0000001 A used in a solid-state device can be written as 0.1 microampere (μA). Some other examples of conversions between units are

$$2200 \text{ ohms } (\Omega) = 2.2 \text{ kilohms } (k\Omega)$$

$$0.083 \text{ watt } (W) = 83 \text{ milliwatts } (mW)$$

$$450,000 \text{ volts } (V) = 450 \text{ kilovolts } (kV)$$

$$2.7 \times 10^6 \text{ ohms } (\Omega) = 2.7 \text{ megohms } (M\Omega)$$

$$3700 \text{ microamperes } (\mu A)$$
$$= 3.7 \text{ milliamperes } (mA)$$

$$6,800,000 \text{ ohms } (\Omega) = 6.8 \times 10^6 \text{ ohms } (\Omega)$$
$$= 6.8 \text{ megohms } (M\Omega)$$

Notice that in all the above examples the conversion is made by moving the decimal point either three places or *multiples of three* places. If you are converting from a smaller unit to a larger unit, the decimal point is moved to the left. Remember that going from *micro* to *milli* is going from a smaller to a larger unit. When converting from a larger unit to a smaller unit, you shift the decimal point to the right.

It is important that you become very familiar with converting from one unit to another. A repair manual for an electric device may list a resistor as 2.2 kΩ. When the technician goes to order a replacement resistor, the parts manufacturer may list the resistor as 2200 Ω. The technician must make the conversion.

History of Electronics

James Watt
The unit of electric power, the watt, is named for James Watt. One watt equals 1 joule of energy transferred in 1 second.

EXAMPLE 2-11

What is the efficiency of a device that requires 0.8 kW of input to deliver 720 W of output?

Given: $P_{in} = 0.8$ kW
$P_{out} = 720$ W

Find: % eff.

Known: % eff. $= \dfrac{P_{out}}{P_{in}} \times 100$

1 kW $= 1000$ W

Solution: $P_{in} = 0.8$ kW $= 800$ W

% eff. $= \dfrac{720 \text{ W}}{800 \text{ W}} \times 100$

$= 0.9 \times 100 = 90\%$

Answer: Efficiency $= 90\%$

In example 2-11, we could have left P_{in} in kW and converted P_{out} to 0.72 kW. Then 0.72 kW ÷ 0.8 kW would also equal 0.9 because the kW units would cancel out just as the W units did.

Self-Test

Answer the following questions.

60. Complete the conversions:
    a. 120 millivolts = _____ volt
    b. 3800 ohms = _____ kilohms
    c. 490 microamperes = _____ ampere
    d. 5.6×10^5 ohms = _____ megohm
    e. 6000 millicoulombs = _____ coulombs

61. Complete the following conversions:
    a. 53 mA = _____ A
    b. 4.7 kΩ = _____ Ω
    c. 0.4 V = _____ mV

2-25 Special Units and Conversions

In this book, standard metric units are used whenever practical. However, there are some areas of electrical work in which nonmetric units are so common that we must consider them.

As noted earlier, electric power companies sell their electric energy by the kilowatthour. A *kilowatthour* (kWh) = 1000 watthours (Wh). An hour = 3600 s. Therefore, a watthour = 3600 wattseconds. And finally, a kilowatthour = 3,600,000 wattseconds or joules. As you can see, the joule would be a small unit of energy for a power company to use. Consider, for instance, that many homes use more than 1000 kWh of energy every month. Remember that the wattsecond and watthour are perfectly usable units of energy, even though the base unit is joule.

EXAMPLE 2-12

How much energy is used by a 1200-W heater in 4 hours of continuous operation?

Given: $P = 1200$ W
$t = 4$ hours

Find: Energy

Known: $W = Pt$

Solution: 1200 W × 4 hours = 4800 Wh

Answer: Energy used = 4800 watthours or 4.8 kilowatthours

Kilowatthour (kWh)

The output power of an electric motor (and many mechanical devices) is specified in *horsepower* (hp) rather than in watts. One *horsepower* is equal to *746 W*. When figuring the efficiency of a motor, you must first convert the output power to watts.

Horsepower (hp)

hp = 746w

EXAMPLE 2-13

What is the efficiency of a ¾-hp motor that requires an input of 1000 W of electric power?

Given: Power input = 1000 W
Power output = ¾ hp

Find: Efficiency

Known: % eff. = $\dfrac{P_{out}}{P_{in}} \times 100$

1 hp = 746 W
¾ hp = 0.75 hp

Solution: $P_{out} = 0.75 \text{ hp} \times 746 \dfrac{\text{W}}{\text{hp}}$

= 559.5 W

% eff. = $\dfrac{559.5 \text{ W}}{1000 \text{ W}} \times 100$

= 55.95

Answer: Efficiency = about 56 percent

 Self-Test

Answer the following questions.

62. Name two ways to condense a large value of an electrical quantity, such as 3,600,000 ohms.

63. What is horsepower?

64. How much power is required by a ½-hp motor that is 62 percent efficient?

65. How many J of energy are used by a 2.3 kW device in one hour and 20 minutes?

Chapter 2 Summary and Review

1. Charge is the electrical property of electrons and protons.
2. One coulomb (C) is the charge possessed by 6.25×10^{18} electrons.
3. Current is the movement of charge in a specified direction.
4. Current can flow in solids, gases, liquids, and a vacuum.
5. In solids the current carriers are electrons.
6. In gases the current carriers are both electrons and ions.
7. In liquids the current carriers are ions—both positive and negative.
8. Current travels at approximately the speed of light.
9. Individual current carriers (electrons) travel much slower than the speed of light.
10. Direct current (dc) never reverses direction.
11. Alternating current (ac) periodically reverses direction.
12. A liquid containing ions is an electrolyte.
13. Voltage is a potential energy difference between two points.
14. Polarity indicates whether a point is negative or positive.
15. Sources of voltage include generators, batteries, thermocouples, solar cells, and crystals.
16. Resistance is opposition to current.
17. Resistance converts electric energy to heat energy.
18. Conductors are materials that have low resistance.
19. Silver, copper, and aluminum (in that order) are the best conductors.
20. Insulators do not allow any current to flow (for practical purposes).
21. An ohm (Ω) is 1 volt per ampere (V/A). It is the resistance of a specified column of mercury at a specified temperature.
22. The resistance of an object is determined by the resistivity, length, cross-sectional area, and temperature of the material.
23. Resistance is directly proportional to length.
24. Resistance is inversely proportional to cross-sectional area.
25. Most conductors have a positive temperature coefficient.
26. A temperature coefficient of resistance specifies the number of ohms of change per million ohms per degree Celsius (abbreviated ppm/°C).
27. Resistivity (specific resistance) of a material is the resistance of a specified-size cube of the material.
28. Resistors are devices used to control current.
29. Power is the rate of doing work or converting energy.
30. A kilowatthour (kWh) is a unit of energy.
31. Horsepower is a unit of power.
32. Table 2-3 is a summary of the units and symbols used throughout this book.
33. Symbols for electric quantities are italicized.
34. Abbreviations for electric base units and prefixes are not italicized.
35. Most of these symbols and abbreviations use capital letters, but a few use lowercase letters.

Table 2-3	Symbols and Abbreviations		
Quantity	Symbol	Base Unit	Abbreviation
Charge	Q	Coulomb	C
Current	I	Ampere	A
Time	t	Second	s
Voltage	V	Volt	V
Energy	W	Joule	J
Resistance	R	Ohm	Ω
Power	P	Watt	W

Related Formulas

$$A = \frac{C}{s}$$

$$W = V \times Q$$

$$\Omega = \frac{V}{A}$$

$$W = \frac{J}{s}$$

$$W = Pt$$

$$\% \text{ eff.} = \frac{P_{out}}{P_{in}} \times 100$$

$$hp = 746 \text{ W}$$

$$R = \frac{KL}{A}$$

Chapter Review Questions

For questions 2-1 to 2-11, determine whether each statement is true or false.

2-1. Neutrons can be current carriers in an electric circuit. (2-3)

2-2. One wattsecond is equal to 1 joule. (2-21)

2-3. 5 microamperes = 0.005 ampere. (2-24)

2-4. Semiconductors generally have five valence electrons. (2-15)

2-5. The liquid in a battery is called an electrolyte. (2-5)

2-6. A potential energy difference between two points is called power. (2-8)

2-7. Copper is a better conductor than aluminum. (2-13)

2-8. $3.2 \times 10^3 = 3200$ (2-23)

2-9. $6 \times 10^8 \times 3 \times 10^{-6} = 1.8 \times 10^3$ (2-23)

2-10. 0.420 watts = 42 milliwatts (2-24)

2-11. Most conductors have a negative temperature coefficient. (2-17)

For questions 2-12 to 2-18, choose the letter that best completes each sentence.

2-12. A joule per second defines a (2-21)
   a. Volt
   b. Ampere
   c. Watt
   d. Ohm

2-13. A coulomb per second defines a (2-7)
   a. Volt
   b. Ampere
   c. Watt
   d. Ohm

2-14. Electric energy is converted into heat energy by (2-12)
   a. Voltage
   b. Charge
   c. Power
   d. Resistance

2-15. Charge has base units of (2-2)
   a. Coulombs
   b. Watts
   c. Protons
   d. Amperes

2-16. Voltage has base units of (2-9)
   a. Ohms
   b. Amperes
   c. Watts
   d. Volts

2-17. Resistance has base units of (2-16)
   a. Ohms
   b. Amperes
   c. Watts
   d. Volts

2-18. Current has base units of (2-7)
   a. Ohms
   b. Amperes
   c. Watts
   d. Volts

Answer the following questions.

2-19. What are the four factors that determine the resistance of an object? (2-16)

2-20. What is the symbol for each of the following terms?
   a. Volt (2-9)
   b. Ampere (2-7)
   c. Resistance (2-12)
   d. Ohm (2-16)
   e. Watt (2-21)
   f. Charge (2-2)
   g. Milliampere (2-24)
   h. Horsepower (2-25)

Chapter Review Problems

2-1. How much energy does a motor use in 6 hours if it requires 650 watts to operate the motor? (2-21)

2-2. What is the percent efficiency of a 1-horsepower motor that requires 1.2 kilowatts to operate? (2-22)

2-3. What is the power rating of a television receiver that uses 162,000 joules to operate for 1.25 hours? (2-21)

2-4. An electronic amplifier produces 60 W of output power. If it is 45 percent efficient, how much power input power does it require? (2-22)

2-5. A 12.6-V battery forces 5 coulombs through a load. How much energy does the load convert? (2-9)

2-6. A lamp requires 0.30 kilowatthour to operate for 90 minutes. What is the power rating of the lamp? (2-21)

2-7. How many seconds are required by a 120-W device to convert 3 Wh of energy? (2-21)

2-8. It requires 3 s for 12.6 C to enter the positive terminal of a battery in an electric circuit. How much current is flowing in the circuit? (2-7)

2-9. If a 24-V battery can provide 48 kilojoules of energy, how much charge is stored in it? (2-9)

2-10. What is the percentage of efficiency of a motor that requires 7.5×10^3 W while it is producing 5 hp? (2-22)

2-11. The resistivity of aluminum, at 20°C, is 2.62×10^{-6} Ω·cm. What is the resistance of an aluminum conductor, at 20°C, when the conductor is 20 meters long and has a cross-sectional area of 1.5×10^{-1} cm²? (2-18)

Critical Thinking Questions

2-1. Explain why efficiency can be calculated using either P_{out} and P_{in} or W_{out} and W_{in}.

2-2. Why does the temperature of a solid conductor increase when the conductor is carrying current?

2-3. Why does a long conductor have more resistance than a short conductor of the same material, cross-sectional area, and temperature?

2-4. Which base unit is represented by the following:

$$\frac{\text{Newton-meter} \times \text{ampere}}{\text{Coulomb}}$$

Explain how you arrived at your answer.

2-5. Prove that a joule per coulomb is equal to a watt per ampere.

2-6. Do you think horsepower is as meaningful a unit of power as the watt? Why?

2-7. What is the power rating of a light that is 20 percent efficient and produces 27 Wh of light energy in 18 minutes? Show the conversions and formulas you used.

2-8. The prefix *mega* multiplies the base unit by 1 million. What do the prefixes *giga* and *tera* multiply the base unit by?

2-9. The heater motor in a car takes 900 C of Q from the car's 12.6-V battery in 5 minutes of operation. How many J of W does the heater motor convert in 15 minutes of operation? How much I is the battery providing?

Answers to Self-Tests

1. An electric charge is the electrical property exhibited by protons and electrons.
2. Q
3. 6.25×10^{18}
4. C
5. electron, ion
6. electron
7. F

8. F
9. F
10. The electron travels from atom to atom. The individual electron moves down the wire slowly.
11. Alternating current periodically reverses its direction, while direct current flows continuously in the same direction.

12. ampere
13. An ampere is equal to 1 coulomb per second.
14. A
15. a. $I = 8A$
    b. $Q = 6\,C$
16. 4 A
17. voltage
18. volt
19. V
20. Electromotive force
21. V
22. copper sulfate
23. Kinetic energy is energy being converted or energy in use. Potential energy is energy at rest or stored energy.
24. Because potential energy differences can exist only between two points. The reference point (floor) must be specified.
25. Because voltage is a potential energy difference. This means that voltage always exists between two points.
26. The volt is equal to 1 joule per coulomb.
27. Polarized means that an electric device has a positive and a negative terminal.
28. battery—chemical energy, crystal (microphone, cartridge)—mechanical energy, generator —mechanical energy, thermocouple—heat energy, solar cell—light energy
29. 20 V
30. Resistance
31. Electric
32. insulators
33. four
34. ohm
35. Ω
36. volt
37. resistivity of the material from which the object is made, length of the object, cross-sectional area of the object, and temperature of the object.
38. The statement means that the resistance of the material decreases 250 ohms for every million ohms for each degree Celsius that the temperature increases above 20°C.
39. conductors: iron, silver, aluminum, and copper; insulators: rubber, paper, glass, and mica
40. carbon
41. Resistivity is the characteristic resistance of a material. It is the resistance of a specified-size cube of

the material. The cube is usually defined as being either 1 cm, 1 m, or 1 ft on each side.
42. ohm-centimeter
43. A resistor is a physical device designed to provide a controlled amount of resistance.
44. 0.52 Ω
45. F
46. F
47. T
48. T
49. F
50. T
51. **Given:** $P = 850\ \text{W}, t = 4\ \text{min}$
    Find: W
    Known: $W = Pt$
                60 seconds = 1 min
    Solution: $t = 4\ \text{min} \times 60\ \frac{\text{s}}{\text{min}} = 240\ \text{s}$
                $W = 850 \times 240 = 204{,}000\ \text{J}$
    Answer: Energy = 204,000 joules
52. **Given:** $W = 6000\ \text{J}, t = 50\ \text{min}$
    Find: P
    Known: $P = \dfrac{W}{t}$
    Solution: $t = 50\ \text{min} \times 60\ \frac{\text{s}}{\text{min}} = 3000\ \text{s}$
                $P = \dfrac{6000}{3000} = 2\ \text{W}$
    Answer: Power = 2 watts
53. **Given:** $P_{\text{in}} = 1200\ \text{W}, P_{\text{out}} = 800\ \text{W}$
    Find: % eff.
    Known: % eff. $= \dfrac{P_{\text{out}}}{P_{\text{in}}} \times 100$
    Solution: % eff. $= \dfrac{800}{1200} \times 100 = 66.7$
    Answer: % eff. = 66.7%
54. a. 1.8×10^2 c. 2×10^6
    b. 4.2×10^4
55. a. 3100   c. 2460
    b. 10,000
56. a. 0.0001   c. 0.00063
    b. 0.00281   d. 630
57. a. 1×10^{-7} or 10^{-7}
    b. 2.8×10^{-2}
    c. 7.2×10^{-3}
    d. 1×10^3 or 10^3
58. a. 8×10^{11} d. 2×10^{-5}
    b. 3.92×10^{-1} e. 6.4×10^4
    c. 2.2×10^6
59. $\dfrac{(3 \times 10^{-3}) + (4 \times 10^{-2})}{(2 \times 10^{-2}) - (1.2 \times 10^{-3})} = \dfrac{4.3 \times 10^{-2}}{1.88 \times 10^{-2}} = 2.287$

60. a. 0.12 V
    b. 3.8 kΩ
    c. 0.00049 A

    d. 0.56 MΩ
    e. 6 C

61. a. 0.053 A
    b. 4700 Ω

    c. 400 mV

62. By using powers of 10 (3.6×10^6 ohms) or by using multiple units (3.6 megohms)

63. Horsepower is a nonmetric unit of power. It is equal to 746 watts.

64. 601.6 W

65. 11.04 MJ

Basic Circuits, Laws, and Measurements

Learning Outcomes

This chapter will help you to:

3-1 *Specify* the four essential parts of a complete circuit.

3-2 *Identify* circuit components and symbols.

3-3 *Use* electrical laws and formulas to solve electrical problems involving I, V, R, P, and W, time and cost.

3-4 *Connect* and read panel meters and multimeters to measure resistance, voltage, and current.

You are now familiar with electrical quantities and units. Now you are ready to explore the circuits, laws, and devices used to control and measure these quantities and units.

3-1   Circuit Essentials

Many complete electric circuits contain six parts:

1. An *energy source* to provide the voltage needed to force current (electrons) through the circuit
2. *Conductors* through which the current can travel
3. *Insulators* to confine the current to the desired paths (conductors, resistors, etc.)
4. A *load* to control the amount of current and convert the electric energy taken from the energy source
5. A *control device,* often a switch, to start and stop the flow of current
6. A *protection device* to interrupt the circuit in case of a circuit malfunction

The first four of the above six parts are essential parts. All complete circuits use them. The control device (item 5) and/or the protection device (item 6) will be omitted in some circuits. Item 6 is the most often omitted item. A complete electric circuit has an uninterrupted path for current (electrons) to flow from the negative terminal of the energy source through the load and control device to the positive terminal of the energy source.

The simplest electric circuit contains only one load, one voltage source, and one control device. It is sometimes referred to as a *simple circuit* to distinguish it from more complex circuits. A single-cell flashlight is an example of a simple circuit. Figure 3-1(*a*) shows (in cross section) the construction details of such a flashlight.

Simple circuit

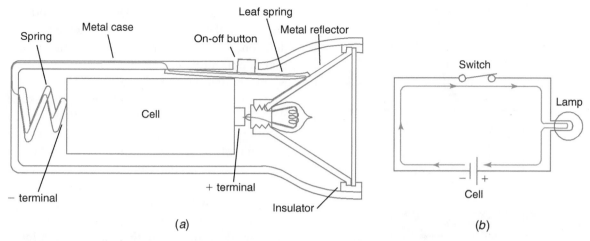

Fig. 3-1 Flashlight. (a) Cross-sectional drawing. (b) Schematic diagram.

Electron flow (current) in the flashlight circuit can be traced by referring to Fig. 3-1. Electrons leave the negative end of the cell, travel through the spring, the metal case, the leaf spring of the switch, the metal reflector, and the light bulb, and back to the positive end of the cell. Note that the spring, case, and reflector are conductors for the flashlight circuit. It is quite common in electric devices for structural parts of the device to serve also as conductors. For example, both an automobile frame and the chassis of many electronic devices serve as circuit conductors.

3-2 Circuit Symbols and Diagrams

In describing an electric circuit, you will find it more convenient to use symbols to represent electric components than to draw pictures of the components. A resistor and the symbol used to represent it are shown in Fig. 3-2. This same symbol is used for all fixed resistors regardless of the material from which they are made. Other common electric components and their symbols are shown in Fig. 3-3. It should be noticed that there is no symbol to distinguish insulated from noninsulated conductors. Insulation is assumed to be present wherever it is needed to keep components and conductors from making

electric contact. In constructing an electric circuit that is described by electrical symbols, the circuit builder must determine where insulation is needed.

A drawing that uses only symbols to show how components are connected together is called a *schematic diagram*. A schematic diagram shows only how the parts are electrically interconnected. The physical size and the mechanical arrangements of the parts are in no way indicated. Also, accessories, such as battery or lamp holders, are not indicated. Unless the schematic diagram is accompanied by some form of pictorial drawing, the physical arrangement of the electric components is left to the discretion of the circuit builder.

Schematic diagram

History of Electronics

Thomas Edison
One of the inventions Thomas Edison is well known for is the light bulb. Through his experiments, he discovered that electrons are emitted when the filament of an incandescent light bulb is heated by passing an electric current through it. (*Encyclopedia of Electronics, Gibilisco* and Sclater, McGraw-Hill, 1990.)

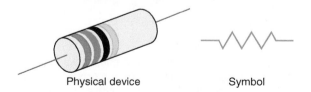

Physical device   Symbol

Fig. 3-2 Resistor and its symbol.

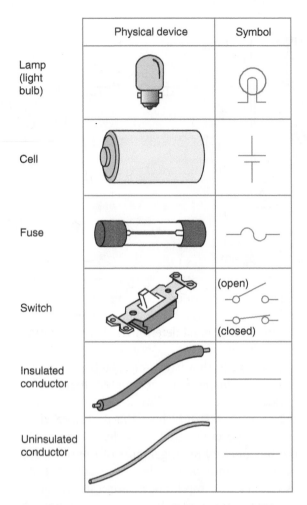

Physical device		Symbol
Lamp (light bulb)		
Cell		
Fuse		
Switch		(open) / (closed)
Insulated conductor		———
Uninsulated conductor		———

Fig. 3-3 Electric components and their symbols.

(which has an excess of electrons) around the circuit to the positive terminal (which has a deficiency of electrons). Therefore, the direction of current flow is as indicated in Fig. 3-1(b). Many books (especially those dealing with transistors, digital circuits, and other solid-state devices) use conventional current flow, which assumes that current flows out of the positive terminal of the source and into the negative terminal of the source. Any circuit or system can be analyzed using either electron-flow direction or conventional-flow direction. Many people find it easier to visualize, and remember, the electron-flow direction than the conventional direction of current.

The *common ground symbol* shown in Fig. 3-4 is often used in schematic diagrams—especially in schematic diagrams for complex circuits and systems. The symbol does not represent any specific electric component. Rather, it represents a common electric point in an electric circuit (or an electric system) that is a common connecting point for many components. For example, the metal frame or chassis of an automobile is the common ground for the many electric circuits used in the automobile. Usually, the negative terminal of the car battery is

Direction of current flow

A schematic diagram of the flashlight illustrated in Fig. 3-1(a) is shown in Fig. 3-1(b). Current can also be traced in the schematic diagram. The line with the arrowheads indicates the *direction of current flow*. The electrons (current) flow from the negative end of the cell through the closed switch and the lamp and back to the positive end of the cell. The conductor between the negative terminal of the cell and the switch in Fig. 3-1(b) represents the spring and the metal case in Fig. 3-1(a). The contact between the positive end of the cell and the light bulb in Fig. 3-1(a) is represented by the long line in the schematic diagram. Notice that lines in a schematic do not necessarily indicate a wire, but they do indicate a path for current to flow through.

In this text, we will always use the direction of electron flow as the direction of current flow. For a voltage source like the cell in Fig. 3-1(b), electrons flow out of the negative terminal

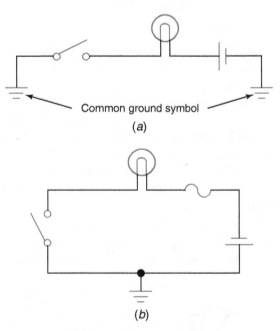

Common ground symbol
(a)

(b)

Fig. 3-4 Common grounds. (a) A conductive path exists between the common grounds. (b) The switch and the positive terminal are connected to the common ground.

connected only to the frame or chassis. Then, any circuit that needs an electric connection to the negative terminal of the battery can be physically (and electrically) connected to any convenient spot on the frame or chassis. This idea is illustrated in Fig. 3-4(*a*), where the two ground symbols tell us that there is a conductive path between them. Figure 3-4(*b*) shows another way to indicate that a circuit is connected to a common ground.

The *electrical values* of the components used in the circuit can also be included on the schematic diagram. This is done in one of two ways. In the first method, shown in Fig. 3-5(*a*), the values of the components are printed beside the symbols for the components. In the second method, shown in Fig. 3-5(*b*), an identifying letter or symbol is printed beside each component and the values of the components are given in an accompanying parts list.

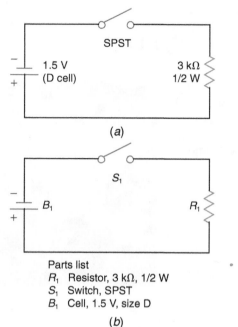

(a)

Electrical values

Parts list
R_1 Resistor, 3 kΩ, 1/2 W
S_1 Switch, SPST
B_1 Cell, 1.5 V, size D

(b)

Fig. 3-5 Specifying component values. (*a*) Values given on the diagram. (*b*) Values given in a parts list.

Self-Test

Answer the following questions.

1. What are the six parts of a complete circuit?
2. True or false. Component values are always given on a schematic diagram.
3. True or false. The conductors in an electric circuit are always insulated wire conductors.
4. Draw the symbol for a lamp, a resistor, and a conductor.
5. What is a schematic diagram?
6. True or false. The chassis or frame of a device can serve as a conductor for more than one circuit.
7. True or false. A schematic diagram is used to show both the electrical and the mechanical layout of a circuit.
8. Which part of a complete circuit is most often omitted?

3-3 Calculating Electrical Quantities

You May Recall

... that in previous chapters you learned how to use relationships between electrical quantities to calculate other electrical quantities.

In this section we work with more relationships of electrical quantities. These new relationships emphasize quantities which can be easily measured or are commonly specified by manufacturers of electrical products.

Ohm's Law

The relationship between current (*I*), voltage (*V*), and resistance (*R*) was discovered by a German scientist named Georg Ohm. This relationship is named *Ohm's law* in his honor. Ohm found that the current in a circuit varies directly with the voltage when the resistance is kept constant. While keeping the resistance

Ohm's law

Georg Simon Ohm

The unit of measure for resistance (ohm) is named for German physicist Georg Simon Ohm. Ohm is also known for his development of Ohm's law:

Voltage = current × resistance

constant, Ohm varied the voltage across the resistance and measured the current through it. In each case, when he divided the voltage by the current, the result was the same. In short, this is Ohm's law, which can be stated as "The current is directly proportional to the voltage and inversely proportional to the resistance."

Written as a mathematical expression, Ohm's law is

$$\text{Current } (I) = \frac{\text{voltage } (V)}{\text{resistance } (R)} \quad \text{or} \quad I = \frac{V}{R}$$

$I = \dfrac{V}{R}$

This equation allows you to determine the value of the current when the voltage and the resistance are known.

Of course, Ohm's law can be rearranged to solve for either resistance or voltage. The rearranged relationships are

$$\text{Resistance } (R) = \frac{\text{voltage } (V)}{\text{current } (I)} \quad \text{or} \quad R = \frac{V}{I}$$

$R = \dfrac{V}{I}$

and

$$\text{Voltage } (V) = \text{current } (I) \times \text{resistance } (R)$$

or

$$V = IR$$

$V = IR$

An aid to remembering the Ohm's law relationships is shown in the divided circle of Fig. 3-6. To use the aid, just cover the quantity you want to find and perform the multiplication or division indicated. Cover the V of Fig. 3-6, and the remainder of the circle indicates I times R. Thus, voltage (V) equals current (I) times resistance (R). Cover the R, and the remainder of the circle shows voltage (V) divided by the current (I). Finally, if

the current (I) is covered, the indicated operation is to divide the voltage (V) by resistance (R).

EXAMPLE 3-1

How much current (I) flows in the circuit shown in Fig. 3-7?

Given:	Voltage (V) = 2.8 volts (V)
	Resistance (R) = 1.4 kilohm (1.4 kΩ)
Find:	Current (I)
Known:	$I = \dfrac{V}{R}$, 1.4 kΩ = 1400 Ω
Solution:	$I = \dfrac{2.8 \text{ V}}{1400 \text{ Ω}}$
	= 0.002 ampere (A)
Answer:	The current is 0.002 A.

EXAMPLE 3-2

A lamp has a resistance of 96 ohms. How much current flows through the lamp when it is connected to 120 volts?

Given:	$R = 96 \text{ Ω}$
	$V = 120 \text{ V}$
Find:	I
Known:	$I = \dfrac{V}{R}$
Solution:	$I = \dfrac{120 \text{ V}}{96 \text{ Ω}} = 1.25 \text{ A}$
Answer:	The current through the lamp equals 1.25 A.

Notice in the preceding examples that the answers for the current are in their base units. This is because both voltage and resistance are also in base units. Remember that an ohm is

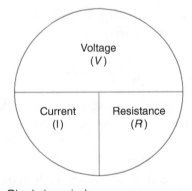

Fig. 3-6 Ohm's law circle.

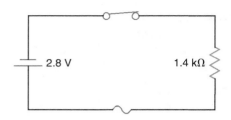

Fig. 3-7 Circuit diagram for example 3-1.

defined as 1 volt per ampere. Therefore, Ohm's law can be expressed in base units as follows:

$$1 \text{ ampere} = \frac{1 \text{ volt}}{1 \text{ volt/ampere}}$$

This expression reduces to 1 ampere = 1 ampere, which shows that proper units were used.

EXAMPLE 3-3

The manufacturer specifies that a certain lamp will allow 0.8 ampere of current when 120 volts is applied to it. What is the resistance of the lamp?

Given:	Current (I) = 0.8 ampere (A)
	Voltage (V) = 120 volts (V)
Find:	Resistance (R)
Known:	$R = \dfrac{V}{I}$
Solution:	$R = \dfrac{120 \text{ V}}{0.8 \text{ A}} = 150 \text{ ohms } (\Omega)$
Answer:	The resistance of the lamp is 150 Ω.

EXAMPLE 3-4

How much voltage is required to cause 1.6 amperes in a device that has 30 ohms of resistance?

Given:	$R = 30 \ \Omega$
	$I = 1.6 \text{ A}$
Find:	V
Known:	$V = IR$
Solution:	$V = 1.6 \text{ A} \times 30 \ \Omega = 48 \text{ V}$
Answer:	The voltage applied to the device must be 48 V.

EXAMPLE 3-5

The current flowing through a 10-kΩ resistor is 35 mA. What is the potential energy difference (voltage) across the resistor?

Given:	$R = 10 \text{ k}\Omega = 10,000 \ \Omega$
	$I = 35 \text{ mA} = 0.035 \text{ A}$
Find:	V
Known:	$V = IR$
Solution:	$V = 0.035 \text{ A} \times 10,000 \ \Omega$
	$= 350 \text{ V}$
Answer:	The voltage across the resistor is 350 V.

Notice in example 3-5 that I in mA times R in kΩ would also yield V in V because m (10^{-3}) times k (10^3) cancel each other.

Calculating Power

You May Recall

. . . that we previously learned how to find the amount of power when energy and time were known.

Now we are going to work with the relationship between current, voltage, and power. Since current and voltage are easily measured quantities, you will be using this relationship quite often in electrical work.

In circuits where the load is entirely resistance, power is equal to current times voltage. Expressed as a formula, we have

$$\text{Power } (P) = \text{current } (I) \times \text{voltage } (V)$$

or

$$P = IV$$

P = IV

Power is in its base unit of watts when voltage is in volts and current is in amperes.

EXAMPLE 3-6

What is the power input to an electric heater that draws 3 amperes from a 120-volt outlet?

Given:	Current (I) = 3 amperes (A)
	Voltage (V) = 120 volts (V)
Find:	Power (P)
Known:	$P = IV$
Solution:	$P = 3 \text{ A} \times 120 \text{ V}$
	$= 360 \text{ watts (W)}$
Answer:	The power input to the electric heater is 360 W.

By rearranging the power formula, we can solve for current when power and voltage are known. The formula is

$$\text{Current } (I) = \frac{\text{power } (P)}{\text{voltage } (V)} \quad \text{or} \quad I = \frac{P}{V}$$

$$I = \frac{P}{V}$$

This formula is used to find the current in a conductor feeding a load of specified power.

The voltage needed to provide a specified current and power can be found using:

$$\text{Voltage } (V) = \frac{\text{power } (P)}{\text{current } (I)}$$

or

$$V = \frac{P}{I}$$

$$V = \frac{P}{I}$$

EXAMPLE 3-7

How much current flows through a 120-volt, 500-watt lamp?

Given: Voltage (V) = 120 volts (V)
Power (P) = 500 watts (W)

Find: Current (I)

Known: $I = \dfrac{P}{V}$

Solution: $I = \dfrac{500 \text{ W}}{120 \text{ V}} = 4.17 \text{ A}$

Answer: The current flowing through the lamp is 4.17 A.

$$P = \frac{V^2}{R}$$

EXAMPLE 3-8

The heating element in a clothes dryer is rated at 4 kilowatts (kW) and 240 volts (V). How much current does it draw?

Given: P = 4 kW = 4000 W
V = 240 V

Find: I

Known: $I = \dfrac{P}{V}$

Solution: $I = \dfrac{4000 \text{ W}}{240 \text{ V}} = 16.7 \text{ A}$

Answer: The current drawn by the clothes dryer is 16.7 A.

$$P = I^2 R$$

Using Ohm's law and the power formula, we can find the power when resistance and current or voltage are known.

EXAMPLE 3-9

Find the power used by the resistor in the circuit shown in Fig. 3-8.

Given: V = 1.5 V
R = 10 Ω

Find: P

Known: $P = IV, I = \dfrac{V}{R}$

Solution: $I = \dfrac{1.5 \text{ V}}{10 \text{ }\Omega} = 0.15 \text{ A}$
$P = 0.15 \text{ A} \times 1.5 \text{ V} = 0.225 \text{ W}$

Answer: The power used (dissipated) by the resistor is 0.225 W.

The procedure used in example 3-9 is a two-step process. If Ohm's law and the power formula are combined, the procedure can be shortened to one step. The combining of Ohm's law and the power formula yields two formulas. From Ohm's law we know that

$$I = \frac{V}{R}$$

Substituting for I in the basic power formula $(P = IV)$, we find that

$$P = \frac{V}{R} \times V$$

$$P = \frac{V^2}{R}$$

Thus, we can solve for the power if we know the voltage and the resistance.

From Ohm's law we also know that

$$V = I \times R$$

Again, substituting for V in the basic power formula, we see that

$$P = I \times I \times R$$

$$P = I^2 R$$

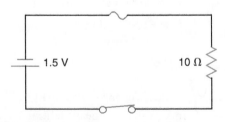

Fig. 3-8 Circuit diagram for examples 3-9, 3-10, and 3-13.

Thus, given the current and the resistance (or the voltage and the resistance), we can solve for the power with just one formula.

Example 3-10 is solved using one of the above formulas.

EXAMPLE 3-10

Find the power dissipated (used) by the resistor in Fig. 3-8 using the appropriate power formula.

Given:	$V = 1.5$ V
	$R = 10\ \Omega$
Find:	P
Known:	$P = \dfrac{V^2}{R}$
Solution:	$P = \dfrac{(1.5)^2}{10} = \dfrac{1.5 \times 1.5}{10}$
	$= 0.225$ W
Answer:	The power dissipated by the resistor is 0.225 W.

EXAMPLE 3-11

How much power is dissipated when 0.2 ampere of current flows through a 100-ohm resistor?

Given:	$R = 100\ \Omega$
	$I = 0.2$ A
Find:	P
Known:	$P = I^2R$
Solution:	$P = 0.2^2 \times 100$
	$= 0.2 \times 0.2 \times 100$
	$= 0.04 \times 100$
	$= 4$ W
Answer:	The resistor uses 4 W of power.

Calculating Energy

You May Recall

. . . from a previous chapter that energy is equal to power times time.

You have just learned that power is equal to current times voltage. Thus, energy can also be determined by knowing current, voltage, and time.

Energy can also be determined if resistance and voltage or current are known. Of course, it is always necessary to know the amount of time the power is being used.

EXAMPLE 3-12

How much energy is converted by a device that draws 1.5 amperes from a 12-volt battery for 2 hours?

Given:	$I = 1.5$ A
	$V = 12$ V
	$t = 2$ hours (h)
Find:	Energy (W)
Known:	$W = Pt, P = IV$
Solution:	$P = 1.5$ A $\times$ 12 V $= 18$ W
	$W = 18$ W $\times$ 2 h $= 36$ Wh
Answer:	The energy is 36 Wh.

EXAMPLE 3-13

In Fig. 3-8, how much energy is taken from the battery by the resistor if the switch remains closed for 30 minutes?

Given:	$V = 1.5$ V, $R = 10\ \Omega$
	$t = 30$ min
Find:	W
Known:	$W = Pt$
	$P = \dfrac{V^2}{R}$
	1 joule = 1 wattsecond
	1 min = 60 s
Solution:	$P = \dfrac{(1.5)^2}{10} = \dfrac{2.25}{10}$
	$= 0.225$ W
	$W = 0.225$ W $\times$ 1800 s
	$= 405$ Ws
	$= 405$ joules (J)
Answer:	The energy drawn from the cell is 405 J.

Calculating Cost

The cost of electric energy can be determined from the amount of energy used and the cost rate. The cost rate is usually specified in *cents per kilowatthour.* Specifying the cost rate of energy in cents per kilowatthour is like

Cents per kilowatthour

specifying the cost rate of gasoline in cents per gallon. The cost of anything is equal to the total quantity times the cost rate. If potatoes cost 10¢ per pound (cost rate) and you buy 10 pounds, the cost is $1. That is, the cost is equal to the rate times the quantity. The cost of electric energy is

Cost = rate × energy

$$\text{Cost} = \text{rate} \times \text{energy}$$
$$= \text{cents per kilowatthour}$$
$$\times \text{kilowatthours}$$

EXAMPLE 3-14

What is the cost of 120 kWh of energy if the rate is 6¢ per kWh?

Given:	$W = 120$ kWh
	Rate = 6¢ per kWh
Find:	Cost
Known:	Cost = rate × energy
Solution:	Cost = 6¢ per kWh
	× 120 kWh
	= 720¢ = $7.20
Answer:	The cost is $7.20.

EXAMPLE 3-15

What is the cost of operating a 100-watt lamp for 3 hours if the rate is 6¢ per kWh?

Given:	$P = 100$ W
	$t = 3$ h
	Rate = 6¢ per kWh
Find:	Cost
Known:	Cost = rate × energy, $W = Pt$
Solution:	$W = 100$ W × 3 h
	= 300 Wh = 0.3 kWh
	Cost = 6¢ per kWh
	× 0.3 kWh
	= 1.8¢
Answer:	It costs 1.8¢ to operate the lamp for 3 hours.

EXAMPLE 3-16

An electric iron operates from a 120-volt outlet and draws 8 amperes of current. At 9¢ per kWh, how much does it cost to operate the iron for 2 hours?

Given:	$V = 120$ V
	$I = 8$ A
	$t = 2$ h
	Rate = 9¢ per kWh
Find:	Cost
Known:	Cost = rate × energy
	$W = Pt, P = IV$
Solution:	$P = 8$ A × 120 V = 960 W
	$W = 960$ W × 2 h
	= 1920 Wh
	= 1.92 kWh
	Cost = 9¢ per kWh
	× 1.92 kWh
	= 17.3¢
Answer:	The cost is 17.3¢.

Notice the order in which the "Known" in example 3-16 is listed. The procedure outlined below was used in developing the order in which the "Known" formulas are listed.

1. Write the formula needed to solve for the quantity to be found. Look at the quantities to the right of the equals sign in this formula. If one of these quantities is not listed in the "Given" row, write the formula needed to find it ($W = Pt$).
2. Look at the right-hand half of the formula just written ($W = Pt$). If a quantity there is not listed in the "Given" row, write the formula needed to find it ($P = IV$).
3. Look at the right-hand half of ($P = IV$). The "Given" row lists both (I and V). The problem can now be solved.

The above procedure should always be used to reduce complex problems into a series of simple steps.

 *Self-Test*

Answer the following questions.

9. Express the answer to example 3-6 in kilowatts.

10. Express the answer to example 3-10 in milliwatts.

11. Refer to Fig. 3-5(*a*). How much current flows when the switch is closed?

12. What is the resistance of a semiconductor device that allows 150 milliamperes of current when 600 millivolts of voltage is connected to it?

13. What is the power of a hair dryer that operates from 120 volts and draws 4.5 amperes of current?

14. How much current is required by a 1000-watt toaster that operates from a 120-volt outlet?

15. What is the power rating of an automobile tape deck that requires 2.2 amperes from a 12.6-volt battery?

16. A miniature lamp has 99 ohms of resistance when lit. It operates from a 6.3-volt battery. What is the power rating of the lamp?

17. An electric heater draws 8 amperes from a 240-volt source. How much energy does it convert in 9 hours?

18. The current through a 100-ohm resistor is 200 milliamperes. How much energy does the resistor convert to heat in 10 minutes?

19. An electric water heater has a 3000-watt heating element. The element is on for 3 hours. What is the cost if the rate is 4¢ per kilowatthour?

20. A string of Christmas tree lights draws 0.5 amperes from a 120-volt outlet. At 8¢ per kilowatthour, how much does it cost to operate the lights for 40 hours?

3-4 Measuring Electrical Quantities

Most electrical quantities are measured with a device called a *meter.* Voltage is measured with a *voltmeter,* current with an *ammeter,* resistance with an *ohmmeter,* and power with a *wattmeter.*

Panel Meters

A meter that measures only one of the above quantities is called a *panel meter.* Panel meters are often permanently connected (wired) into a circuit to provide continuous monitoring of an electrical quantity. The meters may be either analog or digital.

ANALOG PANEL METER

An analog panel meter is shown in Fig. 3-9. With this meter, the needle can point anywhere on the scale. When an analog meter is read, the reading is generally taken to the nearest minor-division mark. If the pointer is halfway between marks, it is read as a half-division. Before reading a meter scale, you must figure out the value of each division of the scale. Look at the scale on the meter in Fig. 3-9. Notice that there is a heavy line halfway between 4 and 6. This heavy line represents five units. Now count the number of divisions (called *minor divisions*) between 4 and 5. Since there are five minor divisions between 4 and 5, each minor division has a value of 0.2 unit. Therefore, each minor division on the

scale represents 0.2 ampere. Suppose the needle (pointer) on the meter in Fig. 3-9 is pointing to the second mark to the right of the 8 mark. The meter would be indicating 8.4 amperes.

DIGITAL PANEL METER

The digital panel meter (DPM) shown in Fig. 3-10 eliminates the need to decide which mark is closest to the pointer. There is no guesswork in trying to decide whether the meter reading is 184.8 or 184.9.

Digital meters are usually specified by the number of digits in their readout. When the

Voltmeter

Ammeter

Ohmmeter

Wattmeter

Panel meter

Fig. 3-9 Panel meter.

Fig. 3-10 Digital panel meter.

most significant (leftmost) digit can be only a 0 or a 1, it is counted as a *half-digit*. The 200-volt DPM in Fig. 3-10 is a 3½-digit meter. Even though it is called a 200-volt meter, it can measure a maximum of 199.9 volts.

Multimeters

Often a single meter serves as a voltmeter, an ammeter, and an ohmmeter. Meters that are capable of measuring two or more electrical quantities are known as *multimeters*. Multimeters use the same basic mechanism to indicate the amount of a quantity as panel meters. However, the multimeter also includes some circuitry (switches, resistors, etc.) inside its housing or case.

ANALOG MULTIMETER

Like most analog multimeters, the meter in Fig. 3-11 has multiple scales printed on its face and can measure current, voltage, and resistance. An analog meter that can measure these three quantities is often called a *volt-ohm-milliammeter (VOM)*.

Although the various analog multimeters may look different, all of them have *functions, ranges,* and *scales. Function* refers to the *quantity being measured. Range* refers to the *amount* of the quantity that can be measured. Which *scale* of the meter is used depends on both the function and the range to which the meter is

set. Proper use of the multimeter involves selection of the correct function, range, and scale. Once you understand the relationships between (1) function and scale and (2) range and scale, using any multimeter is possible.

The VOM (multimeter) in Fig. 3-11 has five functions (ac voltage, dc voltage, dc current, resistance, and continuity). The function switch (on the left side of the meter) has four positions. One position (AC VOLTS ONLY) is used for measuring ac voltage. Two positions (−DC and +DC) are used for measuring dc voltage, dc current, and resistance. Which quantity is being measured in either of these two positions is determined by the position of the range switch (located in the center of the meter). The fourth position (note symbol) is used for checking continuity. In this position, a tone is emitted whenever there is a low-resistance path for current between the test-probe tips.

Except for the resistance function, the range indicates the *maximum* amount of a quantity that can be measured on a given range setting. For the resistance function, the range indicates the amount by which the ohm scale is to be multiplied.

The ohm scale on the meter in Fig. 3-11 is different from most voltage and current scales in three ways. First, it is reverse-reading. Second, it is nonlinear. Third, the number of minor divisions between the heavy lines (and the numbered lines) is not the same throughout the scale. Therefore, the value of a minor division varies across the scale.

The bottom scale on the VOM in Fig. 3-11 is a dB (decibel) scale. A dB is 1/10 of a bel, which is a unit of relative power or change in power. For example, the power output of an audio amplifier relative to the power input of the amplifier can be expressed in bels or decibels. The decibel is a nonlinear (logarithmic) unit that expresses changes in power levels in much the same way that the human ear perceives changes in the audio power output of a speaker. To the ear, a change in audio power from 0.2 W to 2 W is perceived as just as large a change as changing from 2 W to 20 W. Both represent a power change of 10 dB.

A VOM does not measure power. However, since $P = V^2/R$, voltage levels can be used to determine dB as long as the measured voltage is measured across the same value of resistance as was used in developing the dB scale on the meter. The value of resistance, as well as the

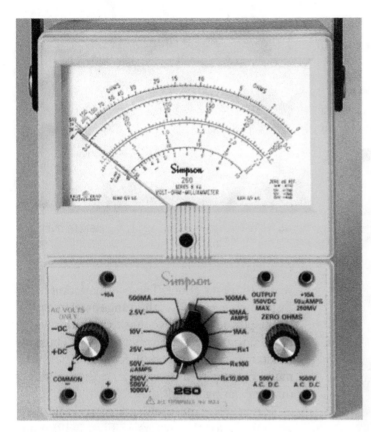

Fig. 3-11 Multimeter (VOM).

power dissipated by the resistor at 0 dB, is given on the face of the meter and/or in the manual supplied with the meter. A common reference is 1 mW dissipated by a 600-Ω resistor.

The dB scale is calibrated for the smallest ac voltage range. A chart on the meter face tells one how many dB to add to the scale reading when on larger ac voltage ranges.

EXAMPLE 3-17

The meter in Fig 3-11 is set to the R × 100 range. The needle points to the second mark to the left of the 20 mark. How much resistance is the meter indicating?

Solution: Between 20 and 30 there are five divisions; each division is worth 2 units (30 − 20 = 10; 10 ÷ 5 = 2). The scale indicates 24 units (20 + 2 + 2). Since the meter is on the R × 100 range

$$R = 24 \times 100 \ \Omega$$
$$= 2400 \ \Omega$$

Answer: The meter indicates 2400 Ω.

When more than one scale is available for a given function and range, select the scale that ends in a number equal to the range or a power of 10 of the range.

EXAMPLE 3-18

Refer to Fig. 3-11. Assume that the function switch is on + DC, the range switch is on 25 V, and the needle points one division left of the 200 mark on the 0 to 250 scale.

Solution: The heavy mark between 150 and 200 represents 175. Between 175 and 200 there are five divisions; so, each division is worth 5 units. The scale reading is, therefore, 195. To make the scale fit the range, the scale must be divided by 10. Thus, the measured voltage is 19.5 V (195 ÷ 10 = 19.5).

Answer: The meter indicates 19.5 V dc.

EXAMPLE 3-19

Assume the same conditions as in example 3-18 except that the range is 2.5 V. How much voltage does the meter indicate?

Solution: Read the same scale as in example 3-18. Now the scale reading of 195 must be divided by 100 to make the 250 scale match the 2.5-V range. Therefore, the measured voltage is 1.95 V (195 ÷ 100 = 1.95).

Answer: The voltage is 1.95 V dc.

Notice in Fig. 3-11 that there is a separate (slightly nonlinear) scale for the 2.5 V ac range. Having a separate scale for the lowest ac voltage range is very common for analog multimeters.

Another common feature of multimeters is having two or more ranges for one setting of the range switch (see Fig 3-11). These additional ranges are selected by inserting the test leads in the appropriate (labeled) jacks.

DIGITAL MULTIMETER

With *digital multimeters (DMMs)*, like those shown in Figs. 3-12 and 3-13, there are no

Fig. 3-12 Digital multimeters. These are handheld (portable) units. (*a*) Note the combined function and range switch. (*b*) This meter is autoranging.

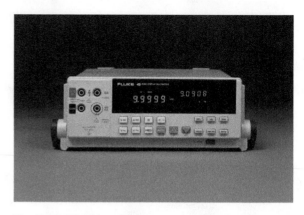

Fig. 3-13 Digital multimeter with separate push-button switches for function and range.

scales to worry about. The decimal point in the display changes when the range is changed so the display reading never has to be multiplied or divided to obtain the correct reading. Thus, the resistance ranges on the DMM tell the maximum resistance a range can measure rather than indicating a multiplier.

The operator must select the correct function and ranges for the meters in Figs. 3-12(*a*) and 3-13. The meter in Fig. 3-13 uses separate push buttons to select the functions and the ranges, and the meter in Fig. 3-12(*a*) uses a common rotary switch. More sophisticated DMMs automatically select the correct range, so the operator must only ensure that the meter is on the correct function. Meters of this type [see Fig. 3-12(*b*)] are said to be *autoranging*.

Using Multimeters

All electrical quantities to be measured are applied to the meter through test leads. A typical test lead is illustrated in Fig. 3-14. For most measurements, the test leads plug into the two jacks in the lower right-hand corner of the meter shown in Fig. 3-11. A black lead goes in the COMMON − jack, and a red lead in the + jack. The black lead is negative and the red lead positive when the function switch is in the +DC position. When the function switch is in the −DC position, the black lead (COMMON −) is positive. Of course, the red lead also reverses polarity. The +DC position should be used except when special measurements are being made. In this way, the red lead is positive. Red is the color often used to indicate positive in electric circuits.

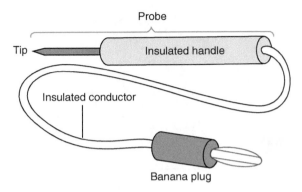

Fig. 3-14 Test lead.

The DMM has no +DC and −DC switch positions. If reverse polarity is applied to a DMM, a minus (−) sign appears ahead of the digits. The magnitude of the indicated value is still correct.

Measuring Resistance

When resistance is measured, the VOM must first be ohms-adjusted. The *ohms-adjust control* is labeled "zero OHMS" in Fig. 3-11. This control is rotated until the meter indicates zero on the ohms scale *when the test-lead tips are touching each other.* The ohms-adjust control must be adjusted for each range of the ohms function. Thus, every time the range is changed, the leads must be touched together and the meter adjusted to zero. No such adjustment is needed with the DMM.

The ohmmeter function of any multimeter uses a cell, battery, or power supply inside the meter housing. That is, it has its own source of energy. Therefore, any other energy source must be disconnected from any circuit in which resistance is to be measured. Never measure the resistance of a load when power (the energy source) is connected to the circuit. Doing this damages the ohmmeter. Figure 3-15(*a*) illustrates the correct technique for measuring the resistance of a lamp. Notice in Fig. 3-15(*b*) the symbols used for the ohmmeter and the test-lead connection.

The procedure used in measuring resistance is as follows:

1. Remove power from the circuit.
2. Select an appropriate range in the ohms function. The appropriate range is the one that gives the best resolution.
3. When using the VOM, short (touch) the test leads. Turn the ohms-adjust control until the pointer reads 0 ohms.
4. Connect, or touch, the test leads to the terminals of the device whose resistance is to be measured. Except for some electronic components, the polarity of the ohmmeter leads is unimportant.

When measuring resistance, do not touch the metal parts of the test leads with your hands. If you do, you will be measuring your body's resistance as well as the circuit's resistance. This will not harm you, but you may not obtain the correct resistance reading for the circuit.

Ohms-adjust control

Measuring Voltage

Voltage measurements are the easiest electrical measurements to make. They are also the most common. Voltage measurements are made with power connected to the circuit. Remember, ALWAYS follow recommended safety procedures when working with energized circuits. The

Voltage measurements

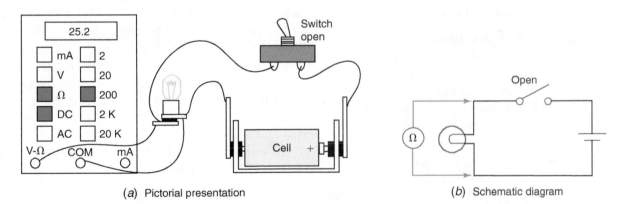

(*a*) Pictorial presentation   (*b*) Schematic diagram

Fig. 3-15 Measuring resistance. Notice that the power source is disconnected from the load by the open switch.

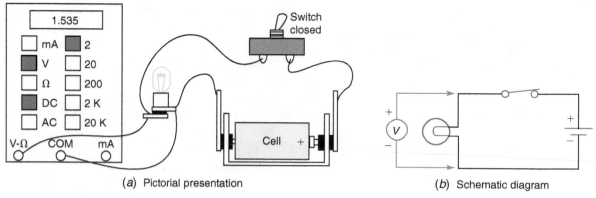

(a) Pictorial presentation   (b) Schematic diagram

Fig. 3-16 Lamp voltage being measured. The switch must be closed.

following procedure is used to make voltage measurements:

1. Select the correct voltage function (ac or dc) for the type of voltage used in the circuit.
2. Select a range that is greater than the expected voltage.
3. Determine the polarity of the voltage to be measured by looking at the schematic diagram or at the battery terminals. This step is omitted when measuring alternating current because the polarity reverses every fraction of a second.
4. Connect the negative (black) lead of the multimeter to the negative end of the voltage to be measured. Touch (or connect) the positive (red) lead of the meter to the positive end of the voltage. In other words, observe polarity when measuring voltage with a multimeter or voltmeter. If you do not, the meter pointer of the VOM may bend when it tries to rotate counterclockwise.

Figure 3-16 shows the correct connections for measuring dc voltage. In this figure, the meter is measuring the voltage across the lamp. Notice that the switch is in the closed position. If the switch were open, there would be no voltage across the lamp and the meter would indicate 0 volts. A load has a voltage across it only when current is flowing through it. If the meter were connected across the cell, as in Fig. 3-17, the meter would indicate the cell's voltage regardless of whether the switch were open or closed. As long as the switch is closed, the meter

Current measurements

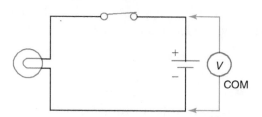

Fig. 3-17 Cell voltage being measured. The switch may be either open or closed.

reading is the same in Figs. 3-16 and 3-17. That is, the voltage output of the cell appears across the lamp when the switch is closed.

Measuring Current

Current measurements are made much less frequently than either resistance or voltage measurements. This is because the circuit usually has to be physically interrupted to insert the meter. In Fig. 3-18, the circuit has been physically interrupted by disconnecting one end of the lead between the cell and the lamp. The meter is then connected between the end of the wire and the lamp. As shown in Fig. 3-19, the meter can just as well be connected on either side of the switch. All three meter locations in Fig. 3-19 are correct, and all three locations yield the same results. Remember, the meter must be inserted into the circuit so that the circuit current flows through the meter as well as the load.

In using the current function of a DMM or a VOM, follow these steps:

1. Select the current function.
2. Select a range that is greater than the expected current.

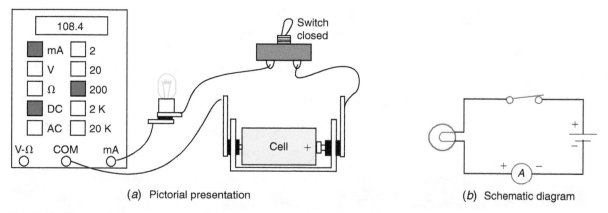

(a) Pictorial presentation   (b) Schematic diagram

Fig. 3-18 Current being measured. Current must flow through both the meter and the load.

3. Physically interrupt the circuit.
4. Observing polarity, connect the DMM or VOM between the points created by the interruption. Correct polarity can be determined by tracing current (electron flow). As indicated in Fig. 3-19, current enters the negative terminal and leaves the positive terminal of the meter.

A meter can be easily damaged, and so it should be used with care. Reverse polarity on a VOM may bend the pointer. However, most damage to multimeters and ammeters occurs when they are incorrectly connected to the circuit. If you connect an ammeter, or a multimeter on the current function, into a circuit in the same way you connect a voltmeter, you may destroy the meter. Even if the meter is not completely destroyed, its accuracy will be greatly decreased.

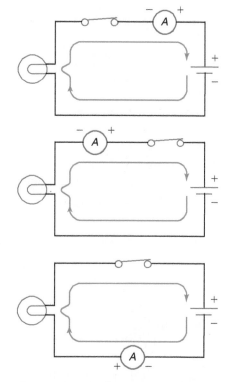

Fig. 3-19 Ammeter location. The location of the ammeter with respect to the lamp or the switch does not change the amount of current.

Self-Test

Answer the following questions.

21. In Fig. 3-9 the needle rests three divisions to the right of the 6 mark. How much current is the meter indicating?
22. Refer to Fig. 3-11. The needle points one mark to the right of 20. The range is set on R × 100. What is the resistance?

23. Refer to Fig. 3-11. Assume that the meter is on the DC function and the 100-mA range. The needle is pointing to 20 on the third scale from the top. How much current is indicated by the meter?
24. True or false. The resistance function of the DMM should be adjusted whenever a new range is selected.

25. True or false. A 10-volt DPM that can indicate a maximum voltage of 9.99 volts is a 2½-digit meter.
26. True or false. Most DMMs are autofunctioning.
27. Summarize the procedure for measuring resistance with the VOM.
28. Summarize the procedure for measuring voltage.
29. Summarize the procedure for measuring current.
30. True or false. Of the possible misuses of an ammeter, reverse polarity is usually the most damaging.
31. Why does a VOM have both a +DC and a −DC function?
32. True or false. Many multimeters have a power function.

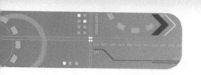

Chapter 3 Summary and Review

Summary

1. Electric components are represented by symbols in schematic diagrams; sometimes their electrical values are also included on the diagrams.
2. Sometimes the physical structure of a device also serves as a conductor in a circuit.
3. Current is directly proportional to voltage and inversely proportional to resistance.
4. A panel meter measures only one quantity.
5. Multimeters measure several quantities.
6. VOMs have functions, ranges, and scales.
7. The ohmmeter function of the VOM must be adjusted every time the range is changed.
8. No ohms adjustment is required with a DMM.
9. Polarity must be observed when measuring dc voltage.
10. Polarity must be observed when measuring direct current.
11. Ammeters are easily damaged by incorrect connections to a circuit.
12. Multimeters may have separate switches for selecting function and range.
13. Many digital meters have automatic range and polarity switching.
14. A readout digit that will always be either a 0 or a 1 is specified as a half-digit.

Related Formulas

$$V = IR$$

$$I = \frac{V}{R}$$

$$R = \frac{V}{I}$$

$$P = IV$$

$$P = I^2R$$

$$P = \frac{V^2}{R}$$

$$W = Pt$$

$$\text{Cost} = \text{rate} \times \text{energy}$$

Chapter Review Questions

For questions 3-1 to 3-12, determine whether each statement is true or false.

3-1. The metal chassis of an electric device is often used as a conductor. (3-1)

3-2. Current is directly proportional to voltage and inversely proportional to resistance. (3-3)

3-3. A watt is equal to 1 volt divided by 1 ampere. (3-3)

3-4. A 60-watt, 120-volt lamp requires less current than a 40-watt, 120-volt lamp. (3-3)

3-5. A 60-watt, 120-volt lamp has less resistance than a 100-watt, 120-volt lamp. (3-3)

3-6. An electric circuit must be physically interrupted to measure voltage in the circuit. (3-4)

3-7. An ohmmeter contains its own energy source. (3-4)

3-8. Always observe polarity when measuring direct current. (3-4)

3-9. Correct polarity for an ammeter requires that current (electrons) enter the positive terminal of the meter. (3-4)

3-10. All DMMs are autoranging. (3-4)

3-11. A VOM has only one ohmmeter scale. (3-4)

3-12. The function switch on the VOM is set in the −DC position to make the common negative jack the negative terminal of the VOM. (3-4)

Answer the following questions.

3-13. List the four essential parts of an electric circuit. (3-1)

3-14. Draw the schematic symbols for the following components: (3-2)
  a. Lamp
  b. Conductor
  c. Resistor
  d. Cell
  e. Open switch

3-15. Measuring which quantity (voltage, current, or resistance) requires physically interrupting the circuit? (3-4)

3-16. Measuring which quantity (voltage, current, or resistance) requires removing power from the circuit or component? (3-4)

3-17. The range of a digital meter is 0000 to 1999. What is the digit rating of this meter? (3-4)

3-18. What name is often used to refer to an analog multimeter that can measure V, I, and R? (3-4)

3-19. How do the resistance range markings on the VOM differ from those on the DMM?

3-20. Why doesn't the readout of a DMM have to be multiplied or divided by some number when the range is changed?

Chapter Review Problems

3-1. What is the resistance of a lamp which draws 240 milliamperes when connected to a 12.6-volt battery? (3-3)

3-2. How much current does a 500-W lamp draw from a 120-V source? (3-3)

3-3. How much power does a heater require if it draws 11 amperes from a 240-volt circuit? (3-3)

3-4. A toaster draws 5 amperes from a 120-volt outlet. How much energy does it use in 2 hours? (3-3)

3-5. How much would it cost to operate the toaster in problem 3-4 if energy costs 12¢ per kWh? (3-3)

3-6. How much voltage is required to force 40 milliamperes of current through a 1-kilohm resistor? (3-3)

3-7. What is the cost per kWh if it costs $1.91 to operate a 240-V heater that draws 6 A for 12 hours? (3-3)

3-8. What is the resistance of a 150-W, 120-V lamp? (3-3)

3-9. An analog multimeter is on the DC voltage function and the 5-V range. On the 0 to 50 scale the needle stops at 38. What is the value of the voltage being measured? (3-4)

3-10. A digital multimeter is on the DC voltage function and the 50-V range. The digital readout indicates 04.340. What is the value of the voltage being measured? (3-4)

Critical Thinking Questions

3-1. Using base unit equivalents, prove that current in amperes multiplied by voltage in volts yields power in watts.

3-2. Some of the example circuits in this chapter include fuses. Yet, the fuses were ignored in calculating resistance, voltage, and current. Does this mean that fuses have no resistance? Explain.

3-3. From the way an ammeter is used to measure current in a circuit, would you conclude that it has a very high or a very low internal resistance? Why?

3-4. Assume that you are purchasing a multimeter for personal use. Would you buy a DMM or VOM? Why?

3-5. It costs $1.20 to operate a heater for 6 hours on a 240-V source. What is the resistance of the heater if energy costs 8¢/kWh?

3-6. Prove that energy will be in joules when current is in amperes, time is in seconds, and the formula is $W = I^2Rt$.

3-7. Explain how you could measure the current in Fig. 3-15 without disconnecting any circuit conductors.

3-8. When you measure the resistance of a 60-W, 120-V lamp, it is approximately 17 Ω. Using its rated wattage and voltage, calculate its resistance. Explain the large difference between the measured and calculated values.

3-9. Explain what will happen to the power of a circuit if the source voltage is doubled and the resistance is not changed

Answers to Self-Tests

1. source of energy, conductors, insulators, load, control device, and protection device
2. F
3. F
4.

    Lamp   Resistor   Conductor

5. A schematic diagram is a drawing which uses symbols for electric components. It indicates the connections between the components.
6. T
7. F
8. protection device
9. 0.36 kW
10. 225 mW
11. **Given:** $V = 1.5$ V, $R = 3$ kΩ $= 3000$ Ω

    Find: I

    Known: $I = \dfrac{V}{R}$

    Solution: $I = \dfrac{1.5\text{ V}}{3000\text{ Ω}} = 0.0005$ A

    Answer: The current is 0.0005 A, or 0.5 mA.
12. **Given:** $I = 150$ mA $= 0.15$ A

             $V = 600$ mV $= 0.6$ V

    Find: R

    Known: $R = \dfrac{V}{I}$

    Solution: $R = \dfrac{0.6\text{ V}}{0.15\text{ A}} = 4$ Ω

    Answer: The resistance is 4 Ω.
13. **Given:** $V = 120$ V, $I = 4.5$ A

    Find: P

    Known: $P = IV$

Solution: $P = 4.5$ A $\times$ 120 V $= 540$ W

Answer: The dryer is a 540-W dryer.

14. **Given:** $P = 1000$ W, $V = 120$ V

    Find: I

    Known: $I = \dfrac{P}{V}$

    Solution: $I = \dfrac{1000\text{ W}}{120\text{ V}} = 8.3$ A

    Answer: The current for the 1000-W toaster is 8.3 A.
15. **Given:** $I = 2.2$ A, $V = 12.6$ V

    Find: P

    Known: $P = IV$

    Solution: $P = 2.2$ A $\times$ 12.6 V $= 27.72$ W

    Answer: The tape deck requires 27.72 W.
16. **Given:** $R = 99$ Ω, $V = 6.3$ V

    Find: P

    Known: $P = \dfrac{V^2}{R}$

    Solution: $P = \dfrac{(6.3)^2}{99} = 0.4$ W

    Answer: The power rating of the lamp is 0.4 W.
17. **Given:** $I = 8$ A, $V = 240$ V, $t = 9$ h

    Find: W

    Known: $W = Pt$, $P = IV$

    Solution: $P = 8$ A $\times$ 240 V $= 1920$ W

             $W = 1920$ W $\times$ 9 h

             $= 17{,}280$ Wh $= 17.28$ kWh

    Answer: The energy converted is 17.28 kWh.
18. **Given:** $R = 100$ Ω, $I = 200$ mA $= 0.2$ A,

             $t = 10$ min $= 600$ s

    Find: W

    Known: $W = Pt$, $P = I^2R$, Ws $=$ J

Solution: $P = (0.2)^2 \times 100 = 4\,\text{W}$
$W = 4\,\text{W} \times 600\,\text{s}$
$= 2400\,\text{Ws} = 2400\,\text{J}$

Answer: The resistor converts 2400 J of electric energy to heat energy.

19. **Given:** $P = 3000\,\text{W}$, $t = 3\,\text{h}$, rate = 4¢ per kWh

    Find: Cost

    Known: Cost = rate × W, $W = Pt$

    Solution: $W = 3000\,\text{W} \times 3\,\text{h} = 9000\,\text{Wh}$
    $= 9\,\text{kWh}$
    Cost = 4¢ per kWh × 9 kWh
    $= 36¢$

    Answer: The cost is 36¢.

20. **Given:** $I = 0.5\,\text{A}$, $V = 120\,\text{V}$, rate = 8¢ per kWh, $t = 40\,\text{h}$

    Find: Cost

    Known: Cost = rate × W, $W = Pt$, $P = IV$

    Solution: $P = 0.5\,\text{A} \times 120\,\text{V} = 60\,\text{W}$
    $W = 60\,\text{W} \times 40\,\text{h} = 2400\,\text{Wh}$
    $= 2.4\,\text{kWh}$
    Cost = 8¢ per kWh × 2.4 kWh
    $= 19.2¢$

    Answer: The cost is 19.2¢.

21. 6.6 amperes
22. 1900 ohms (19 × 100)
23. 40 milliamperes
24. F
25. F
26. F
27. Remove the power from the circuit, select the ohms function, select the correct range, and connect the test leads to the device or component being measured. If a VOM is used, the meter must also be ohms-adjusted.
28. Select the correct voltage function, select the correct range, determine the polarity of the voltage, and—observing polarity—connect the meter to the circuit.
29. Select the current function, select the correct range, interrupt the circuit, and—observing polarity—connect the meter between the interrupted points.
30. F
31. They are provided so that test lead polarities can be reversed without changing the lead connections to the meter.
32. F

CHAPTER 4

Circuit Components

Learning Outcomes

This chapter will help you to:

4-1 *Understand* the types, symbols, ratings, and limitations of chemical cells and batteries.

4-2 *Become familiar with* the chemical reaction, discharging, charging, specific gravity, and safe-handling of lead-acid cells and batteries.

4-3 *List* four major characteristics of nickel-cadmium cells.

4-4 *Specify and explain* four undesirable characteristics of carbon-zinc and zinc-chloride cells.

4-5 *Compare* the secondary alkaline-manganese dioxide cell and the nickel-cadmium cell in terms of charge retention, operating temperature, internal resistance, cycle life, and cell voltage during discharge.

4-6 *Remember* that mercuric oxide cells have stable output voltage, can operate at high temperature, and have good energy-to-volume and energy-to-weight ratios.

4-7 *Compare* silver oxide, nickel-cadmium, and mercuric oxide cells in terms of magnitude and consistency of output voltage during discharge.

4-8 *Realize* that secondary lithium cells have higher output voltage, more stable output voltage during discharge, wider range of operating temperature, and higher energy-to-weight ratio than other types of secondary cells.

4-9 *Describe* how a fuel cell produces an output voltage.

4-10 *Discuss* how a miniature flasher lamp operates.

4-11 *List and discuss* categories, ratings, and color coding of resistors.

4-12 *Explain* why switches have different current and voltage ratings for ac and dc.

4-13 *Discuss* the factors that determine the maximum current an insulated wire can safely carry.

4-14 *Describe* the operation of a resettable fuse.

In Chap. 3, you learned about the requirements for a complete electric circuit. In this chapter, you will learn about some of the components (parts) used to construct electric circuits.

4-1 Batteries and Cells

Although there are many types of cells and batteries, such as the solar cell mentioned in a previous chapter, electric chemical cells and batteries are by far the most common type. Unless otherwise specified, the terms *cell* and *battery* refer to the electric chemical type.

Batteries and cells are chemical devices that provide dc voltage. Since voltage is a potential energy difference, batteries and cells are also known as sources of electric energy. And, since power is the rate of using energy, these devices are also called *power sources*. They are used to power electric devices.

Terminology and Symbols

A *cell* is an electrochemical device consisting of two electrodes made of different materials and an electrolyte. The chemical reaction between the *electrodes* and the *electrolyte* produces a voltage.

A *battery* consists of two or more cells electrically connected together and packaged as a single unit. Although technically a battery has two or more cells, the term *battery* is often used to indicate either a single cell or a group of cells.

The schematic symbols for a cell and a battery are shown in Fig. 4-1. The voltage rating of the cell or battery is specified next to the symbol. The short line in the symbol is always the negative terminal. Polarity indicators (negative and positive signs) are not always shown with the symbol.

Cells and batteries are classified as either primary or secondary. *Primary cells* are cells that are not rechargeable. That is, the chemical

Cell

Electrodes

Electrolytes

Battery

Primary cells

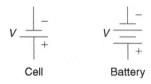

Cell   Battery

Fig. 4-1 Cell and battery symbols.

Ampere-hours

reaction that occurs during discharge is not easily reversed. When the chemicals used in the reactions are all converted, the cell is fully discharged. It must then be replaced by a new cell. Included in the primary cell category are the following types: carbon-zinc, zinc chloride, most alkaline, silver oxide, mercury, and some lithium cells.

Secondary cells

Secondary cells may be discharged and recharged many times. The number of discharge/charge cycles a cell can withstand depends on the type and size of the cell and on the operating conditions. The number of cycles will vary from fewer than 100 to many thousands. Secondary cells include the following types: rechargeable alkaline, lead-acid, nickel metal hydride, nickel-cadmium, nickel-iron, and lithium-ion cells.

Cells and batteries are also classified as either dry or wet. Historically, a *dry cell* was one that had a paste or gel electrolyte. It was semisealed and could be used in any position. With newer designs and manufacturing techniques, it is possible to completely (hermetically) seal a cell. With complete seals and chemical control of gas buildup, it is possible to use liquid electrolyte in dry cells. Today the term *dry cell* refers to a cell that can be operated in any position without electrolyte leakage.

Dry cell

Wet cell

Wet cells are cells that must be operated in an upright position. These cells have vents to allow gases generated during charge or discharge to escape. The most common wet cell is the lead-acid cell.

Internal resistance

Cell (or battery) capacity

Ratings of Cells and Batteries

Cell (or battery) capacity refers to the amount of energy the cell can provide under specified conditions. The conditions specified are temperature, current drain, discharge schedule (minutes per hour), and final output voltage when discharge is complete. The amount of energy available varies widely as these specified conditions change. For example, a typical flashlight cell provides approximately twice as much energy when the discharge rate

is changed from 300 to 50 mA. For the same cell, the capacity decreases about 30 percent when the temperature is decreased from 21 to 5°C. In general, a cell (or battery) provides the most energy when the temperature is high and the discharge schedule, current drain, and final voltage are low.

The capacity of a cell (or battery) is expressed in *ampere-hours* (Ah). When the voltage of the cell is specified, ampere-hour becomes a unit of energy. The energy of a cell is

$$W = Pt$$

Since $P = IV = VI$, we have

$$W = VIt$$

If t is given in hours and I is in amperes, then the energy of a cell is expressed in watthours.

EXAMPLE 4-1

How much energy is stored in a fully charged 12.6-V battery rated at 90 Ah?

Given:	$V = 12.6$ V
	Capacity $= 90$ Ah
Find:	W
Known:	$W = VIt$
Solution:	$W = 12.6$ V $\times$ 90 Ah
	$= 1134$ Wh
Answer:	The stored energy is 1134 watthours.

Internal Resistance

The output voltage from a cell varies as the load on the cell changes. *Load on a cell* refers to the amount of current drawn from the cell. As the load increases, the voltage output drops, and vice versa. The change in output voltage is caused by the *internal resistance* of the cell. Since materials from which the cell is made are not perfect conductors, they have resistance. Current flowing through the external circuit also flows through the internal resistance of the cell. According to Ohm's law, a current flowing through a resistance (either external or internal) results in a voltage drop ($V = IR$). Any voltage developed across the internal resistance is not available at the terminals of the cell. The voltage at the terminals is the voltage due to the chemical reactions *minus the voltage dropped across the internal resistance.* The terminal voltage of a cell, therefore, depends on both the internal resistance of the cell and the amount of load current.

EXAMPLE 4-2

What is the terminal voltage of a 6-V battery with an internal resistance of 0.15 Ω when the load draws a current of 3.4 A?

Given: No-load terminal voltage
$(V_{NLT}) = 6$ V
Internal resistance (R_i)
$= 0.15$ Ω
Load current $(I_L) = 3.4$ A

Find: Loaded terminal voltage (V_{LT})

Known: $V_{LT} = V_{NLT} - (I_L \times R_i)$

Solution: $V_{LT} = 6$ V $- (3.4$ A $\times 0.15$ Ω$)$
$= 5.49$ V

Answer: The terminal voltage of the loaded battery is 5.49 V.

In many applications, the changes in cell terminal voltage are so small that they make no practical difference. In other applications, the changes are very noticeable. For example, while an automobile engine is starting up, the battery output voltage changes from about 12.6 to 8 V.

As a cell is discharged, its internal resistance increases. Therefore, its output voltage decreases for a given value of load current.

Polarization

When a primary cell discharges, gas ions (such as hydrogen) are created around the positive electrode. The gas ions polarize the cell and reduce the cell's terminal voltage. If the gas ions are allowed to build up, the polarization can become so great that the cell is useless. Therefore, cells include a *depolarizing agent* in their chemical composition. The depolarizing agent is a chemical compound that reacts with the polarizing gas to remove it.

Depolarizing agent

 Self-Test

Answer the following questions.

1. What is a cell?
2. What is a battery?
3. Describe the basic structure of a cell.
4. Define the following terms:
   a. Primary cell
   b. Secondary cell
   c. Dry cell
   d. Wet cell
   e. Ampere-hour
5. List six types of secondary cells.
6. List six types of primary cells.

7. Internal resistance causes a battery's terminal voltage to _____ .
8. The development of hydrogen ions on the positive electrode of a cell is called _____ .
9. How much energy is stored in a fully charged 9-V battery rated at 3 Ah ?
10. The terminal voltage of a battery is 4.5 V with no load. What is the battery's internal resistance if the terminal voltage drops to 4.2 V when the load current is 200 mA?

4-2 Lead-Acid Cells

The lead-acid cell is a very common secondary cell. It is the power source for the electric system of most cars, trucks, boats, and tractors. It can provide the large currents (hundreds of amperes) needed to crank internal-combustion engines.

A lead-acid cell produces about 2.1 V. Higher voltages are obtained by connecting cells together to form batteries. A 12-V automobile

battery actually has a nominal voltage of 12.6 V because it contains six cells.

The structure of a lead-acid cell is shown in Fig. 4-2. Notice that the cell terminals connect to a *group of plates*. The groups are tied together (both electrically and mechanically) at the top. The meshing together of the positive and negative groups effectively provides one large negative and one large positive plate.

Group of plates

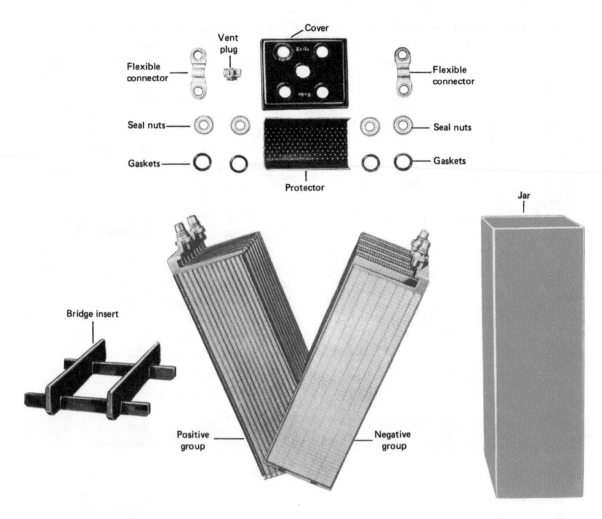

Fig. 4-2 Structure of a lead-acid cell.

Sulfuric acid
(H_2SO_4)

Lead sulfate
($PbSO_4$)

Chemical Action

When a load is connected to the cell of Fig. 4-3(*a*), current flows. Electrons leave the negative terminal (plate), flow through the load, and return to the positive terminal of the cell. At the surface of the positive plate shown in Fig. 4-3(*a*), a molecule of lead peroxide (PbO_2) becomes three ions. Notice that each ion of oxygen (O) has two excess electrons whereas the ion of lead (Pb) has a deficiency of only two electrons. Thus, the plate is left with a deficiency of two electrons; that is, it has a positive charge. In the electrolyte, two molecules of *sulfuric acid* (H_2SO_4) provide four hydrogen (H) ions and two sulfate (SO_4) ions. The four hydrogen ions combine with the two oxygen ions to form two molecules of water (H_2O). One of the sulfate ions combines with the lead ion at the positive plate to form a molecule of *lead sulfate* ($PbSO_4$).

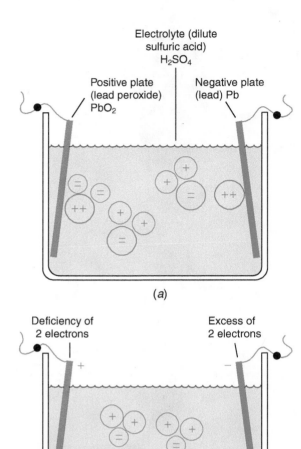

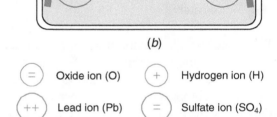

	Oxide ion (O)		Hydrogen ion (H)
$++$	Lead ion (Pb)	$=$	Sulfate ion (SO_4)

Fig. 4-3 Lead-acid cell. (a) Chemical reaction. (b) Results.

On the surface of the negative plate in Fig. 4-3(a), an atom of lead becomes a positive ion. In becoming a positive ion, the lead atom leaves two electrons on the plate. Thus, the plate has a negative charge. The lead ion combines with a sulfate ion from the electrolyte and forms a molecule of lead sulfate. The result of these reactions, shown in Fig. 4-3(b), is that the plates change to lead sulfate and the electrolyte changes to water.

The above-described chemical reaction continues until one of two things happens:

1. The load is removed; the chemical reaction stops. At this time, the charges on the plates and the charges of the electrolyte ions are in a state of equilibrium (balance).
2. All the sulfuric acid solution has been converted to water and/or all the lead and lead peroxide have been converted to lead sulfate. In this case, the battery is fully discharged.

Recharging

A lead-acid cell is recharged by forcing a reverse current through the cell. That is, electrons enter the negative plate and leave the positive plate. When a cell is being charged, all the chemical reactions described above are reversed. The water and lead sulfate are converted back into sulfuric acid, lead, and lead peroxide. Cells are charged by connecting them to a voltage source that is greater than the cell's voltage. In other words, the cell becomes the load rather than the energy source. As the load, the cell is converting electric energy (from the other voltage source) into chemical energy.

The six-cell (12-V) battery in Fig. 4-4 is being recharged. Notice that the negative terminal of the charger is connected to the negative terminal of the battery. Since the charger's voltage is greater than the battery's voltage, current flows in the direction indicated in Fig. 4-4. The exact voltage of the charger depends on the condition of the battery and on the *rate of charge* desired. The charger voltage is adjusted to provide the desired charge current. The manufacturer's recommendation should be followed in determining the rate of charge for a specific type of battery or cell. Too fast a charge rate must be avoided because it will overheat the battery.

Rate of charge

A battery should not be overcharged. Overcharging can weaken the plate structure of the battery. When a battery is overcharged, water from the electrolyte is converted into hydrogen gas and oxygen gas. Having to add more than normal amounts of water to a battery is an indication that it is being overcharged.

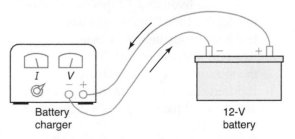

Fig. 4-4 Charging a battery. The charger voltage must exceed the battery voltage.

Many of the newer lead-acid batteries are sealed. There is no provision for adding water. Such batteries require no maintenance other than keeping them clean—especially their terminals.

Specific Gravity

The charge of a battery can be determined by measuring the specific gravity of the electrolyte. The *specific gravity* of a substance is the ratio of its weight to the weight of water. If a substance has a specific gravity of 1.251, it is 1.251 times as heavy as water. Sulfuric acid is heavier than water. Therefore, the more sulfuric acid in the electrolyte, the higher the specific gravity of the electrolyte. Since the amount of sulfuric acid increases as the battery is charged, specific gravity indicates the state of charge of the battery. The specific gravity of a fully charged cell is adjusted at the factory to suit the structure and intended purpose of the cell. For fully charged lead-acid cells, the specific gravity ranges from 1.21 to 1.28. The typical automotive battery is fully charged at a specific gravity of about 1.26.

A lead-acid cell is considered to be completely discharged when the specific gravity drops to 1.12. It should not be left in this discharged state for extended periods. To do so shortens the life of the battery. Discharged cells and batteries need to be protected from low temperature. A completely discharged cell will freeze at about $-9°C$ (16°F). At 50 percent discharge it freezes at $-24°C$ ($-11°F$). When a cell freezes, the electrolyte expands and can break the jar (case) of the cell.

Hydrometer

Specific gravity is measured with a *hydrometer* like the one shown in Fig. 4-5. The hydrometer is used as follows:

1. Squeeze the rubber bulb. Insert the flexible tube into the electrolyte of the cell. Slowly release the rubber bulb, drawing the electrolyte into the hydrometer. When the float in the hydrometer lifts free, remove the flexible tube from the electrolyte and finish releasing the rubber bulb.
2. Read the specific gravity from the float at the top surface of the electrolyte.
3. Reinsert the flexible tube into the cell and slowly squeeze the bulb to return the electrolyte.
4. When finished testing, flush the hydrometer (both inside and out) with clean water.

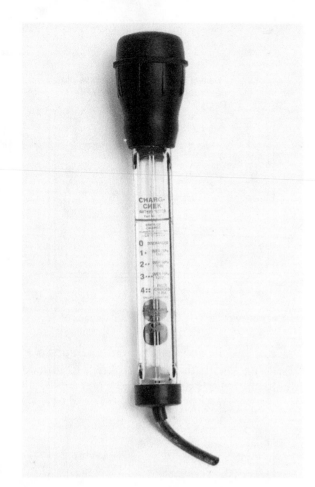

Fig. 4-5 Hydrometer.

When the specific gravity of a cell cannot be raised by charging to within 0.05 of the manufacturer's specification, it is a questionable cell. It will probably completely fail in the near future.

Obviously the condition of a sealed lead-acid cell or battery cannot be tested with a hydrometer. The condition of such a cell or battery can be determined by testing the terminal voltage under a specific load. The load and minimum terminal voltage are specified by the manufacturer. Some large sealed lead-acid cells have a built-in feature (colored balls with different specific gravities) to determine the specific gravity of the electrolyte. When the ball with the highest specific gravity (indicated by its color) floats on the surface of the electrolyte, the cell is charged. It is in a usable condition. When only the ball with the lowest specific gravity surfaces, the cell needs to be charged.

Battery Safety

If not properly handled, lead-acid cells and *batteries can be dangerous.* The acid used in the

Answer the following questions.

11. Describe the chemical reaction that occurs in a lead-acid cell as it is discharged.
12. Describe how a lead-acid battery is recharged.
13. List two precautions to follow to prevent battery damage when charging a battery.
14. Define *specific gravity*.

15. A(n) _____ is used to measure specific gravity.
16. In cold weather a(n) _____ lead-acid battery can freeze.
17. The acid in a lead-acid cell is _____.
18. List the safety rules that apply to handling and charging lead-acid batteries.

electrolyte can cause skin burns and burn holes in clothing. It is extremely harmful to the eyes. Always wear safety glasses when working with lead-acid cells and batteries. If acid does come in contact with the skin or clothing, immediately flush with clean water. Then wash with soap and water, except for the eyes. If the acid is in the eyes, get medical attention immediately after repeated flushing with water. Wash hands in soap and water after handling batteries.

The gases released from charging batteries are explosive. Charge batteries in a ventilated area where there are no sparks or open flames.

4-3 Nickel-Cadmium Cells

Nickel-cadmium cells are dependable, rugged secondary cells. As vented wet cells, they have been in limited use for many years in many types of service. When technological advances made it possible to produce a nickel-cadmium dry cell, the popularity of the cell greatly increased. The dry cell is hermetically sealed and can be operated in any position. Figure 4-6 illustrates the structure of a typical sealed nickel-cadmium cell. These cells are most commonly available in button or cylindrical shape and in capacities ranging from a few milliampere-hours to 5 or 6 Ah.

Sealed nickel-cadmium cells have a life ranging from several hundred to several thousand *charge/discharge cycles*. They have an extremely long *shelf life* (storage life) and can be stored either charged or discharged. They can be stored at temperatures ranging from −40 to 60°C (−40 to 140°F).

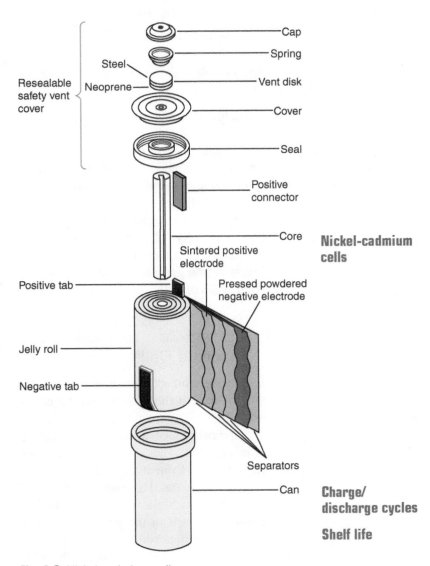

Fig. 4-6 Nickel-cadmium cell.

Nickel-cadmium cells

Charge/ discharge cycles

Shelf life

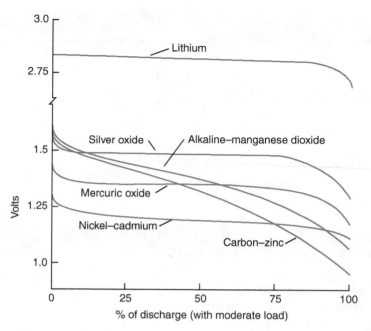

Fig. 4-7 Comparison of various cell voltages as the cells are discharged.

Other notable characteristics of the nickel-cadmium cell are

1. Good low-temperature operating characteristics.
2. High initial cost but low operating cost.
3. Very low internal resistance. Thus, they can provide high currents with little drop in output voltage.
4. Nearly constant output voltage (approximately 1.2 V) as the cell is discharged by a moderate load. See Fig. 4-7 for a general comparison with other types of cells.

4-4 Carbon-Zinc and Zinc Chloride Cells

Carbon-zinc cells

Zinc chloride cells

The *carbon-zinc cell* is the most common of the dry cells. It is also the least expensive of the primary cells.

The composition of a carbon-zinc cell is shown in Fig. 4-8. The zinc can is the negative electrode, and the manganese dioxide is the positive electrode. The carbon rod makes electric contact with the manganese dioxide and conducts current to the positive terminal. However, the carbon rod is not involved in the chemical reaction that produces the voltage. The electrolyte for the chemical system is a solution of ammonium chloride and zinc chloride.

Although extensively used because it is cheap, the carbon-zinc cell has many weaknesses. Some of these weaknesses are

1. Poor low-temperature operating characteristics.
2. Gradual decrease in output voltage as the cell discharges (Fig. 4-7).
3. Low energy-to-weight ratio and low energy-to-volume ratio. Other types of primary cells have ratios two to three times as great as the carbon-zinc cell.
4. Poor efficiency under heavy loads due to high internal resistance.

Carbon-zinc cells and batteries are available in a wide range of sizes and shapes. Batteries with capacities of 30 Ah are available.

Zinc chloride cells are constructed much like carbon-zinc cells, but they use a modified electrolyte system. The electrolyte is a solution of zinc chloride. Because only zinc chloride is used in the electrolyte, the zinc chloride cell has several advantages over the standard carbon-zinc cell. First, the zinc chloride cell is more efficient at higher current drain. Second, the output voltage does not decrease as rapidly as it does in the carbon-zinc cell.

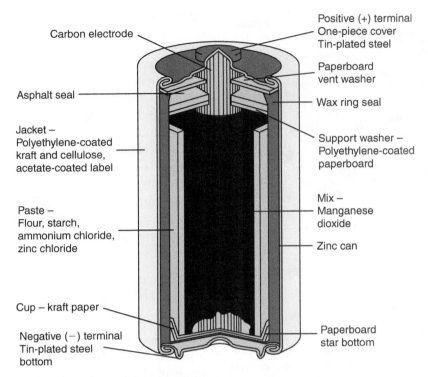

Fig. 4-8 Cutaway view of carbon-zinc cell.

Labels on figure:
- Carbon electrode
- Asphalt seal
- Jacket – Polyethylene-coated kraft and cellulose, acetate-coated label
- Paste – Flour, starch, ammonium chloride, zinc chloride
- Cup – kraft paper
- Negative (–) terminal Tin-plated steel bottom
- Positive (+) terminal One-piece cover Tin-plated steel
- Paperboard vent washer
- Wax ring seal
- Support washer – Polyethylene-coated paperboard
- Mix – Manganese dioxide
- Zinc can
- Paperboard star bottom

4-5 Alkaline–Manganese Dioxide Cells

Alkaline cells use the same electrodes as carbon-zinc cells (zinc and manganese dioxide). However, a solution of potassium hydroxide is used as the electrolyte. There are two types of alkaline cells—primary and secondary (rechargeable).

Alkaline cells produce about 1.5 V. The cell voltage decreases gradually as the cell is discharged. However, the change is not as great as it is with carbon-zinc cells (Fig. 4-7).

The *primary alkaline cell* is more expensive than the carbon-zinc cell. However, it has several advantages. It can be discharged at high current drains and still maintain reasonable efficiency and cell voltage. It can operate effectively at lower temperatures (−30°C versus −7°C for carbon-zinc cells). It has a lower internal resistance and a higher energy-to-weight ratio. An alkaline cell stores at least 50 percent more energy than a carbon-zinc cell of the same size.

Secondary alkaline cells are much cheaper than nickel-cadmium cells. They have better charge retention (ability to remain charged when stored) and a wider operating temperature range than nickel-cadmium cells do. The internal resistance of the two types of cells is comparable. However, the alkaline cell has several shortcomings not found in the nickel-cadmium cell:

1. The cycle life (number of discharge/recharge cycles) is fewer than 75.
2. The cell voltage is not as constant (Fig. 4-7).
3. The cycle life is more dependent on proper discharge and recharge of the cell. Discharging beyond rated capacity and extended overcharging both shorten the cycle life.
4. Charging circuits for optimum cycle life are more complex.

4-6 Mercuric Oxide Cells

Mercuric oxide cells (commonly called mercury cells or batteries) have some distinct advantages over the primary cells discussed above:

1. The cell voltage is more uniform as the cell discharges (Fig. 4-7).
2. The capacity is less dependent on the discharge rate.
3. The energy-to-volume ratio is two to four times as great as that of either the alkaline or the carbon-zinc cell.
4. The energy-to-weight ratio is higher.

Alkaline cells

Primary alkaline cells

Mercuric oxide cells

Secondary alkaline cells

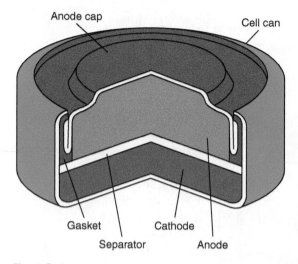

Anode cap   Cell can

Gasket   Cathode

Separator   Anode

Fig. 4-9 Cutaway view of a mercuric oxide cell.

The mercury cell is mechanically rugged, has low internal resistance, and can operate at high temperatures (54°C). Since the output voltage (1.35 V) is so constant over most of its discharge, the mercury cell is used as a *voltage standard*. (A standard is used to check the accuracy of other voltage sources and measuring devices.)

Some of the disadvantages of the mercury cell are its poor low-temperature (0°C) performance and its relatively high cost. Both the initial cost and the operating cost are high compared with costs of carbon-zinc and alkaline cells.

Figure 4-9 shows the internal structure of a mercury cell. Mercury cells and batteries are available in a wide range of capacities (up to 28 Ah) and shapes.

4-7 Silver Oxide Cells

Silver oxide cells are much like mercury cells. However, they provide a higher voltage (1.5 V) and they are made for *light loads*. The loads can be continuous, such as those encountered in hearing aids and electronic watches. Like the mercury cell, the silver oxide cell has good energy-to-weight ratios, poor low-temperature response, and flat output voltage characteristics. The structures of the mercuric and silver oxide cells are very similar. The main difference is that the positive electrode of the silver cell is silver oxide instead of mercuric oxide. Figure 4-7 compares the voltage characteristics of the silver oxide cell with those of other cells.

4-8 Lithium Cells

The original *lithium cells* were primary cells. The newer lithium-ion cells are secondary cells. These cells are available in a variety of sizes and configurations. Compared with other cells, they are quite expensive. Depending on the chemicals used with lithium, the cell voltage is between 2.1 and 3.8 V. Note that this voltage is considerably higher than that of other cells. Lithium cells operate at temperatures ranging from −50 to 75°C. They have a very constant output voltage during discharge. Two of the advantages of lithium cells over other cells are

1. Longer shelf life—up to 10 years
2. Higher energy-to-weight ratio—up to 350 Wh/kg

The recycle life of lithium-ion cells is good. It typically ranges from several hundred to thousands of charge/discharge cycles.

Lithium-ion cells and lithium-ion polymer cells and batteries are being used in digital and video cameras, laptop computers, and other electrical/electronic devices. Because of their higher voltage and higher energy-to-weight and energy-to-volume ratios, they are replacing nickel-cadmium cells in many applications. Although lithium batteries are expensive, they are extensively used in electric and hybrid cars.

4-9 Fuel Cells

The basic principles of fuel cell operation have been known by scientists for over 170 years. However, it has only been in the last two or three decades that researchers and manufacturers have developed the designs and manufacturing techniques to produce practical fuel cells. Although practical, these fuel cells are still quite expensive. This has restricted their widespread use although their high-efficiency, low-noise level, and environmentally clean operation are very desirable.

Figure 4-10 illustrates the major parts of a common fuel cell. It could be the cutaway of a proton exchange membrane fuel cell (PEMFC). In this fuel cell, the electrolyte is a polymer electrolyte membrane. The membrane is made of a porous material that allows protons to pass through it, but will not allow electrons to pass.

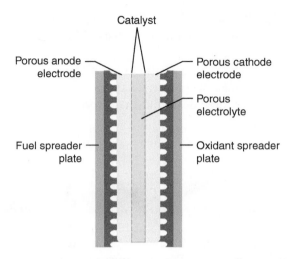

Fig. 4-10 Schematic diagram of a fuel cell. The major components are shown in cross-section.

Labels in figure: Catalyst; Porous anode electrode; Porous cathode electrode; Porous electrolyte; Fuel spreader plate; Oxidant spreader plate

Hydrogen gas (the fuel) is fed into the fuel spreader plate which distributes the gas uniformly across the anode electrode. The anode is a porous, conductive material so the hydrogen passes through it to the surface of the electrolyte membrane, which is coated with a thin, porous layer of a catalyst (platinum). The catalyst causes the hydrogen gas to ionize. When hydrogen ionizes, it provides a free electron (a negative atomic particle) and a proton (a positive atomic particle). The proton migrates through the electrolyte membrane to the cathode electrode. This leaves the anode electrode with a negative charge (an excess electron) and the cathode electrode with a positive charge (an excess proton). An external circuit connected between the anode side of the cell and the cathode side allows the excess electrons at the anode to be forced through the circuit load to the cathode. At the cathode, the electrons and protons provided by the hydrogen atoms and oxygen atoms provided by the oxidant (either oxygen gas or air) combine to form water (H_2O) and heat. The catalyst at the junction of the electrolyte membrane and the cathode electrode aid in forming the water. The water is discharged from the cell at the bottom of the oxidant spreader plate. The chemical reaction at the cathode heats the water to about 80°C.

In broad terms, the action of the proton exchange membrane fuel cell can be summarized this way: The chemical energy from hydrogen and oxygen is converted to electric energy. The by-products of operating the cell are water and heat.

Hydrocarbon fuels, such as methanol, ethanol, and gasoline, can provide the hydrogen for this fuel cell. The hydrocarbon fuel is first processed by a "fuel reformer" to remove the carbon, sulfur, and other elements or compounds that are detrimental to the operation and or longevity of the fuel cell.

An individual fuel cell produces a very small voltage and very little current—thus, very little power. The voltage is increased by stacking the cells and electrically connecting them together so that the voltage of one cell adds to the voltage of adjacent cells. Available current is increased by electrically connecting the cell stacks together so that the currents from all the stacks add together to provide the large current required from the group of cell stacks. The group of fuel cell stacks forms a fuel cell battery. The fuel cell battery is often simple referred to as a fuel cell.

Proton exchange membrane fuel cell batteries typically provide power in the 100-W to 500-kW range. Although expensive, these batteries can be used in a wide variety of applications. And, the cost may decrease if manufacturing techniques improve and increased demand lead to mass production.

There are other types of fuel cells being manufactured, and other types in the research and development stage. They may use a different fuel, and/or electrolyte, and/or catalyst. Some types operate at more than 800°C. Others can produce up to 100 Mega watts (MW) of power output.

〜 Self-Test

Answer the following questions.

19. Can nickel-cadmium cells be damaged by leaving them discharged?

20. Compare the nickel-cadmium and rechargeable alkaline cell in terms of
    a. Voltage during discharge

b. Cycle life

c. Cost

21. Compare alkaline and carbon-zinc cells in terms of

a. Efficiency at high-current loads

b. Voltage at high-current loads

c. Energy storage

22. The _____ cell is the least expensive of the various types of cells.

23. Which cells are expected to provide at least several hundred charge/discharge cycles?

24. The _____ cell has the highest output voltage of the various types of cells.

25. List three desirable characteristics of fuel cells.

4-10 Miniature Lamps and LEDs

Miniature lamps are used extensively in electric and electronic circuits. They are used for indicator lights and instrument illumination in automobiles, aircraft, home appliances, coin-operated machines, and all kinds of electronic instruments. Miniature lamps are also used in flashlights, lanterns, and signal lights. *Light-emitting diodes* (LEDs), *neon glow lamps,* and *incandescent lamps* are commonly used as indicators.

Incandescent Lamps

The heart of the incandescent lamp is the *tungsten filament* (wire), which emits light when heated by an electric current. When the tungsten wire is long, thin, and coiled, as in Fig. 4-11, it is often supported in the center.

Neon glow lamps

Incandescent lamps

Wedge base

Tungsten filament

Screw-type base

Bayonet base

The tungsten filament is enclosed in a glass bulb from which the air is evacuated. Often the air is replaced by an inert gas. As the last of the air is removed from the bulb, the bulb is sealed. It is sealed by melting the glass around the evacuation hole. The seal on the lamp in Fig. 4-11 was made on the small tip of glass at the bottom of the bulb.

The heavy wire leads connecting to the filament carry current to the filament. For this lamp, these leads are also the terminals which make electrical contact with the lamp holder. This configuration of terminal connections is called a *wedge base.* It provides for a simple, compact lamp holder.

The two principal bases used in miniature incandescent lamps are the *screw-type base* and the *bayonet base.* The lamp on the left in Fig. 4-12 has a screw-type base. The screw is stamped on the portion of the base called the *shell.* The next lamp in Fig. 4-12 has a bayonet base. The bayonet-base lamp usually has one or two small projections on the shell base. These projections slide into grooves in the bayonet lamp holder. When the lamp is in place, it is given a slight twist to lock it in the holder. Lamp bases are usually made of brass for good conductivity and long wear.

Figure 4-12 shows a variety of other base and bulb styles used with miniature lamps. All the bases provide two contacts which make connections to the filament. Each base is designed to best fulfill a specific need. For example, the base may provide for exact location of the lamp. This is important when the light is to be focused. Some lamps, like the one on the right in Fig. 4-12, are designed to be soldered into the circuit. Some bases are designed to withstand high temperature. Others are made to withstand severe vibration.

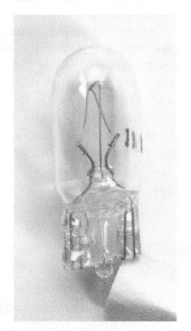

Fig. 4-11 Wedge-base lamp.

Fig. 4-12 Miniature lamps. These lamps are about 2.5 cm long.

Miniature lamps are available in a wide range of voltage and *current ratings*. At rated voltage, a lamp draws its rated current, provides its specified light, and operates for a specified time, usually called the *life* of the lamp.

Operating above *rated voltage* increases the current draw and light output (specified in units called *lumens*). However, operating with overvoltage decreases the life expectancy so greatly that it is not recommended. Operation below rated voltage is permissible. Under this condition, current decreases, light output decreases, and expected life greatly increases. The efficiency in terms of light output per watt input is reduced when operating below rated voltage.

Sometimes the manufacturer specifies the voltage and power rating of a lamp. In this case, the current can be determined by using the power relationship $I = P/V$.

A few miniature lamps are made with built-in flasher units. *Flasher lamps* are used as warning indicators, such as seat belt and emergency brake warnings in automobiles. The flasher action is caused by a *bimetallic strip* located next

EXAMPLE 4-4

What is the power rating of a lamp that draws 240 mA from a 12.6-V source?

Given:	$V = 12.6$ V
	$I = 240$ mA $= 0.24$ A
Find:	P
Known:	$P = IV$
Solution:	$P = 0.24$ A $\times 12.6$ V
	$= 3.024$ W
Answer:	The power rating is 3.024 W.

to the filament. This bimetallic strip is one of the conductors carrying current to the filament, as shown in Fig. 4-13(*a*). When the filament heats the bimetallic strip, the strip bends out, as shown in Fig. 4-13(*b*) and opens the circuit, turning off the lamp. When the lamp is off, the strip cools and straightens out. When the strip returns to its normal shape, it makes contact with the filament again and turns on the lamp.

EXAMPLE 4-3

How much current is drawn by a type-1992 lamp which is rated at 14 V and 35 W?

Given:	Voltage (V) = 14 V
	Power (P) = 35 W
Find:	Current (I)
Known:	$I = \dfrac{P}{V}$
Solution:	$I = \dfrac{35\text{ W}}{14\text{ V}} = 2.5$ A
Answer:	The lamp draws 2.5 A.

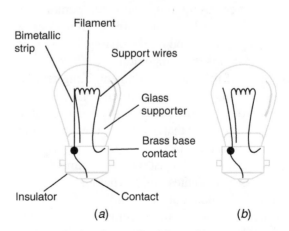

Fig. 4-13 Flasher lamp. The bimetallic strip turns the lamp off and on as it warms up and cools down. (*a*) Lamp on. (*b*) Lamp off.

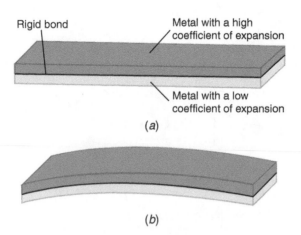

Fig. 4-14 Bimetallic strip. (*a*) At room temperature. (*b*) Above room temperature.

This cycle repeats as long as the lamp is connected to a power source.

Figure 4-14 illustrates the structure of a bimetallic strip, which is composed of two different metals. The two metals are rigidly joined where they meet. The metals have different temperature coefficients of expansion. That is, one of the metals increases its length more rapidly than the other as the temperature increases. This unequal expansion causes the strip to curve when heated, as shown in Fig. 4-14(*b*).

A lamp that flashes is often not a flasher lamp. Instead, it is a regular lamp that is being turned on and off by an electronic circuit or by a remote bimetallic strip unit.

The resistance of a lamp filament when cool is only a fraction of the resistance of the filament when it is hot. Therefore, a lamp draws a large current when it is first turned on. However, the filament comes up to operating temperature in a few milliseconds, so this large current does not last long. The reason for the change in resistance of the lamp is the temperature coefficient of resistance of the tungsten filament. Tungsten, like most metals, has a positive temperature coefficient. Although the coefficient of tungsten is not much higher than that of copper, the temperature change is tremendous.

Measuring the resistance of a lamp with an ohmmeter indicates only whether or not the lamp is burned out. The indicated resistance will be far less than the actual *hot resistance* of the lamp. This is because the current from the meter is too small to bring the lamp up to operating temperature.

Ionization

Metal electrodes

External resistor

Hot resistance

Fig. 4-15 Neon lamp. Note the absence of a filament in the neon lamp.

Neon Glow Lamps

The lamp in Fig. 4-15 is a neon glow lamp. The glass bulb is filled with neon gas. When neon gas is not ionized, it is an insulator. Once it ionizes, it is a fair conductor.

Ionization of the neon in the lamp in Fig. 4-15 occurs between the two *metal electrodes* inside the bulb. If the ionization voltage is direct current, only one electrode glows. If it is alternating current, both electrodes glow. It requires between 70 and 95 V to ionize a neon bulb. After ionization occurs, it requires 10 to 15 percent less voltage, or from 60 to 80 V, to maintain ionization.

Because the internal resistance of an ionized lamp drops to a very low value, a resistor must be connected in series with the lamp. The resistor (Fig. 4-16) limits the current through the lamp to a safe value. In Fig. 4-16 the neon bulb and the resistor are represented by their schematic symbols. Some neon lamps are manufactured with a resistor in the base. In this case no *external resistor* is needed.

Neon lamps are made in a number of different bulb and base styles. Most of the styles are the same as for the incandescent lamps. Neon lamps require very little power to operate. Most are less than ½ W and require no more than 2 mA.

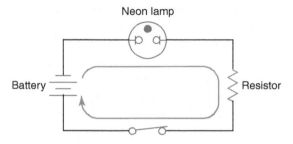

Fig. 4-16 Neon lamp and current-limiting resistor. Without the resistor, the neon lamp would immediately burn out.

Light-Emitting Diodes

The *light-emitting diode* (LED) is often used as an indicator in place of a lamp. It is also used for numeric readouts on calculators and other digital devices. LEDs are small, low-powered, and reliable. They can operate for years without failure. They are available in either a wire-terminal or a socket-type base.

Although many LEDs are red, they are also available in green, blue, and yellow. The color is determined by the combination of materials used in the manufacturing process. The materials used include various compounds of gallium, arsenic, and phosphorus.

An LED can also produce white light by either of two methods. The first method involves combining red, green, and blue elements in a single package. The second method uses only a blue element to excite a phosphor, which emits a white light. White LEDs are more efficient, less subject to vibration failure, and last longer than incandescent lamps. Because they are more expensive than incandescent lamps, their use as an illumination device in place of incandescent lamps is quite limited.

However, as technological and manufacturing advances increase the efficiency and lower the cost of LEDs, their use as illumination devices is rising. Because of their efficiency and longevity, the use of LEDs in battery-operated devices is steadily increasing.

Like neon lamps, LEDs need *current-limiting resistors* to control the current once the LED starts conducting. However, the typical LED needs only about 1.5 V before it starts conducting. Most of these devices can operate on 5 to 40 mA of current. Typically, they are operated at about 20 mA.

The LED operates only on direct current. It is a *polarized device*. The polarity of its leads need to be determined before it is connected to a voltage source.

Working with Miniature Lamps

Some miniature lamps operate at very high temperatures. A quartz lamp can operate with a bulb temperature of 350°C. Wedge-based quartz lamps can operate at temperatures as high as 450°C. At these temperatures, dirt on the surface of the glass can be quite harmful to the lamp. Needless to say, severe burns can result from touching lamps operating at such high temperatures.

If oil and/or dirt from a person's fingers is transferred to the glass bulb of a high-temperature lamp, the glass may break when the lamp heats up. Therefore, never touch the glass portion on this type of lamp with your bare hands when installing or cleaning it.

When connecting wire-terminal lamps, like the neon lamp in Fig. 4-15, you must be careful when bending the wire leads. Do not bend the leads within 3 mm (⅛ in.) of the glass bulb. Also, do not solder within 3 mm of the glass. Either bending or soldering too close to the bulb can damage the metal-to-glass seal.

Light-emitting diode

Current-limiting resistors

Polarized device

ABOUT ELECTRONICS

No More Tangles
Fluctuations in electric power service can cause havoc with sensitive machinery in industries such as the rug industry. Prior to the development of the Westinghouse dynamic voltage restorer (DVR), rug workers had to clean up a mess of yarn on the way to the winder every time there was a brief power outage. The DVR (pictured) works by injecting energy into a power line to fix deviations from sags, swells, and transients that occur because of storms or dust.

Lamp Safety

Following the procedures listed below will reduce the probability of accidents when working with lamps:

1. Turn off the power to a lamp before changing the lamp.
2. Allow a lamp to cool before changing it.

3. Be certain that the *replacement lamp* has the correct voltage. If the voltage of the power source greatly exceeds the voltage rating of the lamp, the lamp may explode.
4. When replacing lamps that fit tightly in their sockets, wear a glove (or wrap the bulb in a heavy cloth) in case the bulb breaks.

Self-Test

Answer the following questions.

26. What material is used in the filament of a miniature lamp?
27. Name two types of miniature lamps used as indicators.
28. What makes a flasher lamp blink off and on?
29. How much current does a 12-V, 3-W lamp draw?
30. The _____ is a polarized indicator.
31. Tungsten has a(n) _____ temperature coefficient.
32. With _____ voltage, only one electrode of neon lamps glows.
33. What is the purpose of the resistor in a neon lamp circuit?
34. What precaution should be observed when using wire-terminal lamps?
35. Why is it dangerous to put a 6-V lamp in a 28-V circuit?
36. What precautions should be observed when working with a quartz lamp?

4-11 Resistors

Zero-ohm resistors

One of the most common and most reliable electric-electronic components is the resistor. The resistor is used as a load, or part of the load, in most electronic circuits. Its major purposes are to control current and divide voltage. Some types of resistors can operate at temperatures as high as 300°C (572°F). Resistors are manufactured in a wide range of resistance values. Resistances of

less than 1 Ω to more than 100 MΩ are available. In fact, devices that have a resistance equivalent to a conductor but have the physical size and shape of a resistor are also manufactured. These devices are known as *zero-ohm resistors*. They are used in place of a short piece of wire (jumper wire) on a circuit board because they are easier to insert by automated machines.

Classification and Symbols

Resistors can be classified into four broad categories: fixed, variable, adjustable, and tapped. Sketches and symbols for these four categories are shown in Fig. 4-17. Notice that the variable and the adjustable resistor use the same symbol.

As shown in Fig. 4-17, there are two types of variable resistors. The *potentiometer* has three terminals. Rotation of the shaft changes the resistances between the middle terminal and the two end terminals.

Most potentiometers are *linear*. That is, one degree of shaft rotation results in the same change of resistance regardless of the shaft

Potentiometer

Typical resistors.

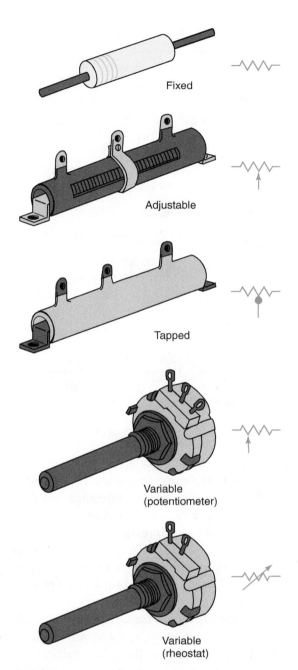

Fig. 4-17 Classification and symbols of resistors.

Fixed

Adjustable

Tapped

Variable
(potentiometer)

Variable
(rheostat)

Fig. 4-18 Potentiometers. The potentiometer on the bottom row has three concentric shafts.

location. Other potentiometers have *nonlinear tapers*. This means that the rate of change of resistance varies as the shaft is rotated. A variety of tapers are available. Tapered potentiometers are sometimes used as volume and tone controls in stereo amplifiers.

Potentiometers are often ganged together so that the resistance in several circuits can be changed simultaneously. Figure 4-18 shows some *ganged potentiometers*. The left potentiometer in Fig. 4-18 is a dual-shaft unit. One shaft controls the ganged potentiometers, and the other shaft controls the switch. A triple-shaft potentiometer

is also shown in Fig. 4-18. One shaft operates the switch, and the other shafts control the potentiometers. Often the same shaft operates both the switch and the potentiometer when the two are ganged together.

Rheostats (Fig. 4-17) have only two terminals. Turning the shaft changes the resistance between these two terminals. Rheostats are used to adjust the current in a circuit to a specified value. Very often potentiometers are used as rheostats by not using one of the end terminals.

Adjustable resistors serve about the same functions as potentiometers and rheostats. However, adjustable resistors are used in high-power circuits. They are used only when infrequent changes in resistance are required. Unlike rheostats and potentiometers, adjustable resistors are not usually adjusted while the circuit is in operation.

A single-tap resistor is illustrated in Fig. 4-17. Multiple-tap resistors are also available. *Tapped resistors,* like adjustable resistors, are mostly used in high-power circuits (greater than 2 W).

Power Ratings

A resistor has a power rating as well as a resistance rating. The *power rating* of a resistor indicates the amount of power the resistor can safely dissipate without being destroyed. As current passes through the resistor, heat is produced. The resistor is converting electric energy into heat energy. If the current were allowed to increase, the heat would burn up the resistor. Thus, some safe level of heat must be specified, and this is what the power rating does.

Rheostats

Adjustable resistors

Tapped resistors

Nonlinear tapers

Power rating

Ganged potentiometers

There is no relationship between the resistance rating and the power rating of a resistor. The same resistance value can be obtained in resistors ranging in power ratings from less than 1 W to many watts. The power rating of a resistor is primarily determined by the physical size of the resistor and the type of materials used. Figure 4-19 shows, from top to bottom, resistors in the ¼-W, ½-W, 1-W, and 2-W size. The three resistors with the squared-off ends are carbon-composition resistors. Those with the rounded ends are metal-film resistors, which, for a given power rating, are slightly smaller than the carbon-composition resistors.

The power rating of a resistor is assigned by the manufacturer under specified conditions. These conditions include free air circulation around the resistor and resistor leads soldered to sizable terminals. Often these conditions are not fully satisfied in the devices that use resistors. Therefore, a resistor with a power rating higher than the value calculated for the circuit is used. As a rule of thumb, many circuit designers specify a power rating twice as great as the calculated value. In most applications this is **Safety factor** a reasonable *safety factor*.

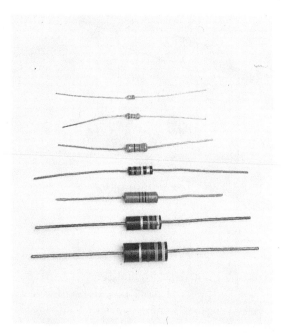

Fig. 4-19 Relative size and power rating of resistors. The physical size of a resistor determines its power rating rather than its resistance.

Resistor Tolerance

It is very difficult to mass-produce resistors that have exactly the same resistance value. Therefore, manufacturers specify tolerances for their resistors. *Tolerance* is specified as a percentage of the nominal (stated) resistance. For example, a 10 percent 1000-Ω resistor can have a resistance anywhere between 900 (1000 − 10 percent of 1000) and 1100 Ω (1000 + 10 percent of 1000).

Common tolerances for resistors are 10, 5, 2, and 1 percent. However, resistors are available with tolerances of less than 0.01 percent. Of course, low-tolerance resistors are more expensive than high-tolerance resistors.

Tolerance

EXAMPLE 4-5

How much power is used (dissipated) by a 1000-Ω resistor when 100 V is connected to it? What power rating would be specified for the resistor?

Given:	Resistance (R) = 1000 Ω
	Voltage (V) = 100 V
Find:	Power (P)
Known:	$P = \dfrac{V^2}{R}$
Solution:	$P = \dfrac{(100\text{ V})(100\text{ V})}{1000\ \Omega}$
	$= \dfrac{10{,}000}{1000} = 10\text{ W}$
Answer:	The calculated power is 10 W. Using the rule of thumb mentioned above, the specified power rating should be twice the calculated value, or
	Resistor rating = 10 W × 2
	= 20 W
	Therefore, use a 20-W resistor.

EXAMPLE 4-6

What is the maximum power a 250-Ω ± 10 percent resistor will dissipate when it is drawing 0.16 A of current?

Given:	Resistor tolerance (T_R) = 10 percent
	Nominal resistance (R_N) = 250 Ω
	Current (I) = 0.16 A
Find:	Power (P)

Types of Resistors

Resistors are grouped according to the type of material or process used to make the resistive element. The major types and their characteristics are listed below.

CARBON-COMPOSITION

The resistance element is made of finely powdered carbon held together with an inert binding material. The resistance of the element is determined by the ratio of carbon to binding material. The size determines the power rating, not the resistance. Carbon composition is used for both fixed and variable resistors. A fixed carbon-composition resistor is shown in Fig. 4-20(*a*). These resistors are reliable and relatively inexpensive. They are available in a wide range of resistances and in power ratings up to 2 W.

WIRE-WOUND

Wire-wound resistors are used for all classes of resistors—variable, fixed, adjustable, and tapped. In fact, nearly all tapped and adjustable resistors are wire-wound and are classified as *power resistors* (above 2 W). The wire for wire-wound resistors is made of alloys such as nickel-chromium or copper-nickel. The wire is wound (evenly spaced) on an insulator form, and the ends are connected to solder-coated copper leads. For fixed resistors, the entire assembly except for the leads is coated with insulating material.

Fixed wire-wound resistors come in many sizes and shapes. Some look just like carbon-composition resistors. Wire-wound resistors are made with power ratings that are in excess of 1000 W.

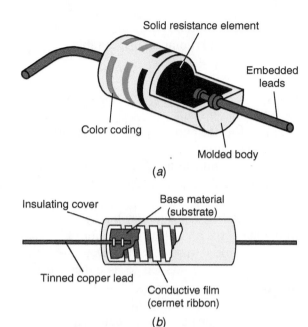

Fig. 4-20 Resistors. The internal structure of various resistors is quite different. (*a*) Carbon-composition resistor. (*b*) Cermet or film resistor.

Wire-wound resistors have low temperature coefficients and good stability and can be made to close tolerances.

CERMET

Cermet is a mixture of fine particles of glass (or ceramic) and finely powdered metal (or oxides), such as silver, platinum, or gold. This mixture, in a paste form, is applied to a base material made of ceramic, glass, or alumina. Leads are attached to the form and to the cermet mixture, and the whole device is fired (baked) in a kiln. The firing fuses everything together. The cermet is put on the substrate in a spiral ribbon. This effectively makes a long, thin resistive element for the resistor. The structure of a cermet (or cermet film, as it is sometimes called) resistor is shown in Fig. 4-20(*b*).

Cermet is used in variable as well as fixed resistors. It has a low temperature coefficient, and it is very stable and rugged. As a potentiometer,

Cermet

Power resistors

ABOUT ELECTRONICS

Resistors Can Become Very Hot! A power resistor can operate at a temperature high enough to cause severe burns if touched before it has had time to cool.

cermet offers fine adjustment and long life. As a fixed resistor, it is compact and durable.

DEPOSITED-FILM

Deposited-film resistors

Deposited-film resistors are made like and look like the resistor in Fig. 4-20(*b*). The film is applied to the substrate by vaporizing (in a vacuum) the film material. The material forms a very thin, uniform film on the substrate. The film is often deposited in the form of a spiral ribbon.

Either carbon or metal may be deposited as a film. One of the nickel-chromium alloys is often used as the metal to be deposited. Deposited-film resistors are often referred to as either *carbon-film* or *metal-film resistors*. These film resistors have characteristics much like those of cermet resistors. They operate well at high frequencies.

Carbon-film resistors

Metal-film resistors

CONDUCTIVE-PLASTIC

Conductive plastics

Conductive plastics are used as the resistive element in some potentiometers. They are made by combining carbon powder with a plastic resin, such as polyester or epoxy. The carbon-resin mixture is applied to a substrate, such as ceramic. Conductive-plastic potentiometers are relatively cheap. They wear very well in applications that require much adjusting of the potentiometer.

Color codes

SURFACE-MOUNT

Many circuit components, especially resistors, capacitors, and solid-state devices, are made without leads for connecting them to tie points or holes on a circuit board trace (conductor). Such components are referred to as *surface-mount devices* (SMDs). They are also called *chip components*. Instead of having leads, these components have conductive terminal pads that make contact with (and are soldered to) the traces on the printed circuit board. In general, an SMD is smaller than its counterpart with leads, and it is also easier to attach to the printed circuit board.

Surface-mount devices

Surface-mount resistors (*chip resistors*) are shown in Fig. 4-21. Because they are very small, their power rating is usually no greater than ¼ W. However, a wide range of resistance values is available.

Chip resistors

The resistance value of a chip is indicated by a three- or four-digit number printed on the chip. With a three-digit number, the first two digits are the digits of the value, and the third (last) digit

Fig. 4-21 Surface-mount resistors. Notice the tinned end terminals.

gives the power of 10 by which the first two digits are to be multiplied. The 391 printed on the chips in Fig. 4-21 indicates a resistance of 390 Ω ($39 \times 10^1 = 390$). With a four-digit number, the first three digits are the digits of the value, and the fourth digit is the power of 10 multiplier. For example, a resistor chip labeled 5363 would have a value of 536,000 Ω (536 kΩ).

Color Codes

The resistance and tolerance of many fixed resistors are specified by color bands around the body of the resistors. Both resistance and tolerance are indicated by the color, number, and position of the bands. See Fig. 4-22 for the values assigned to the various colors. Fig. 4-22 also details the meaning of the location and number of color bands.

As shown in Fig. 4-22, the first band is always closer to one end of the resistor. Since 5 and 10 percent resistors are made only in values with two significant figures, only four color bands are needed to mark them. Three significant figures are common with 1 and 2 percent resistors; therefore, five color bands are needed to identify their resistance and tolerance. Examples of the use of the color code are given in Table 4-1.

To distinguish wire-wound resistors from other types of resistors, many manufacturers make the first color band on wire-wound resistors about twice as wide as the other color bands.

Some resistors have only three color bands. These are resistors with tolerances of ±20 percent. They are not very common anymore.

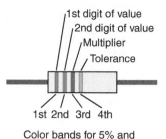

Color	Digit	Multiplier	Tolerance
Black	0	10^0	
Brown	1	10^1	±1%
Red	2	10^2	±2%
Orange	3	10^3	
Yellow	4	10^4	
Green	5	10^5	
Blue	6	10^6	
Violet	7	10^7	
Gray	8	10^8	
White	9	10^9	
Gold		10^{-1}	±5%
Silver		10^{-2}	±10%

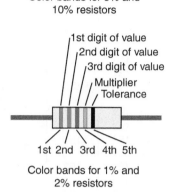

Fig. 4-22 Resistor color code. The 1 and 2 percent resistors have five color bands.

Also, some 5 and 10 percent resistors manufactured to military specifications have five color bands. In this case, the first four bands are read the same as with the four-band system (the fourth band is either gold or silver). The fifth band indicates reliability. The *reliability* of a resistor tells what percentage of the resistors fail within 1000 h.

Careful examination of Fig. 4-22 reveals that the color codes can specify resistances from 0.10 to $999 \times 10^9\ \Omega$. However, as shown in Fig. 4-23, only certain values are commonly available for each tolerance. Figure 4-23 shows only the significant figures. Thus, for resistors with 10 and 5 percent tolerances, resistances of 0.22, 2.2, 22, 220, 2200, and so on, are available. In 2 and 1 percent tolerances, 2.26, 22.6, 226, and so on, are available. Note from Fig. 4-23 that only the 10/100 series is available in all tolerances.

Resistor Networks

Resistor networks are composed of many film resistors, such as cermet, on a single substrate. The resistors and substrate are housed in either a dual in-line package (*DIP*) or a single in-line package (*SIP*) like those in Fig. 4-24. These resistor networks are often used in digital electronic circuits on printed circuit boards.

Special Resistors

A number of special resistors are manufactured. Three of them and their typical applications are discussed on the following two pages.

Resistor networks

Reliability

DIP

SIP

| Table 4-1 | **Color Code Examples** |

	Color of Band				Resistance (Ω)	Tolerance (%)
1st	2d	3d	4th	5th		
Yellow	Violet	Orange	Silver	None	47,000	10
Red	Red	Red	Gold	None	2200	5
Brown	Black	Gold	Gold	None	1	5
Orange	White	Silver	Silver	None	0.39	10
Green	Blue	Red	Red	Brown	56,200	1
Violet	Gray	Violet	Silver	Red	7.87	2

±1%	±2%	±5%	±10%	±1%	±2%	±5%	±10%
100	100	10	10	316	316		
102				324			
105	105			332	332	33	33
107				340			
110	110	11		348	348		
113				357			
115	115			365	365	36	
118				374			
121	121	12	12	383	383		
124				392		39	39
127	127			407	407		
130		13		412			
133	133			422	422		
137				432		43	
140	140			442	442		
143				453			
147	147			464	464		
150		15	15	475		47	47
154	154			487	487		
158				499			
162	162	16		511	511	51	
165				523			
169	169			536	536		
174				549			
178	178			562	562	56	56
182		18	18	576			
187	187			590	590		
191				604			
196	196			619	619	62	
200		20		634			
205	205			649	649		
210				665			
215	215			681	681	68	68
221		22	22	698			
226	226			715	715		
232				732			
237	237			750	750	75	
243		24		765			
249	249			787	787		
255				806			
261	261			825	825	82	82
267				845			
274	274	27	27	866	866		
280				887			
287	287			909	909	91	
294				931			
301	301	30		953	953		
309				976			

Fig. 4-23 Commonly available resistances in various tolerances. Numbers in tolerance columns represent only the significant digits of the resistance values available.

FUSE RESISTOR

Fuse resistors have low resistance (less than 200 Ω) and limited current-carrying capacity. They are wire-wound resistors. These resistors are used in circuits that require high initial currents but low operating currents. If the current remains high, the resistor burns out (opens) and protects the circuit.

THERMAL RESISTOR

These resistors are commonly called *thermistors*. They are resistors which have a very high temperature coefficient. The large changes in resistance for small changes in temperature make the thermistor useful for measuring temperature. The thermistor is also used to compensate for other

Fig. 4-24 Resistor networks. Two views of a DIP are shown on the left. The SIP is on the right.

temperature-induced changes in electronic circuits.

VOLTAGE-SENSITIVE OR VOLTAGE-DEPENDENT RESISTORS (VDR)

Also called *varistors*, these devices are used to protect circuits from very sudden increases in source voltage. If the sudden increase lasts only for a fraction of a second, the varistor limits the voltage to a safe value. These devices are often connected directly across the line input to voltage-sensitive devices such as computers.

Self-Test

Answer the following questions.

37. List the four categories of resistors.
38. A two-terminal variable resistor is called a(n) _____.
39. A three-terminal variable resistor is called a(n) _____.
40. The physical size of a given type of resistor determines the resistor's _____.
41. What are the resistance and tolerance of a resistor with the following color-band combinations:
    a. Brown, black, orange, silver
    b. Red, red, silver, gold
    c. Blue, gray, green
    d. Yellow, violet, black, silver
    e. Gray, red, green, orange, brown
42. What are the minimum and maximum permissible resistances of a 470-Ω, 5 percent resistor?
43. List four resistive materials used in variable resistors.
44. What is the power dissipated by a 270-Ω resistor with 6 V across it?
45. What type of resistor is generally used for high-power applications?

4-12 Switches

There are many types of switches available for electric circuits (Fig. 4-25). They all perform the same basic function of opening or closing circuits. The type used in a given application is often a matter of style and/or convenience of operation. When the switching requirements are complex, the choice narrows to the rotary switch.

Types, Symbols, and Uses

Some of the more common types of small switches are shown in Fig. 4-25. These types are fairly simple switches which usually control only one or two circuit paths. Figure 4-26 shows the schematic symbols of these switches. The dashed line in Fig. 4-26(c) and (d) means that the two poles are mechanically connected

Fig. 4-25 Some common types of switches.

SPST
(Single-pole,
single-throw)

(a)

SPDT
(Single-pole,
double-throw)

(b)

Normally open

Normally closed

Spring-loaded

Rotary switches

DPST
(Double-pole,
single-throw)

(c)

DPDT
(Double-pole,
double-throw)

(d)

Fig. 4-26 Switch symbols and names.

Shorting

Nonshorting

SPDT

Make-before-break

but electrically insulated. Figure 4-27 shows the internal structure of a toggle switch similar to those used as wall switches in the home.

Controlling a lamp from two different locations is illustrated in Fig. 4-28. The circuit uses two single-pole, double-throw (*SPDT*)

switches. As shown, the lamp would be off. Changing the position of either switch turns the lamp on. The lamp can be turned on or off by either switch regardless of the position of the other switch.

Some switches are constructed so that they always return to the same position when released by the operator. These are called *momentary-contact* switches. They can be either *normally open* (*NO*) or *normally closed* (*NC*). Schematic symbols for such switches are shown in Fig. 4-29. This type of switch is sometimes called *spring-loaded* because a spring returns it to its normal position.

Rotary switches are illustrated in Fig. 4-30. They range in complexity from a simple on-off switch to a multideck switch. Figure 4-31 shows the symbols and terminology used for rotary switches.

Rotary switches come in either a *shorting* or a *nonshorting* configuration (Fig. 4-32). The shorting configuration is sometimes referred to as a *make-before-break* switch. This means

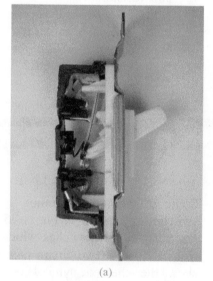

(a)   (b)

Fig. 4-27 Internal structure of a switch. (*a*) Switch on (closed). (*b*) Switch off (open).

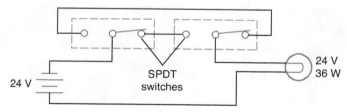

24 V

SPDT
switches

24 V
36 W

Fig. 4-28 Controlling a lamp from two different locations using SPDT switches. The lamp can be turned off or on with either switch.

(a) Normally closed   (b) Normally open

Fig. 4-29 Symbols for momentary-contact switches.

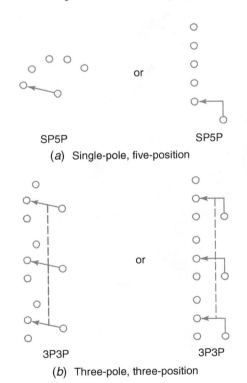

Fig. 4-30 Rotary switches. These switches can provide many switching functions.

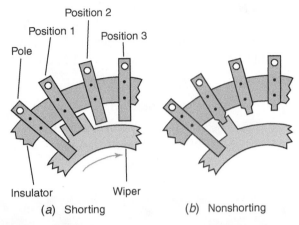

(a) Shorting   (b) Nonshorting

Fig. 4-32 Shorting and nonshorting rotary switches.

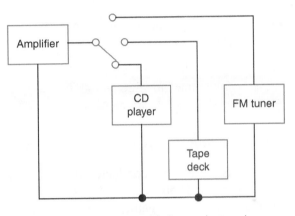

Fig. 4-33 Using a rotary switch to select various input devices.

that the pole makes contact with the new position before it breaks contact with the old position. The wiper in Fig. 4-32(a) comes in contact with position 2 before it loses contact with position 1.

An example of the use of a rotary switch is shown in Fig. 4-33. This switch is a nonshorting switch. In the position shown, the amplifier receives its input from the CD player.

SP5P   or   SP5P

(a) Single-pole, five-position

3P3P   or   3P3P

(b) Three-pole, three-position

Fig. 4-31 Rotary switch symbols and names.

Ratings

Whenever an electric circuit is broken, an arc (ionization of the air) occurs. The greater the current or the voltage, the larger the arc. *Arcing* causes switch contacts to erode away. Therefore, switches have both a voltage and a current rating. Exceeding the manufacturer's ratings shortens the life of the switch. It can also be dangerous to the operator if the ratings are exceeded. An arc can char and break down the insulation in the switch and thus put the operator in contact with the circuit.

Switches are usually given multiple current and voltage ratings. For example, a switch may be rated for 3 A at 250 V ac, 6 A at 125 V ac, and 1 A at 120 V dc. In general, a switch can handle higher alternating currents than direct currents.

Arcing

EXAMPLE 4-7

Determine the minimum voltage and current ratings required for the switches in Fig. 4-28.

Given: Load is a 24-V, 36-W lamp.

Power source is a 24-V battery.

Each switch can either turn off or turn on the lamp.

Find: Minimum I and V ratings for the switches

Known: $P = IV$ therefore $I = P \div V$

Solution: $I_{lamp} = P_{lamp} \div V_{lamp} = 36\text{ W} \div 24\text{ V} = 1.5\text{ A}$

Answer: Each switch must have a minimum dc rating of 1.5 A and 24 V.

Self-Test

Answer the following questions.

46. What do the dashed (broken) lines in a switch symbol indicate?

47. Name at least four common types of switches.

48. Switches are rated for both _____ and _____.

49. When used to describe a switch, NO means _____.

50. A make-before-break switch is also called a(n) _____ switch.

4-13 Wires and Cables

Cable

Wire

Solid conductor

Stranded conductor

A *cable* is a multiple conductor device. The conductors of the cable are insulated from each other and are held together as a single unit. A *wire* is a single conductor. It may or may not be insulated. Typical cables are shown in Fig. 4-34.

A wire may be solid or stranded. A *solid conductor* is made from a single piece of low-resistance material, such as copper, aluminum, or silver. A *stranded conductor* is made from a number of wires twisted or braided together to form a single conductor. Stranded wire is more flexible than solid wire.

Smaller conductors used in electric wire and cable are often coated with a thin layer of solder. Coating a conductor with a thin layer of solder is referred to as tinning the conductor. Bare copper conductors oxidize quite easily. The oxide must be removed before conductors can be joined by soldering. Thus, tinned wires make soldering easier.

Electric Cables

Shielding

Cables may be either shielded or unshielded (Fig. 4-34). The *shielding* helps to isolate the conductors from electromagnetic fields in the vicinity of the cable. Either individual conductors or the whole cable may be shielded.

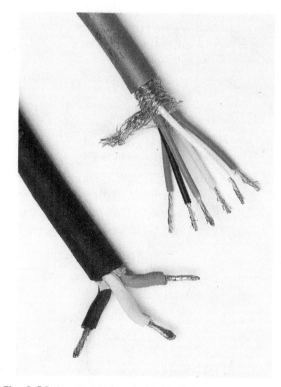

Fig. 4-34 Unshielded and shielded electric cables.

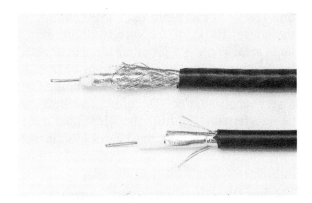

Fig. 4-35 Coaxial cable.

A wide variety of cables are manufactured for general-purpose and audio use. Shielded cables are used for such things as connecting a CD player or tape deck to an amplifier. Unshielded cables are used for such things as extension cords, telephone cords, and doorbell circuits.

Coaxial cable (Fig. 4-35) is a special-purpose, shielded cable used to connect antennas to receivers or transmitters. These cables are used for amateur radio, citizens band radio, and television antenna systems.

Much of the cable used with digital electronic systems is produced in a *flat ribbon* form. Figure 4-36 shows a typical example and also illustrates how this cable can be connected to flat cable plugs.

Power cables carrying very high currents often include a hollow tube through which oil is pumped to cool the cables.

Conductor (Wire) Specifications

The material used in making conductors for electric cables and wires is most often copper. However, aluminum is also used. The size of a round conductor (wire) is specified by a *gage number*. Each gage number represents a different diameter of wire. The diameter of wire is specified in mils. A mil is equal to 0.001 in. Thus, each gage number represents a certain diameter in

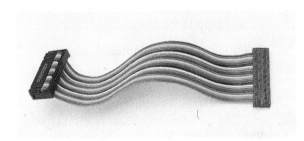

Fig. 4-36 Flat cable.

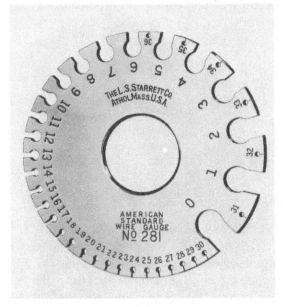

(a)

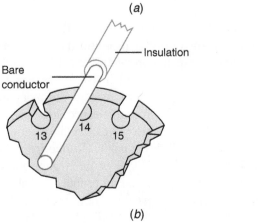

(b)

Fig. 4-37 American wire gage. (*a*) Gage. (*b*) Using a gage.

Coaxial cable

Flat ribbon

mils. Electric wire is measured with the *American wire gage* (AWG), which is the same as the Brown and Sharpe gage used by machinists.

The smaller the gage number, the larger the diameter of the wire. A typical wire gage, such as that shown in Fig. 4-37(*a*), gives both the gage number and the diameter. The diameter is often printed on the reverse side of the gage. Figure 4-37(*b*) illustrates how to use the gage. The width of the slot leading to the hole in the gage indicates the gage size. The hole is larger than the wire and merely allows the wire to move freely through the slot. The slot through which the solid, bare copper wire passes smoothly but snugly represents the wire size. Stranded wire cannot be measured with the wire gage.

The cross-sectional area of a conductor is specified in *circular mils* (cmil or CM). One

American wire gage

Gage number

Circular mils

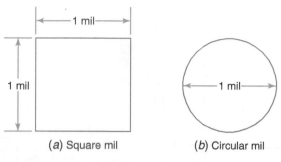

(a) Square mil   (b) Circular mil

Fig. 4-38 Square mil and circular mil. Notice that a circular mil has less cross-sectional area than a square mil.

Maximum temperature rating

circular mil is the area of a circle with a diameter of 1 mil. The cross-sectional area of a conductor is equal to the diameter (in mils) squared. A conductor that is 30 mils in diameter would have a cross-sectional area of 900 cmil ($30 \times 30 = 900$).

Notice from Fig. 4-38 that a circular mil is not equal to a square mil. This could make the calculation of the resistance of a wire a difficult job. However, wire tables which list the resistance per 1000 ft of wire for various gage numbers are readily available. Such a table is provided in Appendix D. This table also gives the diameter and area of wires of different gages. Notice from the wire table that an increase of three gage numbers results in the area being reduced by about half. Of course, the resistance doubles when the cross-sectional area is halved. This leads to a simple method for remembering wire areas and resistances. Number 10 wire is about 10,000 cmil and 1 Ω/1000 ft. Thus, no. 13 wire (an uncommon size) is about 5000 cmil and 2 Ω/1000 ft. Three sizes smaller, no. 16 wire (a common size for extension cords) is about 2500 cmil and 4 Ω/1000 ft. Check these values against those in the table in Appendix D.

The amount of current a copper wire can safely carry depends on the cross-sectional area of the wire and how much heat the wire can dissipate before the insulation is damaged and the metal is affected. This in turn depends on where and how the wire is run. When air is freely circulating around a conductor, it can carry more current than when air circulation is restricted. For example, a no. 12 copper conductor used to wire a house can safely carry 20 A under certain conditions. The same size wire used in a transformer may carry only 4 A. In the house, the wire has about 325 cmil/A, whereas in the

transformer it has more than 1500 cmil/A. In the house, 100 ft of the wire is strung over a large area. There is a lot of air for heat (created by the current in the wire) to transfer into. In the transformer, 100 ft of the wire is coiled into a very small volume. Therefore, less heat can be transferred away from the wire in the transformer, and so the wire must carry less current.

Insulation Specifications

The insulation used on electric wires and cables has a number of important ratings. Each insulating material has a *maximum temperature rating*. This is the maximum continuous temperature to which the insulation should be exposed. Typical maximum temperatures range from 60°C for some thermoplastic compounds to 250°C for extruded polytetrafluoroethylene.

Voltage ratings for cables and wires depend on both the type of insulation and the thickness of the insulation. The voltage rating of an insulated wire is the maximum voltage which the insulation can continuously withstand. Voltage ratings vary from several hundred volts for appliance cords to several hundred thousand volts for high-voltage cables.

Insulating materials are also rated as to their relative ability to resist damage from various chemicals and environments. They are rated on such things as resistance to water, acid, abrasion, flame, and weather (sun and extreme cold, for example).

Insulating materials used on cables and wires include neoprene, rubber, nylon, vinyl, polyethylene, polypropylene, polyurethane, varnished cambric, paper, silicone rubber, and many more.

Electric Wires

Besides building wires, which come in sizes from 18 AWG to as large as 2,000,000 cmil, electric wire is manufactured in an almost endless variety of sizes, styles, and ratings. Some of the common names used in identifying broad groups of electric wires are listed and defined below.

1. *Busbar wire*. An uninsulated, usually tinned, solid copper wire. Available in a wide range of sizes.
2. *Test-prod wire*. Primarily used to make test leads. High-voltage insulation (from 5000 to 10,000 V). Very fine strands for good flexibility.

3. *Hookup or lead wire.* Insulated wire used to connect electric components and devices together. May be either solid or stranded. Usually tinned. Available in a wide variety of sizes and types of insulation.
4. *Magnet wire.* Used in magnetic devices, such as motors, transformers, speakers, and electromagnets. Always insulated, but the insulation is very thin and often has a copper color. Available in a wide variety of sizes and insulating materials.
5. *Litz wire.* A small, finely stranded, low-resistance, insulated wire. Used for making certain coils and transformers for electronic circuits.

Self-Test

Answer the following questions.

51. The cross-sectional area of round conductors is specified in _____.
52. The _____ is used to measure the size of solid round conductors.
53. List at least three common insulating materials used on wires and cables.
54. Where is magnet wire used?
55. When both wires are made from the same materials, will a no. 20 or a no. 18 wire carry more current?

Internet Connection

You may be interested in visiting the website for the American National Standards Institute.

4-14 Fuses and Circuit Breakers

Protection of a circuit from excessive current is provided by a fuse or a circuit breaker. Protection is needed because an excessively high current can damage either the load or the source (or both). Higher than normal currents result from a defective circuit, which has lower than normal resistance.

Shorts and Opens

In electricity and electronics, the term *short* or *short circuit* means that an undesired conductive path exists in a circuit. A partially shorted circuit [Fig. 4-39(*a*)] has only part of the load shorted out (bypassed). Partially shorted circuits have lower than normal resistance. When all the load is shorted out [Fig. 4-39(*b*)], the circuit is said to have a *dead short* or *direct short*. Such circuits have almost no resistance.

Figure 4-39(*c*) shows another way the term *short* is used. When a conductive path develops between the electric part of a device and the frame or housing that encloses the device, the device is said to be shorted or *grounded* to the frame.

The resistors in Fig. 4-39 could just as well be heating elements in a toaster or coffee pot. The battery could be a 120-V outlet. If not properly wired and protected, a shorted circuit can cause a severe (even lethal) shock or start a fire.

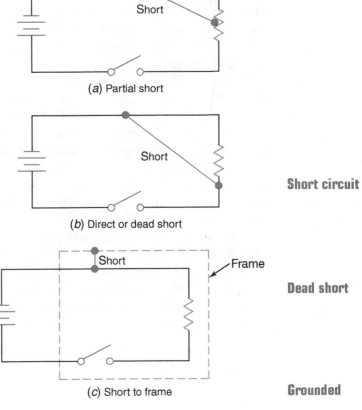

(*a*) Partial short

(*b*) Direct or dead short

Short circuit

Dead short

(*c*) Short to frame

Grounded

Fig. 4-39 Short circuits.

The circuits in Fig. 4-39 could be protected by a fuse or a circuit breaker, as shown in Fig. 4-40. Note the symbols for fuses and circuit breakers.

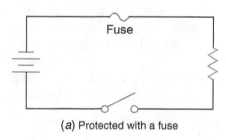

(a) Protected with a fuse

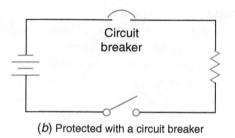

(b) Protected with a circuit breaker

Fig. 4-40 Circuits protected against excessive current.

Open

If a partial short or a dead short of the load occurs in one of the circuits in Fig. 4-40, the fuse or circuit breaker opens. In electricity, *open* means that the path for current has been interrupted or broken. When a fuse opens, we say "The fuse has blown." When a circuit breaker opens, we say "The circuit breaker has tripped."

Traditional Fuses

Fuse

A *fuse* opens when the current through it exceeds its current rating for a period of time. The fuse opens when its conducting element becomes hot enough to melt (Fig. 4-41). The element heats up because it has resistance. A 1-A fuse has approximately 0.13 Ω of resistance. Since $P = I^2R$, more current means more power. More power means more joules (of heat energy) per second converted by the fuse element.

Current rating

A variety of fuses and fuse holders are shown in Fig. 4-42. A glass fuse ($\frac{1}{4}$ in. by $1\frac{1}{4}$ in.) and a panel-mount fuse holder are shown in Fig. 4-42(*a*). Many panel-mount holders do not have the protective screw-on cover shown in this figure. Four subminiature fuses are shown in the upper left part of Figure 4-42(*b*). The lower right part of Fig. 4-42(*b*) shows this type of fuse in the three white fuse holders mounted on a circuit board with other compounds. These fuses are packaged in a tape strip when the fuses are to be surface-mounted (soldered to traces on the circuit board). A miniature radial-lead fuse and a panel-mount holder are shown in Fig. 4-42(*c*). The end of this fuse is clear so that the condition

Voltage rating

of the fuse can be visually determined when the screw-on cover is removed. Indicating panel-mount holders with built-in lamps that glow when the fuse is blown are also available. The fuses pictured in Fig. 4-42(*d*) are pigtail (axial leads) types soldered in place on a printed circuit board. The brownish interior coating of the glass of the lower fuse in Fig. 4-42(*d*) indicates that the fuse is blown.

Fuses have three important ratings, or characteristics, that need to be considered when buying or specifying fuses. These ratings apply to all fuses regardless of size, shape, or style. These three characteristics are (1) a current rating, (2) a voltage rating, and (3) a blowing (fusing) characteristic.

Current ratings of fuses have already been mentioned. The current rating of a fuse is given for specific conditions of air circulation and temperature. These conditions are rarely obtained in typical applications of fuses. Therefore, fuses are usually operated well below their rated current. This practice avoids unnecessary blowing of fuses while still providing short-circuit protection.

Voltage ratings for fuses are necessary for the same reasons that switches have voltage ratings. When a fuse opens, arcing and burning of the element occur. If the current and voltage are high enough, the arcing/burning can continue until the fuse caps and holder are melted. Thus, fuses need voltage ratings. Voltage ratings of fuses are based on a power source capable of providing 10,000 A of current when dead-shorted. A 250-V fuse protects any 250-V power source that can provide no more than 10,000 A. Of course, the current rating of the fuse must be less than the current capabilities of the power source. Many power sources (small batteries and transformers, for example) can provide only a few amperes of current when short-circuited. With limited-current power sources, the voltage

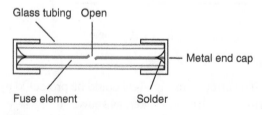

Fig. 4-41 Open fuse.

Fig. 4-42 Fuses and fuse holders. (*a*) Typical small glass fuse and holder. (*b*) Subminiature fuses and circuit board fuse holders. (*c*) Miniature fuse and holder. (*d*) Pigtail fuses.

ratings of fuses are often exceeded. It is fairly common for electronic circuits with more than 400 V to be fused with 250-V fuses.

The *blowing characteristic* of a fuse indicates how rapidly the fuse blows (opens) when subjected to specified overloads. There are three general categories of blowing characteristics of fuses. These are *fast-blow, medium-blow,* and *slow-blow*. These categories are sometimes referred to as *short-time lag, medium-time lag,*

and *long-time lag,* respectively. All three categories respond rapidly (approximately 1 ms) to extreme overloads (more than 10 times the rated current). All three categories respond about the same to very small overloads. At 1.35 times (135 percent) the rated current, they all take a minute or two to open. In between these two extremes, the three categories differ dramatically. For example, at 5 times rated current, a slow-blow may take more than 1 s to blow

Blowing characteristic

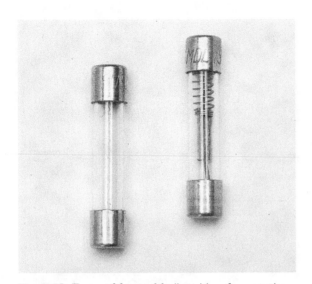

Fig. 4-43 Types of fuses. Medium-blow fuse on the left. Slow-blow fuse on the right.

Fuses can be tested with an ohmmeter. If the fuse is blown, the ohmmeter will indicate an infinite resistance. The meter will show a very low resistance if the fuse is good. Fuses rated at less than ¼ A should not be checked on the $R \times 1$ range. The ohmmeter may provide enough current to blow the fuse. A higher range of the ohmmeter will provide less current and still indicate if the fuse has continuity, that is, if the fuse element still provides a continuous path. When checking a fuse with an ohmmeter, always remove the fuse from the circuit.

Resettable Fuses

A *resettable fuse* (in a surface-mount package) and its schematic symbol are shown in Fig. 4-44. Resettable fuses differ from traditional fuses in a number of ways:

1. The resettable fuse does not open (produce an open circuit); instead it trips. When it trips, its internal resistance greatly increases, and this reduces the circuit current to a value that is safe for the load and/or the source.
2. Once the circuit fault is corrected, this type of fuse resets itself and its internal resistance returns to a low value.
3. After a circuit fault is corrected, the resettable fuse does not need to be replaced the way a traditional fuse does.

Resettable fuses are available with either fast or medium time lag. Like traditional fuses, resettable fuses have both a current and a voltage rating. They are available with either axial or radial leads, as well as in the surface-mount package shown in Fig. 4-44. Notice how small this surface-mount device is.

Circuit Breakers

Circuit breakers have one big advantage over traditional fuses. When they open (trip), they can be reset. Nothing needs to be replaced. Once the overload that caused the breaker to trip has been corrected, the breaker can be reset and used again. There are two types of trip mechanisms: thermal and magnetic.

Thermal circuit breakers are the more common type for such things as small motors, household circuits, and battery chargers. Thermal circuit breakers can have either manual or automatic reset. The automatic reset breaker

while a fast-blow will usually open in less than 1 ms. Under the same conditions, a medium-blow would take about 10 ms.

Fast-blow fuses, sometimes called *instrument fuses,* are used to protect very sensitive devices, such as electric meters. The physical appearance of a fast fuse is the same as that of a medium fuse.

Medium-blow fuses (Fig. 4-43) are general-purpose fuses. They are used when the initial (turn on) current of a device is about the same as the normal operating current.

Slow-blow fuses are used whenever short-duration surge currents are expected. An electric motor is a good example of a device that requires much higher current to start than to continue operating. Its starting current may be five or six times the normal operating current. Therefore, a motor is often fused with a slow-blow fuse. A typical slow-blow fuse is shown in Fig. 4-43. When subjected to an extreme overload (short circuit), the thin part of the element burns open. When the overload is less severe and of long duration, the solder (holding the spring and element together) melts.

Always replace fuses with the style and ratings specified by the manufacturer of the equipment. Do not ignore the voltage ratings of fuses. If a medium or fast fuse is replaced by a slow-blow fuse, the equipment may be damaged before the fuse can respond. When ordering fuses, specify either current, voltage, blowing characteristic, physical dimensions, and style, or current and manufacturer's type number.

(a)

(b)

Fig. 4-44 Resettable fuse. (a) SMD resettable fuse. (b) Symbol for a resettable fuse.

resets after the breaker has cooled. Cooling takes a few minutes. Automatic reset is used where overloads are self-correcting. Thermal breakers use bimetallic strips or disks as their current-sensing element. Therefore, they have long-time lag characteristics.

You May Recall

. . . that the operating principle of a bimetallic strip is illustrated in Fig. 4-14 (page 78) and described in Sec. 4-10.

For manual reset breakers, a mechanical mechanism locks the bimetallic strip in the open position once the breaker is tripped.

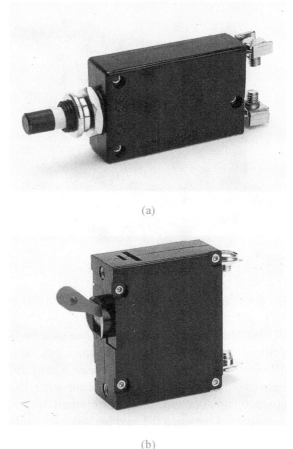

(a)

(b)

Fig. 4-45 Thermal circuit breakers. (a) Push-button reset. (b) Switch/breaker.

Two styles of manual reset thermal breakers are shown in Fig. 4-45. The one in Fig. 4-45(b) is a combination switch/breaker. The breaker is reset by moving the handle all the way to the *off* position and then to the *on* position. The breaker in Fig. 4-45(a) is a typical push-button breaker. When the breaker is tripped, the button extends out so that the white band shows. Pushing the button resets the breaker.

A loose and/or corroded connection on a thermal circuit breaker can generate enough heat to trip a thermal breaker even though the current in the conductor is safely below the rated value of the breaker. Of course, if the faulty connection is some place other than a breaker, the excessive heat can melt insulation and/or start a fire in flammable materials.

The external appearance of a *magnetic circuit breaker* is like that of the thermal breaker. However, its internal structure, principle of operation, and characteristics are quite different. Magnetic breakers can be made with time lags ranging from a few milliseconds to many seconds.

Magnetic circuit breakers

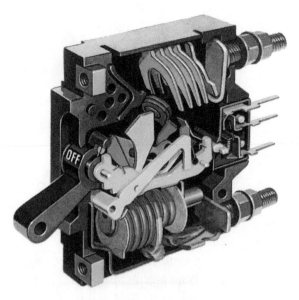

Fig. 4-46 Magnetic circuit breaker.

Figure 4-46 shows a cutaway view of a magnetic circuit breaker in the tripped position. The four shields between the breaker contacts help to reduce the arc that may occur when the contacts are opened. The three small terminals located between the two breaker terminals connect to an SPDT switch. Although the switch is operated by the breaker mechanism, it is electrically isolated from the circuit protected by the breaker. Thus, this switch can be used to control a separate circuit which remotely indicates whether or not the breaker is tripped. The large circuit breakers that protect high-current loads or conductors often have both thermal and magnetic mechanisms. The thermal trip opens when the overload is small and a quick break in the circuit is not needed. A large rush of current, however, causes the breaker to open quickly via the magnetic trip mechanism.

4-15 Other Components

Capacitors, inductors, and transformers are other commonly used circuit components. However, these devices cannot be adequately appreciated until one develops an understanding of alternating current and magnetism. Capacitors, inductors, and transformers are discussed in detail in later chapters.

⎍ Self-Test

Answer the following questions.

56. Define the following terms:
    a. Open circuit
    b. Short circuit
    c. Slow-blow
57. What is an indicating holder?
58. Three major ratings of a fuse are _____, _____, and _____.
59. Fuses protect a circuit from _____.
60. Two types of circuit breakers are the _____ type and the _____ type.
61. What is an instrument fuse?
62. A(n) _____ fuse does not open the circuit when its current rating is exceeded.
63. When a circuit breaker has both a magnetic and a thermal trip mechanism, which trip mechanism provides a quicker response to a large overload current?

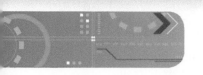

Chapter 4 Summary and Review

Summary

1. Cells are energy or power sources.
2. Cells provide dc voltage.
3. A cell contains an electrolyte and two electrodes.
4. A battery is made from two or more cells.
5. Primary cells are not rechargeable.
6. Secondary cells are rechargeable.
7. Dry cells may use either a paste or a liquid electrolyte.
8. Gases are produced when cells are charged or discharged.
9. The energy storage capacity of a cell or battery is expressed in ampere-hours.
10. The energy available from a battery is dependent on temperature, rate of discharge, and final voltage.
11. Internal resistance causes a cell's voltage to decrease as the current increases.
12. Polarization refers to the buildup of gas ions around an electrode.
13. Common secondary cells are lead-acid, nickel-cadmium, rechargeable alkaline, and lithium-ion.
14. Common primary cells are carbon-zinc, alkaline, mercury, silver oxide, and lithium.
15. When a lead-acid cell is discharging, sulfuric acid is being converted to water.
16. Specific gravity is measured with a hydrometer.
17. Safety glasses should be worn when working with lead-acid cells and batteries.
18. Electrolyte on the skin or clothing should be immediately removed by flushing with lots of water.
19. Carbon-zinc cells are relatively inexpensive.
20. An alkaline cell stores more energy than an equal-size carbon-zinc cell. The former is more efficient at high current drains.
21. Both mercury and silver oxide cells have nearly constant output voltage.
22. Incandescent miniature lamps have tungsten filaments.
23. Miniature lamps have both a current and a voltage rating.
24. Bimetallic strips bend because the two bonded metals have different coefficients of expansion.
25. The resistance of a lamp when hot is greater than its resistance when cold.
26. Only one electrode glows when a neon lamp is operated on direct current.
27. Both a neon lamp and a light-emitting diode (LED) circuit must have a resistor to limit the current through the device.
28. A potentiometer is a variable resistor.
29. A potentiometer can also be used as a rheostat.
30. The power rating of a resistor is independent of its resistance.
31. Common types of resistors are carbon-composition, cermet, wire-wound, deposited-film, and conductive-plastic.
32. Resistance values and tolerances are indicated with a color code on the body of a resistor.
33. Thermistors have high temperature coefficients.
34. Common types of small switches are rotary, toggle, slide, rocker, and push-button.
35. Spring-loaded, momentary-contact switches can be either normally open or normally closed.
36. Rotary switches can handle complex switching involving many circuits.
37. Switches have both a current and a voltage rating.
38. Cables are multiple conductors.
39. Wires are single conductors.
40. Conductors may be stranded or solid.
41. Shielded wires and cables are used to help isolate a conductor's electromagnetic fields.
42. Coaxial cables are used to connect antennas to receivers or transmitters.
43. The American wire gage is the standard used to specify the size of a conductor.
44. The cross-sectional area of a conductor is specified in circular mils.
45. When the gage of a conductor is decreased by three gage numbers, its cross-sectional area doubles. Its resistance is half as much.

46. Insulation used on conductors has both a temperature and a voltage rating.
47. Circuit-breaker mechanisms work on either a magnetic or a thermal principle.
48. Fuses are thermally operated devices.
49. A blown fuse results in an open circuit.
50. Instrument (fast-blow) fuses are used to protect electric meters.
51. Resettable fuses do not produce an open circuit when tripped.
52. SMDs are smaller than traditional components.
53. Fuses and breakers have both current and voltage ratings.
54. The characteristics of various dry cells are given in Table 4-2.

Table 4-2 Dry Cell Characteristics

Cell Type	Rated Voltage	Voltage during Discharge	Cycle Life	Initial Cost	Internal Resistance	Energy-to-Weight Ratio
Carbon-zinc	1.5	Decreases		Low	Med	Med
Alkaline	1.5	Decreases		Med	Low	Med
Mercury	1.35	Almost constant		High	Low	High
Silver oxide	1.5	Almost constant		High	Low	High
Lithium	2.5–3.6	Almost constant		High	Med high	Very high
Rechargeable alkaline	1.5	Decreases	Low	Med	Very low	Low
Nickel-cadmium	1.25	Decreases slightly	High	High	Very low	Low

Chapter Review Questions

For questions 4-1 to 4-21, determine whether each statement is true or false.

4-1. The amount of energy stored in a cell is specified in ampere-hours. (4-1)
4-2. A device which contains two or more cells is called a battery. (4-1)
4-3. All secondary cells are wet cells. (4-1)
4-4. The negative and positive terminals of a cell are produced by polarization of the cell. (4-1)
4-5. Specific gravity is a number which represents the ratio of the weight of a substance to the weight of air. (4-2)
4-6. Lead sulfate is being converted to lead peroxide when a lead-acid cell is being charged. (4-2)
4-7. The specific gravity of a lead-acid cell decreases when the cell is being charged. (4-2)
4-8. The resistance of a lamp is greatest when the lamp is hot. (4-10)
4-9. Copper is commonly used for the filaments of miniature lamps. (4-10)
4-10. Both electrodes glow when direct current is applied to a neon lamp. (4-10)
4-11. A neon lamp circuit uses a resistor only if the voltage in the circuit is more than 20 percent above the lamp's rated voltage. (4-10)
4-12. The control element in a flashing miniature lamp is a bimetallic strip. (4-10)
4-13. Light–emitting diodes operate as well on alternating as on direct current. (4-10)
4-14. The abbreviation NC denotes *no connection* when it is used in switch specifications. (4-12)
4-15. The deposited-film method is usually used to make power resistors. (4-11)
4-16. The greater the resistance of a resistor, the larger its physical size. (4-11)
4-17. The dashed lines between the poles of a switch symbol indicate that the poles are electrically connected to each other. (4-11)
4-18. When a shorting-type switch is rotated, all positions of the switch are momentarily shorted together. (4-11)

4-19. Switches have both a current and a voltage rating. (4-11)

4-20. A circular mil is equal to a square mil. (4-13)

4-21. The larger the diameter of a wire, the larger the gage number of the wire. (4-13)

For questions 4-22 to 4-26, choose the letter that best completes each sentence.

4-22. Cells that cannot be recharged are classified as (4-1)
   a. Primary cells
   b. Secondary cells
   c. Dry cells
   d. Wet cells

4-23. Which one of the following cells is rechargeable? (4-3)
   a. Nickel-cadmium
   b. Mercury
   c. Silver oxide
   d. Carbon-zinc

4-24. Which one of the following cells provides the least constant output voltage? (4-4)
   a. Silver oxide
   b. Lithium
   c. Carbon-zinc
   d. Nickel-cadmium

4-25. A variable resistor with three terminals is a (4-11)
   a. Varistor
   b. Thermostat
   c. Rotary resistor
   d. Potentiometer

4-26. A switch with five poles and four positions would most likely be a (4-12)
   a. Toggle switch
   b. Slide switch
   c. Rotary switch
   d. On-off switch

Answer the following questions:

4-27. What is the nominal resistance of a resistor that is color-banded yellow, violet, gold, gold? (4-11)

4-28. What is the nominal resistance of a resistor that is color-banded red, red, blue, black, brown? (4-11)

4-29. Name two types of circuit breakers. (4-14)

4-30. What are the three electrical ratings used in specifying fuses? (4-14)

4-31. Describe a short circuit. (4-14)

4-32. What type of fuse would you use to protect an electric motor? (4-14)

4-33. Which protection device does not produce an open circuit when tripped? (4-14)

4-34. Will the ampere-hours provided by a cell increase or decrease for each of these deviations from specified conditions? (4-1)
   a. Increased load
   b. Decreased temperature

4-35. What is a variable resistor with two terminals called? (4-11)

4-36. Does an LED or a neon glow lamp use the least power under typical operating conditions? (4-10)

Chapter Review Problems

4-1. What is the minimum resistance of a 4700-Ω, 10 percent resistor? (4-11)

4-2. The terminal voltage of a battery drops from 12.6 V to 12 V when a 4-Ω resistor is connected to it. What is the internal resistance of the battery? (4-1)

4-3. A 1000-Ω resistor is connected to a 31-V source. What power rating should the resistor have? (4-11)

4-4. A 270-Ω $\pm$ 10 percent resistor is drawing 0.2 A from the voltage source. What are the minimum and the maximum power it could be dissipating? (4-11)

4-5. Using the rule of thumb as a safety factor, determine the maximum current a 560-Ω, 1-W resistor should carry. (4-11)

4-6. Determine the amount of energy stored in a 6.3-V battery rated at 150 Ah. (4-1)

4-7. A 9-V battery can provide 3.6×10^4 J when connected to a load that draws 2-A. How long can the battery power a 2-A load? (4-1)

4-1. What are the resistance and power of the load in example 4-1?

4-2. Would you expect the internal resistance of a carbon-zinc cell to increase or decrease with decreasing temperature? Why?

4-3. What assumption must be made about the power sources used in examples 4-2, 4-3, and 4-4? Why?

4-4. Discuss other factors besides those dealt with in this chapter that might be considered in selecting cells and batteries.

4-5. Why do switches have a higher ac than dc rating?

4-6. Why does only one electrode in a neon lamp glow when the power source is direct current?

4-7. What is the maximum power dissipated by a 100-Ω, 5 percent resistor connected to a 50-V source with 5 Ω of internal resistance?

4-8. Determine the minimum voltage across a 50-Ω, 10 percent resistor connected to a 60-V source with 3 Ω of internal resistance.

4-9. A 24-V battery stores 1.728×10^7 J of energy. Determine its Ah rating.

4-10. A battery's terminal voltage changes from 9 V to 8.8 V when the load current changes from 250 mA to 500 mA. Determine the internal resistance of the battery.

Answers to Self-Tests

1. A cell is a device that converts chemical energy to electric energy. A cell is also known as an energy or power source.

2. A battery is an electric device composed of two or more cells.

3. A cell is composed of two electrodes and an electrolyte.

4. a. Primary cell is a cell that is not rechargeable.
   b. Secondary cell is a cell that is rechargeable.
   c. Dry cell is a cell that can be operated in any position. It is sealed.
   d. Wet cell is a vented cell that must be operated in an upright position.
   e. Ampere-hour is a rating of a cell which indicates the amount of energy it stores.

5. rechargeable alkaline, lead-acid, nickel metal hydride, nickel-cadmium, nickel-iron, and lithium-ion.

6. alkaline, carbon-zinc, mercury, silver oxide, zinc chloride, and lithium.

7. decrease

8. polarization

9. 27 Wh

10. 1.5 Ω

11. Lead and lead peroxide are changed to lead sulfate while sulfuric acid is changed to water. This process leaves one plate deficient in electrons and the other plate with an excess of electrons.

12. The battery is recharged by using the battery as a load rather than as a source. Electrons are forced into the negative terminal and out of the positive terminal. The chemical reaction is also reversed.

13. Charge at, or below, the rate recommended by the manufacturer, and do not overcharge the battery.

14. Specific gravity is the ratio of the weight of a substance to the weight of water.

15. hydrometer

16. discharged

17. sulfuric acid

18. Wear safety glasses, avoid spilling the electrolyte, and charge in a spark-free, ventilated area.

19. no

20. a. The nickel-cadmium has a more constant output voltage.
   b. The nickel-cadmium provides 3 to 30 times as many cycles.
   c. The alkaline is much cheaper.

21. a. The alkaline cell is much better for providing high-current loads.
   b. Alkaline cells maintain higher voltage under very heavy current loads.
   c. Alkaline cells store more energy for a given weight.

22. carbon-zinc

23. nickel-cadmium and lithium-ion

24. lithium

25. High-efficiency, low-noise, environmentaly clean operation (by-product is hot water).
26. tungsten
27. incandescent and neon glow
28. A bimetallic strip which acts like a switch as it heats and cools
29. 0.25 A
30. LED
31. positive
32. dc
33. The resistor limits current through the lamp.
34. Do not bend or solder leads within 3 mm of the bulb.
35. The lamp may explode.
36. Do not contaminate the glass enclosure with oil, grease, or dirt (don't touch the glass with bare fingers).
37. variable, adjustable, tapped, and fixed
38. rheostat
39. potentiometer
40. power rating
41. a. 10,000 Ω $\pm$ 10 percent
    b. 0.22 Ω $\pm$ 5 percent
    c. 6,800,000 Ω $\pm$ 20 percent
    d. 47 Ω $\pm$ 10 percent
    e. 825 kΩ $\pm$ 1 percent
42. Minimum resistance is 446.5 Ω, and maximum resistance is 493.5 Ω.
43. carbon-composition, conductive-plastic, cermet, deposited-film, and wire

44. 0.133 W
45. wire-wound
46. The dashed line indicates that two poles are mechanically connected but electrically isolated.
47. rotary, slide, rocker, toggle, and push-button
48. voltage, current
49. normally open
50. shorting
51. circular mils
52. American wire gage
53. vinyl, rubber, neoprene, nylon, asbestos, and other plastic materials
54. electromagnets, motors, generators, speakers, transformers
55. no. 18 wire
56. a. Open circuit means the conductive path for current has been broken.
    b. Short circuit means the circuit has an abnormally low resistance.
    c. Slow-blow means a fuse can stand a short-term overload without blowing.
57. An indicating holder is a fuse holder which lights up when the fuse has blown.
58. current, voltage, blowing characteristic
59. excessive current
60. thermal, magnetic
61. a fast-blow fuse
62. resettable
63. magnetic

CHAPTER 5

Multiple-Load Circuits

Learning Outcomes

This chapter will help you to:

5-1 *Use* subscripts to reference specific components, voltages, currents, etc.

5-2 *Express* the relationships between individual powers and total power in multiple-load circuits.

5-3 *Measure and calculate* individual and total powers, currents, resistances and voltages in a series circuit.

5-4 *Understand* how the internal resistance of the source influences the transfer of power from a source to a load as well as the efficiency of the system.

5-5 *Measure and calculate* individual and total powers, currents, resistance and voltages in a parallel circuit.

5-6 *Learn* the meaning of conductance and its relationship to resistance.

5-7 *Use* your knowledge of series-circuit and parallel-circuit relationships to reduce a series-parallel circuit to a simple single-load equivalent circuit.

5-8 *Determine* the output voltages of an unloaded and a loaded resistor voltage divider circuit and compare these voltages with voltages obtained from a zener diode regulated circuit.

A great majority of electric circuits operate more than one load. Circuits that contain two or more loads are called *multiple-load circuits*. A multiple-load circuit can be a series circuit, a parallel circuit, or a series-parallel circuit.

5-1 Subscripts

Notice the symbols R_1, R_2, and R_3 in Fig. 5-1. The *subscripts* "1," "2," and "3" are used to identify the different-load resistors in the circuit. Two-level subscripts are also used in these types of circuits. For example, I_{R_2} is used to indicate the current flowing through resistor R_2. The symbol V_{R_1} indicates the voltage across resistor R_1. Similarly, P_{R_3} is used to specify the power used by R_3.

In some electrical and electronics literature, only one level of subscript is used. For example, V_{R_1} would be written as V_{R1}. Either way, the reference is to the voltage across resistor R_1.

Electrical quantities for a total circuit are identified by the subscript *T*. Thus, the voltage of the battery of Fig. 5-1 is indicated by the symbol V_T. The symbol I_T indicates the source (battery) current, and R_T represents the total resistance of all the loads combined. The power from the battery or source is identified as P_T.

Subscripts

5-2 Power in Multiple-Load Circuits

The total power taken from a source, such as a battery, is equal to the sum of the powers used by the individual loads. As a formula, this statement is written

$$P_T = P_{R_1} + P_{R_2} + P_{R_3} + \text{etc.}$$

The "+ etc." means that the formula can be expanded to include any number of loads. Regardless of how complex the circuit, the above formula is appropriate.

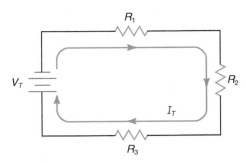

Fig. 5-1 Series circuit. There is only one path for current.

5-3 Series Circuits

A *series circuit* contains two or more loads but only one path for current to flow from the source voltage through the loads and back to the source. Figure 5-1 is an example of a series circuit.

Current in Series Circuits

In Fig. 5-1, the battery current I_T flows through the first load R_1, the second load R_2, and the third load R_3. If 1 A flows through R_1, then 1 A also flows through R_2 and R_3. And, of course, the battery provides 1 A of current. In symbolic form, the current relationship in a series circuit is

$$I_T = I_{R_1} = I_{R_2} = I_{R_3} = \text{etc.}$$

Current in a series circuit can be measured by inserting a meter in series. Since there is only *one path for current,* any part of the circuit can be interrupted to insert the meter. All the

meters in Fig. 5-2 give the same current reading provided the voltage and the total resistance are the same in each case.

Resistance in Series Circuits

The *total resistance* in a series circuit is equal to the sum of the individual resistances around the series circuit. This statement can be written as

$$R_T = R_1 + R_2 + R_3 + \text{etc.}$$

This relationship is very logical if you remember two things: (1) resistance is opposition to current and (2) all of the current has to be forced through all the resistances in a series circuit.

The total resistance R_T can also be determined by Ohm's law if the total voltage V_T and total current I_T are known. Both methods of determining R_T can be seen in the circuit of Fig. 5-3. In this figure, the "2 A" near the ammeter symbol means that the meter is indicating a current of 2 A. The resistance of each resistor is given next to the resistor symbol. Using Ohm's law, the total resistance is

$$R_T = \frac{V_T}{I_T} = \frac{90\text{ V}}{2\text{ A}} = 45\ \Omega$$

Using the relationship for series resistance yields

$$\begin{aligned} R_T &= R_1 + R_2 + R_3 \\ &= 5\ \Omega + 10\ \Omega + 30\ \Omega \\ &= 45\ \Omega \end{aligned}$$

Total resistance

Series circuit

One path for current

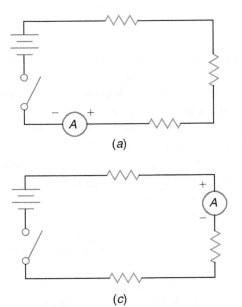

(a)

(c)

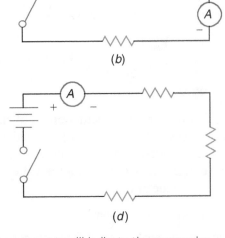

(b)

(d)

Fig. 5-2 Measuring current in a series circuit. The ammeter will indicate the same value of current in any of the positions shown in diagrams (a) to (d).

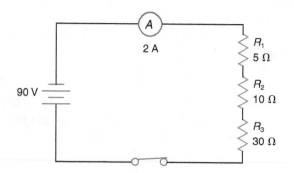

Fig. 5-3 Total resistance can be found by using Ohm's law or by adding individual resistances.

The total resistance of a series circuit can be measured by connecting an ohmmeter across the loads as in Fig. 5-4(a). The power source must be disconnected from the loads. Individual resistances can be measured as shown in Fig. 5-4(b) and (c).

Voltage in Series Circuits

The battery voltage in Fig. 5-3 divides up across the three load resistors. It always divides up so that the sum of the individual load voltages equals the source voltage. That is,

$$V_T = V_{R_1} + V_{R_2} + V_{R_3} + \text{etc.}$$

Kirchhoff's voltage law

This relationship is often referred to as *Kirchhoff's voltage law*. Kirchhoff's law states that "the sum of the voltage drops around a circuit equals the applied voltage."

For Fig. 5-3, this relationship can be verified by using Ohm's law to determine the individual voltages:

$$V_{R_1} = I_{R_1} \times R_1 = 2\,\text{A} \times 5\,\Omega = 10\,\text{V}$$

$$V_{R_2} = I_{R_2} \times R_2 = 2\,\text{A} \times 10\,\Omega = 20\,\text{V}$$

$$V_{R_3} = I_{R_3} \times R_3 = 2\,\text{A} \times 30\,\Omega = 60\,\text{V}$$

$$V_T = V_{R_1} + V_{R_2} + V_{R_3}$$

$$= 10\,\text{V} + 20\,\text{V} + 60\,\text{V}$$

$$= 90\,\text{V}$$

Polarity

Notice from the above calculations that individual voltages are directly proportional to individual resistances. The voltage across R_2 is twice the voltage across R_1 because the value of R_2 is twice the value of R_1.

Voltage Drop and Polarity

Voltage drop

The voltage, or potential energy difference, across a resistor is referred to as a *voltage drop*. We can

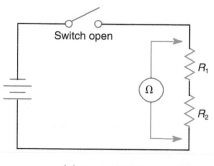

(a) Measuring R_T

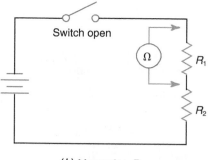

(b) Measuring R_1

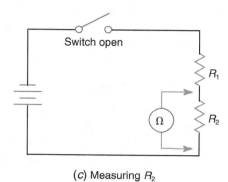

(c) Measuring R_2

Fig. 5-4 Measuring resistance in series.

say that a voltage develops across the resistor. That is, part of the potential energy difference of the source develops, or appears, across each resistor as a smaller potential energy difference. There is a distinction between source voltage and the voltage across the loads. The source voltage provides the electric energy, and the load voltage converts the electric energy into another form.

The voltage drop across a resistor has *polarity*. However, in this case the polarity does not necessarily indicate a deficiency or excess of electrons. Instead, it indicates the direction of current flow and the conversion of electric energy to another form of energy. Current moves through a load resistor from the negative polarity to the positive polarity. This means that electric energy is being converted to another form.

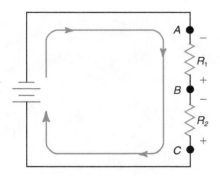

Fig. 5-5 Voltage polarity in a series circuit.

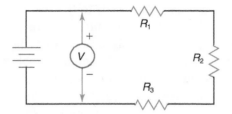

Fig. 5-6 Measuring the total voltage of a series circuit.

The current *through* a battery moves from the positive to the negative polarity. Thus, the battery is providing the electric energy.

Polarity signs are shown on the resistors in Fig. 5-5. The current through R_1 and R_2 flows from the negative end of the resistors to the positive end. Current inside the battery flows from the positive to the negative. However, the external current still flows from the negative terminal to the positive terminal of the battery.

Notice in Fig. 5-5 that point B is labeled both negative (−) and positive (+). This may seem contradictory, but it is not. Point B is *positive with respect to point A*, but *negative with respect to point C*. It is very important to understand the phrase "with respect to." Remember that voltage is defined as "a potential energy difference between two points." Therefore, it is meaningless to speak of a voltage or a polarity at point B. Point B by itself has neither voltage nor polarity. But, with respect to either point A or point C, point B has both polarity and voltage. Point B is positive with respect to point A; this means that point B is at a lower potential energy than point A. It also means that electric energy is being converted to heat energy as electrons move from point A to point B.

Measuring Series Voltages

The total voltage of a series circuit must be measured across the voltage source, as shown in Fig. 5-6. Voltages across the series resistors are also easily measured. The correct connections for measuring series voltages are illustrated in Fig. 5-7. The procedure is to first determine the correct function, range, and polarity on the meter.

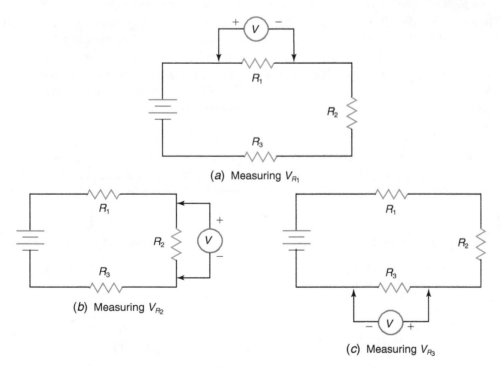

(a) Measuring V_{R_1}

(b) Measuring V_{R_2}

(c) Measuring V_{R_3}

Fig. 5-7 Measuring voltage drops across each resistor in a series circuit.

Then touch the meter leads to the two points at which voltage is to be measured. Voltmeters have a very high internal resistance, so high that connecting them to a circuit has no noticeable effect on many circuits. For now, assume that meters (both ammeters and voltmeters) do not change the circuits in which they are used. Also, assume that the voltage source has insignificant internal resistance.

Open in Series Circuits

Open

In a series circuit, if any part of the circuit is *open*, current stops flowing and voltage and power are removed from all loads. This is one of the weaknesses of a series circuit. For example, when one lamp in a series (such as the old-style holiday tree lights) burns out (opens), all lamps go out.

An easy way to determine which load in a series is open is to measure the individual voltages. The load that is open will have a voltage drop

Shorted

equal to the entire source voltage. In Fig. 5-8(a) a meter connected across either of the good resistors reads 0 V. Since R_2 is open, no current flows in the series circuit. A meter across the open resistor R_2, however, as in Fig. 5-8(b), reads approximately 50 V. When a voltmeter is connected across R_2 as in Fig. 5-8(b), a small current flows through R_1, the voltmeter, and R_3. In most circuits, the internal resistance of the voltmeter is very, very high compared with the other series loads. Therefore, nearly all the battery voltage is developed (and read) across the voltmeter. It should be noted that 50 V is also across R_2, in Fig. 5-8(a). It has to be to satisfy Kirchhoff's voltage law. That is, $V_{R_1} + V_{R_2} + V_{R_3}$ has to equal the 50 V of the source.

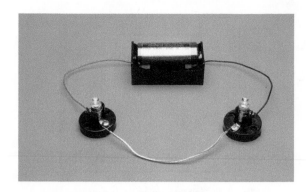

An example of a series circuit.

The diagrams of Fig. 5-8 show the open resistor. However, in a real physical circuit, R_2 may look just like R_1 and R_3. There may be no physical evidence to indicate that it is open. A technician repairing the circuit would have to interpret the voltmeter readings to conclude that R_2 is open.

Shorts in Series Circuits

When one load in a series circuit is *shorted* out, the other loads *may continue to operate*. Or, one of the other loads *may open* because of increased voltage, current, and power. In Fig. 5-9(a), each lamp is rated at 10 V and 1 A. Each lamp has a power rating of 10 W ($P = IV = 1\text{ A} \times 10\text{ V}$). When lamp L_2 shorts out, as shown in Fig. 5-9(b), the 30-V source must divide evenly between the two remaining lamps. This means that each remaining lamp must drop 15 V. Since the lamp voltages increase by 50 percent, the lamp currents also increase by 50 percent to 1.5 A (this assumes the individual lamp resistance does not change). With 15 V and 1.5 A, each lamp has to dissipate 22.5 W. One or both of the lamps will soon burn out (open).

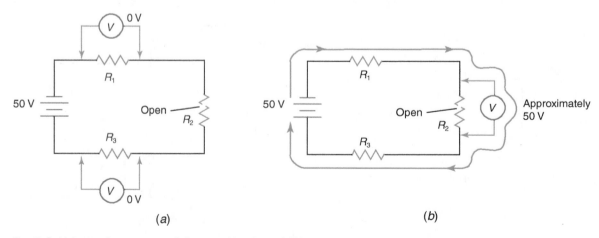

(a)   (b)

Fig. 5-8 Voltage drops across (a) normal loads and (b) open loads.

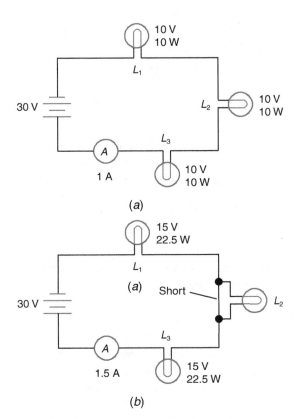

Fig. 5-9 Effects of a shorted series load.

The effects of one shorted load in a series can be summarized as follows:

1. The total resistance decreases.
2. The total current increases.
3. The voltage across the remaining loads increases.
4. The power dissipation of the remaining load increases.
5. The total power increases.
6. The resistance, voltage, and power of the shorted load decrease. If the short is a dead short, these quantities decrease to zero.

Solving Series-Circuit Problems

Proper use of Ohm's law and the series relationships of current, voltage, resistance, and power solves most series problems. In using Ohm's law, you should develop the habit of *using subscripts* for voltage, current, and resistance. Without subscripts, it is easy to forget which voltage, current, or resistance to substitute into the formula. If you are using Ohm's law to find the voltage across R_1, write

$$V_{R_1} = I_{R_1} R_1$$

If you are calculating the total voltage, you should write

$$V_T = I_T R_T$$

EXAMPLE 5-1

As shown in Fig. 5-10, an 8-V, 0.5-A lamp is to be operated from a 12.6-V battery. What resistance and wattage rating are needed for R_1?

Given: $V_T = 12.6\text{ V}$
$V_{L_1} = 8\text{ V}$
$I_{L_1} = 0.5\text{ A}$

Find: R_1, P_{R_1}

Known: $R_1 = \dfrac{V_{R_1}}{I_{R_1}}$
$V_T = V_{L_1} + V_{R_1}$
$I_T = I_{R_1} = I_{L_1}$
$P_{R_1} = V_{R_1} I_{R_1}$

Solution: $I_{R_1} = I_{L_1} = 0.5\text{ A}$
$V_T = V_{L_1} + V_{R_1}$

Therefore,

$V_{R_1} = V_T - V_{L_1}$
$V_{R_1} = 12.6\text{ V} - 8\text{ V} = 4.6\text{ V}$

$R_1 = \dfrac{4.6\text{ V}}{0.5\text{ V}} = 9.2\ \Omega$

$P_{R_1} = V_{R_1} I_{R_1}$

$= 4.6\text{ V} \times 0.5\text{ A}$

$= 2.3\text{ W}$

Answer: The calculated values for R_1 are 9.2 Ω and 2.3 W. Use a 5-W resistor to provide a safety factor.

Using subscripts

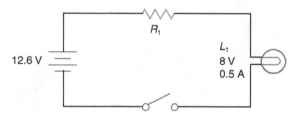

Fig. 5-10 Circuit diagram for example 5-1.

EXAMPLE 5-2

Find the total current and total resistance of the circuit in Fig. 5-11. Also determine the voltage across each resistor.

Voltage-divider equation

Given:
$$V_T = 90 \text{ V}$$
$$R_1 = 35 \text{ }\Omega$$
$$R_2 = 70 \text{ }\Omega$$
$$R_3 = 45 \text{ }\Omega$$

Find: $I_T, R_T, V_{R_1}, V_{R_2}, V_{R_3}$

Known:
$$I_T = \frac{V_T}{R_T}$$
$$R_T = R_1 + R_2 + R_3$$
$$V_{R_1} = I_{R_1} R_1$$
$$I_T = I_{R_1} = I_{R_2} = I_{R_3}$$

Solution:
$$R_T = 35 \text{ }\Omega + 70 \text{ }\Omega + 45 \text{ }\Omega$$
$$= 150 \text{ }\Omega$$
$$I_T = \frac{90 \text{ V}}{150 \text{ }\Omega} = 0.6 \text{ A}$$
$$V_{R_1} = 0.6 \text{ A} \times 35 \text{ }\Omega = 21 \text{ V}$$
$$V_{R_2} = 0.6 \text{ A} \times 70 \text{ }\Omega = 42 \text{ V}$$
$$V_{R_3} = 0.6 \text{ A} \times 45 \text{ }\Omega = 27 \text{ V}$$

Answer:
$$I_T = 0.6 \text{ A}, R_T = 150 \text{ }\Omega$$
$$V_{R_1} = 21 \text{ V}$$
$$V_{R_2} = 42 \text{ V}$$
$$V_{R_3} = 27 \text{ V}$$

Cross-check

After you solve a complex problem, it is a good idea to *cross-check* the problem for mathematical errors. This can usually be done by checking some relationship *not used* in originally solving the problem. The problem of

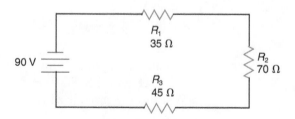

Fig. 5-11 Circuit diagram for example 5-2.

example 5-2 can be cross-checked by using Kirchhoff's voltage law:

$$V_T = V_{R_1} + V_{R_2} + V_{R_3}$$
$$= 21 \text{ V} + 42 \text{ V} + 27 \text{ V}$$
$$= 90 \text{ V}$$

Since 90 V was specified for V_T, the cross-check verifies that at least the sum of the individual voltages is as it should be.

Voltage-Divider Equation

When you want to find the voltage across only one of the resistors in a series circuit, you can use the *voltage-divider equation*. This equation in its general form is

$$V_{R_n} = \frac{V_T R_n}{R_T}$$

where R_n is any one of the resistors in the series circuit. The logic of this equation is obvious if it is written in the form

$$V_{R_n} = \frac{V_T}{R_T} \times R_n = I_T \times R_n = I_{R_n} \times R_n$$

To illustrate the usefulness of the voltage-divider equation, let us use it to solve for V_{R_2} in Fig. 5-11. If we remember that $R_T = R_1 + R_2 + R_3$ for Fig. 5-11, we can write the voltage divider equation as

$$V_{R_2} = \frac{V_T R_2}{R_1 + R_2 + R_3}$$
$$= \frac{90 \text{ V} \times 70 \text{ }\Omega}{35 \text{ }\Omega + 70 \text{ }\Omega + 45 \text{ }\Omega} = 42 \text{ V}$$

Estimations, Approximations, and Tolerances

In a series circuit, the resistor with the most resistance dominates the circuit. That is, the highest resistance drops the most voltage, uses the most power, and has the most effect on the total current. Sometimes one resistor is so much larger than the other resistors that its value almost determines the total current. For example, resistor R_1 in Fig. 5-12 dominates the circuit. Shorting out R_2 in the circuit would increase the current from about 18 mA to only 20 mA. Assume R_1 and R_2 are ± 10 percent resistors. Then R_1 could be as low as 90 kΩ and R_2 as low as 9 kΩ. If this were the case, the current in Fig. 5-12 would still be about

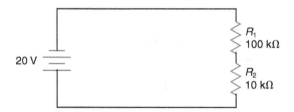

Fig. 5-12 Dominant resistance. Circuit current and power are largely determined by R_1.

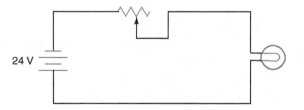

Fig. 5-14 Lamp-intensity control.

20 mA. The presence or absence of R_2 has no more effect on the current than the tolerance of R_1 does. Therefore, a good *estimate* of the current in Fig. 5-12 could be obtained by ignoring R_2. This estimate would be close enough for such things as

1. Determining the power rating needed for the resistors by using $P = I^2R$
2. Determining which range of an ammeter should be used to measure the current
3. Estimating the power required from the battery

When estimating current in a series circuit, ignore the lowest resistance if it is less than the tolerance of the highest resistance.

Applications of Series Circuits

One application of series circuits has already been mentioned—holiday tree lights. Other applications include (1) motor-speed controls, (2) lamp-intensity controls, and (3) numerous electronic circuits.

A simple *motor-speed control* circuit is shown in Fig. 5-13. This type of control is used on very small motors, such as the motor on a sewing machine. With a sewing machine, the variable resistance is contained in the foot control. The circuit provides continuous, smooth control of motor speed. The lower the resistance, the faster the motor rotates. A big disadvantage of this type of motor control is

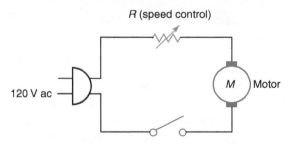

Fig. 5-13 Motor-speed control.

that it is inefficient. Sometimes the resistance converts more electric energy to heat than the motor converts to mechanical energy.

The intensity of an indicator lamp is often controlled by a *variable* series *resistor*. Such circuits are used to illuminate dials and meters on radio and navigation equipment in airplanes. A typical circuit is illustrated in Fig. 5-14. In this circuit, increasing the resistance decreases the total current and the lamp intensity. Notice that the symbol for the variable resistor in Fig. 5-14 is different from the one used in Fig. 5-13. Either symbol is correct. They both are symbols for a *rheostat*.

Estimate

Variable resistor

Rheostat

EXAMPLE 5-3

The rheostat (R) in Fig. 5-14 is adjusted to provide a lamp voltage (V_L) and lamp current (I_L) of 16 V and 0.6 A, respectively. What resistance (R_R) is the rheostat set for, and how much power (P_R) is the rheostat dissipating?

Given: $V_{\text{battery}}\ (V_T) = 24\text{ V}$
$V_L = 16\text{ V}$
$I_L = 0.6\text{ A}$

Find: $R_{\text{rheostat}}\ (R_R)$
$P_{\text{rheostat}}\ (P_R)$

Known: $V_T = V_L + V_R$
$I_T = I_L = I_R$
$P_R = I_R \times V_R$

Solution: $V_R = V_T - V_L = 24\text{ V} - 16\text{ V} = 8\text{ V}$
$I_R = I_L = 0.6\text{ A}$
$R_R = V_R \div I_R = 8\text{ V} \div 0.6\text{ A}$
$\quad = 13.33\ \Omega$
$P_R = I_R \times V_R = 0.6\text{ A} \times 8\text{ V}$
$\quad = 4.8\text{ W}$

Answer: $R_R = 13.33\ \Omega$
$P_R = 4.8\text{ W}$

Motor-speed control

Transistor

A portion of a *transistor* circuit is shown in Fig. 5-15. Resistor R_1 is in series with the collector and emitter of the transistor. The transistor acts like a variable resistor. Its resistance is controlled by the current supplied to the base. Thus, controlling the current to the base controls the current through the transistor. This makes it

Amplify

possible for a transistor to *amplify,* or increase, the small current presented to the base.

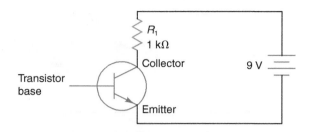

Fig. 5-15 Series resistor in a transistor circuit.

Self-Test

Answer the following questions.

1. Write the symbol for
   a. Voltage across resistor R_4
   b. Source current
   c. Current through resistor R_2
2. True or false. In any multiple-load circuit the total power is the sum of the individual powers.
3. True or false. When resistors are in series, they share a common current.
4. True or false. In a series circuit, the source current is equal to the current through any load.
5. Give two formulas for determining the total resistance of a series circuit.
6. Can the resistance of a resistor in a series circuit be measured without completely removing it from the circuit?
7. Write the formula that shows the relationship between total and individual voltages in a series circuit.
8. Which drops more voltage in a series circuit, a 100-Ω resistor or a 56-Ω resistor?
9. Does a negative-polarity marking on a voltage drop indicate an excess of electrons at that point?

10. A series circuit contains two resistors. One resistor is good, and the other is open. Across which resistor will a voltmeter indicate more voltage?
11. A series circuit contains three resistors: R_1, R_2, and R_3. If R_2 shorts out, what happens to each of the following:
    a. Current through R_1
    b. Total power
    c. Voltage across R_3
12. A 125-Ω resistor (R_1) and a 375-Ω resistor (R_2) are connected in a series to a 100-V source. Determine the following:
    a. Total resistance
    b. Total current
    c. Voltage across R_1
    d. Power dissipated by R_2
13. Refer to Fig. 5-10. Change L_1 to a 7-V, 150-mA lamp. What resistance is now needed for R_1? How much power does the battery have to furnish?
14. If the resistance in Fig. 5-14 decreased, what would happen to the intensity of the lamp?
15. If the transistor in Fig. 5-15 opened, how much voltage would be measured across R_1?
16. If the transistor in Fig. 5-15 shorted, how much current would flow through R_1?

5-4 Maximum Power Transfer

Maximum power transfer

Maximum power transfer refers to getting the maximum possible amount of power from a source to its load. The source may be a battery, and the load a lamp, or the source could be a guitar amplifier and the load a speaker.

Impedance

Maximum power transfer occurs when the source's internal opposition to the current equals the load's opposition to the current. Resistance is one form of opposition to current; *impedance* is another form of opposition. You will learn more about impedance when you

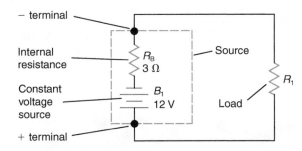

Fig. 5-16 Maximum power transfer occurs when R_B and R_1 are equal.

study alternating current. In dc circuits, maximum power transfer occurs when resistances are matched.

Referring to Fig. 5-16 will aid you in understanding resistance matching and power transfer. In this figure, the internal resistance of the battery is represented by R_B. The battery B_1 represents what is called a *constant voltage source* or an *ideal voltage source*. That is, it represents a voltage source that has no internal resistance. The dotted line around R_B and B_1 means that R_B and B_1 together behave like a real battery. Resistance R_B and battery B_1 do not actually exist as separate components. They cannot be separated. Thus, they are enclosed in dotted lines. Together they form the source. The load is R_1. When R_1 equals R_B, the maximum amount of power is transferred from the source to the load. This statement can be proved by an example. First assign fixed values to R_B and B_1. Then calculate the power dissipated by R_1 for various values of R_1. For example, when R_1 is 9 Ω, the series calculations are

$$R_T = R_1 + R_B = 9\ \Omega + 3\ \Omega = 12\ \Omega$$

$$I_T = \frac{V_T}{R_T} = \frac{12\ \text{V}}{12\ \Omega} = 1\ \text{A}$$

$$P_{R_1} = I_{R_1}{}^2 R_1 = (1\ \text{A})^2 \times 9\ \Omega = 9\ \text{W}$$

$$P_{R_B} = I_{R_B}{}^2 R_B = (1\ \text{A})^2 \times 3\ \Omega = 3\ \text{W}$$

$$P_T = P_{R_1} + P_{R_B}$$
$$= 9\ \text{W} + 3\ \text{W} = 12\ \text{W}$$

EXAMPLE 5-4

A battery (B_1) produces 16 V (V_{NL}) when no load is connected to it. Connecting a 25-Ω load resistor (R_L) to B_1 causes its voltage (V_L) to drop to 15 V. Determine the internal resistance (R_{B_1}) of the battery and the maximum power (P_{max}) that B_1 can deliver to any load resister.

Given: $V_{NL} = 16\ \text{V}$
$V_L = 15\ \text{V}$
$R_L = 25\ \Omega$

Find: R_{B_1}
P_{max}

Known: $I_{B_1} = I_L$
$R_{B_1} = (V_{NL} - V_L) \div I_L$
$I_L = V_L \div R_L$
$R_T = R_L + R_{B_1}$
P_{max} occurs when $R_L = R_{B_1}$
$P = I^2 R$

Solution: $I_{B_1} = I_L = V_L \div R_L = 15\ \text{V} \div 25\ \Omega$
$= 0.6\ \text{A}$

$R_{B_1} = (V_{NL} - V_L) \div I_L$
$= (16\ \text{V} - 15\ \text{V}) \div 0.6\ \text{A}$
$= 1\ \text{V} \div 0.6\ \text{A} = 1.67\ \Omega$

$R_T = R_L + R_{B_1} = 1.67\ \Omega + 1.67\ \Omega$
$= 3.34\ \Omega$

$I_L = V_{NL} \div R_T = 16\ \text{V} \div 3.34\ \Omega$
$= 4.79\ \text{A}$

$P_{max} = I_L{}^2 R_L = (4.79\ \text{A})^2 \times 1.67\ \Omega$
$= 38.32\ \text{W}$

Answer: Internal resistance of the battery is 1.67 Ω.

Maximum power available from the battery is 38.32 W.

Constant voltage source

Calculations for other values of R_1 have been made and recorded in Table 5-1. Notice that the maximum power dissipation occurs when R_1 is 3 Ω.

Table 5-1	Calculated Values for Fig. 5-16				
R_1 (Ω)	R_T (Ω)	I_T (A)	P_{R_1} (W)	P_{R_B} (W)	P_T (W)
1	4	3.00	9.00	27.00	36.00
2	5	2.40	11.52	17.28	28.80
3	6	2.00	12.00	12.00	24.00
4	7	1.71	11.76	8.82	20.57
5	8	1.50	11.25	6.75	18.00
6	9	1.33	10.67	5.33	16.00

Also shown in Table 5-1 are the power dissipated within the source and the total power taken from the source. Notice that when maximum power transfer occurs, the efficiency is only 50 percent. Of the 24 W furnished by the source, only 12 W is used by the load. As the load gets larger, the efficiency improves and the power transferred decreases.

Self-Test

Answer the following questions.

17. When is maximum power transferred from a source to a load?

18. For high efficiency, should the load resistance be equal to, less than, or greater than the internal resistance of the source?

5-5 Parallel Circuits

Parallel circuits

More than one path

Branch

Parallel circuits are multiple-load circuits which have *more than one path* for current. Each different current path is called a *branch*. The circuit in Fig. 5-17 has three branches. Current from the battery splits up among the three branches. Each branch has its own load. As long as the voltage from the source remains constant, each branch is independent of all other branches. The current and power in one branch are not dependent on the current, resistance, or power in any other branch.

In Fig. 5-18, switch S_2 in the second branch controls the lamp in that branch. However, turning switch S_2 on and off does not affect the lamps in the other branches. Figure 5-18 illustrates the way in which the lamps in your home are connected. In a house, the electric circuits are parallel circuits. The fuse in Fig. 5-18 protects all three branches. It carries the current for all branches. If any one branch draws too much current, the fuse opens. Also, if the three branches together draw too much current, the fuse opens. Of course, when the fuse opens, all branches become inoperative.

Voltage in Parallel Circuits

All voltages in a parallel circuit are the same. In other words, the source voltage appears across each branch of a parallel circuit. In Fig. 5-19(*a*), each voltmeter reads the same

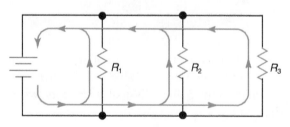

Fig. 5-17 Parallel circuit. There is more than one path for current to take.

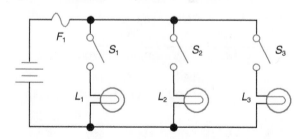

Fig. 5-18 Independence of parallel branches. Opening or closing S_2 has no effect on L_1 or L_3.

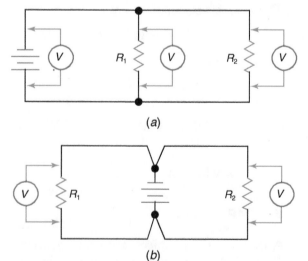

(a)

(b)

Fig. 5-19 Voltage measurement in parallel circuits. All voltmeters will indicate the same value of voltage.

voltage. Rearranging the circuit in Fig. 5-19(*a*) yields the circuit shown in Fig. 5-19(*b*). It is easier to see in Fig. 5-19(*b*) that each branch of the circuit receives the total battery voltage. For a parallel circuit, the relationship of source voltage to load voltage is expressed as

$$V_T = V_{R_1} = V_{R_2} = V_{R_3} = \text{etc.}$$

In a parallel circuit, the voltage measured across the load does not change if the load opens. If R_2 in Fig. 5-19(*a*) were open, the voltmeter across R_2 would still measure the voltage of the battery.

Current in Parallel Circuits

The relationship of the currents in a parallel circuit is as follows:

$$I_T = I_{R_1} + I_{R_2} + I_{R_3} + \text{etc.}$$

In other words, the total current is equal to the sum of the individual *branch currents*. Figure 5-20 shows all the places where an ammeter could be inserted to measure current in a parallel circuit. The current being measured at each location is indicated inside the meter symbol. The total current I_T in Fig. 5-20 splits at junction 1 into two smaller currents—namely I_{R_1} and I_A. At junction 2, I_A splits into currents I_{R_2} and I_{R_3}. Currents I_{R_2} and I_{R_3} join at junction 3 to form I_B. Finally, I_B and I_{R_1} join at junction 4 and form the total current returning to the source.

The various currents entering and leaving a junction are related by *Kirchhoff's current law.*

History of
Electronics

Gustav R. Kirchhoff
German physicist Gustav R. Kirchhoff is best known for his statement of two basic laws of the behavior of current and voltage. Developed in 1847, these laws enable scientists to understand and therefore evaluate the behavior of networks.

This law states that "the sum of the currents entering a junction equals the sum of the currents leaving a *junction*." No matter how many wires are connected at a junction, Kirchhoff's current law still applies. Thus, in Fig. 5-20, the following relationships exist:

$$I_T = I_{R_1} + I_A$$
$$I_A = I_{R_2} + I_{R_3}$$
$$I_{R_2} + I_{R_3} = I_B$$
$$I_{R_1} + I_B = I_T$$
$$I_B = I_A$$

In Fig. 5-21, five wires are joined at a junction. If three of the wires carry a total of 8 A into the junction, the other two must carry a total of 8 A out of the junction. Therefore, the unmarked wire in Fig. 5-21 must be carrying 3 A out of the junction.

Branch currents

Junction

Kirchhoff's current law

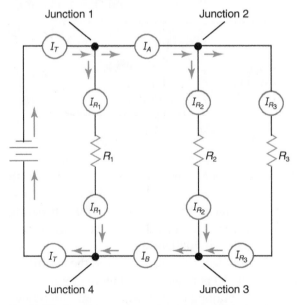

Fig. 5-20 Current measurement in parallel circuits.

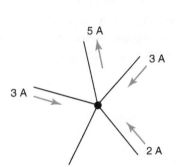

Fig. 5-21 Current at a junction. The unmarked line must carry 3 A of current away from the junction.

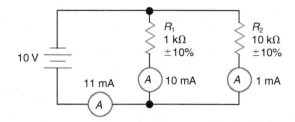

Fig. 5-22 Resistor tolerance and estimated current. Notice that the small resistance carries nearly all the circuit current.

As shown in Fig. 5-22, the branch of a parallel circuit with the lowest resistance dominates the circuit. That is, the lowest resistance takes the most current and power from the source. Ohm's law shows why this is so:

$$I = \frac{V}{R}$$

With V equal for all branches, a low value of R results in a high value of I. As R increases, I decreases. Since V is the same for all branches, the branch with the most current uses the most power ($P = IV$).

In Fig. 5-22, removing the 10-kΩ resistor would reduce the total power and current about 10 percent. Following the logic used for series circuits, engineers have developed a *rule of thumb for parallel circuits*. When *estimating* total current and power, ignore a parallel resistor whose resistance is 10 times higher than that of the other resistor. This rule assumes that the resistors have a 10 percent tolerance. If the tolerance is 5 percent, the rule is modified to read "20 times higher." In Fig. 5-22, ignoring the 10-kΩ resistor results in an estimated total current of 10 mA.

Rule of thumb for parallel circuits

EXAMPLE 5-5

Determine I_{R_1}, I_{R_2}, and I_T for Fig. 5-22 when both resistors are at the high side of their ±10 percent tolerance.

Given: Nominal value of R_1 is 1 kΩ ±10%
Nominal value of R_2 is 10 kΩ ±10%
$V_T = 10$ V

Known: $I_{R_1} = V_T \div R_1$ and $I_{R_2} = V_T \div R_2$
$I_T = I_{R_1} + I_{R_2}$

Solution: $R_1 = 1000\,\Omega + (1000\,\Omega \times 0.10)$
$= 1100\,\Omega$
$R_2 = 10{,}000\,\Omega + (10{,}000\,\Omega \times 0.10) = 11{,}000\,\Omega$
$I_{R_1} = 10\,\text{V} \div 1100\,\Omega = 9.09\,\text{mA}$
$I_{R_2} = 10\,\text{V} \div 11{,}000\,\Omega$
$= 0.91\,\text{mA}$
$I_T = 9.09\,\text{mA} + 0.91\,\text{mA}$
$= 10\,\text{mA}$

Answer: $I_{R_1} = 9.09\,\text{mA}$
$I_{R_2} = 0.91\,\text{mA}$
$I_T = 10\,\text{mA}$

EXAMPLE 5-6

Determine I_{R_1}, I_{R_2}, and I_T for Fig. 5-22 when both resistors are at the low side of their ±10 percent tolerance.

Given: Nominal value of R_1 is 1 kΩ ±10%
Nominal value of R_2 is 10 kΩ ±10%
$V_T = 10$ V

Known: $I_{R_1} = V_T \div R_1$ and $I_{R_2} = V_T \div R_2$
$I_T = I_{R_1} + I_{R_2}$

Solution: $R_1 = 1000\,\Omega - (1000\,\Omega \times 0.10) = 900\,\Omega$
$R_2 = 10{,}000\,\Omega - (10{,}000\,\Omega \times 0.10) = 9000\,\Omega$
$I_{R_1} = 10\,\text{V} \div 900\,\Omega = 11.11\,\text{mA}$
$I_{R_2} = 10\,\text{V} \div 9000\,\Omega = 1.11\,\text{mA}$
$I_T = 11.11\,\text{mA} + 1.11\,\text{mA}$
$= 12.22\,\text{mA}$

Answer: $I_{R_1} = 11.11\,\text{mA}$
$I_{R_2} = 1.11\,\text{mA}$
$I_T = 12.22\,\text{mA}$

Examples 5-5 and 5-6 show that when both resistors are on the high side of their ±10 percent tolerance, the *rule of thumb* for estimating total current provides the correct value of current for the circuit. However when they are both on the low side of their tolerance, the estimated current is 10 percent lower than the total circuit current obtained using the actual resistor values (see Fig. 5-22).

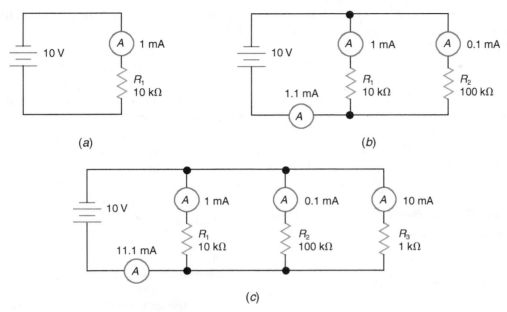

Fig. 5-23 Total resistance in parallel circuits. Adding resistors, as in circuits (b) and (c), increases total current and decreases total resistance.

Resistance in Parallel Circuits

The *total resistance* of a parallel circuit is always less than the *lowest branch resistance*. It may seem illogical at first that adding more resistors to a parallel circuit decreases the total resistance. The logic of the above statements can be shown through the use of Ohm's law and reference to Fig. 5-23. In Fig. 5-23(a) the current in the circuit is

$$I = \frac{V}{R} = \frac{10 \text{ V}}{10,000 \text{ }\Omega}$$

$$= 0.001 \text{ A} = 1 \text{ mA}$$

Adding R_2 in parallel, as in Fig. 5-23(b), does not change either the resistance of R_1 or the voltage across R_1. Therefore, R_1 will still draw 1 mA. The current drawn by R_2 can also be calculated:

$$I_{R_2} = \frac{V_2}{R_2} = \frac{10 \text{ V}}{100,000 \text{ }\Omega}$$

$$= 0.0001 \text{ A} = 0.1 \text{ mA}$$

By Kirchhoff's current law, the total current in Fig. 5-23(b) is 1.1 mA ($I_T = I_{R_1} + I_{R_2}$). Now, if the total voltage is still 10 V and the total current has increased, the total resistance has to decrease. By Ohm's law, the total resistance of Fig. 5-23(b) is

$$R_T = \frac{V_T}{I_T} = \frac{10 \text{ V}}{0.0011 \text{ A}}$$

$$= 9091 \text{ }\Omega = 9.09 \text{ k}\Omega$$

Notice that the addition of a 100-kΩ resistor in parallel with a 10-kΩ resistor *reduces* the total

resistance. Also, notice that the total resistance is less than the lowest (10-kΩ) resistance. In Fig. 5-23(c), a 1-kΩ resistor has been added to the circuit of Fig. 5-23(b). The total resistance of Fig. 5-23(c) is

$$R_T = \frac{V_T}{I_T} = \frac{10 \text{ V}}{0.0011 \text{ A}}$$

$$= 900.9 \text{ }\Omega = 0.901 \text{ k}\Omega$$

Again, notice that R_T has decreased and is less than the lowest resistance (1 kΩ).

In the above examples, the total resistance was calculated by using Ohm's law and the circuit voltage and current. A formula to determine the total resistance directly from the branch resistance can be developed using Ohm's law and the current and voltage relationships in a parallel circuit. From parallel-circuit relationships, $I_1 + I_2 + I_3 +$ etc. can be substituted for I_T, and V_T can be substituted for V_{R_1} or V_{R_2}, etc. Since, by Ohm's law, V_{R_1}/R_1 can be substituted for I_{R_1}, and V_T can be substituted for V_{R_1}, we can substitute V_T/R_1 for I_{R_1}. Now, starting with Ohm's law for a parallel circuit, we can write

$$R_T = \frac{V_T}{I_T} = \frac{V_T}{I_{R_1} + I_{R_2} + I_{R_3} + \text{ etc.}}$$

$$= \frac{V_T}{\dfrac{V_T}{R_1} + \dfrac{V_T}{R_2} + \dfrac{V_T}{R_3} + \text{ etc.}}$$

Total resistance

Lowest branch resistance

Finally, both the numerator and the denominator of the right side of the equation can be divided by V_T to yield

$$R_T = \frac{1}{\frac{1}{R_1} + \frac{1}{R_2} + \frac{1}{R_3} + \text{etc.}}$$

This formula is often referred to as the *reciprocal formula* because the reciprocals of the branch resistances are added, and then the reciprocal of this sum is taken to obtain the total (equivalent) resistance.

Instead of using fractions to calculate the total resistance, you can convert the reciprocals

formula is derived from the reciprocal formula. It is

$$R_T = \frac{R_1 \times R_2}{R_1 + R_2}$$

This formula is sometimes referred to as the product-over-sum formula.

The simpler formula can also be used for circuits containing more than two resistors. The process is to find the *equivalent resistance* of R_1 and R_2 in parallel, label it $R_{1,2}$, and then use this equivalent resistance and R_3 in a second application of the formula to find R_T. As an

Reciprocal formula

Equivalent resistance

EXAMPLE 5-7

What is the total resistance of three resistors—20 Ω, 30 Ω, and 60 Ω—connected in parallel?

Given: $R_1 = 20\ \Omega$
$R_2 = 30\ \Omega$
$R_3 = 60\ \Omega$

Find: R_T

Known: $R_T = \dfrac{1}{\dfrac{1}{R_1} + \dfrac{1}{R_2} + \dfrac{1}{R_3}}$

Solution: $R_T = \dfrac{1}{\dfrac{1}{20} + \dfrac{1}{30} + \dfrac{1}{60}}$

$= \dfrac{1}{\left(\dfrac{6}{60}\right)} = 10\ \Omega$

Answer: The total resistance is 10 Ω.

EXAMPLE 5-8

What is the total resistance of a 27-Ω resistor in parallel with a 47-Ω resistor?

Given: $R_1 = 27\ \Omega$
$R_2 = 47\ \Omega$

Find: R_T

Known: $R_T = \dfrac{R_1 \times R_2}{R_1 + R_2}$

Solution: $R_T = \dfrac{27 \times 47}{27 + 47} = \dfrac{1269}{74}$

$= 17.1\ \Omega$

Answer: The total resistance is 17.1 Ω.

illustration, let us solve example 5-7 using this two-step method.

$$R_{1,2} = \frac{R_1 \times R_2}{R_1 + R_2} = \frac{20 \times 30}{20 + 30} = \frac{600}{50}$$

$$= 12\ \Omega$$

$$R_T = \frac{R_{1,2} \times R_3}{R_{1,2} + R_3} = \frac{12 \times 60}{12 + 60} = \frac{720}{72}$$

$$= 10\ \Omega$$

When all the resistors in a parallel circuit have the same value, the total resistance can be found easily. Just divide the value of a resistor by the number of resistors. That is, $R_T = R/n$, where n is the number of resistors. For example, three 1000-Ω resistors in parallel have a total resistance of

$$R_T = \frac{R}{n} = \frac{1000}{3}$$

$$= 333.3\ \Omega$$

Decimal equivalents

to their *decimal equivalents*. Let us solve the problem in example 5-7 using the decimal equivalents:

$$R_T = \frac{1}{\frac{1}{20} + \frac{1}{30} + \frac{1}{60}}$$

$$= \frac{1}{0.05 + 0.033 + 0.017}$$

$$= \frac{1}{0.100} = 10\ \Omega$$

Two resistors in parallel

$R_T = \dfrac{R}{n}$

When only *two resistors* are *in parallel*, a simplified formula can be used to solve parallel-resistance problems. This simplified

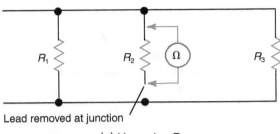

Lead removed at junction

(a) Measuring R_2

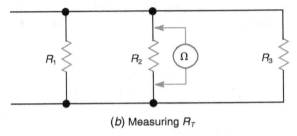

(b) Measuring R_T

Fig. 5-24 Measurement of resistance in parallel.

Two 100-Ω parallel resistors have an equivalent resistance of

$$R_T = \frac{R}{n} = \frac{100}{2} = 50\ \Omega$$

Measuring Resistance in Parallel

The total resistance of a parallel circuit is measured in the same way that it is measured in other types of circuits: the power source is disconnected and the resistance is measured across the points where the power was applied.

To measure an individual resistance in a parallel circuit, one end of the load must be disconnected from the circuit. The correct technique is shown in Fig. 5-24(a). When the load is not disconnected, as illustrated in Fig. 5-24(b), the meter again reads the total resistance.

Solving Parallel-Circuit Problems

Now that we know the relationship between individual and total resistance, current, voltage, and power, we can solve parallel-circuit problems. These relationships, plus Ohm's law and the power formula, allow us to solve most parallel-circuit problems. The formulas listed below will be the "Known" for examples that follow.

Parallel-Circuit Formulas

$$I_T = I_{R_1} + I_{R_2} + I_{R_3} + \text{etc.}$$
$$V_T = V_{R_1} = V_{R_2} = V_{R_3} = \text{etc.}$$

$$R_T = \cfrac{1}{\cfrac{1}{R_1} + \cfrac{1}{R_2} + \cfrac{1}{R_3} + \text{etc.}}$$
$$P_T = P_{R_1} + P_{R_2} + P_{R_3} + \text{etc.}$$
$$I = \frac{V}{R}$$
$$P = IV = I^2R = \frac{V^2}{R}$$

EXAMPLE 5-9

For the circuit in Fig. 5-25, find I_T, R_T, and P_T.

Given: $V_T = 10\ \text{V}$ $R_1 = 100\ \Omega$
$P_{L_1} = 2\ \text{W}$ $I_{R_2} = 0.5\ \text{A}$

Find: I_T, R_T, P_T

Known: Parallel-circuit formulas

Solution: For branch I:

$$I_{L_1} = \frac{P_{L_1}}{V_{L_1}} = \frac{2\ \text{W}}{10\ \text{V}} = 0.2\ \text{A}$$

For branch II:

$$I_{R_1} = \frac{V_{R_1}}{R_1} = \frac{10\ \text{V}}{100\ \Omega} = 0.1\ \text{A}$$

For the total circuit:
$$I_T = I_{L_1} + I_{R_1} + I_{R_2}$$
$$= 0.2\ \text{A} + 0.1\ \text{A} + 0.5\ \text{A}$$
$$= 0.8\ \text{A}$$
$$R_T = \frac{V_T}{I_T} = \frac{10\ \text{V}}{0.8\ \text{A}} = 12.5\ \Omega$$
$$P_T = I_T V_T = 0.8\ \text{A} \times 10\ \text{V}$$
$$= 8\ \text{W}$$

Answer: The total current is 0.8 A. The total resistance is 12.5 Ω. The total power is 8 W.

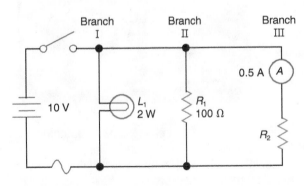

Fig. 5-25 Circuit for example 5-9.

EXAMPLE 5-10

With the data found in example 5-9, find the total resistance using the reciprocal formula.

Solution:
$$R_{L_1} = \frac{V_{L_1}}{I_{L_1}} = \frac{10\text{ V}}{0.2\text{ A}} = 50\ \Omega$$

$$R_2 = \frac{V_{R_1}}{I_{R_2}} = \frac{10\text{ V}}{0.5\text{ A}} = 20\ \Omega$$

$$R_T = \frac{1}{\dfrac{1}{50} + \dfrac{1}{100} + \dfrac{1}{20}}$$

$$= \frac{1}{0.02 + 0.01 + 0.05}$$

$$= \frac{1}{0.08} = 12.5\ \Omega$$

Answer: The total resistance is $12.5\ \Omega$.

Using the reciprocal formula can be a laborious task—especially when the resistances do not have an easily determined common denominator. This task is greatly simplified by using one of the many computer programs available for analyzing or simulating electrical or electronic circuits.

Current-Divider Formula

When you are interested in finding the current through only one of two parallel resistors, you can use the *current-divider formula*. This formula is

$$I_{R_1} = \frac{I_T R_2}{R_1 + R_2} \quad \text{or} \quad I_{R_2} = \frac{I_T R_2}{R_1 + R_2}$$

The current-divider formula can easily be derived by substituting parallel relationships and formulas into the Ohm's law expression for I_{R_1}. The derivation is

$$I_{R_1} = \frac{V_{R_1}}{R_1} = \frac{R_T I_T}{R_1} = \frac{\left(\dfrac{R_1 R_2}{R_1 + R_2}\right) I_T}{R_1}$$

$$= \frac{R_1 R_2 I_T}{(R_1 + R_2) R_1} = \frac{I_T R_2}{R_1 + R_2}$$

We can demonstrate the usefulness of the current-divider formula by solving for I_{R_1} in the circuit shown in Fig. 5-26:

$$I_{R_1} = \frac{I_T R_2}{R_1 + R_2} = \frac{2\text{ A} \times 47\ \Omega}{22\ \Omega + 47\ \Omega} = 1.36\text{ A}$$

Calculating I_{R_1} in Fig. 5-26 without the current divider formula involves the following steps:

1. Calculate R_T using either the reciprocal formula or the product-over-sum formula.

$$R_T = \frac{1}{\dfrac{1}{R_1} + \dfrac{1}{R_2}} = \frac{1}{\dfrac{1}{22\ \Omega} + \dfrac{1}{47\ \Omega}}$$

$$= \frac{1}{0.0455 + 0.0213} = \frac{1}{0.0668}$$

$$= 14.97\ \Omega$$

or
$$R_T = \frac{R_1 \times R_2}{R_1 + R_2} = \frac{22\ \Omega \times 47\ \Omega}{22\ \Omega + 47\ \Omega}$$

$$= \frac{1034\ \Omega}{69\ \Omega} = 14.99\ \Omega$$

2. Calculate V_T using Ohm's law.

$$V_T = I_T \times R_T = 2\text{ A} \times 14.99\ \Omega = 29.98\text{ V}$$

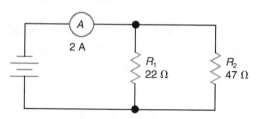

Fig. 5-26 The current-divider formula can solve for I_{R_1} without calculating R_T or V_T.

3. Calculate I_{R_1} using Ohm's law.

$$I_{R_1} = \frac{V_T}{R_1} = \frac{29.99 \text{ V}}{22 \text{ }\Omega} = 1.36 \text{ A}$$

Although the calculated value for I_{R_1} was the same using either procedure, the current-divider formula requires far fewer calculations and is a faster way to determine the value of I_{R_1} when the values of V_T and R_T are unknown.

Applications of Parallel Circuits

An electric system in which one section can fail and other sections continue to operate has parallel circuits. As previously mentioned, the electric system used in homes consists of many parallel circuits.

An automobile electric system uses parallel circuits for lights, heater motor, radio, etc. Each of these devices operates independently of the others.

Individual television circuits are quite complex. However, the complex circuits are connected in parallel to the main power source. That is why the audio section of television receivers can still work when the video (picture) is inoperative.

5-6 Conductance

So far in this text, we have considered only a resistor's opposition to current. However, no resistor completely stops current. Therefore, we could just as well consider a resistor's ability to *conduct* current. Instead of considering a resistor's resistance, we could consider a resistor's conductance. *Conductance* refers to the ability to conduct current. It is symbolized by the letter G. The base unit for conductance is the *siemens*, abbreviated S, in honor of the inventor Ernst Werner von Siemens.

Conductance is the exact opposite of resistance. In fact, the two are mathematically defined as reciprocals of each other. That is,

$$G = \frac{1}{R} \quad \text{and} \quad R = \frac{1}{G}$$

Thus, a 100-Ω resistor has a conductance of $\frac{1}{100}$, or 0.01, siemens (S).

Using the relationship $R = 1/G$ and the series-resistance formula, we can determine total conductance of conductances, in series.

$$R_T = R_1 + R_2 + R_3 + \text{etc.}$$

so

$$\frac{1}{G_T} = \frac{1}{G_1} + \frac{1}{G_2} + \frac{1}{G_3} + \text{etc.}$$

Taking the reciprocal of both sides yields

$$G_T = \frac{1}{\dfrac{1}{G_1} + \dfrac{1}{G_2} + \dfrac{1}{G_3} + \text{etc.}}$$

The formula for total conductance of parallel conductances can be found in a like manner.

$$R_T = \frac{1}{\dfrac{1}{R_1} + \dfrac{1}{R_2} + \dfrac{1}{R_3} + \text{etc.}}$$

Taking the reciprocal of both sides gives

$$\frac{1}{R_T} = \frac{1}{R_1} + \frac{1}{R_2} + \frac{1}{R_3} + \text{etc.}$$

so

$$\frac{1}{\dfrac{1}{G_T}} = \frac{1}{\dfrac{1}{G_1}} + \frac{1}{\dfrac{1}{G_2}} + \frac{1}{\dfrac{1}{G_3}} + \text{etc.}$$

This reduces to

$$G_T = G_1 + G_2 + G_3 + \text{etc.}$$

The relationships between I, V, and G can be determined by taking the reciprocal of both sides of $R = V/I$. This yields $G = I/V$ and $I = GV$ and $V = I/G$. The voltage-divider equation for series conductance can be derived from these relationships as follows:

$$V_{G_n} = \frac{I_{G_n}}{G_n} = \frac{G_T V_T}{G_n} \quad \text{so} \quad V_{G_n} = \frac{G_T V_T}{G_n}$$

For parallel conductance, the current divider formula is derived as:

$$I_{G_n} = G_n V_{G_n} = G_n \times \frac{I_T}{G_T} \quad \text{so} \quad I_{G_n} = \frac{G_n I_T}{G_T}$$

Conductance

Siemens

EXAMPLE 5-11

Determine the individual conductances and the total conductance of a 25-Ω resistor (R_1) and a 50-Ω resistor (R_2) connected in series.

Given: $R_1 = 25 \text{ }\Omega$
$R_2 = 50 \text{ }\Omega$

Find: G_1, G_2, and G_T

Known: $G = 1/R$
$$G_T = \frac{1}{1/G_1 + 1/G_2}$$

Solution: $G_1 = 1/25 = 0.04 \text{ S}$
$G_2 = 1/50 = 0.02 \text{ S}$
$$G_T = \frac{1}{1/0.04 + 1/0.02}$$
$$= \frac{1}{75} = 0.0133 \text{ S}$$

$G = \dfrac{1}{R}$

$R = \dfrac{1}{G}$

Answer: The conductances are as follows: 0.04 S, 0.02 S, and 0.0133 S. Notice that the total conductance could also have been found by determining R_T and taking the reciprocal of it.

EXAMPLE 5-12

Determine the total conductance for the resistors in example 5-11 when they are in parallel.

Given: $G_1 = 0.04$ S
$G_2 = 0.02$ S

Find: G_T

Known: $G_T = G_1 + G_2$

Solution: $G_T = 0.04 + 0.02 = 0.06$ S

Answer: The total conductance is 0.06 S.

EXAMPLE 5-13

Determine the current through G_2 when $G_1 = 0.5$ S, $G_2 = 0.25$ S, $G_3 = 0.20$ S, $I_T = 19$ A, and the conductances are connected in parallel.

Given: $G_1 = 0.5$ S
$G_2 = 0.25$ S
$G_3 = 0.20$ S
$I_T = 19$ A

Find: I_{G_2}

Known: $G_T = G_1 + G_2 + G_3$

$I_{G_2} = \dfrac{G_2 I_T}{G_T}$

Solution: $G_T = 0.5$ S $+ 0.25$ S $+ 0.20$ S
$= 0.95$ S

$I_{G_2} = \dfrac{0.25 \text{ S} \times 19 \text{ A}}{0.95 \text{ S}}$

$= \dfrac{4.75 \text{ SA}}{0.95 \text{ S}} = 5$ A

Answer: The current through G_2 is 5 A.

Self-Test

Answer the following questions.

19. Define *parallel circuit*.
20. How is the voltage distributed in a parallel circuit?
21. How is the current distributed in a parallel circuit?
22. Does the highest or the lowest resistance dominate a parallel circuit?
23. Give two formulas that could be used to find the total resistance of two parallel resistors.
24. True or false. The base unit for conductance is the siemens.
25. True or false. Adding another resistor in parallel increases the total resistance.
26. True or false. The total resistance of a 15-Ω resistor in parallel with a 39-Ω resistor is less than 15 Ω.
27. True or false. The total resistance of two 100-Ω resistors in parallel is 200 Ω.
28. True or false. In a parallel circuit, a 50-Ω resistor dissipates more power than does a 150-Ω resistor.

29. True or false. The resistance of a parallel resistor can be measured while the resistor is connected in the circuit.
30. True or false. Voltage measurements are used to determine whether or not a load is open in a parallel circuit.
31. If one load in a parallel circuit opens, what happens to each of the following?
    a. Total resistance
    b. Total current
    c. Total power
    d. Total voltage
32. Refer to Fig. 5-27. What are the values of I_{R_1} and I_{R_2}?
33. Refer to Fig. 5-27. What is the voltage of B_1?
34. Refer to Fig. 5-27. What is the resistance of R_2?
35. Refer to Fig. 5-27. Determine the following:
    a. Total resistance
    b. Total conductance
    c. Current through R_2
    d. Power dissipated by R_1

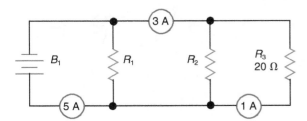

Fig. 5-27 Circuit test questions 32 to 35.

5-7 Series-Parallel Circuits

Some of the features of both the series circuit and the parallel circuit are incorporated into *series-parallel circuits*. For example, R_2 and R_3 in Fig. 5-28(*a*) are in parallel. Everything that has been said about parallel circuits applies to these two resistors. In Fig. 5-28(*d*), R_7 and R_8 are in series. All the series-circuit relationships apply to these two resistors.

Resistors R_1 and R_2 in Fig. 5-28(*a*) are not directly in series because the same current does not flow through each. However, the equivalent of R_2 and R_3 in parallel [$R_{2,3}$ in Fig. 5-28(*b*)] is in series with R_1. Combining R_2 and R_3 is the first step in determining the total resistance of this circuit. The result is $R_{2,3}$. Of course, $R_{2,3}$ is not an actual resistor; it merely represents R_2 and R_3 in parallel. Resistor R_1 is in series with $R_{2,3}$. The final step in finding the total resistance is to combine R_1 and $R_{2,3}$, as shown in Fig. 5-28(*c*). Referring to Fig. 5-28(*a*), assume that $R_1 = 15\ \Omega$, $R_2 = 20\ \Omega$, and $R_3 = 30\ \Omega$. Combining R_2 and R_3 gives

$$R_{2,3} = \frac{R_2 \times R_3}{R_2 + R_3} = \frac{20 \times 30}{20 + 30} = \frac{600}{50} = 12\ \Omega$$

The total resistance is

$$R_T = R_1 + R_{2,3} = 15 + 12 = 27\ \Omega$$

Series-parallel circuits

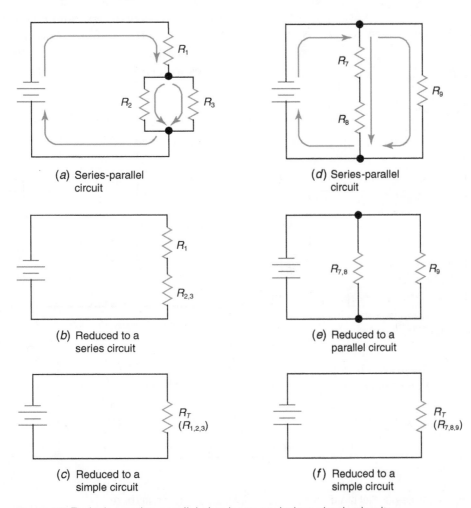

(a) Series-parallel circuit

(b) Reduced to a series circuit

(c) Reduced to a simple circuit

(d) Series-parallel circuit

(e) Reduced to a parallel circuit

(f) Reduced to a simple circuit

Fig. 5-28 Reducing series-parallel circuits to equivalent simple circuits.

Neither R_7 nor R_8 in Fig. 5-28(d) is in parallel with R_9 because neither is across the same voltage source as R_9. Therefore, the first step in reducing this circuit is to combine the series resistors R_7 and R_8. This results in the circuit shown in Fig. 5-28(e). Now the parallel circuit of Fig. 5-28(e) can be reduced, by a parallel-resistance formula, to the simple circuit of Fig. 5-28(f). Refer to Fig. 5-28(d) and let $R_7 = 40 \ \Omega$, $R_8 = 60 \ \Omega$, and $R_9 = 20 \ \Omega$.

$$R_{7,8} = R_7 + R_8 = 40 + 60 = 100 \ \Omega$$

$$R_T = \frac{R_{7,8} \times R_9}{R_{7,8} + R_9}$$

$$= \frac{100 \times 20}{100 + 20} = \frac{2000}{120} = 16.7 \ \Omega$$

A more complex series-parallel circuit is shown in Fig. 5-29(a). Determining the total resistance of the circuit is illustrated in Fig. 5-29(b) through (e). The calculations required to arrive at the total resistance are

$$R_{3,5} = R_3 + R_5 = 50 + 30 = 80 \ \Omega$$

$$R_{3,4,5} = \frac{R_4 \text{ or } R_{3,5}}{2} = \frac{80}{2} = 40 \ \Omega$$

$$R_{2,3,4,5} = R_2 + R_{3,4,5} = 60 + 40 = 100 \ \Omega$$

$$R_T = \frac{R_1 \times R_{2,3,4,5}}{R_1 + R_{2,3,4,5}} = \frac{200 \times 100}{200 + 100}$$

$$= \frac{20{,}000}{300}$$

$$= 66.7 \ \Omega$$

Using Kirchhoff's Laws in Series-Parallel Circuits

Individual currents and voltages in series-parallel circuits can often be determined by Kirchhoff's laws. In Fig. 5-30, current in some of the conductors is given. The rest of the currents can be found using Kirchhoff's current law. Since 0.6 A enters the battery, the same amount must leave. This is the total current I_T. Thus $I_T = 0.6$ A. Since R_3 and R_4 are in series, the current entering R_3 is the same as the current leaving R_4. Therefore, $I_2 = 0.2$ A. The current entering point A is 0.6 A. Leaving that point are I_1 and I_2. From Kirchhoff's current law we know that $I_1 + I_2 = 0.6$ A. But $I_2 = 0.2$ A, and thus $I_1 + 0.2$ A $= 0.6$ A, and $I_1 = 0.4$ A. We can check this result by examining point B. Here $I_1 + 0.2$ A $= 0.6$ A. Since I_1 was found to be 0.4 A, we have 0.4 A $+$ 0.2 A $=$ 0.6 A, which agrees with Kirchhoff's current law.

Kirchhoff's laws in series-parallel circuits

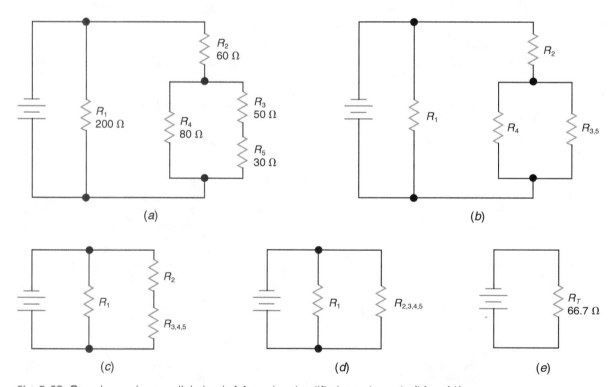

Fig. 5-29 Complex series-parallel circuit (a) can be simplified, as shown in (b) to (d).

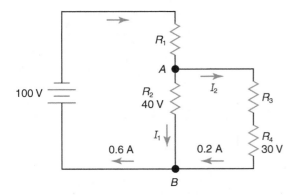

Fig. 5-30 Currents and voltages in series-parallel circuits. Kirchhoff's laws can be used to determine the unknown values.

Some of the voltage drops in Fig. 5-30 are indicated beside the resistor symbols. The unspecified voltage drops can be found by using Kirchhoff's voltage law. In a series-parallel circuit, Kirchhoff's law applies to all loops, or current paths, in the circuit. Thus, for the circuit of Fig. 5-30, we can write two voltage relationships.

$$V_T = V_{R_1} + V_{R_2}$$
$$V_T = V_{R_1} + V_{R_3} + V_{R_4}$$

Studying the above two equations shows that

$$V_{R_2} = V_{R_3} + V_{R_4}$$

In other words, the voltage between points A and B is 40 V, no matter which path is taken. Rearranging the first equation, we can solve for the voltage drop across R_1:

$$V_{R_1} = V_T - V_{R_2} = 100 \text{ V} - 40 \text{ V} = 60 \text{ V}$$

The voltage across R_3 can be found by rearranging the second equation:

$$V_{R_3} = V_T - V_{R_1} - V_{R_4}$$
$$= 100 \text{ V} - 60 \text{ V} - 30 \text{ V}$$
$$= 10 \text{ V}$$

Also, the voltage across R_3 can be found without knowing the voltage across R_1:

$$V_{R_3} = V_{R_2} - V_{R_4} = 40 \text{ V} - 30 \text{ V} = 10 \text{ V}$$

Solving Series-Parallel Problems

Several problems are solved in the examples that follow. These problems illustrate how to use Ohm's law and Kirchhoff's two laws to solve series-parallel problems.

EXAMPLE 5-14

For the circuit of Fig. 5-31(a), determine all unknown currents and voltages.

Given: $V_T = 60 \text{ V}$
            $V_{R_2} = 40 \text{ V}$
            $I_T = 4 \text{ A}$
            $R_3 = 20 \text{ }\Omega$

Find: $V_{R_1}, V_{R_3}, I_{R_1}, I_{R_2}, I_{R_3}$

Known: Ohm's law and Kirchhoff's laws

Solution: $V_{R_3} = V_T - V_{R_2}$
               $= 60 \text{ V} - 40 \text{ V}$
               $= 20 \text{ V}$
               $V_{R_1} = V_T = 60 \text{ V}$
               $I_{R_3} = \dfrac{V_{R_3}}{R_3}$
               $= \dfrac{20 \text{ V}}{20 \text{ }\Omega} = 1 \text{ A}$
               $I_{R_2} = I_{R_3} = 1 \text{ A}$
               $I_{R_1} = I_T - I_{R_2}$
               $= 4 \text{ A} - 1 \text{ A} = 3 \text{ A}$

Answer: $V_{R_1} = 60 \text{ V}, V_{R_3} = 20 \text{ V}, I_{R_1}$
             $= 3 \text{ A}, I_{R_2} = 1 \text{ A}, I_{R_3}$
             $= 1 \text{ A}$

EXAMPLE 5-15

Refer to Fig. 5-31(b). For this circuit compute the resistance of R_3, the power dissipation of R_4, and the voltage across R_1.

Given: $V_T = 100 \text{ V}$
            $I_{R_1} = 0.8 \text{ A}$
            $I_{R_3} = 0.3 \text{ A}$
            $R_2 = 100 \text{ }\Omega$
            $V_{R_4} = 30 \text{ V}$

Find: R_3, P_{R_4}, V_{R_1}

Known: Ohm's law and Kirchhoff's laws

Solution: $I_{R_2} = I_{R_1} - I_{R_3}$
               $= 0.8 \text{ A} - 0.3 \text{ A}$
               $= 0.5 \text{ A}$
               $V_{R_2} = I_{R_2} R_2$
               $= 0.5 \text{ A} \times 100 \text{ }\Omega$
               $= 50 \text{ V}$
               $V_{R_1} = V_T - V_{R_2} - V_{R_4}$
               $= 100 \text{ V} - 50 \text{ V} - 30 \text{ V}$
               $= 20 \text{ V}$

Solving series-parallel problems

$$V_{R_3} = V_{R_2} = 50 \text{ V}$$
$$R_3 = \frac{V_{R_3}}{I_{R_3}} = \frac{50 \text{ V}}{0.3 \text{ A}}$$
$$= 166.7 \text{ }\Omega$$
$$P_{R_4} = I_{R_4} V_{R_4}$$
$$= 0.8 \text{ A} \times 30 \text{ V} = 24 \text{ W}$$

Answer: $R_3 = 166.7 \text{ }\Omega$, $P_{R_4} = 24 \text{ W}$,
$V_{R_1} = 20 \text{ V}$

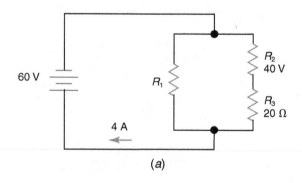

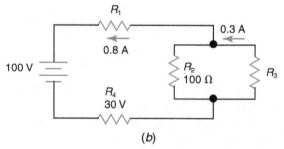

Fig. 5-31 Circuits for examples 5-14 and 5-15.

For Fig. 5-31(a), all the currents and voltages have now been determined. Using Ohm's law and the power formula, we can easily find the resistance and power of each resistor.

Relationships in Series-Parallel Circuits

As in series circuits, the current, the voltage, and the power in series-parallel circuits are dependent on one another. That is, changing any one resistance usually changes all currents, voltages, and powers except the source voltage. For example, increasing the resistance of R_3 in Fig. 5-32(a) from 40 to 90 Ω causes the changes listed below:

1. R_T increases (because $R_{3,4}$ increases).
2. I_T decreases (because R_T increases).
3. V_{R_1} decreases (because $I_T = I_{R_1}$).
4. V_{R_2} increases (because $V_{R_2} = V_T - V_{R_1}$).
5. I_{R_2} increases (because $I_{R_2} = V_{R_2}/R_2$).
6. I_{R_4} decreases (because $I_{R_4} = I_{R_1} - I_{R_2}$).
7. V_{R_4} decreases (because $V_{R_4} = I_{R_4} R_4$).
8. V_{R_3} increases (because $V_{R_3} = V_{R_2} - V_{R_4}$).
9. P_{R_1} decreases, P_{R_2} increases, P_{R_4} decreases, and P_T decreases (because of the I and V changes specified above).
10. P_{R_3} increases (because V_{R_3} increases more than I_{R_3} decreases).

The magnitude of these changes is shown in Fig. 5-32(b).

The changes detailed above occur when any resistor is changed, except in circuits like the one in Fig. 5-29(a). In this circuit, changing R_1 affects only the current and power of R_1 and the battery. This is because R_1 is in parallel with the combination of R_2, R_3, R_4, and R_5 [Fig. 5-29(d)].

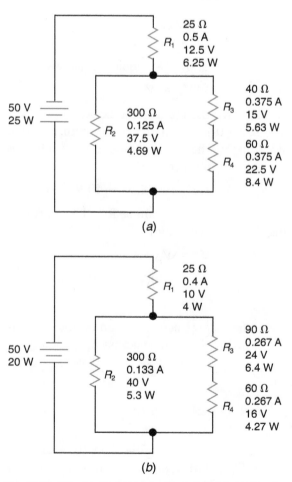

Fig. 5-32 Effects of changing R_3. (a) Original circuit values. (b) Values after R_3 increases.

Answer the following questions.

36. Refer to Fig. 5-29(a). Which resistors, if any, are
    a. Directly in series?
    b. Directly in parallel?

37. Referring to Fig. 5-29(a), determine whether each of the following statements is true or false.
    a. $I_{R_2} = I_{R_3} + I_{R_5}$
    b. $I_{R_2} = I_{R_4} + I_{R_5}$
    c. $I_{R_1} = I_T - I_{R_2}$
    d. $V_{R_3} = V_{R_5}$
    e. $V_{R_4} = V_{R_1} - V_{R_2}$

38. Refer to Fig. 5-31(a) and compute the following:
    a. R_1
    b. P_{R_2}
    c. P_T

39. Refer to Fig. 5-28(a). If R_2 is decreased, indicate whether each of the following increases or decreases:
    a. I_{R_1}
    b. V_{R_3}
    c. P_{R_3}
    d. V_{R_1}

40. Refer to Fig. 5-32(a). Change the value of R_1 to 50 Ω. Then determine the value of each of the following:
    a. R_T
    b. V_{R_1}
    c. I_{R_4}
    d. P_{R_3}

41. For Fig. 5-33, determine the following:
    a. R_1
    b. I_{R_2}
    c. P_{R_3}
    d. I_T
    e. R_T

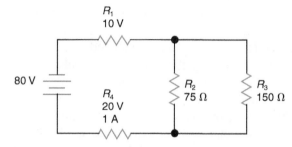

Fig. 5-33 Circuit diagram for test question 41.

5-8 Voltage Dividers and Regulators

A series circuit is an unloaded voltage divider. The circuit in Fig. 5-34 shows how a single-source voltage can provide three voltages. In this figure, the voltage values include a polarity sign, which indicates the voltage at a given point (A, B, or C) with reference to the common ground. The voltages at points A and B are easily calculated using the voltage-divider formula given in section 5-3, Series Circuits. The calculations are:

$$V_A = \frac{V_{B_1} R_3}{R_T} = \frac{50 \text{ V} \times 2 \text{ k}\Omega}{10 \text{ k}\Omega} = 10 \text{ V}$$

$$V_B = \frac{V_{B_1}(R_3 + R_2)}{R_T} = \frac{50 \text{ V} \times 5 \text{ k}\Omega}{10 \text{ k}\Omega} = 25 \text{ V}$$

The problem with this type of voltage divider is that the voltages at both point A and point B will

change when a load is connected to either point A or point B. Loading the divider converts the circuit from a series to a series-parallel circuit. For example, if a 5-kΩ load is connected to point B, the resistance from point B to ground is reduced to

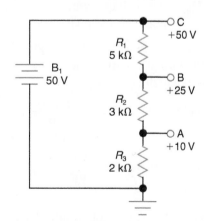

Fig. 5-34 A series voltage divider providing two voltages in addition to the source voltage.

2.5 kΩ and the total resistance reduces to 7.5 kΩ. Now the voltage at point B will be

$$V_B = \frac{50 \text{ V} \times 2.5 \text{ k}\Omega}{7.5 \text{ k}\Omega} = 16.67 \text{ V}$$

and at point A it will reduce to

$$V_A = \frac{16.7 \text{ V} \times 2 \text{ k}\Omega}{5 \text{ k}\Omega} = 6.67 \text{ V}$$

Zener diode

The severity of the voltage changes can be reduced by using smaller resistances for R_1, R_2, and R_3. Of course, this requires more power from the source and reduces the efficiency of the circuit. When the load current is insignificant compared to the current through the divider resistors, then the change in divider voltage will be very small. The current through the divider resistance is often called the bleeder current. The smaller the ratio is of the load current to the bleeder current, the smaller the decrease in voltage will be between the unloaded and the loaded divider.

If one knows the approximate voltage and current requirements of the load (or loads) before the divider is designed, then one can use one's knowledge of series-parallel circuits and design a divider that will provide the desired voltage when the divider is loaded. However, the current requirements of most loads vary as conditions change so the voltage out of the divider will vary.

Percentage of voltage regulation

The severity of voltage variation as a load changes is usually expressed as a *percentage of voltage regulation*. Percentage of voltage regulation is determined with the formula

$$\% \text{ Regulation} = \frac{V_{ML} - V_{FL}}{V_{FL}} \times 100$$

where V_{ML} = the voltage with the minimum load (highest-resistance load)
V_{FL} = the voltage with the maximum load (lowest-resistance load)

The percentage of voltage regulation for Fig. 5-34 with no load (which is V_{ML}) and a 5-kΩ load (which is V_{FL}) is

$$\% \text{ Regulation for point A} = \frac{10 \text{ V} - 6.67 \text{ V}}{6.67 \text{ V}}$$
$$\times 100 = 50\%$$

$$\% \text{ Regulation for point B} = \frac{25 \text{ V} - 16.67 \text{ V}}{16.67 \text{ V}}$$
$$\times 100 = 50\%$$

The divider in Fig. 5-34 would have to be redesigned if it has to provide approximately 25 V to a 5-kΩ load.

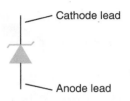

Fig. 5-35 Symbol for a zener diode.

Simple resistor voltage dividers have been replaced by solid-state circuits that divide the source voltage while providing much better voltage regulation than straight resistor dividers. The *zener diode* is a solid-state device that can be used in series with a resistor to make a voltage divider. The symbol for the zener diode is shown in Fig. 5-35. Notice the zener diode, like the LED, is a polarized device and incorrect polarity connections on either device will destroy it. As can be seen in Fig. 5-36(a), the zener diode is operated with its cathode positive with respect to its anode. This mode of operation is called *reverse bias*. With reverse bias, the zener does not allow current to flow until the voltage across the zener reaches a value very close to the voltage rating of the zener. After this critical voltage is reached, the current through the zener can increase to any value less than its rated maximum value and the zener voltage will increase very little. Thus, the zener diode provides good voltage regulation.

Table 5-2 compares the operation of the two divider circuits in Fig. 5-36 in terms of voltage regulation and efficiency as the load resistance varies from its nominal value of 1 kΩ down to 500 Ω and up to 2 kΩ. The data for the zener diode circuit [Fig. 5-36(a)] were obtained using electronic circuit simulation software. The data for the resistor circuit were calculated using the values given in Fig. 5-36(b). Both circuits were designed to provide 5 V across a 1-kΩ load. The calculations needed to determine the data for the resistor circuit with a 500-Ω load are

$$R_{2, \text{ load}} = \frac{1000 \text{ }\Omega \times 500 \text{ }\Omega}{1000 \text{ }\Omega + 500 \text{ }\Omega} = 333.3 \text{ }\Omega$$

$$R_T = 500 \text{ }\Omega + 333.3 \text{ }\Omega = 833.3 \text{ }\Omega$$

$$V_L = \frac{R_{2, \text{ load}} \times V_T}{R_T} = \frac{333.3 \text{ }\Omega \times 10 \text{ V}}{833.3 \text{ }\Omega} = 4.00 \text{ V}$$

$$I_L = \frac{V_L}{R_{\text{load}}} = \frac{4.00 \text{ V}}{500 \text{ }\Omega} = 8 \text{ mA}$$

$$I_T = \frac{R_I}{R_T} = \frac{10 \text{ V}}{833.3 \text{ }\Omega} = 12 \text{ mA}$$

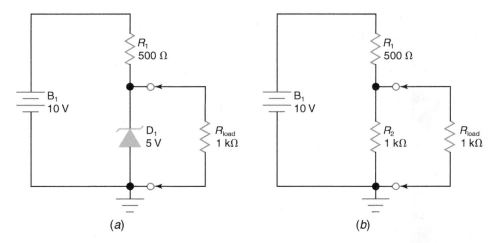

Fig. 5-36 Voltage divider and regulator circuits. The zener diode circuit has very good regulation. (a) Zener divider and regulator. (b) Resistor voltage divider.

$$\% \text{ eff} = \frac{P_{\text{load}}}{P_T} = \frac{4 \text{ V} \times 8 \text{ mA}}{10 \text{ V} \times 12 \text{ mA}} \times 100$$

$$= \frac{32 \text{ mW}}{120 \text{ mW}} \times 100 = 26.7\%$$

Notice in Table 5-2 that the zener circuit is much more efficient than the resistor circuit when the load current is larger than the value the circuits were designed for. However, at a current load less than design value, the resistor circuit is slightly more efficient.

Look carefully at the percentage of voltage regulation column in Table 5-2. It shows the big advantage of the zener circuit over the resistor circuit.

EXAMPLE 5-16

Determine I_{R_1}, $I_{R,\text{load}}$, and I_{D_1} for Fig. 5-36(a) when R_{load} is 1.5 kΩ.

Given: $V_T = 10 \text{ V}$, $V_{D_1} = 5 \text{ V}$, $R_1 = 500 \text{ }\Omega$, $R_{\text{load}} = 1.5 \text{ k}\Omega$.

Find: I_{R_1}, $I_{R,\text{load}}$, I_{D_1}

Known: Ohm's law and Kirchhoff's laws

Solution:
$$V_{R,\text{load}} = V_{D_1} = 5 \text{ V}$$
$$I_{R,\text{load}} = V_{R,\text{load}} \div R_{\text{load}}$$
$$= 5 \text{ V} \div 1.5 \text{ k}\Omega$$
$$= 3.33 \text{ mA}$$
$$V_{R_1} = V_T - V_{D_1}$$
$$= 10 \text{ V} - 5 \text{ V} = 5 \text{ V}$$
$$I_{R_1} = V_{R_1} \div R_1$$
$$= 5 \text{ V} \div 500 \text{ }\Omega = 10 \text{ mA}$$
$$I_{D_1} = I_{R_1} - I_{R,\text{load}}$$
$$= 10 \text{ mA} - 3.33 \text{ mA}$$
$$= 6.67 \text{ mA}$$

Answer: $I_{R_1} = 10 \text{ mA}$
$$I_{R,\text{load}} = 3.33 \text{ mA}$$
$$I_{D_1} = 6.67 \text{ mA}$$

![Self-Test] **Self-Test**

Answer the following questions.

42. What happens to the load voltage of an unloaded resistor voltage divider circuit when a load is connected to it?

43. Calculate the voltage at point B in Fig. 5-34 when a 15-kΩ load is connected to point B.

44. Calculate the percentage of voltage regulation for the conditions given in question 43.

45. What will happen to the load voltage in Fig. 5-36(a) when the load is reduced to 750 Ω?

46. Show the calculations you would use to determine the percentage of efficiency listed in Table 5-2 for the zener circuit with a 2-kΩ load.

Table 5-2 A comparison of two voltage divider circuits

Load resistance	V_L		I_L		I_T		% of V reg.		% of eff.	
	Zener circuit	Resistor circuit	Zener circuit	Resistor circuit	Zener circuit	Resistor circuit	Zener circuit	Resistor circuit	Zener circuit	Resistor circuit
500 Ω	4.92 V	4.00 V	9.84 mA	8.00 mA	10.16 mA	12.00 mA			47.6	26.7
1 kΩ	4.97 V	5.00 V	4.97 mA	5.00 mA	10.05 mA	10.00 mA			24.6	25.0
2 kΩ	4.98 V	5.71 V	2.49 mA	2.86 mA	10.03 mA	8.57 mA			12.4	19.1
Increase 1 kΩ to 2 kΩ							0.2	14.3		
Decrease 1 kΩ to 500 Ω							1.0	25.0		
Increase 500 Ω to 2 kΩ							1.2	42.8		

Chapter 5 Summary and Review

Summary

1. Multiple-load circuits include series, parallel, and series-parallel circuits.
2. Series circuits are single-path circuits.
3. The same current flows throughout a series circuit.
4. The total resistance equals the sum of the individual resistances in a series circuit.
5. The sum of the voltage drops around a circuit equals the total source voltage (Kirchhoff's voltage law).
6. A voltage drop (voltage across a load) indicates that electric energy is being converted to another form.
7. The polarity of a voltage drop indicates the direction of current flow.
8. The voltage across an open series load is usually equal to the source voltage.
9. A shorted load in a series circuit increases the current, voltage, and power of the other loads.
10. When one resistance in a series circuit is smaller than the tolerance of another, the smaller resistance has little effect on the circuit current and power.
11. The highest resistance in a series circuit drops the most voltage.
12. Maximum power transfer occurs when the source resistance equals the load resistance.
13. Conductance is the ability to conduct current. Its symbol is G. Its base unit is siemens (S).
14. Parallel circuits are multiple-path circuits.
15. Each branch of a parallel circuit is independent of the other branches.
16. The same voltage appears across each branch of a parallel circuit.
17. The currents entering a junction must equal the currents leaving a junction (Kirchhoff's current law).
18. The total current in a parallel circuit equals the sum of the branch currents.
19. Adding more resistance in parallel decreases the total resistance.
20. The total resistance in a parallel circuit is always less than the lowest branch resistance.
21. The relationships of both series and parallel circuits are applicable to parts of series-parallel circuits.

Related Formulas

$$P_T = P_{R_1} + P_{R_2} + P_{R_3} + \text{etc.}$$

$$R = \frac{1}{G}$$

For series circuits:

$$I_T = I_{R_1} = I_{R_2} = I_{R_3} = \text{etc.}$$

$$V_T = V_{R_1} + V_{R_2} + V_{R_3} + \text{etc.}$$

$$R_T = R_1 + R_2 + R_3 + \text{etc.}$$

$$G_T = \frac{1}{\dfrac{1}{G_1} + \dfrac{1}{G_2} + \dfrac{1}{G_3} + \text{etc.}}$$

For parallel circuits:

$$I_T = I_{R_1} + I_{R_2} + I_{R_3} + \text{etc.}$$

$$V_T = V_{R_1} = V_{R_2} = V_{R_3} = \text{etc.}$$

$$R_T = \frac{1}{\dfrac{1}{R_1} + \dfrac{1}{R_2} + \dfrac{1}{R_3} + \text{etc.}}$$

$$G_T = G_1 + G_2 + G_3 + \text{etc.}$$

$$R_T = \frac{R_1 \times R_2}{R_1 + R_2}$$

$$R_T = \frac{R}{n}$$

For questions 5-1 to 5-5, supply the missing word or phrase in each statement.

5-1. The total current is equal to the sum of the individual currents in a _____ circuit. (5-3)

5-2. The highest resistance dissipates the least power in a _____ circuit. (5-3)

5-3. The highest resistance drops the _____ voltage in a series circuit. (5-3)

5-4. Adding another parallel resistor _____ the total resistance. (5-5)

5-5. The total voltage is equal to the sum of the voltage drops in a _____ circuit. (5-3)

For questions 5-6 to 5-20, determine whether each statement is true or false.

5-6. To measure the resistance of an individual resistor in a parallel circuit, one end of the resistor must be disconnected from the circuit. (5-5)

5-7. Maximum power transfer occurs when the source resistance is very low compared with the load resistance. (5-4)

5-8. The negative polarity marking on a resistor in a schematic diagram indicates an excess of electrons at that point. (5-3)

5-9. Voltmeters have a very low internal resistance. (5-3)

5-10. In a series circuit, an open load drops no voltage. (5-3)

5-11. If one load in a parallel circuit opens, all the other loads will use more power. (5-5)

5-12. If one load in a series circuit shorts out, all the other loads will use more power. (5-3)

5-13. Adding more resistors to a parallel circuit increases the total power used by the circuit. (5-5)

5-14. The total resistance of parallel resistances is always less than the value of the lowest resistance. (5-5)

5-15. Changing the value of one resistor in a parallel circuit changes the current through all other resistors in that circuit. (5-5)

5-16. The unit of conductance is the siemen. (5-6)

5-17. A voltage drop indicates that some other form of energy is being converted to electric energy. (5-3)

5-18. The direction of current flow determines the polarity of the voltage drop across the resistor. (5-3)

5-19. The lowest resistance in a series circuit dominates the circuit current and power. (5-3)

5-20. Most circuits in a home are series circuits. (5-5)

For questions 5-21 and 5-22, choose the letter that best completes each sentence.

5-21. The total resistance of a 45-Ω resistor and a 90-Ω resistor connected in series is (5-3)

a. 30 Ω

b. 45 Ω

c. 67.5 Ω

d. 135 Ω

5-22. The total resistance of a 30-Ω resistor and a 60-Ω resistor connected in parallel is (5-5)

a. 20 Ω

b. 30 Ω

c. 45 Ω

d. 90 Ω

Answer the following questions.

5-23. If R_2 in Fig. 5-39 is increased to 1200 Ω, does V_{R_1} increase, decrease or remain the same? (5-7)

5-24. What would happen to G_T in Fig. 5-38 if a fourth parallel resistor were added? (5-6)

5-25. What would happen to G_T in Fig. 5-37 if a fourth series resistor were added? (5-6)

5-26. For the circuit in Fig. 5-34, is it possible for P_{R_2} or P_{R_3} to be larger than P_{R_1} under any load conditions? Explain your answer. (5-8)

5-27. Which circuit in Fig. 5-36 has the better voltage regulation? (5-8)

5-1. For the circuit in Fig. 5-37, compute the following: (5-3)
   a. V_{R_1}
   b. R_2
   c. P_T
   d. G_3

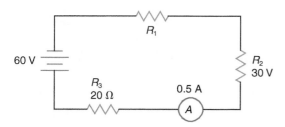

Fig. 5-37 Circuit for chapter review problem 5-1.

5-2. For the circuit in Fig. 5-38, compute the following: (5-5)
   a. I_{R_1}
   b. R_1
   c. I_{R_2}
   d. G_T

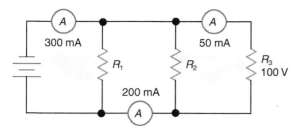

Fig. 5-38 Circuit for chapter review problem 5-2.

5-3. For the circuit in Fig. 5-39, compute the following: (5-7)
   a. V_{R_3}
   b. I_{R_1}
   c. V_{R_1}

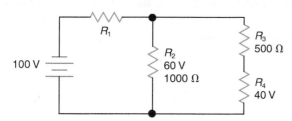

Fig. 5-39 Circuit for chapter review problem 5-3.

5-4. If I_T in Fig. 5-38 is changed to 280 mA, what is the value of: (5-5)
   a. R_1
   b. P_T

5-5. Change the value of R_3 in Fig. 5-37 to 40 Ω and then determine: (5-3)
   a. R_1
   b. R_2
   c. R_T

5-6. Change the voltage across R_2 in Fig. 5-39 to 80 V and then determine: (5-7)
   a. R_1
   b. R_T
   c. P_T

5-7. How much current does a 3-kΩ load draw when connected to point A in Fig. 5-34? (5-8)

5-8. Using the data in Table 5-2, determine I zener when the load resistance is 2 kΩ. (5-8)

5-9. How much power does the zener use under the conditions given in problem 5-8 above? (5-8)

5-10. How much power is used by R_1 in Fig. 5-36(a) when the load resistance is 1 kΩ? Use data given in Table 5-2. (5-8)

5-11. Determine I_{D_1}, I_T, and P_{D_1} for the circuit in Fig. 5-40.

5-12. Determine the efficiency of the circuit in Fig. 5-40.

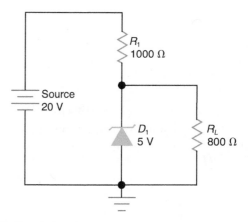

Fig. 5-40 Circuit for chapter review problems 5-11 and 5-12.

Critical Thinking Questions

5-1. List and explain several applications of parallel circuits not mentioned in this chapter.

5-2. For Fig. 5-11, assume that all resistors have a tolerance of ±5 percent and that the source has an internal resistance of 9.2 Ω. Determine the maximum total power dissipated by the loads.

5-3. Is it desirable to have an amplifier's output impedance (opposition) equal to the speaker's impedance? Why?

5-4. Is it desirable to have a 100-kW generator's internal resistance equal to the resistance of its load? Why?

5-5. Derive the reciprocal formula for parallel resistances from the formula for parallel conductances.

5-6. Are measured circuit voltages more likely to be inaccurate in a series or a parallel circuit? Why?

5-7. Discuss the results of shorting out R_2 in Fig. 5-11 when the power ratings for R_1, R_2, and R_3 are 25 W, 50 W, and 25 W respectively.

5-8. Without disconnecting either end of any of the loads in Fig. 5-41, how could you determine whether one or more of the loads is outside its tolerance range?

5-9. When only a voltmeter is available, how could you determine whether any of the loads in Fig. 5-41 were open?

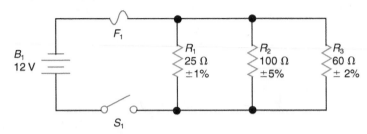

Fig. 5-41 Circuit for critical thinking questions 5-8 and 5-9.

5-10. Determine the efficiency of a 24-V battery with 0.04 Ω of internal resistance when it is supplying 150 A to a load.

5-11. In Fig. 5-34, what value of resistor is needed for R_2 and R_3 to provide the specified voltages at points A and B when a 5-kΩ load is connected to point A and a 10-kΩ load is connected to point B?

Answers to Self-Tests

1. a. V_{R_4}
   b. I_T
   c. I_{R_2}
2. T
3. T
4. T
5. $R_T = R_1 + R_2 +$
   $R_3 +$ etc.
   $R_T = V_T/I_T$
6. Yes.
7. $V_T = V_{R_1} + V_{R_2} + V_{R_3} +$ etc.
8. 100-Ω resistor
9. No.
10. across the open resistor
11. a. increases
    b. increases
    c. increases
12. a. 500 Ω

b. 0.2 A
c. 25 V
d. 15 W
13. R_1 is 37.3 Ω. The battery has to furnish 1.89 W.
14. It would increase.
15. 0 V
16. 9 mA
17. when load resistance is equal to source resistance
18. greater than
19. A parallel circuit is one that has two or more loads and two or more independent current paths.
20. The total (source) voltage is applied to each load:
    $V_T = V_{R_1} = V_{R_2} = V_{R_3}$
21. The total current divides among the branches of the circuit:
    $I_T = I_{R_1} + I_{R_2}$
    $+ I_{R_3} +$ etc.
22. lowest

23. $R_T = \dfrac{1}{\dfrac{1}{R_1} + \dfrac{1}{R_2}}$

    $R_T = \dfrac{R_1 \times R_2}{R_1 + R_2}$

24. T

25. F

26. T

27. F

28. T

29. F

30. F

31. a. increases

   b. decreases

   c. decreases

   d. remains the same

32. $I_{R_1} = I_T - (I_{R_2} + I_{R_3})$

      $= 5\,\text{A} - 3\,\text{A}$

      $= 2\,\text{A}$

   $I_{R_2} = (I_{R_2} + I_{R_3}) - I_{R_3}$

      $= 3\,\text{A} - 1\,\text{A}$

      $= 2\,\text{A}$

33. $V_{B_1} = V_{R_3} = I_{R_3} R_3$

      $= 1\,\text{A} \times 20\,\Omega$

      $= 20\,\text{V}$

34. $V_{R_2} = V_{R_3} = 20\,\text{V}$

   $R_2 = \dfrac{V_{R_2}}{I_{R_2}} = \dfrac{20\,\text{V}}{2\,\text{A}}$

      $= 10\,\Omega$

35. a. $4\,\Omega$

   b. 0.25 S

   c. 2 A

   d. 40 W

36. a. R_3 and R_5

   b. none

37. a. F

   b. T

   c. T

   d. F

   e. T

38. a. $20\,\Omega$

   b. 40 W

   c. 240 W

39. a. increases

   b. decreases

   c. decreases

   d. increases

40. a. $125\,\Omega$

   b. 20 V

   c. 0.3 A

   d. 3.6 W

41. a. $10\,\Omega$

   b. 0.67 A

   c. 16.7 W

   d. 1 A

   e. $80\,\Omega$

42. The voltage decreases. The magnitude of the decrease is a function of the ratio of bleeder current to load current.

43. 21.4 V

44. 16.8 %

45. It will decrease slightly. The data in Table 5-2 indicate it will decrease less than 0.05 V.

46. $P_{\text{in}} = 10\,\text{V} \times 10\,\text{mA}$

      $= 100\,\text{mW}$

   $P_{\text{out}} = P_{\text{load}}$

      $= 4.98\,\text{V} \times 2.49\,\text{mA} = 12.4\,\text{mW}$

   $\%\,\text{eff} = \dfrac{12.4\,\text{mW}}{100\,\text{mW}}$

      $\times 100 = 12.4\,\%$

Complex-Circuit Analysis

Learning Outcomes

This chapter will help you to:

6-1 *Solve* simultaneous equations using a variable eliminations technique or a determinant method.

6-2 *Use* Kirchhoff's voltage law to write a set of loop equations for a complex circuit. Demonstrate how these equations can solve for circuit values.

6-3 *Use* Kirchhoff's current law to write a set of loop equations for a complex circuit. Demonstrate how these equations can solve for circuit values.

6-4 *Solve* multiple-source complex circuits using the superposition theorem. Does this method require simultaneous equations, variable elimination techniques, determinates, or node voltage techniques?

6-5 *Discuss* the significance of viewing a source voltage as an ideal voltage source or an equivalent-circuit voltage source.

6-6 *View* any circuit as a two-terminal voltage network, i.e., an equivalent-circuit voltage source.

6-7 *Understand* how, and why, a voltage source can be viewed as, and converted to, a current source.

6-8 *View* any circuit as a two-terminal current network, i.e., an equivalent-circuit current source.

6-9 *Realize* that once a circuit is viewed as a two-terminal network and the network has been reduced to either an equivalent circuit voltage source or current source, it is easy to find the current and voltage of any load connected to the two terminals.

Some circuits are arranged so that the loads are neither directly in series nor directly in parallel. Such circuits are called *complex circuits or networks*. Some complex circuits (networks) include two or more voltage sources. Reducing complex circuits to a single equivalent resistance, or determining individual currents and voltages for the circuit, requires techniques beyond those covered in Chap. 5.

6-1 Simultaneous Equations

Equations with two or more unknown variables are readily solvable if there are at least as many independent equations relating the variables as there are variables. Such sets, or groups, of equations are known as *simultaneous equations*. We can use the *variable elimination* method for solving these types of equations. This method is simple and straightforward in that it involves only three algebraic operations: (1) multiplying by a constant, (2) adding equations, and (3) substituting numerical value for variables.

Simultaneous equations

Variable elimination

> ### You May Recall
>
> . . . that in solving problems in previous chapters, you have been performing all these operations except the second.

Adding Equations

The addition of equations can most readily be illustrated with a few examples. Suppose you want to add the equation $A = 4$ to the equation $B = 7$. The first step is to align the variables A and B vertically so that A is in one column and B in another column:

$$
\begin{array}{r}
A \quad\quad = 4 \\
B = 7 \\
\hline
A + B = 11
\end{array}
$$

Notice that the addition of two unequal variables yields A plus B, not A times B.

Next, let us add $2A + B = 13$ to $2B = 6$. Again, like variables are aligned vertically:

$$\begin{aligned} 2A + \quad B &= 13 \\ 2B &= 6 \\ \hline 2A + 3B &= 19 \end{aligned}$$

Answer the following questions.

1. Add $2A + 5B = 43.2$ to $4A + B = 16.4$

2. Add $3R - 2S - 4T = -23$ to $R + 3S - 2T = 48$

3. Add $-2A + 3B = 10$ to $-3B + 4C = -2$

Finally, let us add $2A - 6B + 3C = -25$ to $A + B - 2C = 2$:

$$\begin{aligned} 2A - 6B + 3C &= -25 \\ A + \quad B - 2C &= 2 \\ \hline 3A - 5B + \quad C &= -23 \end{aligned}$$

Variable Elimination with Two Variables

Solving simultaneous equations with *two unknown variables* requires two independent equations, each of which shows a relationship between the unknown variables. The procedure involves multiplying each of the terms of one of the equations by a *constant* to obtain a new, third, equation. The new equation and the equation that was not multiplied are then added together. The constant is selected so that when the two equations are added, one of the variables cancels out and does not appear in the sum. For example, $2A$ in one equation cancels $-2A$ in the other equation. The constant is obtained by dividing a term in the first equation by the corresponding term in the second equation and then multiplying the answer by -1. If each of the terms in the second equation is then multiplied by the constant, a term in the new equation will be equal in magnitude but opposite in sign to the corresponding term in the first equation. When the two equations are added, the variables will cancel each other. Only one variable will remain, and its value can then be calculated.

Let us try this process by solving the two-variable simultaneous equations $6A + 5B = 45$

and $2A + 3B = 19$. First divide $6A$ by $2A$ and multiply by -1 to get the constant:

$$\frac{6A}{2A} \times (-1) = 3 \times (-1) = -3$$

Next, multiply the second equation by the constant -3:

$$-3(2A + 3B) = -3(19)$$
$$-6A - 9B = -57$$

Then add the first equation to the new equation that is the result of the above multiplication:

First equation:	$6A + 5B = 45$
New equation:	$-6A - 9B = -57$
Sum of the two equations:	$-4B = -12$

Solving for B yields

$$B = \frac{-12}{-4} = 3$$

We can also eliminate a variable without multiplying the constant by -1. We must then subtract rather than add the equations. In the example just completed, the constant would be 3 and the new equation would be

$$6A + 9B = 57$$

Now, subtracting the new equation from the first equation yields

$$\begin{aligned} 6A + 5B &= 45 \\ 6A + 9B &= 57 \\ \hline -4B &= -12 \end{aligned}$$
$$B = 3$$

Two unknown variables

Constant

Now we can go back to either the first equation or the second and solve for A by substituting the value of B just found:

$$6A + 5B = 45$$
$$6A + 5(3) = 45$$
$$6A + 15 = 45$$
$$6A = 30$$
$$A = 5$$

Finally, we can *check for* any arithmetic *error* by substituting the values for A and B into the equation we did not use in the above step:

$$2A + 3B = 19$$
$$2(5) + 3(3) = 19$$
$$10 + 9 = 19$$
$$19 = 19 \text{ check}$$

EXAMPLE 6-1

Determine the values of A and B for $4A + 3B = 36.9$ and $A + 2B = 17.1$.

Given:	$4A + 3B = 36.9$
	$A + 2B = 17.1$
Find:	A and B
Known:	Variable-elimination method
Solution:	$\dfrac{4A}{A} \times (-1) = -4$
	$-4(A + 2B) = -4(17.1)$

New equation:	$-4A - 8B = -68.4$
First equation:	$4A + 3B = 36.9$
New equation:	$-4A - 8B = -68.4$
	$-5B = -31.5$
	$B = 6.3$
	$4A + 3(6.3) = 36.9$
	$4A + 18.9 = 36.9$
	$4A = 18$
	$A = 4.5$
Answer:	A equals 4.5, and B equals 6.3.

 Self-Test

Answer the following questions.

4. Solve for X and Y when $3X + 4.2Y = 38.1$ and $2.5X + 3Y = 28.75$.

5. Solve for C and D when $4C + 5D = 42$ and $-3C + 7D = 26$.

6. Solve for A and B when $3A - 4B = 6$ and $-2A - 2B = -18$.

Variable Elimination with Three Variables

Independent equations

Three variables

Solving for three unknown variables requires three *independent equations* involving the *three variables*. Suppose the three equations are

(1) $\quad 2X + 4Y - 5Z = -1$

(2) $\quad -4X - 2Y + 6Z = -2$

(3) $\quad 3X + 5Y + 7Z = 96$

The step-by-step procedure for solving for X, Y, and Z is as follows:

1. Combine any two of the three equations so that any one of the three variables is eliminated. For example, let us combine Eqs. 1 and 3 to eliminate variable Y:

Find the constant: $\dfrac{4Y}{5Y} \times (-1) = -0.8$

Multiply: $-0.8(3X + 5Y + 7Z) = -0.8(96)$

$$-2.4X - 4Y - 5.6Z = -76.8$$

Add:

Eq. 1:	$2X + 4Y - 5Z = -1$
New equation:	$-2.4X - 4Y - 5.6Z = -76.8$
	$-0.4X \qquad -10.6Z = -77.8$

To simplify this equation and eliminate the decimal numbers, we multiply the equation by 5:

$$-2X - 53Z = -389$$

2. Combine any other two of the three original equations to eliminate the same variable as eliminated in step 1. Let us use Eqs. 1 and 2 and again eliminate Y by the same procedure:

Constant: $\quad \dfrac{4Y}{-2Y} \times (-1) = 2$

Multiply: $\quad 2(-4X - 2Y + 6Z) = 2(-2)$
$$-8X - 4Y + 12Z = -4$$

Add:
Eq. 1: $\qquad\qquad 2X + 4Y - 5Z = -1$
New Eq. 2: $\qquad \underline{-8X - 4Y + 12Z = -4}$
$$-6X \qquad\quad + 7Z = -5$$

3. Combine the equations derived in steps 1 and 2 to eliminate one of the two remaining variables. Let us eliminate X and solve for Z:

Constant: $\quad \dfrac{-6X}{-2X} \times (-1) = -3$

Multiply: $\quad -3(-2X - 53Z) = -3(-389)$
$$6X + 159Z = 1167$$

Add: $\qquad\qquad -6X + 7Z = \quad -5$
$$\underline{6X + 159Z = 1167}$$
$$166Z = 1162$$
$$Z = 7$$

4. Substitute the value for Z found in step 3 into the equation derived in either step 1 or step 2 to find the value of X. Using the equation of step 1 yields

$$-2X - 53Z = -389$$
$$-2X - 53(7) = -389$$
$$-2X = -389 + 371$$
$$2X = 18$$
$$X = 9$$

5. Substitute the values of Z and X from steps 3 and 4 into any one of the three original equations and solve for Y. Let us use Eq. 1:

$$2(9) + 4Y - 5(7) = -1$$
$$18 + 4Y - 35 = -1$$
$$4Y - 17 = -1$$
$$4Y = 16$$
$$Y = 4$$

All five steps are needed only if all three equations contain all three variables. If either one or two of the equations contain only two of the three variables, then step 2 can be omitted.

Determinant Method of Solving Simultaneous Equations with Two Unknowns

Another method of solving simultaneous equations involves setting up a number of *square matrixes* and calculating a *determinant* value for each matrix. Before developing the details of setting up a matrix and calculating its determinant, it is necessary to describe or define some of the terms we will be using. A square matrix is a group of related numbers arranged

Square matrixes

Determinant

EXAMPLE 6-2

Determine the values of A, B, and C for the given equations:

Given: $\qquad 2A + 4C = 40$
$$-4B + 3C = -4$$
$$5A + 3B - 7C = -15$$

Find: $\qquad A$, B, and C
Known: $\quad$ Variable-elimination technique

Solution: $\quad \dfrac{-4B}{3B} \times (-1) = \dfrac{4}{3}$

$$\tfrac{4}{3}(5A + 3B - 7C) = \tfrac{4}{3}(-15)$$

$$6.6A + 4B - 9.3C = -20$$
$$-4B + 3C = -4$$
$$\underline{6.6A + 4B - 9.3C = -20}$$
$$6.6A \qquad - 6.3C = -24$$
$$10(6.6A) - 10(6.3C) = 10(-24)$$
$$66A - 63C = -240$$

$$\dfrac{66A}{2A} \times (-1) = -33$$

$$-33(2A + 4C) = -33(40)$$
$$-66A - 132C = -1320$$
$$66A - 63C = -240$$
$$\underline{-66A - 132C = -1320}$$
$$-195C = -1560$$
$$C = 8$$
$$2A + 4(8) = 40$$
$$2A + 32 = 40$$
$$2A = 8$$
$$A = 4$$
$$-4B + 3(8) = -4$$
$$-4B + 24 = -4$$
$$-4B = -28$$
$$B = 7$$

Answer: $\quad A$ equals 4, B equals 7, and C equals 8.

in an equal number of rows and columns. A determinant is a single number that is calculated from the data in the matrix. In other words, a determinant reduces the data in the matrix to a single number that can be used to replace the matrix in a formula that determines the value for an unknown variable. In an equation, the number in front of the unknown variable is the *coefficient* of that variable. The number on the other side of the equation is called the *constant* of the equation. Thus, for the first equation in example 6-1, the coefficient of variable A is 4, for variable B it is 3, and the constant for the equation is 36.9.

A generalized (2 × 2) matrix like the type used in solving for two unknown variables in two simultaneous equations is

$$\begin{vmatrix} a_1 & b_1 \\ a_2 & b_2 \end{vmatrix}$$

In this matrix, a_1 is the coefficient of the first unknown variable in the first equation, b_1 is the coefficient of the second unknown variable in the first equation, and a_2 and b_2 are the corresponding coefficients from the second equation. The straight lines enclosing the coefficients indicate this is a matrix for which a determinant can be calculated.

The determinant for a (2 × 2) matrix is found by multiplying a_1 by b_2 and a_2 by b_1 and then subtracting the a_2b_1 product from the a_1b_2 product. Sometimes arrows are added to a matrix as an aid to remembering which diagonal product is subtracted from the other diagonal product. The upward dashed-line arrow indicates that the a_2b_1 product is to be subtracted from the a_1b_2 product symbolized by the downward solid-line arrow.

$$\begin{vmatrix} a_1 & b_1 \\ a_2 & b_2 \end{vmatrix}$$

The matrix for the coefficients given in example 6-1 would be:

$$\begin{vmatrix} 4 & 3 \\ 1 & 2 \end{vmatrix}$$

For these coefficients, the coefficient determinant (D) would be

$$D = (4 \times 2) - (1 \times 3) = 5.$$

To determine the value of the A variable in example 6-1, first construct a (2 × 2) matrix in which the coefficients a_1 and a_2 of the A variable are replaced by the constants C_1 and C_2 from equations one and two, respectively. Then the determinant (D_A) for this matrix can be calculated thus:

$$D_A = \begin{vmatrix} C_1 & b_1 \\ C_2 & b_2 \end{vmatrix} = \begin{vmatrix} 36.9 & 3 \\ 17.1 & 2 \end{vmatrix}$$

$$= (36.9 \times 2) - (17.1 \times 3)$$

$$= 73.8 - 51.3 = 22.5$$

Finally, the value of unknown variable A can be calculated with the following formula:

$$A = \frac{D_A}{D} = \frac{22.5}{5} = 4.5$$

Notice that this is the same value obtained by the variable elimination method used in example 6-1. To find the value of the B variable in example 6-1, construct another matrix in which the coefficients b_1 and b_2 of the B variable are replaced by the constants C_1 and C_2 from equations one and two, respectively. Then the determinant (D_B) for this matrix can be calculated thus:

$$D_B = \begin{vmatrix} a_1 & C_1 \\ a_2 & C_2 \end{vmatrix} = \begin{vmatrix} 4 & 36.9 \\ 1 & 17.1 \end{vmatrix}$$

$$= (4 \times 17.1) - (1 \times 36.9)$$

$$= 68.4 - 36.9 = 31.5$$

Finally, the value of unknown variable B can be calculated with the following formula:

$$B = \frac{D_B}{D} = \frac{31.5}{5} = 6.3$$

Again, notice this is the same value obtained in example 6-1.

Determinant Method of Solving Simultaneous Equations with Three Unknowns

When three unknown variables (and thus three equations) are to be solved, we need to construct four (3 × 3) matrixes and calculate the determinant for each one. A generalized (3 × 3) matrix is:

$$\begin{vmatrix} a_1 & b_1 & c_1 \\ a_2 & b_2 & c_2 \\ a_3 & b_3 & c_3 \end{vmatrix}$$

If this were a matrix for coefficients, then a_1 and b_1 and c_1 would be the coefficients for the first, second, and third unknown variables in the first equation. With a (3 × 3) matrix, we need to calculate six diagonal products where each diagonal has three coefficients. The sum of three of these products (namely $a_3b_2c_1 + b_3c_2a_1 + c_3a_2b_1$) is then subtracted from the sum of the other products (namely $a_1b_2c_3 + b_1c_2a_3 + c_1a_2b_3$). This process yields the determinant (D) of the coefficient matrix. Determining which coefficient to use in each diagonal product is easier if the first two columns of the matrix are repeated on the right-hand side of the matrix and upward and downward diagonal arrows are then drawn through the coefficients. This is illustrated below where the significance of the type of arrow is the same as it was for a (2 × 2) matrix.

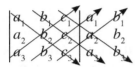

Now that we know how to work with (3 × 3) matrixes, we can use determinants to find the value of the unknown variables A, B, and C in example 6-2.

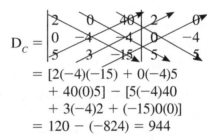

$$D = \begin{vmatrix} 2 & 0 & 4 \\ 0 & -4 & 3 \\ 5 & 3 & -7 \end{vmatrix}$$

$$= [2(-4)(-7) + 0(3)5 + 4(0)3] - [5(-4)4 + 3(3)2 + (-7)0(0)]$$
$$= 56 - (-62) = 118$$

$$D_A = \begin{vmatrix} 40 & 0 & 4 \\ -4 & -4 & 3 \\ -15 & 3 & -7 \end{vmatrix}$$

$$= [40(-4)(-7) + 0(3)(-15) + 4(-4)3] - [(-15)(-4)4 + 3(3)40 + (-7)(-4)0]$$
$$= 1072 - 600 = 472$$

$A = D_A \div D = 472 \div 118 = 4$

$$D_B = \begin{vmatrix} 2 & 40 & 4 \\ 0 & -4 & 3 \\ 5 & -15 & -7 \end{vmatrix}$$

$$= [2(-4)(-7) + 40(3)5 + 4(0)(-15)] - [5(-4)4 + (-15)3(2) + (-7)0(40)]$$
$$= 656 - (-170) = 826$$

$B = D_B \div D = 826 \div 118 = 7$

$$D_C = \begin{vmatrix} 2 & 0 & 40 \\ 0 & -4 & -4 \\ 5 & 3 & -15 \end{vmatrix}$$

$$= [2(-4)(-15) + 0(-4)5 + 40(0)5] - [5(-4)40 + 3(-4)2 + (-15)0(0)]$$
$$= 120 - (-824) = 944$$

$C = D_C \div D = 944 \div 118 = 8$

Notice that the determinant method yielded the same value for A, B, and C (4, 7, and 8, respectively) as did the variable elimination method.

The procedures used in the determinant method are known as Cramer's rule, which was developed by Gabriel Cramer (1704–1752) a Swiss mathematician. If you want to develop a little understanding of why the determinant method solves unknown variables, refer to Appendix J.

Self-Test

Answer the following questions.

7. Solve for R, S, and T when

$$2R - 4S + 3T = 49$$
$$-3R + 5S = -52$$
$$6R - 2S + 4T = 52$$

8. Solve for X, Y, and Z when

$$2X - Y + 2Z = 5.6$$
$$2X + 3Y - Z = 17.1$$
$$-3X + 4Y + 5Z = 29.4$$

9. Solve for C, D, and E when

$$4C - 5D + 3E = -9$$
$$2C - 3D = 4$$
$$8D - 5E = -18$$

6-2 Loop-Equations Technique

Now that you are familiar with simultaneous-equation techniques, you can use them to solve complex electric networks. The equations you will be working with are called *loop equations*. They are derived by applying Kirchhoff's voltage law to the various loops of an electric circuit.

Either *multiple-source* or *single-source circuits* can be solved by the *loop-equations* technique. This technique allows one to find every voltage and every current in the circuit.

Single-Source Circuits

A complex circuit with a single-source voltage is shown in Fig. 6-1. This circuit configuration is called a *bridge circuit.* It is a popular configuration for measuring temperature indirectly. In such an application R_5 represents the internal resistance of an electric meter and R_4 is a thermistor. The scale of the meter is calibrated in temperature units rather than in current or voltage units.

A study of Fig. 6-1 shows that none of the resistors are directly in series or directly in parallel. Thus, there is no place to start using series rules, parallel rules, or Ohm's law. This complex circuit can be solved by developing loop equations using Kirchhoff's voltage law.

A logical way to determine the loops in a circuit is to trace possible paths of current flow, as shown in Fig. 6-2. Without knowing resistor values, it is impossible to determine whether I_1 (loop 1) is in the direction shown in Fig. 6-2(a) or in that shown in Fig. 6-2(b). The direction of the current depends upon the polarity of the voltage between A and B. It does not matter which direction one selects when analyzing circuits by loop equations. If the wrong direction is selected, then the value of I_1 will be *negative*. If I_1

turns out to be negative, then we know that our assumption about current direction was wrong and we have to reverse the polarity markings of the voltage across R_5. Notice that three loops (current paths) are required in Fig. 6-2 to include every resistor in one or more loops. You cannot use the two loops shown in Fig. 6-2(a) and (b) because a single source would be trying to force current through R_5 in both directions. Thus, three loop equations will have to be

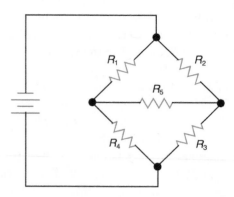

Fig. 6-1 Bridge circuit. None of the resistors are in series or parallel.

derived and then solved simultaneously. To aid in deriving the loop equations, the three loops of Fig. 6-2(a), (c), and (d) have been combined on a single diagram in Fig. 6-3. Also, values have been assigned to the resistors and the voltage source in Fig. 6-3 so that we can calculate the values for all currents and voltages. Notice in Fig. 6-3 that R_1 has two currents (I_1 and I_2) flowing through it. Both currents are flowing in the same direction; therefore, the voltage drop across R_1 is

$$V_{R_1} = R_1(I_1 + I_2) = R_1 I_1 + R_1 I_2$$
$$= 2 \text{ k}\Omega \, (I_1) + 2 \text{ k}\Omega \, (I_2)$$

The same procedure is applicable to V_{R_3}.

We are now ready to derive the loop equations for the *three loops* of Fig. 6-3. We use Kirchhoff's voltage law, which states that "the algebraic sum of the voltage drops around a loop of a circuit is equal to the source voltage." For loop 1 we can write

$$V_T = V_{R_1} + V_{R_5} + V_{R_3}$$

By substituting known and equivalent values, we get

$$9 \text{ V} = 2 \text{ k}\Omega \, (I_1 + I_2) + 4 \text{ k}\Omega \, (I_1)$$
$$+ 5 \text{ k}\Omega \, (I_1 + I_3)$$

By expanding we get

$$9 \text{ V} = 2 \text{ k}\Omega \, (I_1) + 2 \text{ k}\Omega \, (I_2) + 4 \text{ k}\Omega \, (I_1)$$
$$+ 5 \text{ k}\Omega \, (I_1) + 5 \text{ k}\Omega \, (I_3)$$

By collecting terms, the equation for loop 1 becomes

$$9 \text{ V} = 11 \text{ k}\Omega \, (I_1) + 2 \text{ k}\Omega \, (I_2) + 5 \text{ k}\Omega \, (I_3)$$

(loop 1)

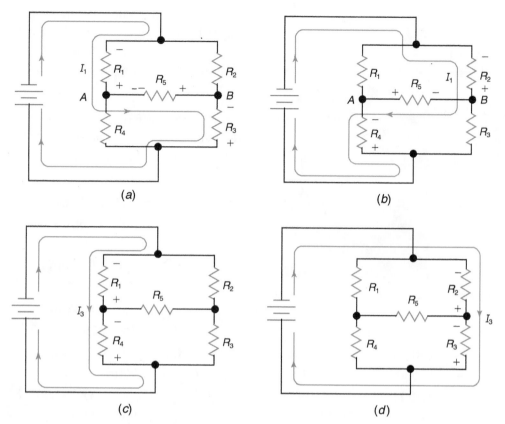

(a) (b) (c) (d)

Fig. 6-2 Three loops for analyzing a bridge circuit using Kirchhoff's voltage law. (a) Loop 1 — assuming that point A is negative with respect to point B. (b) Loop 1 — assuming that point A is positive with respect to point B. (c) Loop 2. (d) Loop 3.

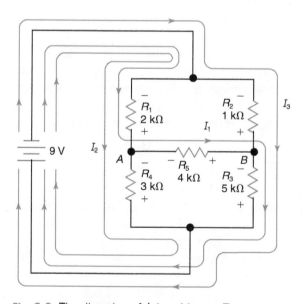

Fig. 6-3 The direction of I_1 is arbitrary. Two currents must be used in calculating the voltage drops across R_1 and R_3.

Using the same procedure for loop 2 yields

$$V_T = V_{R_1} + V_{R_4}$$
$$9\text{ V} = 2\text{ k}\Omega\,(I_1 + I_2) + 3\text{ k}\Omega\,(I_2)$$
$$= 2\text{ k}\Omega\,(I_1) + 2\text{ k}\Omega\,(I_2) + 3\text{ k}\Omega\,(I_2)$$
$$= 2\text{ k}\Omega\,(I_1) + 5\text{ k}\Omega\,(I_2)$$

(loop 2)

And for loop 3 the result is

$$V_T = V_{R_2} + V_{R_3}$$
$$9\text{ V} = 1\text{ k}\Omega\,(I_3) + 5\text{ k}\Omega\,(I_1 + I_3)$$
$$= 1\text{ k}\Omega\,(I_3) + 5\text{ k}\Omega\,(I_1) + 5\text{ k}\Omega\,(I_3)$$
$$= 5\text{ k}\Omega\,(I_1) + 6\text{ k}\Omega\,(I_3)$$

(loop 3)

Our next step is to solve the three equations simultaneously using the variable-elimination technique we learned in the preceding section. We start by adding the equations for loops 1 and 2 to eliminate the variable I_2:

$$\frac{5\text{ k}\Omega\,(I_2)}{2\text{ k}\Omega\,(I_2)} \times (-1) = -2.5$$

History of
Electronics

Charles Wheatstone
Named for Charles Wheatstone, the Wheatstone bridge is a circuit used for measuring unknown resistances (see Sec. 15-8 for more about the Wheatstone bridge).

$$-2.5 \ (9 \ \text{V}) = -2.5 \ [11 \ \text{k}\Omega \ (I_1) + 2 \ \text{k}\Omega \ (I_2)$$
$$+ 5 \ \text{k}\Omega \ (I_3)]$$
$$-22.5 \ \text{V} = -27.5 \ \text{k}\Omega \ (I_1) - 5 \ \text{k}\Omega \ (I_2)$$
$$- 12.5 \ \text{k}\Omega \ (I_3)$$

$$\underline{\quad 9 \ \text{V} = -2 \ \text{k}\Omega \ (I_1) + 5 \ \text{k}\Omega \ (I_2) \quad}$$
$$-13.5 \ \text{V} = -25.5 \ \text{k}\Omega \ (I_1) - 12.5 \ \text{k}\Omega \ (I_3)$$

(loops 1 and 2)

Next, let us add loops 1 and 2 to loop 3 so that we can eliminate the variable I_3 and solve for I_1:

Assumed polarity

$$\frac{-12.5 \ \text{k}\Omega \ (I_3)}{6 \ \text{k}\Omega \ (I_3)} \times (-1) = 2.083$$

$$2.083 \ (9 \ \text{V}) = 2.083 \ [5 \ \text{k}\Omega \ (I_1) + 6 \ \text{k}\Omega \ (I_3)]$$
$$18.75 \ \text{V} = 10.416 \ \text{k}\Omega \ (I_1) + 12.5 \ \text{k}\Omega \ (I_3)$$
$$\underline{-13.5 \ \text{V} = -25.5 \ \text{k}\Omega \ (I_1) - 12.5 \ \text{k}\Omega \ (I_3)}$$
$$5.25 \ \text{V} = -15.083 \ \text{k}\Omega \ (I_1)$$

$$I_1 = \frac{-5.25 \ \text{V}}{15.083 \ \text{k}\Omega} = -0.348 \ \text{mA}$$

Now we can solve for I_3 by substituting the value of I_1 into the equation for loop 3:

$$9 \ \text{V} = (5 \ \text{k}\Omega) \ (-0.348 \ \text{mA}) + 6 \ \text{k}\Omega \ (I_3)$$

$$I_3 = \frac{9 \ \text{V} + 1.74 \ \text{V}}{6 \ \text{k}\Omega} = 1.79 \ \text{mA}$$

Finally, we solve for I_2 by substituting the value of I_1 into the equation for loop 2:

$$9 \ \text{V} = (2 \ \text{k}\Omega) \ (-0.348 \ \text{mA}) + 5 \ \text{k}\Omega \ (I_2)$$

$$I_2 = \frac{9 \ \text{V} + 0.696 \ \text{V}}{5 \ \text{k}\Omega} = 1.939 \ \text{mA}$$

Notice that the value of I_1 is negative. Thus, we assumed the wrong direction for I_1. Therefore, we must change the polarity of the voltage drop across R_5.

Now that we know the values of I_1, I_2, and I_3, we can calculate the current through and the voltage across each resistor. Referring to Fig. 6-3 and our values of I_1, I_2, and I_3, we can write

$$I_{R_1} = I_1 + I_2 = -0.348 \ \text{mA} + 1.939 \ \text{mA}$$
$$= 1.591 \ \text{mA}$$
$$I_{R_2} = I_3 = 1.79 \ \text{mA}$$
$$I_{R_3} = I_1 + I_3 = -0.348 \ \text{mA} + 1.79 \ \text{mA}$$
$$= 1.442 \ \text{mA}$$
$$I_{R_4} = I_2 = 1.939 \ \text{mA}$$
$$I_{R_5} = I_1 = -0.348 \ \text{mA}$$
$$I_T = I_1 + I_2 + I_3 = -0.348 \ \text{mA}$$
$$+ 1.939 \ \text{mA} + 1.79 \ \text{mA} = 3.381 \ \text{mA}$$
$$V_{R_1} = I_{R_1} R_1 = (1.591 \ \text{mA})(2 \ \text{k}\Omega) = 3.182 \ \text{V}$$
$$V_{R_2} = I_{R_2} R_2 = (1.79 \ \text{mA})(1 \ \text{k}\Omega) = 1.79 \ \text{V}$$
$$V_{R_3} = I_{R_3} R_3 = (1.442 \ \text{mA})(5 \ \text{k}\Omega) = 7.21 \ \text{V}$$
$$V_{R_4} = I_{R_4} R_4 = (1.939 \ \text{mA})(3 \ \text{k}\Omega) = 5.817 \ \text{V}$$
$$V_{R_5} = I_{R_5} R_5 = (-0.348 \ \text{mA})(4 \ \text{k}\Omega)$$
$$= -1.392 \ \text{V}$$
$$R_T = \frac{V_T}{I_T} = \frac{9 \ \text{V}}{3.381 \ \text{mA}} = 2.662 \ \text{k}\Omega$$

The negative sign for V_{R_5} again tells us that the *assumed polarity* of the voltage across R_5 in Fig. 6-3 is incorrect.

After finishing an involved problem like the one above, it is worthwhile to check for errors. One way to do this is to substitute the voltages calculated into the original loop equations and see whether the equations balance. For instance,

$$V_T = V_{R_1} + V_{R_5} + V_{R_3}$$
$$= 3.182 \ \text{V} - 1.392 \ \text{V} + 7.21 \ \text{V}$$
$$= 9 \ \text{V}$$

In analyzing the bridge circuit, we arbitrarily chose the direction of the current through the center resistor (R_5). Although choosing the wrong direction makes no difference in the final results, such an error can be easily avoided. Suppose we wish to know the direction of current through R_5 in Fig. 6-4(a). The direction can be found by temporarily removing R_5, as shown in Fig. 6-4(b), and determining the polarity of point A with respect to point B. Removal of R_5 makes the circuit a series-parallel

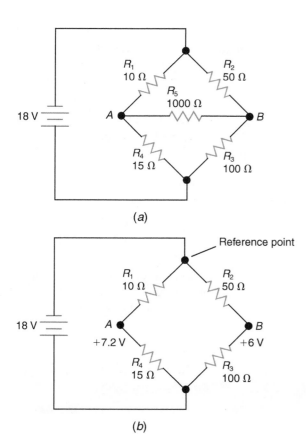

(a)

Reference point

(b)

Fig. 6-4 Determining the direction of current through R_5. (a) Current direction through R_5 unspecified. (b) Point A is positive with respect to point B.

circuit with two series resistors in each parallel branch. Inspection of Fig. 6-4(b) shows that, with respect to the negative terminal of the power source, point A must be more positive than point B because the ratio R_1/R_4 is greater than the ratio R_2/R_3. Thus, point A is positive with respect to point B. The calculations used to determine the values shown in Fig. 6-4(b) are

$$V_{R_1} = \frac{R_1 V_T}{R_1 + R_4} = \frac{(10\ \Omega)(18\ V)}{10\ \Omega + 15\ \Omega}$$

$$= \frac{180\ \Omega \cdot V}{25\ \Omega} = 7.2\ V$$

$$V_{R_2} = \frac{R_2 V_T}{R_2 + R_3} = \frac{(50\ \Omega)(18\ V)}{50\ \Omega + 100\ \Omega}$$

$$= \frac{900\ \Omega \cdot V}{150\ \Omega} = 6\ V$$

Although connecting R_5 back into the circuit of Fig. 6-4(b) changes the voltage distributions, the relative polarities remain the same. Current flows from point B to point A in Fig. 6-4(a).

EXAMPLE 6-3

For Fig. 6-2, assume $R_1 = 1.5\ k\Omega$, $R_2 = 2\ k\Omega$, $R_3 = 5\ k\Omega$, $R_4 = 2.5\ k\Omega$, $R_5 = 1\ k\Omega$ and $V_T = 12\ V$.

Determine : 1. Will the left end of R_5 be negative or positive with respect to the right end of R_5?
2. What are the values of I_{R_5} and V_{R_5}?

Given: $R_1 = 1.5\ k\Omega$,
$R_2 = 2\ k\Omega$,
$R_3 = 5\ k\Omega$,
$R_4 = 2.5\ k\Omega$,
$R_5 = 1\ k\Omega$,
$V_T = 12\ V$

Find: Polarity of R_5 and the values of I_{R_5} and V_{R_5}

Known: Simultaneous-equations techniques and loop-equations techniques

Solution: Polarity of R_5:
Remove R_5, then

$$V_{R_4} = \frac{R_4 \times V_T}{R_1 + R_4} = \frac{2.5\ k\Omega \times 12\ V}{1.5\ k\Omega + 2.5\ k\Omega} = \frac{30\ k\Omega \cdot V}{4\ k\Omega}$$
$$= 7.5\ V\ (-\ \text{on the top end})$$

$$V_{R_3} = \frac{R_3 \times V_T}{R_2 + R_3} = \frac{5\ k\Omega \times 12\ V}{2\ k\Omega + 5\ k\Omega} = \frac{60\ k\Omega \cdot V}{7\ k\Omega}$$
$$= 8.57\ V\ (-\ \text{on the top end})$$

Therefore the left end of R_5 is + with respect to the right end.

Value of I_{R_5} and V_{R_5}:
Reconnect R_5 and have loop 1 (I_1) go through R_2, R_5, and R_4; loop 2 through R_1 and R_4; and loop 3 through R_2 and R_3. Now, develop the three loop equations.

$12\ V = 2\ k\Omega\ (I_1 + I_3)$
  $+ 1\ k\Omega\ (I_1) + 2.5\ k\Omega\ (I_1 + I_2)$ (loop 1)
Now, expand:
$12\ V = 2\ k\Omega\ (I_1) + 2\ k\Omega\ (I_3) + 1\ k\Omega\ (I_1)$
  $+ 2.5\ k\Omega\ (I_1) + 2.5\ k\Omega\ (I_2)$ (loop 1)
Now, collect terms:
$12\ V = 5.5\ k\Omega\ (I_1) + 2.5\ k\Omega\ (I_2)$
  $+ 2\ k\Omega\ (I_3)$ (loop 1)
$12\ V = 1.5\ k\Omega\ (I_2) + 2.5\ k\Omega\ (I_1 + I_2)$
  $= 1.5\ k\Omega\ (I_2) + 2.5\ k\Omega\ (I_1) + 2.5\ k\Omega\ (I_2)$
  (loop 2)
Now, collect terms:
$12\ V = 2.5\ k\Omega\ (I_1) + 4\ k\Omega\ (I_2)$
  (loop 2)

$$12\,V = 2\,k\Omega\,(I_1 + I_3) + 5\,k\Omega\,(I_3)$$
$$= 2\,k\Omega\,(I_1) + 2\,k\Omega\,(I_3) + 5\,k\Omega\,(I_3)$$
(loop 3)

Now, collect terms:
$$12\,V = 2\,k\Omega\,(I_1) + 7\,k\Omega\,(I_3) \qquad \text{(loop 3)}$$

Multiply loop 2 by $\dfrac{2.5\,k\Omega\,(I_{2_3})}{4\,k\Omega\,(I_{2_1})} \times -1$
$$= -0.625$$
$$-7.5\,V = -1.5625\,k\Omega\,(I_1) + (-2.5\,k\Omega\,[I_2])$$
loop 2

Now, add loop 1 with loop 2 to eliminate I_2;
$$12\,V = 5.5\,k\Omega\,(I_1) + 2.5\,k\Omega\,(I_2)$$
$$+\, 2\,k\Omega\,(I_3) \qquad\qquad \text{loop 1}$$
$$4.5\,V = 3.9375\,k\Omega\,(I_1) + 2\,k\Omega\,(I_3)$$
loop 1 + 2

Multiply loop 1 + 2 by $\dfrac{7\,k\Omega(I_{3_3})}{2\,k\Omega(I_{3_1})} \times -1$
$$= -3.5,$$

Then add loop 1 + 2 with loop 3 to eliminate I_3;
$$-15.75\,V = -13.78\,k\Omega\,(I_1) + (-7\,k\Omega[I_3])$$
loop 1 + 2
$$12\,V = 2\,k\Omega\,(I_1) + 7\,k\Omega\,(I_3) \qquad\quad \text{loop 3}$$
$$-3.75\,V = -11.78\,k\Omega\,(I_1) \quad \text{loop 1 + 2 + 3}$$

Now, solve for I_1
$$I_1 = \frac{-3.76\,V}{-11.78\,k\Omega} = 0.318\,mA$$

Finally, solve for I_{R_5} and V_{R_5}
$$I_{R_5} = I_1 = 0.318\,mA$$
$$V_{R_5} = I_{R_5} \times R_5 = 0.318\,mA \times 1\,k\Omega$$
$$= 0.318\,V$$

Answer: The left end of R_5 is positive with respect to the right end. And, $I_{R_5} = 0.318\,mA$ and $V_{R_5} = 0.318\,V$.

Self-Test

Answer the following questions.

10. Write the three loop equations for Fig. 6-3 assuming the direction of I_1 shown in Fig. 6-2(b).
11. Do the equations you wrote for question 10 yield the same absolute values of I_1, I_2, and I_3 as those we obtained using Fig. 6-3?
12. Do the equations of question 10 yield the same values for the current through and voltage across each of the resistors as those we obtained using Fig. 6-3?
13. Change the values in Fig. 6-4(a) to $R_1 = 40\,\Omega$, $R_2 = 70\,\Omega$, $R_3 = 80\,\Omega$, $R_4 = 80\,\Omega$, and $R_5 = 500\,\Omega$. Will the current now flow from point A to point B or from point B to point A?
14. Does the value of R_5 in Fig. 6-4(a) have any effect on the polarity of the voltage between points A and B?

Multiple-source circuits

Direction of current

Draw the loops

Multiple-Source Circuits

Several voltage sources can be applied to the same circuit components. When some, or all, of the components have no series or parallel relationship to other components, a complex multiple-source circuit exists. Such circuits are encountered in electronic systems and must be analyzed.

Look at the circuit of Fig. 6-5(a) and note that the resistors are neither in series nor in parallel. Therefore, we can use loop equations to solve the circuit. Two loops are required to include all the resistors in at least one loop. The traditional way to *draw the loops*

is shown in Fig. 6-5(b). Notice in Fig. 6-5(b) that the two loops show current through R_3 in opposite directions, but we know that this is not possible. Current will be in one direction or the other. This way of drawing loops is just a convention that implies that the direction of current could be either way. Although the traditional way is correct, the true direction of current flow is as shown in Fig. 6-5(c) or 6-5(d). Which direction the current takes through R_3 depends on whether B_1 or B_2 dominates the circuit. We could easily find the *direction of current* through R_3 by temporarily removing R_3 from the circuit and determining

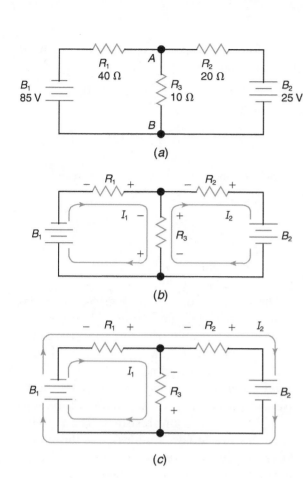

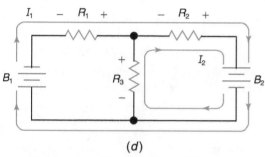

Fig. 6-5 Alternate sets of loops for a complex circuit. Any set will lead to the correct solution of circuit values. (a) Circuit to be analyzed. (b) Traditional loops. (c) Current path if B_1 dominates. (d) Current path if B_2 dominates.

the polarity of the voltage between points A and B. However, there is no need to determine the true direction because the final results using any set of loops will give the current direction. If the loops of Fig. 6-5(b) are used, the direction will be that of the highest current (I_1 or I_2) and the magnitude will be the difference between the two currents. If the loops of Fig. 6-5(c) are used, the direction will

be as indicated if the calculated value of I_1 is positive. If I_1 turns out negative when using the loops of Fig. 6-5(c), then the true current paths are those of Fig. 6-5(d). If the Fig. 6-5(d) loops are used, the polarity of I_2 determines the direction of the current.

Let us verify the equivalency of the three sets of loops by determining the circuit values using each set. For Fig. 6-5(b) the equations and calculations are

$$85 \text{ V} = 40 \ \Omega \ (I_1) + 10 \ \Omega \ (I_1) - 10 \ \Omega \ (I_2)$$
$$\text{(loop 1)}$$
$$25 \text{ V} = -10 \ \Omega \ (I_1) + 10 \ \Omega \ (I_2) + 20 \ \Omega \ (I_2)$$
$$\text{(loop 2)}$$

Collect terms:

$$85 \text{ V} = 50 \ \Omega \ (I_1) - 10 \ \Omega \ (I_2) \qquad \text{(loop 1)}$$
$$25 \text{ V} = -10 \ \Omega \ (I_1) + 30 \ \Omega \ (I_2) \qquad \text{(loop 2)}$$

Determine the constant and eliminate I_1:

$$\frac{50 \ \Omega \ (I_1)}{-10 \ \Omega \ (I_1)} \times (-1) = 5$$

$$125 \text{ V} = -50 \ \Omega \ (I_1) + 150 \ \Omega \ (I_2)$$
$$\underline{85 \text{ V} = \quad 50 \ \Omega \ (I_1) - 10 \ \Omega \ (I_2)}$$
$$210 \text{ V} = 140 \ \Omega \ (I_2)$$

Determine I_2:

$$I_2 = \frac{210 \text{ V}}{140 \ \Omega} = 1.5 \text{ A}$$

Determine I_1:

$$25 \text{ V} = -10 \ \Omega \ (I_1) + (30 \ \Omega)(1.5 \text{ A})$$
$$= -10 \ \Omega \ (I_1) + 45 \text{ V}$$
$$-20 \text{ V} = -10 \ \Omega \ (I_1)$$
$$I_1 = \frac{20 \text{ V}}{10 \ \Omega} = 2 \text{ A}$$

Determine the current through each resistor:

$$I_{R_3} = I_1 - I_2 = 2 \text{ A} - 1.5 \text{ A} = 0.5 \text{ A}$$

(Note that I_{R_3} flows from point A to point B and that point A is negative with respect to point B.)

$$I_{R_2} = I_2 = 1.5 \text{ A}$$
$$I_{R_1} = I_1 = 2 \text{ A}$$

For Fig. 6-5(c) the procedure is

$$85 \text{ V} = 40 \ \Omega \ (I_1) + 40 \ \Omega \ (I_2) + 10 \ \Omega \ (I_1)$$
$$\text{(loop 1)}$$

$$85 \text{ V} = 40 \ \Omega \ (I_1) + 40 \ \Omega \ (I_2)$$
$$+ 20 \ \Omega \ (I_2) - 25 \text{ V} \qquad \text{(loop 2)}$$

(Note that the B_2 voltage is negative because its polarity is the reverse of that of the voltage drops across R_1 and R_2.) Collect terms:

$$85 \text{ V} = 50 \text{ } \Omega \text{ } (I_1) + 40 \text{ } \Omega \text{ } (I_2) \qquad \text{(loop 1)}$$
$$110 \text{ V} = 40 \text{ } \Omega \text{ } (I_1) + 60 \text{ } \Omega \text{ } (I_2) \qquad \text{(loop 2)}$$

Determine the constant and eliminate I_2:

$$\frac{60 \text{ } \Omega \text{ } (I_2)}{40 \text{ } \Omega \text{ } (I_2)} \times (-1) = -1.5$$

$$-127.5 \text{ V} = -75 \text{ } \Omega \text{ } (I_1) - 60 \text{ } \Omega \text{ } (I_2)$$
$$\underline{110 \text{ V} = 40 \text{ } \Omega \text{ } (I_1) + 60 \text{ } \Omega \text{ } (I_2)}$$
$$-17.5 \text{ V} = -35 \text{ } \Omega \text{ } (I_1)$$

Determine I_1:

Series-aiding

$$I_1 = \frac{17.5 \text{ V}}{35 \text{ } \Omega} = 0.5 \text{ A}$$

Determine I_2:

Series-opposing

$$85 \text{ V} = 50 \text{ } \Omega \text{ } (0.5 \text{ A}) + 40 \text{ } \Omega \text{ } (I_2)$$
$$60 \text{ V} = 40 \text{ } \Omega \text{ } (I_2)$$

$$I_2 = \frac{60 \text{ V}}{40 \text{ } \Omega} = 1.5 \text{ A}$$

Determine the current through each resistor:

$$I_{R_3} = I_1 = 0.5 \text{ A}$$
$$I_{R_2} = I_2 = 1.5 \text{ A}$$
$$I_{R_1} = I_1 + I_2 = 0.5 \text{ A} + 1.5 \text{ A} = 2 \text{ A}$$

These values agree with those for Fig. 6-5(b).
For Fig. 6-5(d) the calculations and results are

$$85 \text{ V} = 40 \text{ } \Omega \text{ } (I_1) + 20 \text{ } \Omega \text{ } (I_1)$$
$$+ 20 \text{ } \Omega \text{ } (I_2) - 25 \text{ V} \quad \text{(loop 1)}$$
$$25 \text{ V} = 10 \text{ } \Omega \text{ } (I_2) + 20 \text{ } \Omega \text{ } (I_1)$$
$$+ 20 \text{ } \Omega \text{ } (I_2) \quad \text{(loop 2)}$$

Collect terms:

$$110 \text{ V} = 60 \text{ } \Omega \text{ } (I_1) + 20 \text{ } \Omega \text{ } (I_2)$$
$$25 \text{ V} = 20 \text{ } \Omega \text{ } (I_1) + 30 \text{ } \Omega \text{ } (I_2)$$

Determine the constant and eliminate I_1:

$$\frac{60 \text{ } \Omega \text{ } (I_1)}{20 \text{ } \Omega \text{ } (I_1)} \times (-1) = -3$$

$$-75 \text{ V} = -60 \text{ } \Omega \text{ } (I_1) - 90 \text{ } \Omega \text{ } (I_2)$$
$$\underline{110 \text{ V} = 60 \text{ } \Omega \text{ } (I_1) + 20 \text{ } \Omega \text{ } (I_2)}$$
$$35 \text{ V} = -70 \text{ } \Omega \text{ } (I_2)$$

Determine I_2:

$$I_2 = \frac{-35 \text{ V}}{70 \text{ } \Omega} = -0.5 \text{ A}$$

Thus, Fig. 6-5(d) shows the wrong direction for I_{R_3}. Determine I_1:

$$25 \text{ V} = 20 \text{ } \Omega \text{ } (I_1) + 30 \text{ } \Omega \text{ } (-0.5 \text{ A})$$
$$40 \text{ V} = 20 \text{ } \Omega \text{ } (I_1)$$

$$I_1 = \frac{40 \text{ V}}{20 \text{ } \Omega} = 2 \text{ A}$$

Determine the current through each resistor:

$$I_{R_3} = I_2 = -0.5 \text{ A}$$
$$I_{R_2} = I_1 + I_2 = 2 \text{ A} - 0.5 \text{ A} = 1.5 \text{ A}$$
$$I_{R_1} = I_1 = 2 \text{ A}$$

Again, these current values agree with the previously calculated values.

The voltage sources in Fig. 6-5 are connected so that they are *series-aiding*. That is, the negative terminal of B_2 is connected to the positive terminal of B_1. Multiple-source circuits can also be connected so that the sources are *series-opposing*. Furthermore, both the negative and positive source terminals can be separated by one or more resistors. The possible variations are almost limitless; yet the procedures and rules for working with the circuits remain the same.

EXAMPLE 6-4

Determine the voltages across R_1, R_2, and R_3 in Fig. 6-6(a) using the loop-equations technique.

Given: Circuit diagram and values of Fig. 6-6(a).

Find: V_{R_1}, V_{R_2}, and V_{R_3}

Known: Kirchhoff's voltage law and simultaneous equations

Solution: Determine loops and polarities as shown in Fig. 6-6(b). Derive loop equations:

$$3 \text{ V} = 4 \text{ } \Omega \text{ } (I_1) + 9 \text{ } \Omega \text{ } (I_1) + 9 \text{ } \Omega \text{ } (I_2)$$
$$\text{(loop 1)}$$
$$24 \text{ V} = 9 \text{ } \Omega \text{ } (I_1) + 9 \text{ } \Omega \text{ } (I_2) + 6 \text{ } \Omega \text{ } (I_2)$$
$$\text{(loop 2)}$$

Collect terms:

$$3 \text{ V} = 13 \text{ } \Omega \text{ } (I_1) + 9 \text{ } \Omega \text{ } (I_2) \quad \text{(loop 1)}$$
$$24 \text{ V} = 9 \text{ } \Omega \text{ } (I_1) + 15 \text{ } \Omega \text{ } (I_2) \quad \text{(loop 2)}$$

Add, eliminate I_2, and solve for I_1:

$$\frac{9 \Omega \text{ } (I_2)}{15 \text{ } \Omega \text{ } (I_2)} \times (-1) = -0.6$$

$$-14.4\,\text{V} = -5.4\,\Omega\,(I_1) - 9\,\Omega\,(I_2)$$
$$\underline{3\text{V} = 13\,\Omega\,(I_1) + 9\,\Omega(I_2)}$$
$$-11.4\text{V} = 7.6\,\Omega\,(I_1)$$

$$I_1 = \frac{-11.4\,\text{V}}{7.6\,\Omega} = -1.5\,\text{A}$$

Solve for I_2:
$$24\,\text{V} = 9\,\Omega\,(-1.5\,\text{A}) + 15\,\Omega\,(I_2)$$
$$I_2 = \frac{37.5\,\text{V}}{15\,\Omega} = 2.5\,\text{A}$$

Solve for the currents through the resistors:
$$I_{R_1} = I_1 = -1.5\,\text{A}$$
$$I_{R_2} = I_1 + I_2 = -1.5\,\text{A} + 2.5\,\text{A} = 1\,\text{A}$$
$$I_{R_3} = I_2 = 2.5\,\text{A}$$

Solve for the voltages across the resistors:
$$V_{R_1} = I_{R_1}R_1$$
$$= (-1.5\,\text{A})(4\,\Omega) = -6\,\text{V}$$
$$V_{R_2} = I_{R_2}R_2$$
$$= (1\,\text{A})(9\,\Omega) = 9\,\text{V}$$
$$V_{R_3} = I_{R_3}R_3 = (2.5\,\text{A})(6\,\Omega) = 15\,\text{V}$$

Answer:
$$V_{R_1} = -6\,\text{V}$$
$$V_{R_2} = 9\,\text{V}$$
$$V_{R_3} = 15\,\text{V}$$

In example 6-4, the negative sign for V_{R_1} tells us that the polarity for V_{R_1} in Fig. 6-6(b) is incorrect. The correct polarities and current directions are shown in Fig. 6-6(c). Note that B_1 is being charged because electrons are entering its negative terminal. Battery B_2 is furnishing all the power for the circuit.

As a final illustration of a multiple-source circuit, let us solve one that involves three loops. The general procedures are the same as those used for the bridge circuits except that this time we have two voltage sources. In this example, let us omit all units from the formulas for the sake of simplicity. Our answers for I will still be in amperes as long as circuit resistances are in ohms and circuit voltages are in volts.

EXAMPLE 6-5

Determine the current through each of the resistors in Fig. 6-7(a).

Given: Circuit diagram and values of Fig. 6-7(a).

Find: $I_{R_1}, I_{R_2}, I_{R_3}, I_{R_4}, I_{R_5}$

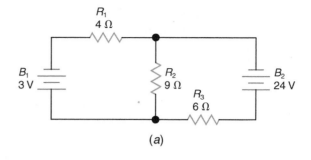

(a)

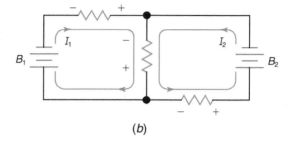

(b)

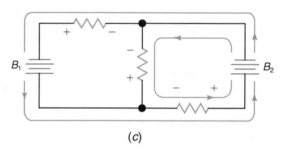

(c)

Fig. 6-6 Circuits for example 6-4. Battery B_1 is being charged. (a) Given circuit and values. (b) Loops used in circuit solution. (c) True polarities and current directions.

Known: Kirchhoff's laws and three-loop techniques

Solution: Specify loops and polarities [Fig. 6-7(b)]. Write loop equations:
$$44 = 10I_1 + 10I_2 + 20I_1 \quad \text{(loop 1)}$$
$$44 = 10I_1 + 10I_2 + 15I_2$$
$$\quad + 12I_2 + 12I_3 \quad \text{(loop 2)}$$
$$18 = 30I_3 + 12I_3 + 12I_2 \quad \text{(loop 3)}$$

Collect terms:
$$44 = 30I_1 + 10I_2 \quad \text{(loop 1)}$$
$$44 = 10I_1 + 37I_2 + 12I_3 \quad \text{(loop 2)}$$
$$18 = 12I_2 + 42I_3 \quad \text{(loop 3)}$$

Add loops 1 and 2 and eliminate I_1:
$$\frac{30I_1}{10I_1} \times (-1) = -3$$
$$-132 = -30I_1 - 111I_2 - 36I_3$$
$$\underline{44 = 30I_1 + 10I_2}$$
$$-88 = -101I_2 - 36I_3 \quad \text{(loops 1 and 2)}$$

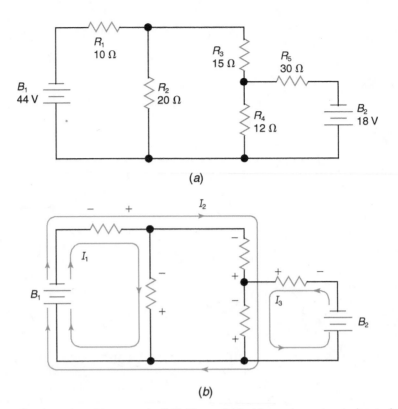

(a)

(b)

Fig. 6-7 Three-loop circuit analyzed in example 6-5. The selected loops turned out also to be the true current paths. (a) Given circuit and values. (b) Assumed loops and resulting polarities.

Add loops 1 and 2 to loop 3, eliminate I_3, and solve for I_2:

$$\frac{42I_3}{-36I_3} \times (-1) = 1.16$$

$$\begin{array}{r} -102.6 = -117.83I_2 - 42I_3 \\ 18 = 12I_2 + 42I_3 \\ \hline -84.6 = -105.83I_2 \\ I_2 = 0.8 \text{ A} \end{array}$$

Solve for I_1 and I_3:
(from loop 1 equation) $44 = 30I_1 + 10(0.8)$
$$I_1 = 1.2 \text{ A}$$
(from loop 3 equation) $18 = 12(0.8) + 42I_3$
$$I_3 = 0.2 \text{ A}$$
Solve for currents through each resistor:
$$I_{R_1} = I_1 + I_2 = 1.2 \text{ A} + 0.8 \text{ A} = 2 \text{ A}$$

$I_{R_2} = I_1 = 1.2 \text{ A}$
$I_{R_3} = I_2 = 0.8 \text{ A}$
$I_{R_4} = I_2 + I_3 = 0.8 \text{ A} + 0.2 \text{ A} = 1 \text{ A}$
$I_{R_5} = I_3 = 0.2 \text{ A}$

Answer: $I_{R_1} = 2 \text{ A}$,
$ I_{R_2} = 1.2 \text{ A}$,
$ I_{R_3} = 0.8 \text{ A}$,
$ I_{R_4} = 1 \text{ A}, I_{R_5} = 0.2 \text{ A}$

Note in example 6-5 that the true direction of the currents is as shown in Fig. 6-7(b) because I_1, I_2, and I_3 all came out positive. Thus, the polarity of the voltage drops in Fig. 6-7(b) is also correct.

Self-Test

Answer the following questions.

15. True or false. There can be more than one correct set of loop equations for a multiple-source circuit.

16. True or false. When the calculated value of a current is negative, the wrong loops have been used in analyzing the circuit.

17. Determine I_{R_1}, I_{R_2}, I_{R_3}, and V_{R_3} for the circuit shown in Fig. 6-8.

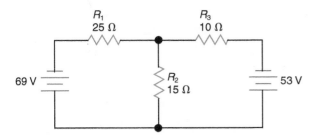

Fig. 6-8 Multiple-source circuit for Self-Test question 17.

6-3 Node Voltage Technique

In Sec. 6-2, we used Kirchhoff's voltage law to develop equations that enabled us to solve current values in a complex circuit. In this section, we will use Kirchhoff's current law to develop equations that allow us to determine the voltage at one or more *nodes* in a complex circuit.

Any point in a circuit where two or more components are joined is called a node. Minor nodes join only two components. Nodes joining three or more components are referred to as major nodes. In Fig. 6-9, all the nodes are labeled with capital letters (A through D). The node voltage technique requires that one node be designated as the reference node. Select the node that connects to the voltage sources for the reference node. The reference node in Fig. 6-9 is node A. All other node voltages are referenced to this node, so all other node voltages will have a polarity as well as a numerical value. Thus, in Fig. 6-9, $V_B = +25$ V, $V_C = +$ unknown, and $V_D = +15$ V. If a node voltage is specified without a polarity sign, it is understood to be a + value; that is, $V_D = 15$ V means $V_D = +15$ V.

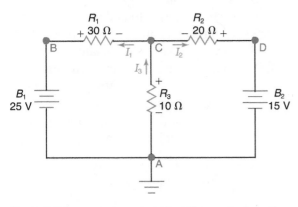

Fig. 6-9 The ground symbol identifies node A as the reference node.

After selecting the reference node, assign current directions to all nodes that have three or more component connections and an unknown voltage. Remember that the reference node voltage is zero so current direction assignments are not needed at this node. Therefore, current directions are assigned only at node C in Fig. 6-9. Also, the polarity of the voltage drops across the resistors (as dictated by the assigned current directions) are marked on Fig. 6-9. An assumed (assigned) current direction that is incorrect will be detected by a calculated current that has a negative value. Also, a calculated resistor voltage will have the wrong polarity to satisfy Kirchhoff's voltage law when applied to the various loops in the circuit. However, the calculated numerical (absolute) values will be correct.

Using Kirchhoff's current law and the assigned current direction, we can write the following equations for node C in Fig. 6-9:

$$I_3 = I_2 + I_1$$

Next, we can use Ohm's law and Kirchhoff's voltage law to express these currents in their $V \div R$ form.

$$I_3 = \frac{V_C}{R_3} = \frac{V_C}{10\ \Omega}$$

Should the expression for the voltage across R_2 be $V_C - V_D$ or $V_D - V_C$? This question is answered by inspecting the assigned current direction through R_2 in Fig. 6-9. For current to flow from node C to node D, node D has to be more positive (or less negative) than node C. Therefore, the expression for V_{R_2} should be $V_D - V_C$. Two general rules to follow when deciding which node voltage to subtract from are:

1. When both node voltages are positive, subtract the less positive node from the more positive node.
2. When both node voltages are negative, subtract the less negative node from the more negative node.

$$I_2 = \frac{V_{R_2}}{R_2} = \frac{V_D - V_C}{R_2} = \frac{15\ V - V_C}{20\ \Omega}$$

$$I_1 = \frac{V_{R_1}}{R_1} = \frac{V_B - V_C}{R_1} = \frac{25\ V - V_C}{30\ \Omega}$$

Now, by substituting these current-equivalent values into the current formula, we obtain the formula:

$$\frac{V_C}{10\,\Omega} = \frac{15\text{ V} - V_C}{20\,\Omega} + \frac{25\text{ V} - V_C}{30\,\Omega}$$

The only unknown in the above formula is V_C so we can rearrange it and solve for V_C as follows. By expanding to get

$$\frac{V_C}{10\,\Omega} = \frac{15\text{ V}}{20\,\Omega} - \frac{V_C}{20\,\Omega} + \frac{25\text{ V}}{30\,\Omega} - \frac{V_C}{30\,\Omega}$$

By rearranging to get

$$\frac{V_C}{30\,\Omega} + \frac{V_C}{20\,\Omega} + \frac{V_C}{10\,\Omega} = \frac{15\text{ V}}{20\,\Omega} + \frac{25\text{ V}}{30\,\Omega}$$

By finding a common denominator

$$\frac{2V_C}{60\,\Omega} + \frac{3V_C}{60\,\Omega} + \frac{6V_C}{60\,\Omega} = \frac{45\text{ V}}{60\,\Omega} + \frac{50\text{ V}}{60\,\Omega}$$

By collecting terms

$$\frac{11V_C}{60\,\Omega} = \frac{95\text{ V}}{60\,\Omega}$$

By cross multiplying

$$660\,\Omega \cdot V_C = 5700\,\Omega\text{V}$$

By dividing by $660\,\Omega$

$$V_C = 8.64\text{ V}$$

Since V_C is positive, the direction of I_3 is correct in Fig. 6-9. Now that we know all node voltages, we can calculate the values for the three currents.

When determining the voltage drop across a resistor, use the absolute value (absolute values have no + or − signs) of the node voltages. We can see why this is necessary if we consider the case where $V_B = -20$ V and $V_C = -12$ V. If we use absolute values and subtract V_C from V_B, we get 20 V − 12 V = 8 V, which is the correct value. But, using signed values we would get $(-20\text{ V}) - (-12\text{ V}) = -8$ V, which indicates an error because a voltage drop cannot be a signed value.

$$I_3 = \frac{V_C}{R_3} = \frac{8.64\text{ V}}{10\,\Omega} = 0.864\text{ A}$$

$$I_2 = \frac{V_D - V_C}{R_2} = \frac{15\text{ V} - 8.64\text{ V}}{20\,\Omega}$$
$$= \frac{6.36\text{ V}}{20\,\Omega} = 0.318\text{ A}$$

$$I_1 = \frac{V_B - V_C}{R_2} = \frac{25\text{ V} - 8.64\text{ V}}{30\,\Omega}$$
$$= \frac{16.36\text{ V}}{30\,\Omega} = 0.545\text{ A}$$

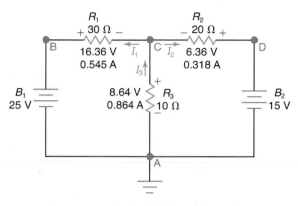

Fig. 6-10 The calculated voltages and currents have been added to the circuit shown in Fig. 6-9. They verify that Kirchhoff's laws are satisfied.

We can check our work for calculation errors by substituting the above current values into the original current formula:

$$I_3 = I_2 + I_1$$
$$0.864\text{ A} = 0.318\text{ A} + 0.545\text{ A} = 0.863\text{ A}$$

Finally, use the calculated resistor voltages, which have been added to Fig. 6-10, and Kirchhoff's voltage law to see that assigned current directions and voltage polarities are correct.

Several comments need to be made about the procedures used in solving for V_C in Fig. 6-9.

1. The units (Ω and V) were used throughout the calculations so that we could see why the correct unit for V_C was the volt. Now that we know our answer will be in volts when the resistances are in ohms, we will omit the units in solving other circuits by node-voltage analysis.
2. Instead of finding a common denominator before collecting terms, we could have converted all terms to decimal equivalents. This approach is often simpler than finding a common denominator.
3. When there is a convenient common denominator for the equation, it is not necessary to expand the equation. We can simply multiply the equation by the common denominator. For example:

$$\frac{V_C}{10} = \frac{15 - V_C}{20} + \frac{25 - V_C}{30}$$

Multiplied by 60 is

$$6V_C = 45 - 3V_C + 50 - 2V_C$$

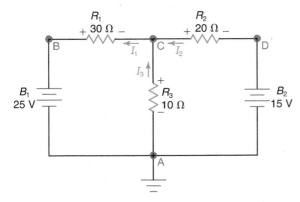

Fig. 6-11 This is Fig. 6-9 with the assumed direction of I_2 reversed. This also reverses the polarity of the voltage across R_2.

and collecting terms gives

$$11\,V_C = 95$$
$$V_C = 8.64 \text{ V}$$

Incorrect Current Direction

Figure 6-11 is the same as Fig. 6-9 except that the direction of I_2 and the polarity on R_2 have been reversed. Analyzing this circuit will illustrate the results of assigning an incorrect current direction. We can use the same procedure as we used for Fig. 6-9.

$$I_1 = I_2 + I_3$$
$$\frac{V_B - V_C}{30} = \frac{V_C - V_D}{20} + \frac{V_C}{10}$$

Notice that the expression for the voltage across R_2 is $V_C - V_D$.

$$\frac{25 - V_C}{30} = \frac{V_C - 15}{20} + \frac{V_C}{10}$$

$$\frac{25}{30} - \frac{V_C}{30} = \frac{V_C}{20} - \frac{15}{20} + \frac{V_C}{10}$$

$$-\frac{V_C}{20} - \frac{V_C}{10} - \frac{V_C}{30} = -\frac{15}{20} - \frac{25}{30}$$

$$-\frac{3\,V_C}{60} - \frac{6\,V_C}{60} - \frac{2\,V_C}{60} = -\frac{45}{60} - \frac{50}{60}$$

$$-\frac{11\,V_C}{60} = -\frac{95}{60}$$

$$660\,V_C = 5700$$
$$V_C = 8.64 \text{ V}$$

$$I_1 = \frac{25 - 8.64}{30} = 0.545 \text{ A}$$

$$I_2 = \frac{8.64 - 15}{20} = -0.318 \text{ A}$$

The − sign for I_2 tells us the assigned direction for I_2 in Fig. 6-11 is incorrect.

$$I_3 = \frac{8.64}{10} = 0.864 \text{ A}$$

Checking calculations:

$$I_1 = I_2 + I_3$$
$$0.545 \text{ A} = -0.318 \text{ A} + 0.864 \text{ A}$$

Although this equation is out of balance by 0.001 A, this small error is the result of rounding off answers in the caldulations used to obtain the three current values entered into the formula. Calculating voltage drops:

$$V_{R_1} = 25 - 8.64 = 16.36 \text{ V}$$
$$V_{R_2} - 8.64 - 15 = -6.36 \text{ V}$$

The − sign indicates the voltage polarity across R_2 is incorrect.

$$V_{R_3} = V_C = 8.64 \text{ V}$$

The above calculated voltage drops and currents are shown in Fig. 6-12. Notice that the polarity shown for the voltage across R_2 is correct for the assigned direction of current but is incorrect to satisfy Kirchhoff's voltage law. Compare the voltage and current values in Fig. 6-10 and Fig. 6-12. Notice that the absolute values are identical. The only differences in the values are the negative signs for the R_2 values. Now we can see that selecting the wrong direction for a current does not require additional or repeat calculations. Correcting the error just

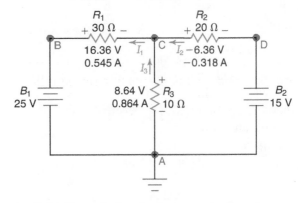

Fig. 6-12 The calculated voltages and currents have been added to the circuit shown in Fig. 6-11. Notice that Kirchhoff's laws are satisfied (the − signs have no meaning when referring to current and voltage drop).

requires dropping the − signs, changing a current arrow, and changing the polarity markings on one component.

EXAMPLE 6-6

Using the node voltage technique, determine the voltage at the junction of the three resistors and the current through each resistor in Fig. 6-13.

Given: Circuit diagram and values of Fig. 6-13.

Find: I_{R_1}, I_{R_2}, I_{R_3} and one node voltage.

Known: Kirchhoff's laws and node-voltage technique.

Solution: Label nodes, select and mark current directions, and label resistor-voltage polarities on Fig. 6-13. These are shown in Fig. 6-14(a).

Write equations:

$$I_3 = I_2 + I_1$$
$$\frac{V_B}{R_3} = \frac{V_C - V_B}{R_2} + \frac{V_A - V_B}{R_1}$$

Enter known values:

$$\frac{V_B}{6} = \frac{-20 - V_B}{8} + \frac{-60 - V_B}{8}$$

Multiply by common denominator (48):

$$8V_B = -120 - 6V_B + (-360) - 6V_B$$

Collect terms and solve V_B:

$$20V_B = -480$$
$$V_B = -24 \text{ V}$$

Calculate the currents (use absolute values):

$$I_3 = \frac{24}{6} = 4 \text{ A}$$

$$I_2 = \frac{20 - 24}{8} = \frac{-4}{8} = -0.5 \text{ A}$$

$$I_1 = \frac{60 - 24}{8} = \frac{36}{8} = 4.5 \text{ A}$$

Note: The negative value of I_2 tells us that the assumed direction of I_2 and the polarity shown on R_2 in Fig. 6-14 are incorrect and need to be reversed. However, the absolute values are correct.

Answer: $V_B = -24 \text{ V}$
$I_{R_1} = 4.5 \text{ A}$
$I_{R_2} = 0.5 \text{ A}$
$I_{R_3} = 4 \text{ A}$

Notice in example 6-6 that battery B_2 is receiving energy from battery B_1. Saying it another way, battery 2 is being recharged if it is rechargeable.

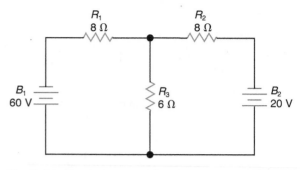

Fig. 6-13 Circuit for example 6-6.

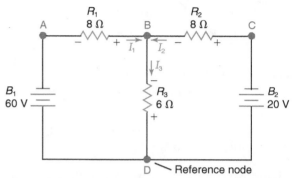

(a) Nodes, current direction, and polarities added to circuit in Fig. 6-13.

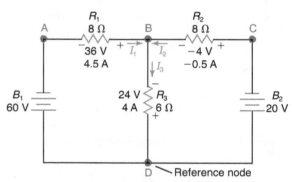

(b) Negative values associated with R_2 show that I_2 flows in the opposite direction and the polarity for R_2 is reversed.

Fig. 6-14 Circuit additions for Fig. 6-13 used in example 6-6.

Node voltage analysis can be used on circuits with more than one unknown node voltage. We can analyze a circuit with two unknown node voltages by writing equations for each node using the same methods and rules we used for circuits with one unknown node voltage. This will give us two independent equations with the two unknown voltages in each equation. Then we can use the simultaneous-equation method (developed in Sec. 6-1) to solve for the two unknowns.

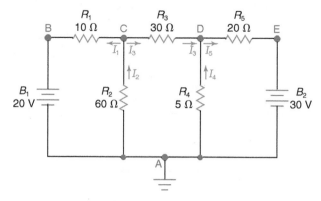

Fig. 6-15 Circuit with two unknown node voltages.

We will use the circuit in Fig. 6-15 to illustrate the procedures for analyzing circuits with two unknown node voltages. Start by assigning current directions at the two nodes with unknown voltages (see Fig. 6-15). Next, write the equations for node C.

$$I_2 = I_1 + I_3$$

$$\frac{V_C}{R_2} = \frac{V_B - V_C}{R_1} + \frac{V_D - V_C}{R_3}$$

$$\frac{V_C}{60} = \frac{20 - V_C}{10} + \frac{V_D - V_C}{30}$$

Multiply by 60

$$V_C = 120 - 6V_C + 2V_D - 2V_C$$

$$9V_C = 120 + 2V_D \qquad \text{(equation #1)}$$

Next, write the equation for node D.

$$I_5 = I_3 + I_4$$

$$\frac{V_E - V_D}{R_5} = \frac{V_D - V_C}{R_3} + \frac{V_D}{R_4}$$

$$\frac{30 - V_D}{20} = \frac{V_D - V_C}{30} + \frac{V_D}{5}$$

Multiply by 60

$$90 - 3V_D = 2V_D - 2V_C + 12V_D$$

$$2V_C = -90 + 17V_D \qquad \text{(equation # 2)}$$

Now multiply equation #1 by -2 and equation #2 by 9, and then add the results to eliminate V_C.

$$-18V_C = -240 - 4V_D$$

$$18V_C = -810 + 153V_D$$

$$0 = -1050 + 149V_D$$

$$V_D = 7.047 \text{ V}$$

Next use this value of V_D in equation #1 and solve for V_C.

$$9V_C = 120 + (2 \times 7.047) = 134.094$$

$$V_C = 14.899 \text{ V}$$

Finally, we can use these values of V_C and V_D to solve for the currents.

$$I_1 = \frac{20 \text{ V} - 14.899 \text{ V}}{10 \ \Omega} = 0.510 \text{ A}$$

$$I_2 = \frac{14.899 \text{ V}}{60 \ \Omega} = 0.248 \text{ A}$$

$$I_3 = \frac{7.047 \text{ V} - 14.899 \text{ V}}{30 \ \Omega} = -0.262 \text{ A}$$

$$I_4 = \frac{7.047 \text{ V}}{5 \ \Omega} = 1.409 \text{ A}$$

$$I_5 = \frac{30 \text{ V} - 7.047 \text{ V}}{20 \ \Omega} = 1.148 \text{ A}$$

Again, the negative answer for I_3 tells us that the indicated current direction for I_3 is incorrect and must be reversed. This, of course, will also reverse the polarity of the voltage drop across R_3.

Self-Test

Answer the following questions.

18. True or false. A junction of two components should not be referred to as a node.

19. What law is used to write the first equation needed to determine an unknown node voltage?

20. True or false. For circuits like the one in Fig. 6-9, solving the node C voltage allows one to determine all other circuit currents and voltages.

21. True or false. When the node voltage technique is used, an incorrect direction of current is detected by checking with Kirchhoff's current law.

22. Using node-voltage analysis, determine the node B voltage for the circuit in Fig. 6-16.

23. Does the current through R_2 in Fig. 6-16 flow left to right or right to left?
24. Do the signed values of the currents calculated for Fig. 6-15 satisfy Kirchhoff's current law?
25. After the direction of I_3 is changed and the minus sign for I_3 is dropped, do the currents in Fig. 6-15 satisfy Kirchhoff's current law?
26. Is Kirchhoff's voltage law satisfied when the assumed current directions and the absolute current values are used to calculate the voltage drops in Fig. 6-15?

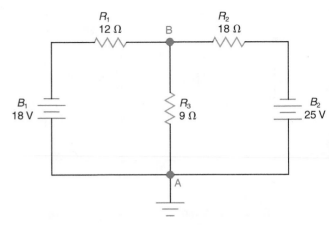

Fig. 6-16 Circuit for Self-Test questions 22 and 23.

6-4 Superposition Theorem

Algebraically add

Superposition theorem

Another approach to analyzing multiple-source circuits involves the use of the *superposition theorem*. The general idea of this theorem is that the overall effect of all sources on a circuit is the sum of effects of the individual sources. Since each voltage source has an effect on the circuit currents, the circuit currents are analyzed once for each voltage source. Then, the currents resulting from each source are added together to arrive at the final circuit currents. Once the currents are known, it is a simple matter of applying Ohm's law to find the voltage drops in the circuit.

The specific steps used in applying the superposition theorem are as follows:

1. If the internal resistances of the voltage sources are very small compared with the external resistance in the circuit, which is the case in most electronic circuits, replace every voltage source but one with a *short circuit* (conductor). If the internal resistance is a significant part (1 percent or more) of the total circuit resistance, replace the source with a resistor equal to its internal resistance.
2. Calculate the magnitude and direction of the current through each resistor in the temporary circuit produced by step 1. (Use the procedures developed in an earlier chapter for series-parallel circuits when making these calculations.)
3. Repeat steps 1 and 2 until each voltage source has been used as the *active source*.

Short circuit

Active source

4. *Algebraically add* all the currents (from step 2) through a specific resistor to determine the magnitude of the current through that resistor in the original circuit. The direction of the current is the direction of the dominant current through the resistor.

Now, let us apply these four steps to the circuit of Fig. 6-17(a) so that we can observe in detail how the superposition theorem works. Applying step 1 by shorting out B_1 yields the temporary circuit shown in Fig. 6-17(b). The currents shown in Fig. 6-17(b) result from following step 2. The calculations involved are

$$R_{1,2} = \frac{R_1 R_2}{R_1 + R_2}$$

$$= \frac{(40\ \Omega)(10\ \Omega)}{40\ \Omega + 10\ \Omega} = 8\ \Omega$$

$$R_T = R_{1,2} + R_3$$

$$= 8\ \Omega + 30\ \Omega = 38\ \Omega$$

$$I_{R_3} = I_T$$

$$= \frac{V_T}{R_T}$$

$$= \frac{38\ V}{38\ \Omega}$$

$$= 1\ A$$

$$I_{R_2} = \frac{I_T R_1}{R_1 + R_2}$$

$$= \frac{(1\ A)(40\ \Omega)}{40\ \Omega + 10\ \Omega}$$

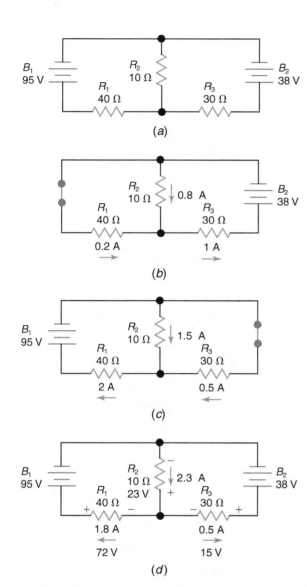

Fig. 6-17 Using the superposition theorem to analyze a multiple-source circuit. Currents and voltage drops for the original circuit (a) are shown in (d). (a) Original circuit. (b) Shorting B_1 puts R_1 and R_2 in parallel. (c) Shorting B_2 puts R_2 and R_3 in parallel. (d) Final results.

$$= 0.8 \text{ A}$$
$$I_{R_1} = I_{R_3} - I_{R_2}$$
$$= 1 \text{ A} - 0.8 \text{ A}$$
$$= 0.2 \text{ A}$$

Going on to step 3 leads us back to steps 1 and 2 for the second time. The second time through steps 1 and 2 results in the circuit and current values shown in Fig. 6-17(c). The currents were calculated by using the procedure outlined for Fig. 6-17(b). We have now used each of the sources as the active source, and so we move on to step 4. Following step 4 leads to the currents shown in Fig. 6-17(d). These currents show the magnitude and direction of the currents in the original circuit. They were obtained by algebraically adding the appropriate currents from Fig. 6-17(b) and (c). When the currents are in opposite directions, the smaller current is assigned a minus sign because the larger current shows the direction of current in the original circuit. The calculations are:

$$I_{R_1} = 2 \text{ A} - 0.2 \text{ A} = 1.8 \text{ A}$$
$$I_{R_2} = 1.5 \text{ A} + 0.8 \text{ A} = 2.3 \text{ A}$$
$$I_{R_3} = 1 \text{ A} - 0.5 \text{ A} = 0.5 \text{ A}$$

Also shown in Fig. 6-17(d) are the voltage drops, which were calculated with Ohm's law. We can use these voltage drops and Kirchhoff's voltage law to verify that our analysis is correct. For example,

$$95 \text{ V} = 23 \text{ V} + 72 \text{ V} = 38 \text{ V} - 15 \text{ V} + 72 \text{ V}$$
$$38 \text{ V} = 23 \text{ V} + 15 \text{ V} = 95 \text{ V} - 72 \text{ V} + 15 \text{ V}$$

The use of the superposition theorem allows us to analyze Fig. 6-17(a) without solving simultaneous equations. However, it does require repeated use of series-parallel techniques.

Self-Test

Answer the following questions.

27. Change R_1 in Fig. 6-6(a) to 8 Ω and then use the superposition theorem to find I_{R_1}, I_{R_2}, and I_{R_3}.

28. Can the superposition theorem be used with single-source circuits?

29. True or false. Application of the superposition theorem always leads to the correct direction of currents through all resistors.

Three-Loop Circuits

The superposition theorem can be applied to *three-loop circuits* like the one shown in Fig. 6-7. For example, shorting out B_1 sets up a series-parallel circuit where R_T can be found by applying these formulas:

$$R_{1,2} = \frac{R_1 R_2}{R_1 + R_2}$$

$$R_{1,2,3} = R_{1,2} + R_3$$

$$R_{1,2,3,4} = \frac{(R_{1,2,3})(R_4)}{R_{1,2,3} + R_4}$$

$$R_T = R_{1,2,3,4} + R_5$$

The remainder of the analysis of Fig. 6-7 by the superposition theorem is left up to you. You can check your answers against the answers given in example 6-5.

So far we have not examined any circuits that involve three sources. Such circuits are easily solved using either the loop techniques of Sec. 6-2 or the superposition theorem. We

will analyze a *three-source circuit* using the superposition theorem.

EXAMPLE 6-7

Using the superposition theorem, calculate the current through every resistor in Fig. 6-18(a).

Given: Circuit diagram and values of Fig. 6-18(a).

Find: I_{R_1}, I_{R_2}, I_{R_3}, and I_{R_4}

Known: Superposition theorem, series-parallel circuit techniques

Solution: Short out B_2 and B_3 and solve for all currents:

$$R_{1,2} = \frac{R_1 R_2}{R_1 + R_2}$$

$$= \frac{(10)(15)}{10 + 15} = 6\ \Omega$$

$$R_{3,4} = \frac{R_3 R_4}{R_3 + R_4}$$

$$= \frac{(12)(20)}{12 + 20} = 7.5\ \Omega$$

$$R_T = R_{1,2} + R_{3,4}$$
$$= 6 + 7.5 = 13.5\ \Omega$$

$$I_T = \frac{V_T}{R_T} = \frac{55\ \text{V}}{13.5\ \Omega}$$

$$= 4.074\ \text{A}$$

$$I_{R_1} = \frac{I_T R_2}{R_1 + R_2}$$

$$= \frac{(4.074)(15)}{10 + 15} = 2.444\ \text{A} \downarrow$$

The arrow indicates current direction.

$$I_{R_2} = I_T - I_{R_1}$$
$$= 4.074 - 2.444 = 1.630\ \text{A} \downarrow$$

$$I_{R_3} = \frac{I_T R_4}{R_3 + R_4}$$

$$= \frac{(4.074)(20)}{12 + 20} = 2.546\ \text{A} \downarrow$$

$$I_{R_4} = I_T - I_{R_3}$$
$$= 4.074 - 2.546 = 1.528\ \text{A} \downarrow$$

Next, short out B_1 and B_2 and solve all for currents:

$$R_{1,2,4} = \frac{1}{\dfrac{1}{R_1} + \dfrac{1}{R_2} + \dfrac{1}{R_4}}$$

$$R_{1,2,4} = \frac{1}{\dfrac{1}{10} + \dfrac{1}{15} + \dfrac{1}{20}} = 4.615\ \Omega$$

$$R_T = R_{1,2,4} + R_3 = 4.615 + 12$$
$$= 16.615\ \Omega$$

$$I_T = \frac{V_T}{R_T} = \frac{58\ \text{V}}{16.615\ \Omega} = 3.491\ \text{A}$$

$$V_{R_3} = I_{R_3} R_3 = I_T R_3 = (3.491\ \text{A})(12\ \Omega)$$
$$= 41.892\ \text{V}$$

$$V_{R_1} = V_{R_2} = V_{R_4} = V_T - V_{R_3}$$
$$= 58 - 41.892 = 16.108\ \text{V}$$

$$I_{R_1} = \frac{V_{R_1}}{R_1} = \frac{16.108\ \text{V}}{10\ \Omega} = 1.611\ \text{A} \uparrow$$

$$I_{R_2} = \frac{V_{R_2}}{R_2} = \frac{16.108\ \text{V}}{15\ \Omega} = 1.074\ \text{A} \uparrow$$

$$I_{R_3} = I_T = 3.491\ \text{A} \uparrow$$

$$I_{R_4} = \frac{V_{R_4}}{R_4} = \frac{16.108\ \text{V}}{20\ \Omega} = 0.805\ \text{A} \downarrow$$

Next, short out B_1 and B_3 and again solve for all currents:

$$R_{1,3,4} = \frac{1}{\dfrac{1}{R_1} + \dfrac{1}{R_3} + \dfrac{1}{R_4}} = \frac{1}{\dfrac{1}{10} + \dfrac{1}{12} + \dfrac{1}{20}}$$

$$= 4.286\ \Omega$$

$$R_T = R_{1,3,4} + R_2 = 19.286\ \Omega$$

$$I_T = \frac{V_T}{R_T} = \frac{30\ \text{V}}{19.286\ \Omega} = 1.556\ \text{A}$$

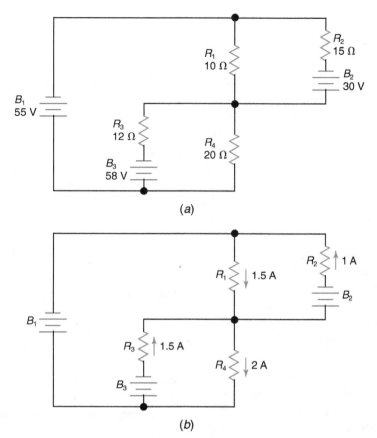

Fig. 6-18 Three-source circuit for example 6-7. Magnitude and direction of current are shown in (b). (a) Given circuit and values. (b) Final results of analysis.

$$V_{R_2} = I_{R_2}R_2 = I_T R_2 = (1.556 \text{ A})(15 \ \Omega)$$
$$= 23.34 \text{ V}$$
$$V_{R_1} = V_{R_3} = V_{R_4} = V_T - V_{R_2} = 30 - 23.34$$
$$= 6.66 \text{ V}$$
$$I_{R_1} = \frac{V_{R_1}}{R_1} = \frac{6.66 \text{ V}}{10 \ \Omega} = 0.666 \text{ A} \downarrow$$
$$I_{R_2} = I_T = 1.556 \text{ A} \uparrow$$
$$I_{R_3} = \frac{V_{R_3}}{R_3} = \frac{6.66 \text{ V}}{12 \ \Omega} = 0.555 \text{ A} \uparrow$$
$$I_{R_4} = \frac{V_{R_4}}{R_4} = \frac{6.66 \text{ V}}{20 \ \Omega} = 0.333 \text{ A} \uparrow$$

Finally, we must algebraically add the individual currents:

$$I_{R_1} = \quad 2.444 - 1.611 + 0.666 = 1.5 \text{ A} \downarrow$$
$$I_{R_2} = -1.630 + 1.074 + 1.556 = 1 \text{ A} \uparrow$$
$$I_{R_3} = -2.546 + 3.491 + 0.555 = 1.5 \text{ A} \uparrow$$
$$I_{R_4} = \quad 1.528 + 0.805 - 0.333 = 2 \text{ A} \downarrow$$

Answer: $I_{R_1} = 1.5 \text{ A}, I_{R_2} = 1 \text{ A}$
$I_{R_3} = 1.5 \text{ A}, I_{R_4} = 2 \text{ A}$

We could cross-check the answers to example 6-7 with Kirchhoff's current law and/or Kirchhoff's voltage law. The cross-check is left up to you.

Self-Test

Answer the following questions.

30. Apply Kirchhoff's current law to the junction of R_1, R_3, R_4, and B_2 in Fig. 6-18.

31. How much current does B_1 provide in Fig. 6-18?

6-5 Voltage Sources

There are two ways of looking at a voltage source. It can be considered a *constant voltage source* (also called an *ideal voltage source*), or it can be considered a constant voltage source in series with some internal resistance. This latter approach is referred to as an *equivalent-circuit voltage source*. The way in which we view a voltage source depends on both the nature of the source and the nature of the circuit to which the source is connected. Actual voltage sources are not true constant voltage sources; however, for practical applications, many voltage sources are assumed to be constant voltage sources.

If a constant (ideal) voltage source were possible, it would have no internal resistance; therefore, the terminal voltage would be constant regardless of the amount of current drawn from the source. The *open-circuit terminal voltage* (V_{oc}) of Fig. 6-19(a) would be the same as the *closed-circuit terminal voltage* of Fig. 6-19(b). Of course, such a voltage source does not exist because all voltage sources have some internal resistance. The conductors in the coils of a generator have some resistance, the chemicals in the lead-acid cell have some resistance, and so forth. Thus, the terminal voltage of all voltage sources must decrease when the current drawn from the source is increased. Viewing a battery as a constant voltage source when analyzing circuits introduces some error.

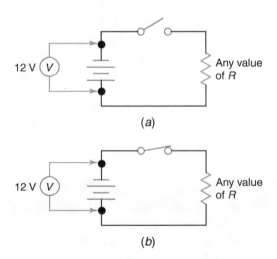

(a)

(b)

Fig. 6-19 An ideal voltage source. Such voltage sources do not actually exist. (a) Open-circuit terminal voltage (V_{oc}) equals 12 V. (b) Closed-circuit (loaded) terminal voltage equals 12 V.

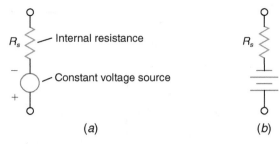

(a)   (b)

Fig. 6-20 Equivalent circuit of a voltage source. (a) Equivalent circuit using a constant voltage source symbol. (b) Alternative circuit using a battery symbol.

In most cases, however, the error is less than that caused by rounding off numbers when doing the calculations. With but one exception (Sec. 5-4), we have assumed that all of our circuits are powered by constant voltage sources. (The battery symbol, with a voltage specified, implies a constant voltage source.) This is standard practice as long as the internal resistance of the source is less than 1 percent of the total circuit resistance.

When the *internal resistance* of a voltage source becomes a significant part of the total circuit resistance, we must take it into account when analyzing the circuit. To do this, we replace the simple battery symbol with its true equivalent circuit. That is, we view the source as an equivalent-circuit voltage source. The equivalent circuit of a voltage source is shown in Fig. 6-20(a). In this figure, the circle with the polarity markings represents the constant voltage source. Its value is the open-circuit voltage (V_{oc}) of the voltage source. It can be thought of as the voltage produced by the chemicals from which a cell is constructed. The voltage produced by the chemical reaction in the cell is indeed independent of the current. The internal resistance (R_S) represents the resistance of the chemicals in the cell. When current flows through the cell, this resistance drops part of the voltage produced by the chemical reaction. When dc circuits are being analyzed, the polarized circle of Fig. 6-20(a) is sometimes replaced by a battery symbol, as shown in Fig. 6-20(b). The small circles at the end of the equivalent circuits in Fig. 6-20 represent the terminals of the voltage source. Sometimes they are omitted.

EXAMPLE 6-8

A voltage source has an open-circuit voltage of 30 V. Its internal resistance is 1 Ω. Calculate the terminal voltage when this source is connected to a 10-Ω load.

Given: $V_{oc} = 30$ V, $R_S = 1$ Ω, $R_L = 10$ Ω
("R_L" stands for "load resistor")

Find: Closed-circuit terminal voltage for a specified load resistance

Known: Equivalent circuit for a source and voltage-divider formula

Solution: Draw the schematic diagram as shown in Fig. 6-21. Then use the voltage-divider equation to determine V_{RL}:

$$V_{RL} = \frac{(30 \text{ V})(R_L)}{R_L + R_S} = \frac{(30 \text{ V})(10 \text{ Ω})}{10 \text{ Ω} + 1 \text{ Ω}}$$

Fig. 6-21 Circuit diagram for example 6-8.

$= 27.27$ V

Since R_L is connected directly to the source terminals, the closed-circuit terminal voltage equals V_{RL}.

Answer: The terminal voltage with a 10-Ω load is 27.27 V.

Self-Test

Answer the following questions.

32. What determines whether a voltage source should be assumed to be constant?

33. Whether we view a voltage source as constant or as an equivalent circuit (constant voltage source in series with its internal resistance), the voltage of the constant voltage source is equal to the _____.

34. True or false. Batteries are never assumed to be a constant voltage source.

6-6 Thevenin's Theorem

Any of the circuits we have been working with can be viewed as a *two-terminal network*. For example, the circuit in Fig. 6-22(a) can be assigned two terminal locations as shown in Fig. 6-22(b). Between these two terminals a voltage exists; however, the amount of voltage depends on the amount of current we draw from the terminals. Thus, we can view everything to the left of the terminals, as shown in Fig. 6-22(c), as a voltage source. Of course, this voltage source must be viewed as an equivalent-circuit voltage source because its internal resistance is quite large. The location of the two terminals in a circuit is entirely up to the person analyzing the circuit. The location of the terminals for the circuit of Fig. 6-22(a) could just as well have been as shown in Fig. 6-22(d).

Thevenin's theorem provides us with an easy way to develop an equivalent circuit of a two-terminal network. This equivalent circuit will, of course, be an equivalent-circuit voltage source like that shown in Fig. 6-20(a). Applying Thevenin's theorem to a two-terminal network requires only two steps:

1. Determine the open-circuit voltage between the terminals. This can be done either by calculating the voltage using the schematic diagram or by measuring the voltage if the circuit actually exists.

2. Determine the internal resistance of the equivalent-circuit voltage source after replacing all voltage sources in the original circuit with resistances equal to their internal resistance. When the original voltage sources are viewed as ideal sources,

Thevenin's theorem

Two-terminal network

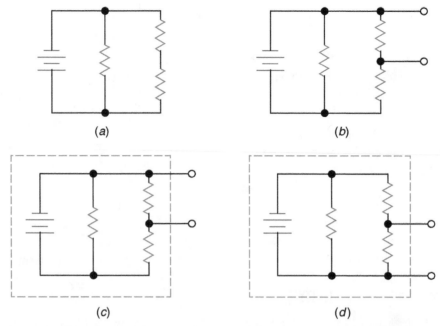

Fig. 6-22 Viewing a circuit as a two-terminal voltage source. (*a*) Original circuit. (*b*) Two terminals added. (*c*) Voltage source with large internal resistance. (*d*) Alternative location of terminals.

they are short-circuited. After the original voltage sources are replaced by conductors, the internal resistance (total resistance between the terminals) can be either calculated or measured.

Equivalent circuits developed by Thevenin's theorem are often called *Thevenin equivalent circuits;* the original circuit is said to have been *thevenized.* The constant voltage source determined in step 1 is usually labeled V_{TH} instead of V_{oc}, and the internal resistance of step 2 is R_{TH} instead of R_S.

Let us try an example using Thevenin's theorem.

EXAMPLE 6-9

Determine the Thevenin equivalent circuit for the circuit in Fig. 6-23(*a*) when the terminals are considered to be the two ends of R_2.

Given: Circuit diagram and values in Fig. 6-23(*a*)
Find: V_{TH} and R_{TH}
Known: Thevenin's theorem
Solution: Calculate V_{TH} (note that V_{TH} is the same as V_{R_2}):

$$V_{TH} = \frac{V_{B_1} R_2}{R_1 + R_2}$$

$$= \frac{(30 \text{ V})(1000 \text{ }\Omega)}{500 \text{ }\Omega + 1000 \text{ }\Omega} = 20 \text{ V}$$

Calculate R_{TH} (note that shorting out V_{B_1} puts R_1 in parallel with R_2):

$$R_{TH} = \frac{R_1 R_2}{R_1 + R_2}$$

$$= \frac{(500 \text{ }\Omega)(1000 \text{ }\Omega)}{500 \text{ }\Omega + 1000 \text{ }\Omega} = 333.3 \text{ }\Omega$$

Answer: The Thevenin equivalent circuit is drawn in Fig. 6-23(*b*).

$$V_{TH} = V_{oc} = 20 \text{ V}$$
$$R_{TH} = R_S = 333.3 \text{ }\Omega$$

Any load connected to the terminals of the original circuit in Fig. 6-23(*a*) has the same current, voltage, and power as the same load connected to the Thevenin equivalent circuit of Fig. 6-23(*b*). We can prove this statement by applying a 1000-Ω load to each circuit, as shown in Fig. 6-23(*c*) and (*d*). For the loaded Thevenin equivalent circuit [Fig. 6-23(*d*)], the calculations are

$$V_{R_L} = \frac{V_{TH} R_L}{R_{TH} + R_L} = \frac{(20 \text{ V})(1000 \text{ }\Omega)}{333.3 \text{ }\Omega + 1000 \text{ }\Omega}$$

$$= 15 \text{ V}$$

$$I_{R_L} = \frac{V_{R_L}}{R_L} = \frac{15 \text{ V}}{1000 \text{ }\Omega} = 15 \text{ mA}$$

Thevenin equivalent circuits

Thevenized

V_{TH}

R_{TH}

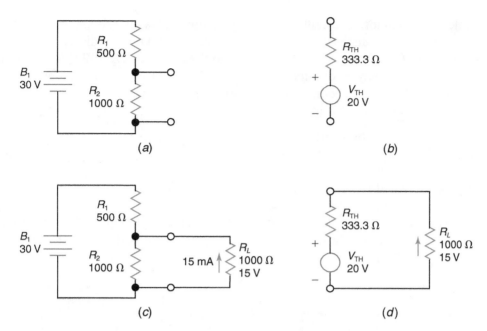

Fig. 6-23 Thevenin equivalent circuit without and with a load. (a) Original circuit (voltage divider). (b) Thevenin equivalent circuit. (c) Original circuit with a load added. (d) Thevenin equivalent circuit with a load added.

For the loaded original circuit shown in Fig. 6-23(c), the calculations are

$$R_{2,L} = \frac{R_2 R_L}{R_2 + R_L} = \frac{(1000\ \Omega)(1000\ \Omega)}{1000\ \Omega + 1000\ \Omega}$$
$$= 500\ \Omega$$

$$V_{RL} = V_{R2,L} = \frac{V_{B_1} R_{2,L}}{R_1 + R_{2,L}} = \frac{(30\ \text{V})(500\ \Omega)}{500\ \Omega + 500\ \Omega}$$
$$= 15\ \text{V}$$

$$I_{RL} = \frac{V_{RL}}{R_L} = \frac{15\ \text{V}}{1000\ \Omega} = 15\ \text{mA}$$

From the above calculations, you can see that the calculations for solving the loaded original circuit are more difficult than for solving the loaded Thevenin equivalent circuit. The level of difficulty becomes more pronounced as the original circuit becomes more complex. Now you can see the advantage of thevenizing a circuit if you need to know the load voltage for many different values of R_L.

The preceding illustration shows how to thevenize a circuit and then determine the effects of adding a load resistance. Next, we will see how to use a Thevenin equivalent circuit to analyze a complex circuit. The general procedure involves selecting one resistor in the complex circuit to act as a load and then thevenizing

the remainder of the circuit. The specific steps for using Thevenin's theorem to analyze a complex circuit are as follows:

1. Select one of the resistors in the complex circuit to be removed from the circuit. This resistor will later be added back to the Thevenin equivalent circuit and treated like the load resistor as in example 6-9. If possible, select this resistor so that the remainder of the original circuit can be analyzed using series, parallel, or series-parallel rules.
2. Analyze the remainder of the circuit to determine the open-circuit voltage (V_{TH}) at the terminals created by step 1.
3. Analyze the remainder of the circuit to determine the internal resistance (R_{TH}).
4. Draw the Thevenin equivalent circuit (using the values obtained in steps 2 and 3) and load it with the resistor removed in step 1.
5. Determine the voltage across and the current through the resistor added in step 4.
6. Insert the resistor removed in step 1 back into the original circuit. The current and voltage of the reinserted resistor are the same in the original circuit as in the loaded Thevenin equivalent circuit of step 5.

7. If possible, use Kirchhoff's current and voltage laws to calculate the other currents and voltages of the complex circuit. Steps 6 and 7 may be omitted if you are interested only in the current and voltage of the resistor removed in step 1.

Let us illustrate these steps by evaluating the complex circuit of Fig. 6-24(a). The result of applying step 1 is shown in Fig. 6-24(b), where terminals have been added to the points from which R_3 was removed. Removal of R_3 results in a simple series circuit with two opposing source voltages. Of course, the larger voltage

(B_2) dominates the circuit and forces current through B_1 in a reverse direction. Applying step 2 to the circuit in Fig. 6-24(b) leads to the voltages specified in Fig. 6-24(c), which were calculated as follows:

$$V_T = B_2 - B_1 = 19 \text{ V} - 10 \text{ V} = 9 \text{ V}$$

$$V_{R_2} = \frac{V_T R_2}{R_1 + R_2} = \frac{(9 \text{ V})(20 \ \Omega)}{10 \ \Omega + 20 \ \Omega} = 6 \text{ V}$$

$$V_{TH} = B_2 - V_{R_2} = 19 \text{ V} - 6 \text{ V} = 13 \text{ V}$$

As called for in step 3, the circuit of Fig. 6-24(b) was modified to arrive at the circuit in

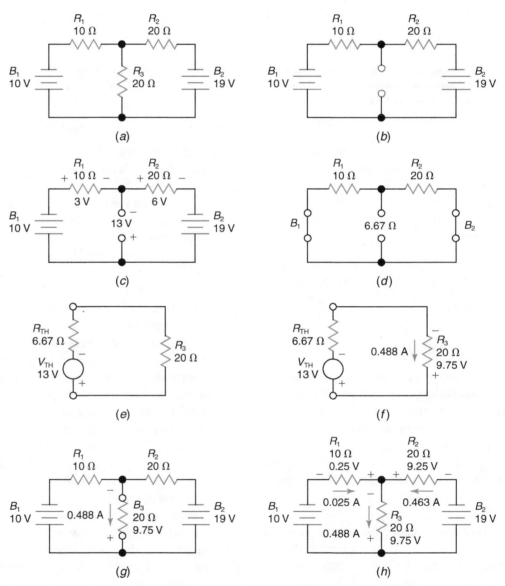

Fig. 6-24 Analyzing a complex circuit by thevenizing the circuit. (a) Original complex circuit. (b) Step 1— remove R_3. (c) Step 2—calculate V_{TH}. (d) Step 3—calculate R_{TH}. (e) Step 4—loaded Thevenin equivalent circuit. (f) Step 5—calculate V_{R_3} and I_{R_3}. (g) Step 6—replace R_3. (h) Step 7—calculate other currents and voltages.

Fig. 6-24(d) so that R_{TH} could be calculated. As seen from the two terminals, R_1 and R_2 in Fig. 6-24(d) are in parallel. Thus, R_{TH} is

$$R_{TH} = \frac{R_1 R_2}{R_1 + R_2} = \frac{(10\ \Omega)(20\ \Omega)}{10\ \Omega + 20\ \Omega} = 6.67\ \Omega$$

Next, step 4 tells us to draw the Thevenin equivalent circuit for the circuit of Fig. 6-24(b) and load it with R_3. This step is illustrated in Fig. 6-24(e). Then, as called for in step 5 and illustrated in Fig. 6-24(f), the voltage and current associated with R_3 are calculated:

$$V_{R_3} = \frac{V_{TH} R_3}{R_{TH} + R_3} = \frac{(13\ V)(20\ \Omega)}{6.67\ \Omega + 20\ \Omega}$$

$$= 9.75\ V$$

$$I_{R_3} = \frac{V_{R_3}}{R_3} = \frac{9.75\ V}{20\ \Omega} = 0.488\ A$$

Following through with step 6 yields the conditions shown in Fig. 6-24(g). Step 7 yields the values of additional voltages and currents shown in Fig. 6-24(h), calculated as follows:

$$V_{R_1} = B_1 - V_{R_3} = 10\ V - 9.75\ V = 0.25\ V$$

$$I_{R_1} = \frac{V_{R_1}}{R_1} = \frac{0.25\ V}{10\ \Omega} = 0.025\ A$$

$$V_{R_2} = B_2 - V_{R_3} = 19\ V - 9.75\ V = 9.25\ V$$

$$I_{R_2} = I_{R_3} - I_{R_1} = 0.488\ A - 0.025\ A$$

$$= 0.463\ A$$

You may wonder why the last sentence in step 1 starts out "If possible." Studying Fig. 6-25 should reveal that the removal of any one resistor still leaves a complex circuit. Such circuits can still be thevenized, but one of the methods we have learned for analyzing a complex circuit must be used to calculate R_{TH} or V_{TH} or both.

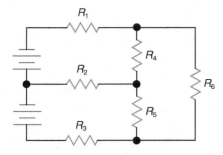

Fig. 6-25 Removal of any one resistor still leaves a complex circuit for which R_S and V_{oc} cannot be determined by applying series-parallel rules.

Note that step 7 also starts out "If possible." The bridge circuit of Fig. 6-4(a) shows that knowing the current and voltage for only one of five resistors does not allow solving the other currents and voltages by using series-parallel rules or Kirchhoff's laws. All currents and voltages for the bridge circuit could be found by applying Thevenin's theorem to the circuit two times. With the first application, find V_{R_5} and I_{R_5}. With the second application, find the current and voltage for any one of the other four resistors. Then Kirchhoff's laws can be used to find the remaining currents and voltages.

As mentioned above, thevenizing a circuit sometimes requires the use of other techniques, such as loop equations, to determine V_{TH}.

EXAMPLE 6-10

Use Thevenin's theorem to find the voltage across each resistor in Fig. 6-26(a).

Given:	Complex circuit and values of Fig. 6-26(a)
Find:	V_{R_1}, V_{R_2}, V_{R_3}, V_{R_4}
Known:	Thevenin's theorem and loop-equations technique
Solution:	Remove R_4 as in Fig. 6-26(b). Define loops [Fig. 6-26(c)] and solve for I_1:

$$58 = 12I_1 + 10I_1 - 10I_2 + 55 \quad \text{(loop 1)}$$
$$3 = 22I_1 - 10I_2 \quad \text{(loop 1)}$$
$$30 = 15I_2 + 10I_2 - 10I_1 \quad \text{(loop 2)}$$
$$30 = -10I_1 + 25I_2 \quad \text{(loop 2)}$$

$$\frac{25I_2}{-10I_2} \times (-1) = 2.5 \quad \text{(constant)}$$

$$7.5 = 55I_1 - 25I_2 \quad \text{(loop 1)}$$
$$\underline{30 = -10I_1 + 25I_2} \quad \text{(loop 2)}$$
$$37.5 = 45I_1 \quad \text{(loops 1 and 2)}$$
$$I_1 = 0.83\ A$$

Find V_{TH}:

$$V_{R_3} = I_1 R_3 = (0.83\ A)(12\ \Omega) = 10\ V$$
$$V_{TH} = B_3 - V_{R_3} = 58\ V - 10\ V = 48\ V$$

Find R_{TH} [Fig. 6-26(d) shows that R_1, R_2, and R_3 are in parallel]:

$$R_{TH} = \frac{1}{\dfrac{1}{10} + \dfrac{1}{15} + \dfrac{1}{12}} = \frac{60}{15} = 4\ \Omega$$

Construct the Thevenin equivalent circuit. Load it with R_4, as in Fig. 6-26(e), and solve for V_{R_4}:

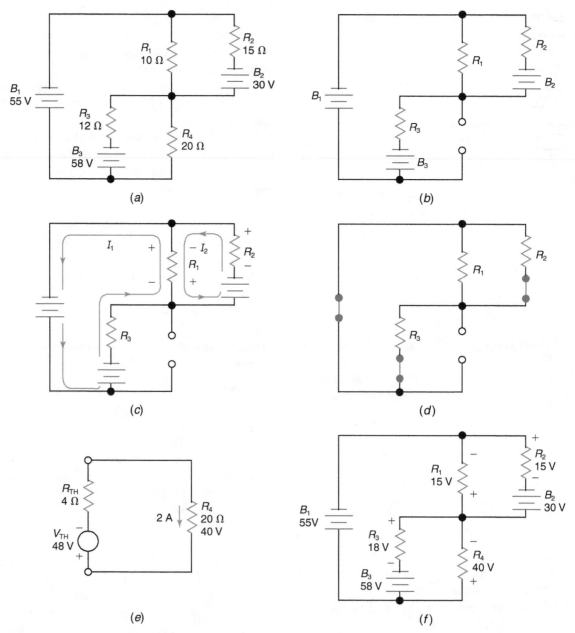

Fig. 6-26 Complex circuit analyzed in example 6-10. (a) Original circuit. (b) R_4 removed. (c) Loops used to find V_{TH}. (d) Finding R_{TH}. (e) Equivalent circuit with load. (f) Final results.

$$V_{R_4} = \frac{V_{TH}R_4}{R_{TH} + R_4} = \frac{(48 \text{ V})(20 \text{ }\Omega)}{4 \text{ }\Omega + 20 \text{ }\Omega} = 40 \text{ V}$$

Finally, solve for other voltage drops. The polarities of these voltage drops are shown in Fig. 6-26(f):

$$V_{R_3} = B_3 - V_{R_4} = 58 \text{ V} - 40 \text{ V} = 18 \text{ V}$$
$$V_{R_1} = B_1 - V_{R_4} = 55 \text{ V} - 40 \text{ V} = 15 \text{ V}$$
$$V_{R_2} = B_2 - V_{R_1} = 30 \text{ V} - 15 \text{ V} = 15 \text{ V}$$

Answer: $V_{R_1} = 15 \text{ V}, V_{R_2} = 15 \text{ V},$
$V_{R_3} = 18 \text{ V}, V_{R_4} = 40 \text{ V}$

The circuit analyzed in example 6-10 is the same circuit analyzed with the superposition theorem in example 6-7. In example 6-7 we find the individual currents, whereas in example 6-10 we find the individual voltages. Of course, application of Ohm's law solves the voltages in example 6-7 and the currents in example 6-10. Note that the number of calculations involved is fewer with the Thevenin method than with the superposition method. Once we have the Thevenin equivalent circuit of Fig. 6-26(e), we can easily find the voltage distribution for any value of R_4. By contrast, changing the value of R_4 when using the superposition method requires that the complete procedure be repeated to find the new currents.

 Self-Test

Answer the following questions.

35. Change the value of R_1 in Fig. 6-17(a) to 15 Ω and calculate all voltage drops using Thevenin's theorem.

36. If you remove R_3 from Fig. 6-7(a), can you
    a. Find R_{TH} without using the superposition theorem or loop analysis?
    b. Find V_{TH} without using the superposition theorem or loop analysis?

37. If you use Thevenin's theorem to find I_{R_3} and V_{R_3} in Fig. 6-7(a), can you then find the other voltage drops with Kirchhoff's voltage law?

38. Which resistor(s) can be removed in Fig. 6-7(a) so that all voltages can be found without using the superposition theorem or loop analysis and by using Thevenin's theorem only once?

6-7 Current Source

Up to this point we have viewed the source that powers a circuit as a voltage source—either as a constant (ideal) voltage source or as an equivalent-circuit voltage source. We can also consider the source that powers a circuit to be a *current source*. The current source, too, can be viewed as a *constant current source* or as an equivalent-circuit current source.

By definition, a constant current source is a source which supplies the same amount of current to its terminals regardless of the terminal voltage. This definition implies that the parallel internal resistance of the current source is infinite. The symbol for a constant current source is drawn in Fig. 6-27(a). The arrow in the symbol indicates the direction of the constant current.

Like the constant voltage source, the constant current source does not really exist. All current sources have some finite internal resistance. However, it is possible to construct a current source with such a large internal resistance that it can be treated as a constant current source for many practical applications. In fact, a transistor is sometimes viewed as a constant current source.

The *equivalent-circuit model* of a current source is shown in Fig. 6-27(b). It consists of a constant current source in parallel with the internal resistance of the source. This model of a current source is the one we are interested in because it is very useful in analyzing complex circuits. Therefore, let us see how we can develop an equivalent-circuit current source.

Figure 6-28 illustrates why we can view a source as either a voltage source or a current source. Figure 6-28(a), (b), and (c) shows a voltage source under three conditions: open-circuited, short-circuited, and loaded. Figure 6-28(d), (e), and (f) shows a current source under the same three conditions. Notice that the end results are the same whether we view the source as a current source or as a voltage source. In Fig. 6-28(d), the constant current source provides 1 A of current which must all flow through

Current source

Constant current source

Equivalent-circuit model

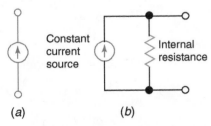

(a)   (b)

Fig. 6-27 Current sources. (a) Ideal constant current source. (b) Equivalent circuit of a current source.

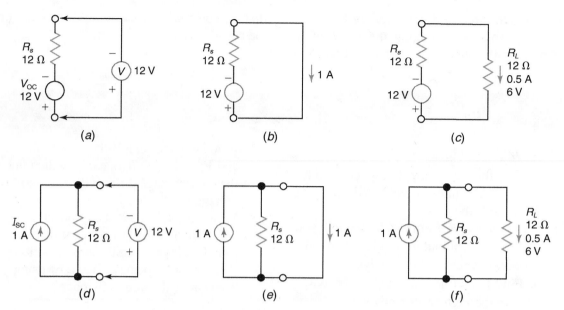

Fig. 6-28 Equivalency of voltage and current sources. R_S is in parallel for a current source and in series for a voltage source. (*a*) Voltage source—open-circuit voltage. (*b*) Voltage source—short-circuit current. (*c*) Voltage source—load current and voltage. (*d*) Current source—open-circuit voltage. (*e*) Current source—short-circuit current. (*f*) Current source—load current and voltage.

the internal resistance of the source since there is no load. This 1 A of current through the 12 Ω of resistance provides 12 V of open-circuit voltage. In Fig. 6-28(*e*), the 1 A of the constant current is routed around the 12 Ω of internal resistance R_s and through the external short circuit. In Fig. 6-28(*f*) the 1 A from the constant current source is split between the internal resistance and the external resistance. When the ratio of the current split is not obvious, use the current-divider formula for parallel circuits:

$$I_{RL} = \frac{I_{constant} R_S}{R_L + R_S}$$

Three very important points should be noted from Fig. 6-28.

1. The internal resistance of a source is the same whether the source is treated as a voltage source or as a current source.
2. The *short-circuit current* of a voltage source is the same as the constant current of the current source. Thus, the abbreviation I_{sc} is sometimes used for the constant current source.
3. The open-circuit voltage of a current source is the same as the constant voltage of the voltage source.

Short-circuit current

Keeping these three points in mind, it is very easy to convert a voltage source to a current source and vice versa.

EXAMPLE 6-11

Convert the voltage source shown in Fig. 6-29(*a*) to a current source.

Given: Voltage source and values of Fig. 6-29(*a*)

Find: Constant current and equivalent-circuit current source

Known: Relationships between voltage source and current source

Solution: Short-circuit the voltage source to find the constant current [Fig. 6-29(*b*)]:

$$I_{sc} = \frac{V_{oc}}{R_S} = \frac{24\text{ V}}{4\text{ Ω}} = 6\text{ A}$$

The R_S of the two sources will be the same.

Answer: The constant current is 6 A, and the current source is as shown in Fig. 6-29(*c*).

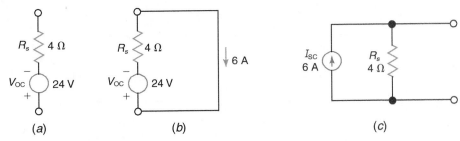

Fig. 6-29 Circuit for example 6-11—converting a voltage source to a current source. (a) Voltage source. (b) Short-circuit current. (c) Current source.

 *Self-Test*

Answer the following questions.

39. A current source has an internal resistance of 2 Ω and a constant current of 7 A.
    a. If this current source were converted to a voltage source, what would the value of the constant voltage be?

b. If this current source were loaded with a 12-Ω resistor, how much current would flow through the resistor?

40. Determine the constant current for the voltage source in Fig. 6-23(b).

6-8 Norton's Theorem

Norton's theorem provides a way of converting any two-terminal circuit into an *equivalent-circuit current source*. The conversion can be accomplished by following the four steps listed below and illustrated in Fig. 6-30:

1. Select two points on the circuit to be used as the terminals of the current source. This is done in Fig. 6-30(b).
2. Calculate or measure the current available from the terminals when the terminals are short-circuited as shown in Fig. 6-30(c). This short-circuit current is labeled I_{sc} in Fig. 6-30(c), and it is calculated

$$I_{sc} = \frac{B_1}{R_1} = \frac{16\text{ V}}{10\text{ }\Omega} = 1.6\text{ A}$$

3. Calculate R_S (the internal resistance) using the technique used for Thevenin's theorem. As shown in Fig. 6-30(d), R_1 and R_2 are in parallel, and so R_S is 8 Ω.
4. Using I_{sc} from step 2 and R_S from step 3, draw the equivalent-circuit current source shown in Fig. 6-30(e). This current source is called a Norton equivalent circuit. Thus, I_{sc} is labeled I_N and R_S is labeled R_N in Fig. 6-30(e).

Figure 6-30(f) illustrates the results of connecting a 4.8-Ω load to the Norton equivalent circuit of Fig. 6-30(e). The load current I_{RL} is calculated using the current-divider equation:

$$I_{RL} = \frac{I_N R_N}{R_L + R_N} = \frac{(1.6\text{ A})(8\text{ }\Omega)}{4.8\text{ }\Omega + 8\text{ }\Omega} = 1\text{ A}$$

It is left to you to verify that connecting a 4.8-Ω load in parallel with R_2 in Fig. 6-30(a) yields the same value for I_{RL} and V_{RL}.

Norton's theorem can also be used to analyze complex circuits. The procedure is the same as with Thevenin's theorem except that the circuit is reduced to a current source after a resistor is removed. Thus, instead of determining the open-circuit voltage (V_{TH}) at the terminals, we determine the short-circuit current (I_N) at the terminals. Let us try an example to illustrate the use of Norton's theorem in analyzing a complex circuit.

Norton's theorem

Equivalent-circuit current source

EXAMPLE 6-12

Using Norton's theorem, find the voltage across and current through R_2 in Fig. 6-31(a)

Given: Values and diagram of Fig. 6-31(a)

Find: V_{R_2} and I_{R_2}

(continued)

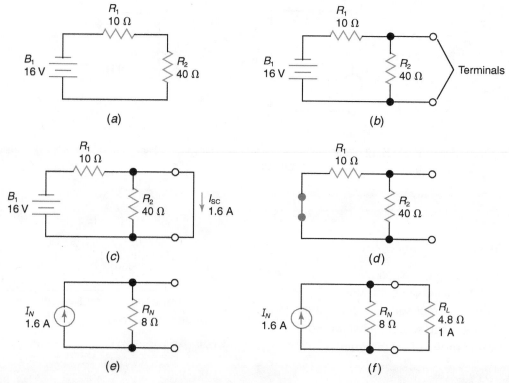

Fig. 6-30 Converting a circuit to an equivalent-circuit current source. The equivalent circuit is a constant current source in parallel with an internal resistance. (a) Original circuit. (b) Source terminals added. (c) Short-circuit current. (d) Internal resistance ($R_S = 8\ \Omega$). (e) Norton equivalent circuit (equivalent-circuit current source). (f) Norton equivalent circuit with a load connected.

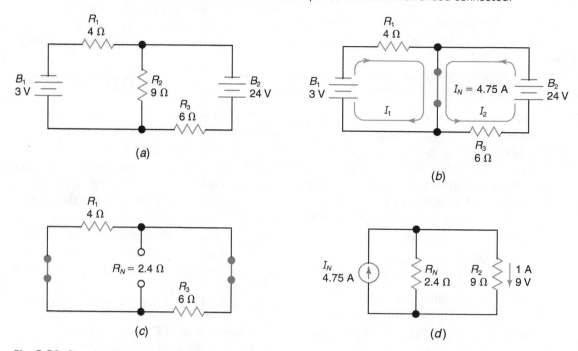

Fig. 6-31 Circuits for example 6-12.

Known: Norton's theorem

Solution: Remove R_2 and calculate I_N as indicated in Fig. 6-31(b):

$$I_1 = \frac{B_1}{R_1} = \frac{3\ \text{V}}{4\ \Omega} = 0.75\ \text{A}$$

$$I_2 = \frac{B_2}{R_3} = \frac{24\ \text{V}}{6\ \Omega} = 4\ \text{A}$$

$I_N = I_1 + I_2 = 0.75 \text{ A} + 4 \text{ A} = 4.75 \text{ A}$

Replace B_1 and B_2 with short circuits as in Fig. 6-31(c) and calculate R_N:

$$R_N = \frac{R_1 R_2}{R_1 + R_2} = \frac{24}{10} = 2.4 \ \Omega$$

Draw the Norton equivalent, load it with R_2 as in Fig. 6-31(d), and calculate V_{R_2} and I_{R_2}:

$$I_{R_2} = \frac{I_N R_N}{R_N + R_2} = \frac{(4.75 \text{ A})(2.4 \ \Omega)}{2.4 \ \Omega + 9 \ \Omega} = 1 \text{ A}$$

$$V_{R_2} = I_{R_2} R_2 = (1 \text{ A})(9 \ \Omega) = 9 \text{ V}$$

Answer: $I_{R_2} = 1$ A, and $V_{R_2} = 9$ V

The circuit used in example 6-12 is the same circuit used in example 6-4 and solved with

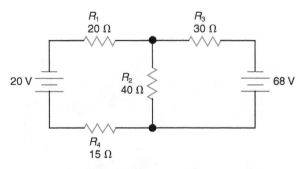

Fig. 6-32 Complex circuit for Self-Test questions 41, 42, and 43.

loop equations. Notice that the results for V_{R_2} and I_{R_2} are the same in both cases. Of course, we can use Kirchhoff's laws to find the remaining currents and voltages in Fig. 6-31(a) after we know V_{R_2} and I_{R_2}.

 ## Self-Test

Answer the following questions.

41. Using Norton's theorem, calculate I_{R_2} and V_{R_2} for the circuit of Fig. 6-32.
42. Can you solve for the other values of current and voltage in Fig. 6-32 using only

Kirchhoff's laws, Ohm's law, and the values of I_{R_2} and V_{R_2} found in question 41?

43. Change the value of R_3 in Fig. 6-32 to 60 Ω. Then use Norton's theorem to find I_{R_3} and V_{R_3}. Finally, determine I_{R_4} and V_{R_1} using Kirchhoff's and Ohm's laws.

6-9 Comparison of Techniques

Most circuits can be analyzed by two or more of the techniques covered in this chapter. In deciding which techniques to use, ask yourself two questions.

1. Which technique will work?
2. Which technique will involve the fewest and simplest calculations to obtain the desired information?

Answering these two questions is not too difficult if you keep in mind the major characteristics of the various techniques, which are as follows:

Loop equations:

1. Can be used with either single-source or multiple-source circuits
2. Solve for all currents and voltages in the circuit

Superposition theorem:

1. Applicable only to multiple-source circuits
2. Solves for all currents and voltages in the circuit
3. Uses only series-parallel rules and procedures

Node voltage technique:

1. Applicable to multiple-source circuits
2. Solves for all currents and voltages
3. Can solve for more than one unknown node voltage
4. May involve simultaneous equations

Thevenin's theorem:

1. Can be used with either single-source or multiple-source circuits
2. May not solve for all values of current and voltage

Comparison of Techniques

Superposition theorem

Loop equations

Thevenin's theorem

4. Requires only one application of the current-divider formula to determine the current through any value of load resistor

From the above summary you can tell that Thevenin's and Norton's theorems have much in common.

Since Thevenin's equivalent circuit is a voltage source and Norton's equivalent circuit is a current source, we can also convert one to the other. The appropriate formulas are

$$R_N = R_{TH}$$

$$I_N = \frac{V_{TH}}{R_{TH}}$$

$$V_{TH} = I_N R_N$$

You May Recall

. . . that Sec. 6-7 points out the relationships between current sources and voltage sources and illustrates how to convert from one to the other.

3. May require other techniques to develop the Thevenin equivalent circuit
4. Requires only a single calculation using the voltage division formula to determine the voltage across any value of load resistor

Norton's theorem

Norton's theorem:

1. Can be used with either single-source or multiple-source circuits
2. May not solve for all values of current and voltage
3. May require other techniques to develop the Norton equivalent circuit

An example will illustrate why we may want to convert from a Thevenin equivalent circuit to a Norton equivalent circuit. Suppose we want to know the current through R_5 in Fig. 6-4(a) when R_5 is any one of 10 different values. This is an easy task if we have the Norton equivalent circuit derived from viewing R_5 as the load resistor. But try calculating I_N for Fig. 6-4(a); it is not easy. However, calculating V_{TH} and R_{TH} requires only series-parallel procedures. Thus, for this circuit, it is best to find the Thevenin equivalent circuit and then convert it to a Norton equivalent circuit.

 Self-Test

Answer the following questions.

44. What is the value of the Thevenin resistance in Fig. 6-31(d)?

45. Which theorem cannot be used with a single-source circuit?

Chapter 6 Summary and Review

Summary

1. Complex circuits (networks) cannot be analyzed using only series-parallel rules and procedures.
2. Equations with more than one variable can be solved by simultaneous-equation techniques.
3. To solve simultaneous equations, there must be as many independent equations as there are unknown variables.
4. Loop equations are derived by applying Kirchhoff's voltage law to each loop of a circuit.
5. The loop-equations technique can solve for all voltages and currents in a circuit.
6. The loop-equations technique can be applied to either single-source or multiple-source circuits.
7. Node voltage techniques can be used when one or more major node voltages are unknown.
8. When the calculated value of current has a negative sign, the direction of current originally assumed is incorrect.
9. The superposition theorem can be used to analyze complex multiple-source circuits without using simultaneous equations. This technique can solve for all currents and voltages in the circuit.
10. When the superposition theorem is used, all but one voltage source is replaced by its internal resistance.
11. In many circuits the internal resistance of the voltage source can be assumed to be 0 Ω, that is, the voltage source can be viewed as a constant (ideal) voltage source.
12. An equivalent-circuit voltage source consists of a constant voltage source (V_{oc} or V_{TH}) in series with an internal resistance (R_S or R_{TH}).

13. If an electric circuit is viewed as a two-terminal network, Thevenin's theorem provides a way of reducing the circuit to an equivalent-circuit voltage source.
14. Thevenin's theorem can be used to find the current and voltage associated with a single resistor in a complex circuit.
15. A constant current source assumes that current from the terminals of the source remains the same for all values of load resistance. Furthermore, the constant current source is assumed to have an infinite internal resistance.
16. An equivalent-circuit current source is a constant current source (I_{sc}) in parallel with an internal resistance (R_S).
17. For a given circuit, the internal resistances of the equivalent-circuit current source and the equivalent-circuit voltage source are the same.
18. Norton's theorem provides a way of reducing any two-terminal circuit to an equivalent-circuit current source.
19. Norton's theorem can be used to determine the voltage across and the current through an individual resistor in a complex circuit.
20. Neither Thevenin's nor Norton's theorem always leads to the easy solution of all voltages and currents in a circuit.
21. Either Thevenin's or Norton's theorem can lead to easy determination of I_{RL} and V_{RL} when the value of R_L is changed.

Related Formulas

$$I_{sc} = \frac{V_{oc}}{R_S}$$

$$V_{oc} = I_{sc}R_S$$

$$I_N = \frac{V_{TH}}{R_{TH}}$$

$$V_{TH} = I_N R_N$$

For questions 6-1 to 6-4, supply the missing word or phrase in each statement.

6-1. A complex circuit can also be referred to as a _____. (6-1)

6-2. A loop equation is written by using _____ law. (6-2)

6-3. A negative value of calculated current indicates that the assumed current direction was _____. (6-2)

6-4. _____ independent equations are required to solve for two unknown variables. (6-1)

For questions 6-5 to 6-9, determine whether each statement is true or false.

6-5. Each equation in a set of equations must contain all unknown variables. (6-1)

6-6. Application of the superposition theorem allows you to determine all voltages and currents in a dual-source complex circuit without using simultaneous equations. (6-3)

6-7. A constant current source provides a constant voltage at its terminals under all load conditions. (6-7)

6-8. For a given circuit, R_N of the Norton equivalent circuit always equals R_{TH} of the Thevenin equivalent circuit. (6-9)

6-9. Norton's theorem reduces a circuit to an equivalent-circuit voltage source. (6-8)

Answer the following questions.

6-10. In a particular complex circuit, three loops are needed to include all resistors in at least one loop. How many loop equations are required to analyze the circuit? (6-2)

6-11. Refer to question 6-10. How many unknown variables will there be in the set of equations? (6-2)

6-12. Which theorem cannot be used on a single-source circuit? (6-3)

6-13. A power source is represented by an equivalent-circuit current source and an equivalent-circuit voltage source. Will the two equivalent circuits have the same value of (6-7)

a. Internal resistance

b. Open-circuit voltage

c. Short-circuit current

Chapter Review Problems

6-1. Add $3A + 5B = 14$ to $-A + 2B - 3C = -7$. (6-1)

6-2. Solve for A and B when $3A + 8B = 30$ and $2A - 4B = -8$. (6-1)

6-3. A current source has an internal resistance of $4\ \Omega$ and a constant current of 9 A. (6-7)

a. What is its open-circuit voltage?

b. What is its terminal voltage when loaded with an $8\text{-}\Omega$ resistor?

6-4. If you convert the current source in problem 6-3 to a voltage source, what will the value of the constant voltage source be? (6-7)

6-5. If you convert a Thevenin equivalent circuit with $V_{TH} = 16$ V and $R_{TH} = 2$ kΩ to a Norton equivalent circuit, what will the values of I_N and R_N be? (6-9)

6-6. Solve for A, B, and C when $2A + 4B - 3C = 10$, $-A - 3B + 2C = 12$, and $3A + 2B + 4C = 20$. (6-1)

6-7. For the circuit in Fig. 6-33(a), compute the following by applying Thevenin's theorem: (6-6)

a. V_{R_3}

b. I_{R_3}

6-8. For the circuit in Fig. 6-33(b), compute the following by using loop equations: (6-2)

a. V_{R_3}

b. V_{R_4}

c. I_{R_1}

d. I_{R_2}

6-9. For the circuit in Fig. 6-33(c), compute the following by using the superposition theorem: (6-3)

a. I_{R_1}

b. I_{R_2}

c. I_{R_3}

d. V_{R_4}

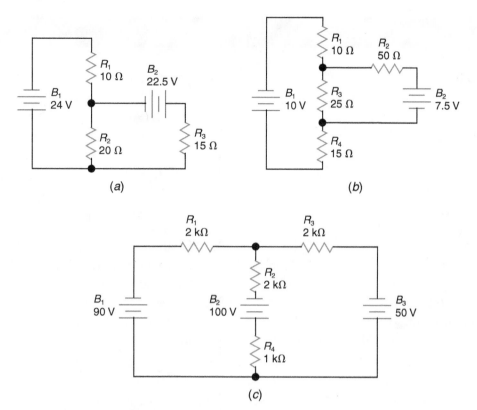

(a)

(b)

(c)

Fig. 6-33 Circuit for chapter review problems 6-7 to 6-11.

6-10. For the circuit in Fig. 6-33(*a*), compute the following by applying Norton's theorem: (6-8)
   a. I_{R_2}
   b. V_{R_2}

6-11. Write the loop equations for Fig. 6-33(*a*). (6-2)

Critical Thinking Questions

6-1. When loop techniques are used on a two-source complex circuit, is it possible for all calculated currents to be negative? Explain.

6-2. Is it possible to have a complex circuit in which no current flows in one of the resistors? Explain.

6-3. Would a 6-V carbon-zinc battery rated at 1 Ah be more like a constant voltage source than a 6-V carbon-zinc battery rated at 10 Ah? Why?

6-4. Compare the efficiency of an equivalent-circuit voltage source to an equivalent-circuit current source.

6-5. Using the values from example 6-5, prove that $V_{R_1} + V_{R_2} = B_1$ in Fig. 6-7.

6-6. Using the values from example 6-5, prove that $B_1 = V_{R_1} + V_{R_3} + V_{R_5} + B_2$.

1. $6A + 6B = 59.6$
2. $4R + S - 6T = 25$
3. $-2A + 4C = 8$
4. $X = 4.3, Y = 6$
5. $C = 3.81, D = 5.35$
6. $A = 6, B = 3$
7. $R = 4, S = -8, T = 3$
8. $X = 2.3, Y = 5.2, Z = 3.1$
9. $C = -5.19, D = -4.79, E = -4.07$
10. $9\text{ V} = 8\text{ k}\Omega\ (I_1) +$
    $3\text{ k}\Omega\ (I_2) + 1\text{ k}\Omega\ (I_3)$ (loop 1)
    $9\text{ V} = 3\text{ k}\Omega\ (I_1) + 5\text{ k}\Omega\ (I_2)$ (loop 2)
    $9\text{ V} = 1\text{ k}\Omega\ (I_1) + 6\text{ k}\Omega\ (I_3)$ (loop 3)
11. No; I_2 and I_3 are different.
12. Yes; however, V_{R_5} is positive with the equations from question 10.
13. point A to point B
14. no
15. T
16. F
17. $I_{R_1} = 1.2$ A,
    $I_{R_2} = 2.6$ A,
    $I_{R_3} = 1.4$ A,
    $V_{R_3} = 14$ V
18. F
19. Kirchhoff's current law
20. T
21. F
22. -11.56 V
23. right-to-left
24. yes
25. yes
26. no
27. $I_{R_1} = 0.983$ A,
    $I_{R_2} = 1.207$ A,
    $I_{R_3} = 2.190$ A
28. no
29. T
30. $I_{R_3} + I_{R_1} = I_{R_4} + I_{B_2}$,
    or 1.5 A $+ 1.5$ A $=$
    2 A $+ 1$ A
31. 0.5 A
32. the amount of internal resistance R_S relative to the external-circuit resistance R_L
33. open-circuit voltage V_{oc}
34. F
35. $V_{R_1} = 57$ V, $V_{R_2} = 38$ V, $V_{R_3} = 0$ V
36. a. yes
    b. yes
37. no
38. R_1 or R_5
39. a. 14 V
    b. 1 A
40. 0.06 A
41. $I_{R_2} = 0.816$ A,
    $V_{R_2} = 32.7$ V
42. yes
43. $I_{R_3} = 0.729$ A,
    $V_{R_3} = 43.7$ V,
    $I_{R_4} = 0.122$ A,
    $V_{R_1} = 2.44$ V
44. $2.4\ \Omega$
45. superposition theorem

Magnetism and Electromagnetism

Learning Outcomes

This chapter will help you to:

7-1 *List* four elements often found in alloys used in manufactured magnets.

7-2 *Visualize* magnetic forces, flux, fields and poles of various shapes of magnets.

7-3 *Predict* direction of the flux around a current-carrying conductor and direction of the force between two parallel current-carrying conductors.

7-4 *Explain* why some materials are nonmagnetic and others are magnetic and why some magnetic materials make permanent magnets and others make temporary magnets.

7-5 *Understand* how a current-carrying coil can magnetize either permanent-magnetic material or temporary-magnetic material.

7-6 *Define* magnetomotive force (mmf).

7-7 *Determine* when a magnetic material is saturated.

7-8 *Discuss* ways to demagnetize a permanent magnet.

7-9 *Name* the flux remaining in a temporary magnetic material after removal from a magnetic field.

7-10 *Explain* why magnetic materials are attracted to magnets.

7-11 *Discuss* how a magnetic shield can protect an object made of magnetic material from an adjacent magnetic field.

7-12 *List* the necessary conditions for a magnetic field to induce voltage in a conductor by "generator action" and by "transformer action."

7-13 *Define* magnetic quantities and units and use them to solve magnetic problems.

7-14 *Discuss* the four factors that control the strength of an electromagnet.

7-15 *Name* the two components of a dc motor that change the magnetic polarity of the armature at the appropriate time.

Electricity and magnetism cannot be separated. Wherever an electric current exists, a magnetic field also exists. Magnetism, created by an electric current, operates many devices, such as transformers, motors, and loudspeakers.

7-1 Magnetism and Magnets

Magnetism is a force field that acts on some materials but not on other materials. Physical devices that possess this force are called *magnets*. *Lodestone* (an iron compound) is a natural magnet that was discovered centuries ago.

The magnets we use today are all manufactured. They are made from various alloys containing elements like copper, nickel, aluminum, iron, and cobalt. These magnets are much, much stronger than the natural lodestone magnet.

Magnetism

Lodestone

7-2 Magnetic Fields, Flux, and Poles

The force of magnetism is referred to as a *magnetic field*. This field extends out from the magnet in all directions, as illustrated in Fig. 7-1.

Magnetic field

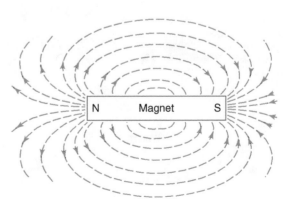

Fig. 7-1 Magnetic field of a magnet. The flux is most dense at the poles.

North pole

South pole

Flux

Like magnetic poles

In this figure the lines extending from the magnet represent the magnetic field.

The invisible lines of force that make up the magnetic field are known as magnetic *flux* (ϕ). The lines of force in Fig. 7-1 represent the flux. Where the lines of flux are dense, the magnetic field is strong. Where the lines of flux are sparse, the field is weak. The lines of flux are most dense at the ends of the magnet; therefore, the magnetic field is strongest at the ends of the magnet.

The picture in Fig. 7-2(*a*) shows how the magnetic flux and magnetic field are distributed around a magnet. The setup for the picture in Fig. 7-2(*a*) consisted of sprinkling a uniformly distributed layer of iron filings on a thin sheet of clear plastic

[shown in Fig. 7-2(*b*)] and then placing the plastic over the magnet. Notice in Fig. 7-2(*a*) that the ends of the magnet possess a magnetic field strong enough to pull the surrounding iron filings to them. Also notice in this figure that the less dense areas of flux arrange the iron filings to indicate the lines of flux.

The arrows on the lines in Fig. 7-1 indicate the direction of the flux. Lines of force are always assumed to leave the *north pole* (N) and enter the *south pole* (S) of a magnet. *North pole* and *south pole* refer to the polarity of the ends of a magnet. When a magnet is suspended on a string and allowed to rotate, its ends will point north and south. The end of the magnet that seeks (or points to) the earth's north magnetic pole is the magnet's north pole. Assuming that flux (lines of force) leaves the north pole and enters the south pole was an arbitrary decision. However, assigning direction to the lines aids in understanding the behavior of magnetism.

Like magnetic poles repel each other. The two north poles of Fig. 7-3(*a*) and the two south poles of Fig. 7-3(*b*) create a repelling force. A picture of the magnetic field created by two magnets with like poles facing each other is provided in Fig. 7-3(*c*). Notice that the iron

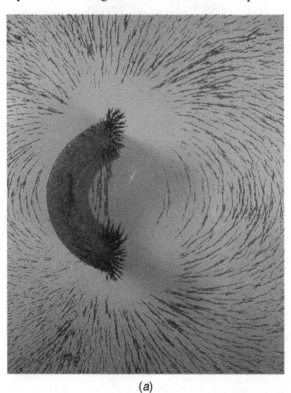

(a)

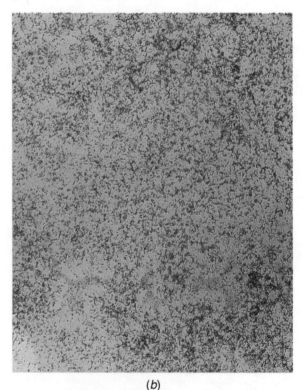

(b)

Fig. 7-2 Distribution of iron filings in a magnetic field. (*a*) Iron filings arranged by the magnetic field of a magnet. (*b*) Distribution of iron filings before being exposed to a magnetic field.

Flux lines

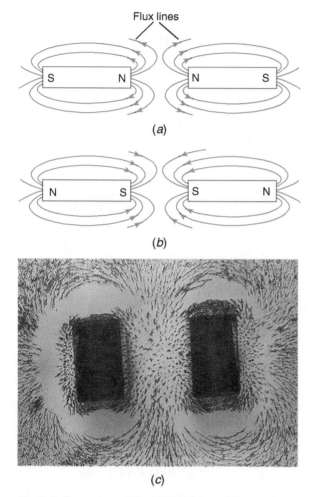

(a)

(b)

(c)

Fig. 7-3 Repulsion of like poles. The repelling poles can both be north poles, as in (a), or south poles, as in (b). (c) A picture of the magnetic field created by like poles.

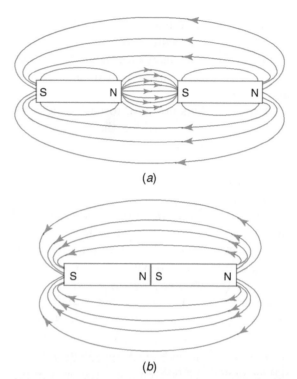

(a)

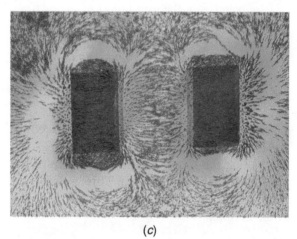

(b)

(c)

Fig. 7-4 Attraction of unlike poles. The attraction shown in (a) is maximized when the poles are touching as in (b). The distribution of iron filings around unlike poles is pictured in (c).

filings between the magnets do not indicate flux lines between the like poles. The closer the poles are, the more they repel each other. The force of repulsion between magnetic poles varies inversely as the square of the distance between them. That is, if the distance is doubled, the force becomes one-fourth as great. Or, if the distance is halved, the force becomes four times as great.

Unlike magnetic poles (Fig. 7-4) create a force of attraction. This force also varies inversely as the square of the distance between the poles. As shown in Fig. 7-4(a), much of the flux of the two magnets joins together to form the force of attraction. The picture in Fig. 7-4(c) shows the result of the heavy concentration of flux between the unlike facing poles of the two magnets. When the two poles

touch, essentially all the flux joins together [Fig. 7-4(b)]. When joined together, the two magnets behave as a single magnet. They create a single magnetic field.

Magnetic lines of force (flux) are assumed to be continuous loops. Although not shown in Fig. 7-4, the flux lines continue on through the magnet. They do not stop at the poles. In fact, the poles of the magnet are merely the areas where most of the flux leaves the magnet and enters the air. If a magnet is broken in

Unlike magnetic poles

half (Fig. 7-5), two new poles are created. The one magnet becomes two magnets. A magnet can be broken into many pieces, and each piece becomes a new magnet with its own north pole and south pole.

A magnet does not have to have poles. All the flux lines can be within the magnet. This idea is illustrated in Fig. 7-6. In Fig. 7-6(*a*), a typical horseshoe magnet is shown. The dark lines represent the flux within the magnet. In Fig. 7-6(*b*) the horseshoe magnet has been bent around to form a circle with a small gap. And finally, in Fig. 7-6(*c*) the gap has been closed to form a circle. Now all the flux (and the magnetic field) is confined within the magnet. The magnet has no poles.

It should be noted that a circular magnet can be (and often is) made with poles. In this case, the flux runs parallel to the hollow core of the magnet, and the flat surfaces of the magnet are the north and south poles.

The patterns of the iron filings in the pictures in Fig. 7-7 show that the flux is most dense on the flat surface (one of the poles) of the circular magnet. The picture in Fig. 7-7(*a*) is a top view of a circular magnet that is covered with a sheet of clear plastic with iron filings on top of the plastic. The iron filings were uniformly distributed on the plastic before the plastic was placed on the magnet. The edge-view picture in Fig. 7-7(*b*) shows that the filings on the flat surface of the magnet are pointing upward. This verifies that the flat surface is one of the poles of this circular magnet.

Fig. 7-5 Creation of poles. Each time the magnet is broken, a new pair of poles is created.

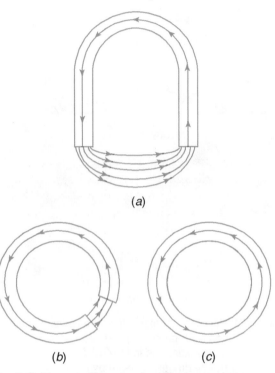

Fig. 7-6 Magnet without poles. The horseshoe magnet (*a*) can be formed into a circle with a gap (*b*) and maintain poles. Eliminating the gap (*c*) eliminates the poles.

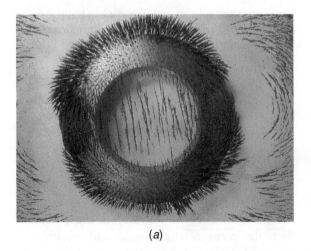

(*a*)

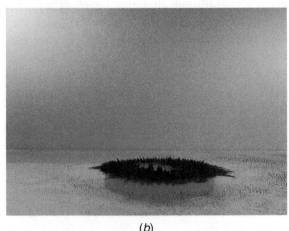

(*b*)

Fig. 7-7 Poles on a circular (ring) magnet. (*a*) Distribution of iron filings around a pole of a circular magnet. (*b*) Edge view of the magnet and filings shown in (*a*).

Answer the following questions.

1. Define *magnetism*.
2. Define *magnetic field*.
3. The invisible lines of force in a magnetic field are called _____.
4. A magnetic north pole is repelled by a magnetic _____ pole.
5. Magnetic flux flows from the _____ pole to the _____ pole.
6. Halving the distance between two poles will _____ the force between the poles.
7. The flux from a north pole will join the flux from a(n) _____ pole.
8. Are lines of flux continuous loops?
9. Does breaking a magnet in half destroy its magnetic field?
10. Do all magnets have a north pole?

7-3 Electromagnetism

So far our discussion has centered around the magnetic field and flux possessed by a magnet. However, magnetic fields are also created by electric current. The current-carrying conductor in Fig. 7-8 has a magnetic field around it. The field is always at right angles (perpendicular) to the direction of current. Although the field is shown in only five places along the conductor, it actually exists as a continuous field for the entire length. Notice in Fig. 7-8 that the magnetic field has no poles. The flux exists only in the air. However, the flux still has an assumed direction, just as it does in the circular magnet of Fig. 7-6(*c*). The *direction of* the *flux* around a conductor can be determined by using what is called the *left-hand rule*. Grasp the conductor with your left hand so that your thumb points in the direction of current. Your fingers then indicate the direction of the flux.

Often the direction of current and flux is indicated as shown in Fig. 7-9. In this figure, you are looking at the ends of current-carrying conductors. Of course, the end of a round conductor looks like a circle. An x in the circle, as in Fig. 7-9(*a*), means that the current is going into the paper. A way to remember this convention is to visualize an arrow or a dart as indicating the direction of current. If the dart were going into the paper, you would see the tail end with feathers arranged in an x shape. If the dart were coming toward you (out of the paper), you would see the front, pointed end. Thus, the dot in a circle represents current flowing toward you.

The strength of the magnetic field around a conductor is determined by the amount of current flowing through the conductor. The strength at some fixed distance from the conductor is directly proportional to the current. Doubling the current doubles the strength of the magnetic field.

Direction of flux

Left-hand rule

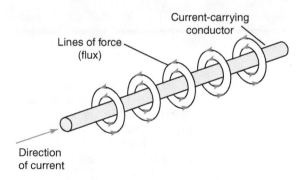

Fig. 7-8 Magnetic field around a conductor. The flux is perpendicular to the direction of current.

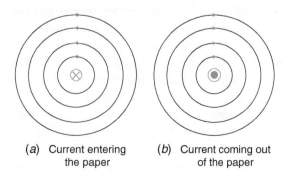

(*a*) Current entering the paper   (*b*) Current coming out of the paper

Fig. 7-9 Indicating direction of current.

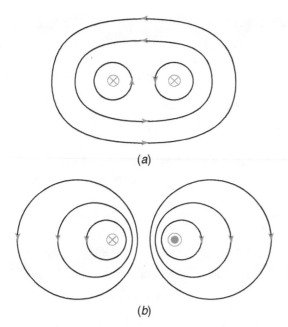

(a)

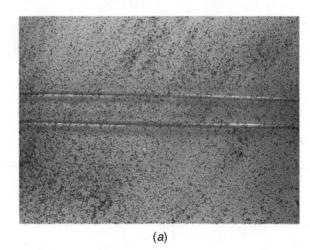

(a)

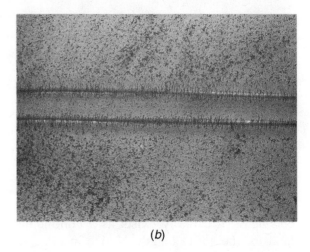

(b)

(b)

Fig. 7-10 Forces between parallel conductors. (a) Attraction of conductors. (b) Repulsion of conductors.

Force between Conductors

Parallel conductors

Two *parallel,* current-carrying *conductors* attract each other if the currents in the conductors are in the same direction. This is because the fields of the two conductors join together, as shown in Fig. 7-10(a). When fields join together, there is a force of attraction. This is the same type of attraction that occurs between unlike poles (Fig. 7-4).

When the currents in parallel conductors are not in the same direction, a force of repulsion exists. The magnetic fields in Fig. 7-10(b) are not able to join together because they are opposing each other. Since lines of flux cannot cross each other and cannot share the same space, they repel each other.

The pictures in Fig. 7-11 depict the magnet fields around parallel conductors with currents in the opposite directions. In Fig. 7-11(a), the iron filings are randomly distributed because no current has passed through the conductors. After a current flow of about 10 A, filings have gathered around the conductor as shown in Fig. 7-11(b). The clustering of the filings can be seen more clearly in Fig. 7-11(c) where the conductors have been removed without disturbing the iron filings. Notice the scarcity of filings between the conductors and along the outer edge of both conductors after current has passed through the conductors.

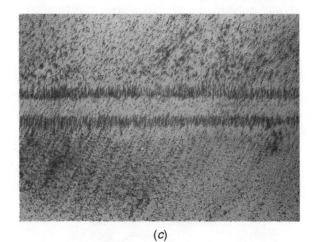

(c)

Fig. 7-11 Magnetic field around parallel conductors when currents are in opposite directions. (a) Distribution of iron filings before current flows through the conductors. (b) Iron filings are attracted to the conductors by a 10-A current flowing in the conductors. (c) Iron fillings' distribution after removing the conductors without disturbing the filings.

Fig. 7-12 Distribution of iron filings around current carrying conductor.

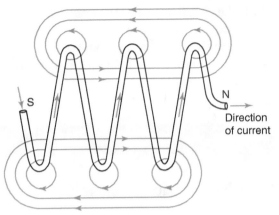

(a) Coil with space between turns

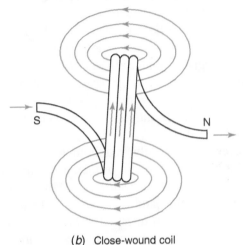

(b) Close-wound coil

Fig. 7-13 Magnetic fields of coils.

Internet Connection

Career and other information may be found at the website for the Institute of Electrical and Electronic Engineers (IEEE).

Figure 7-12 shows patterns produced in fine iron particles by the current flow in the ground cable (shown in the picture) of a MIG arc welder. This welder produces direct current in the 60-A to 120-A range. This welding current flows through the ground cable. The cable moved a little whenever the ground clamp or the object being welded was moved. Thus, we see multiple groupings of the iron particles.

Coils

The magnetic field of a single conductor is too weak for many applications. A stronger field can be created by combining the fields associated with two or more conductors. This is done by coiling a conductor as shown in Fig. 7-13. A conductor formed into this shape is called a *coil*. Forming a coil out of a conductor creates an *electromagnet*. Notice that the coils in Fig. 7-13 have poles at the ends of the coil where the flux enters and leaves the center of the coil.

The *polarity of a coil* can be determined by again applying the left-hand rule. This time, wrap your fingers around the coil in the direction current flows through the turns of the coil. Your thumb will then point to the north pole (it also indicates direction of the flux). Reversing the direction of the current reverses the polarity of an electromagnet.

In Fig. 7-13(a) the three turns of the coil have considerable space between them. This spacing of the turns allows some of the flux to loop only a single conductor. The result is a weakening of the magnetic field at the poles of the magnet. This loss of flux at the poles can be minimized by *close-winding* the coil, as shown in Fig. 7-9(b). In this figure all the flux of the individual turns combines to produce one stronger magnetic field.

Coil

Electromagnet

Polarity of a coil

Close-winding

Self-Test

Answer the following questions.

11. Magnetic fields around a conductor are _____ to the conductor.

12. True or false. The magnetic field around a straight conductor has poles.

13. Is the direction of the flux around a conductor clockwise or counterclockwise when the current is flowing away from you?
14. When a wire is twisted into circular turns, it is called a(n) _____.
15. Refer to Fig. 7-10(*b*). Is the north pole above or below the two wires?
16. Increasing the space between the turns of a coil _____ the pole flux.
17. True or false. Lines of flux can cross each other.

7-4 Magnetic Materials

Domains

Magnetic materials

Materials that are attracted by magnetic fields (and materials from which magnets can be made) are called *magnetic materials*. The most common magnetic materials are iron, iron compounds, and alloys containing iron or steel. These magnetic materials are also called *ferromagnetic materials*. (*Ferro* is a prefix that means "iron.") A few materials, such as nickel and cobalt, are slightly magnetic. They are attracted by strong magnets. Compared with iron, however, they are only weakly magnetic.

Unmagnetized

Materials that are not attracted by magnets are called *nonmagnetic materials*. Most materials, both metallic and nonmetallic, are in this category. A magnet does not attract metals like copper, brass, aluminum, silver, zinc, and tin. Nor does a magnet attract nonmetals like wood, paper, leather, plastic, and rubber. A nonmagnetic material does not stop magnetic flux. Flux goes through nonmagnetic materials about as readily as it goes through air.

Magnetized

Theory of Magnetism

It is well known, and easily demonstrated, that an electric current produces a magnetic field. Since current is nothing but the movement of electric charges, any moving charge should create a magnetic field. The electrons of an atom or molecule possess electric charges, and they are in motion. Therefore, they should have a magnetic field associated with them. In the molecules of nonmagnetic materials, each electron is paired with another electron spinning in the opposite direction. This causes the magnetic field of one electron to be canceled by the opposite field of the other electron. The result is that the molecule ends up with no overall magnetic field. In magnetic materials, more electrons spin in one direction than in the other. Thus, not all the magnetic fields are canceled, and a

Permanent magnet

molecule of the material possesses a weak magnetic field. These molecules then group together to form small *domains* that have a magnetic field with definite poles.

Figure 7-14 represents the theoretical internal structure of a piece of magnetic material. In this figure, magnetic domains are represented by small rectangles. The domains are arranged randomly. They are also moving about randomly in the material. The magnetic fields of the randomly arranged domains cancel each other. Overall, the piece of material in Fig. 7-14 has no magnetic field or poles. It is an *unmagnetized* piece of magnetic material.

Temporary and Permanent Magnets

When a magnetic material is put in the magnetic field of a magnet, it becomes *magnetized*. As shown in Fig. 7-15, all the domains are aligned in an orderly fashion. The magnetic field of one domain supports the field of the next domain. The magnetic material in Fig. 7-15 became a magnet when it was magnetized. Whether it is a *temporary* or a *permanent magnet* depends on how it reacts when removed from the original magnetic field. If most of the domains remain aligned, as shown in Fig. 7-15, the magnetic material becomes a *permanent magnet*. Many alloys of iron, especially those that contain more than 0.8 percent carbon, become permanent

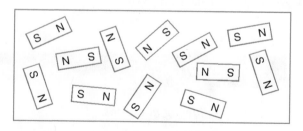

Fig. 7-14 Random arrangement of magnetic domains in an unmagnetized material.

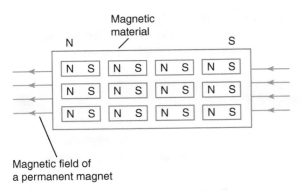

Magnetic material

Magnetic field of a permanent magnet

Fig. 7-15 Orderly arrangement of domains in a magnetized material.

magnets. Most tools, such as screwdrivers, pliers, and hacksaw blades, contain more than 0.8 percent carbon. They can become permanent magnets capable of attracting other magnetic materials. Most permanent magnets are made of alloys (such as alnico) which can be highly magnetized. *Alnico* magnets are composed of iron, cobalt, nickel, aluminum, and copper. Ceramic materials also make strong permanent magnets. Permanent magnets are used to make door catches and the magnetic poles for loudspeakers, electric meters, and motors.

Materials such as pure iron, ferrite, and *silicon steel* make *temporary magnets*. When these materials are removed from a magnetic field, their domains immediately revert to the random arrangement of Fig. 7-14. They lose almost all their magnetism. They no longer attract other magnetic materials. Temporary magnetic materials are used in great quantities in motors, generators, transformers, and electromagnets.

7-5 Magnetizing Magnetic Materials

Magnetic materials can be magnetized by the magnetic fields of either a permanent magnet or an electromagnet. Magnetizing with an electromagnet is shown in Fig. 7-16(a). When the switch is closed, the field from the coil magnetizes the tool steel in the screwdriver. Since tool steel makes a permanent magnet, the screwdriver remains magnetized after the switch is opened. Now the screwdriver can be used to attract other magnetic materials, such as steel screws [Fig. 7-16(b)].

When the material in the center of a coil is silicon steel or *ferrite,* a temporary magnet is made. Refer to Fig. 7-17(a), where a temporary

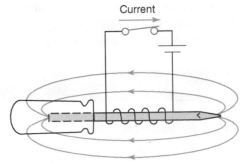

Current

(a) Magnetizing a screwdriver

(b) The screwdriver becomes a permanent magnet

Fig. 7-16 Making a permanent magnet.

Alnico

Silicon steel

Temporary magnets

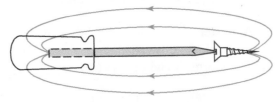

Ferrite rod

Current

(a) Magnetizing a ferrite rod (temporary magnet)

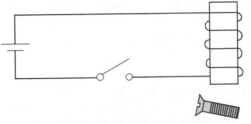

(b) When current stops, ferrite rod loses its magnetism

Fig. 7-17 Making a temporary magnet.

magnet is shown attracting a temporary-magnetic material. It continues to attract the material as long as current flows. However, as shown in Fig. 7-17(b), the magnetic field disappears the instant the switch is opened. The screw is no longer attracted to the ferrite

Ferrite

rod. However, if the screw were made from a permanent-magnet material, it would not separate from the rod when the switch was opened. Instead, the screw's permanent magnetic field would attract the ferrite rod, and the two would remain in contact.

⎍⋏⋏⎍ Self-Test

Answer the following questions.

18. The most strongly magnetic element is _____.

19. _____ steel is used for temporary magnets.

20. A cluster of molecules that has a small magnetic field is called a(n) _____.

21. Transformers and motors use _____ steel.

22. True or false. Most permanent magnets are made from an alloy rather than from a single element.

23. What happens to the flux in a ferrite rod in a coil when the electric circuit is opened?

24. What happens to the magnetic flux in a carbon-steel rod which is in a coil when the circuit is opened?

7-6 Magnetomotive Force

Magnetomotive force (mmf)

The effort exerted in creating a magnetic field (and flux) is called *magnetomotive force (mmf)*. Increasing either the number of turns or the current in the coil of Fig. 7-17 increases the mmf. It also increases the flux *if* the ferrite rod can support more flux.

7-7 Saturation

A magnetic material is saturated when an increase in mmf no longer increases the flux in the material. Increasing the coil current (or turns) in the coil shown in Fig. 7-17(*a*) increases the flux until the ferrite rod saturates. Once the rod has saturated, increasing either the current or the number of turns no longer increases the flux.

7-8 Demagnetizing

A permanent magnet can be partially demagnetized by hammering on it. It can also be demagnetized by heating it to a very high temperature. However, neither of these methods is very practical for demagnetizing many materials and devices. Most demagnetizing is done with a coil and an ac source, as shown in Fig. 7-18(*a*). Alternating current periodically reverses its direction of flow. It is also constantly varying in value [Fig. 7-18(*b*)]. In other words, alternating current starts at zero current, slowly increases

to maximum current, and then decreases back to zero current. At this time it reverses direction, increases to a maximum, and decays back to zero. This sequence is repeated rapidly (every $\frac{1}{60}$ s in the ac system in your home) as long as the circuit is connected. This change in value and direction of the current creates a magnetic field that changes in mmf and polarity. The changes in mmf [Fig. 7-18(*c*)]

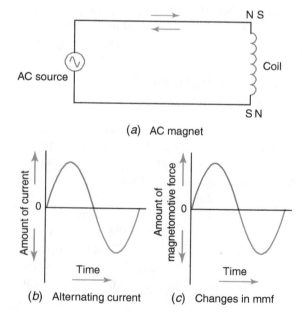

(*a*) AC magnet

(*b*) Alternating current   (*c*) Changes in mmf

Fig. 7-18 Electromagnetism with alternating current. The polarity of the magnetic field periodically reverses.

follow the changes in current. Both changes (current and mmf) are indicated in Fig. 7-18(a) by the arrows and the pole markings.

When a permanent-magnet material is put into the coil of Fig. 7-18(a), it is magnetized. It is magnetized first in one polarity and then in the reverse polarity. Just as the polarity is reversing, there is an instant when no magnetic field exists. If the switch were opened at that exact instant, the magnet would be demagnetized. However, the chances of opening the switch at that instant are very small. It is very likely that the switch would be opened at some other instant, when a magnetic field was present. Then the magnet would still be magnetized. The strength of the magnet depends on the strength of the field at the instant the switch is opened.

To demagnetize with the circuit of Fig. 7-18, the switch must remain closed as the magnet is removed. As the magnet is removed, it is being magnetized first in one polarity and then in the other. However, each reversal of polarity yields a weaker magnet. This is because the field of the coil gets weaker as the magnet is moved farther away. When the permanent-magnet material has been moved several feet away from the coil, it is completely demagnetized. The switch to the coil can then be opened.

An alternative approach to removing the magnet from the coil connected to an ac source is to slowly reduce the magnitude of the alternating current until it reaches zero. As the current decays to zero, so does the strength of the magnet.

A soldering gun like the one shown in Fig. 7-19 can be used to either magnetize or demagnetize. The soldering tip and connecting leads are a single-turn coil. The coil has a large

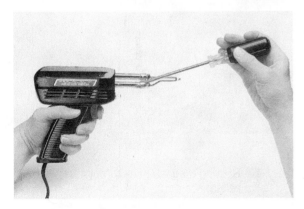

Fig. 7-19 Magnetizing and demagnetizing with a soldering gun. The soldering gun tip is a one-turn coil.

alternating current flowing in it when the gun is on. To magnetize the screwdriver shown in Fig. 7-19, just turn the gun on and off with the screwdriver located as shown. To demagnetize, turn the gun on and pull the screwdriver away from the gun. Then turn off the gun.

7-9 Residual Magnetism

The flux that remains in a temporary magnet after it is removed from a magnetic field is called *residual magnetism*. All magnetic materials retain some flux after they have been exposed to a magnetic field. Permanent-magnet materials retain a high magnetism. The ideal permanent magnet would retain all the flux. The ideal temporary magnet would retain no flux.

Temporary-magnet materials have very low residual magnetism. In most temporary magnets, the magnetic field is created by current-carrying conductors. When the current stops, the temporary-magnet material does not become completely unmagnetized. A small residual magnetism always remains. However, the magnetic field of the residual magnetism is very weak.

Residual magnetism causes ac-operated magnetic devices to heat up. It is one of the reasons devices like motors and transformers are not 100 percent efficient.

7-10 Reluctance

The opposition to magnetic flux is called *reluctance*. The reluctance of an object depends upon both the material and the dimensions of the material. Nonmagnetic materials have about the same (within 1 percent) reluctance as air does. On the other hand, air has 50 to 5000 times as much reluctance as common magnetic materials do. For example, silicon steel used in motors and transformers has about $\frac{1}{3000}$ as much reluctance as air.

When offered a choice, flux takes the lowest-reluctance path. In Fig. 7-20(a), the flux extends away from the poles into the surrounding air. However, in Fig. 7-20(b), the flux concentrates in the small piece of iron. This is because the iron has much, much less reluctance. The insertion of the iron between the poles concentrates the flux into a smaller volume. If the iron in Fig. 7-20(b) were replaced by a piece of lead, the flux pattern would be the same as it is in Fig. 7-20(a).

Residual magnetism

Reluctance

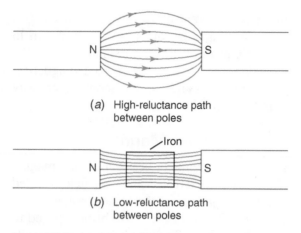

(a) High-reluctance path between poles

(b) Low-reluctance path between poles

Iron

N   S

N   S

Fig. 7-20 Reluctance. Iron has less reluctance than air does.

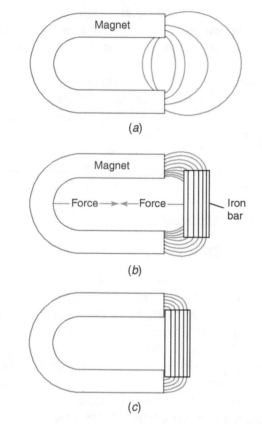

Magnet

(a)

Magnet

Force →   ← Force

Iron bar

(b)

(c)

Fig. 7-21 Attraction of magnetic materials. The flux in (a) is concentrated and shortened by the magnetic material in (b) and (c).

Magnetic materials are attracted to a magnet because of their low reluctance. Magnetic flux lines try to take the shortest path between the poles of a magnet. They also try to take the path of lowest reluctance. The flux lines can do both if they can pull the surrounding magnetic material to the magnet. This idea is illustrated in Fig. 7-21. The normal lines of force around

the poles of a horseshoe magnet are shown in Fig. 7-21(a). In Fig. 7-21(b), the flux lines have bent to take the path of low reluctance through the iron. There is a force trying to shorten the lines of force. This resembles the force exerted by a stretched rubber band. If either the bar or the magnet is free to move, the two will come together, as shown in Fig. 7-21(c). Now the flux has both the shortest path and the lowest-reluctance path.

The pictures in Fig. 7-22 show the iron filing distribution around a semicircular magnet with and without a magnetic material between the two poles. Figure 7-22(b) shows that nearly all of the flux is passing through the low reluctance of the temporary magnetic bar.

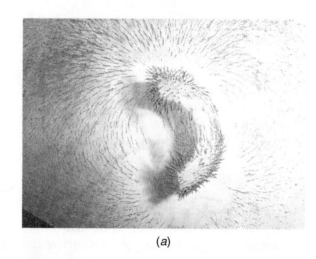

(a)

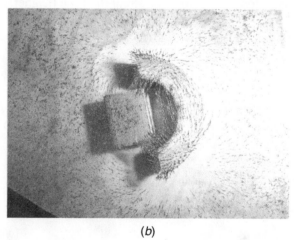

(b)

Fig. 7-22 Routing magnetic flux through a low-reluctance path. (a) A picture illustrating lines of force around a semicircular magnet. (b) Nearly all of the lines of force appear to be going through the bar of magnetic material.

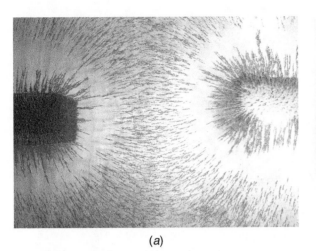

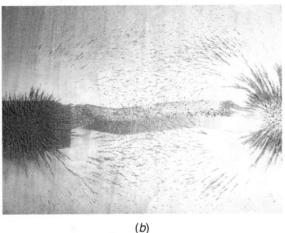

(a)　　　　　　　　　　　　　　　　(b)

Fig. 7-23 Guiding flux through a low-reluctance iron path. (*a*) Flux pattern between north and south poles with no iron guide. (*b*) Flux pattern between north and south poles with an iron guide.

7-11　Magnetic Shields

Magnetic flux can be bent, distorted, and guided by low-reluctance materials inserted in the magnetic field. For example, most of the flux in Fig. 7-23(*b*) is being bent and guided through the crooked iron bar. *Magnetic shields* make use of the tendency of flux to distort and follow the path of lowest reluctance. A magnetic shield is just a material of very low reluctance. The material is put around the object which is to be protected from any magnetic field in its vicinity. Figure 7-24 illustrates how a magnetic object, such as a watch movement, can be protected by a shield.

Magnetic shields

Self-Test

Answer the following questions.

25. The force that creates magnetic flux is a(n) _____.
26. When an increase in mmf causes no increase in the flux in a material, the material is _____.
27. True or false. Opening and closing the switch controlling an ac coil will demagnetize the material inserted in the coil.
28. True or false. A dc coil can be used to demagnetize a permanent-magnet material.
29. Magnetism remaining in a material after the material has been removed from a magnetic field is called _____.
30. Which will have more residual magnetism, a permanent-magnet material or a temporary-magnet material?
31. Why are magnetic materials attracted to a magnet?
32. True or false. Magnetic shields are made from nonmagnetic materials.

7-12　Induced Voltage

We have seen that a current-carrying conductor produces a magnetic field. It is equally true that a magnetic field can induce a voltage (and thus a current) in a conductor. When a conductor moves across lines of flux (Fig. 7-25), a voltage is induced in the conductor. The polarity of the *induced voltage* depends on the direction of motion of the conductor and on the direction of the flux. Changing the direction of either changes the polarity of the induced voltage.

Induced voltage

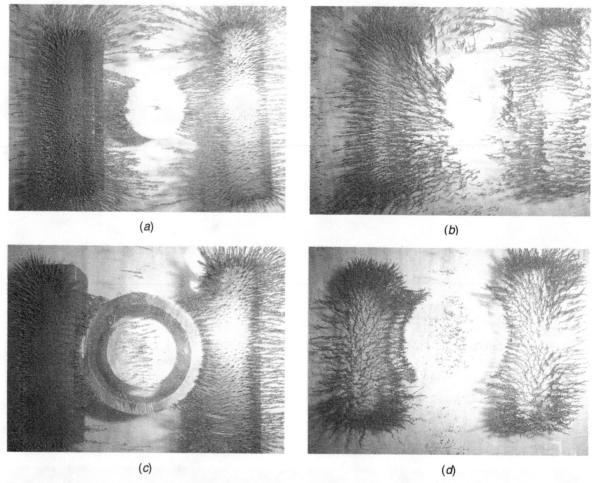

Fig. 7-24 A magnetic shield provides a low-reluctance path to direct the flux around the magnetic object to be protected. (*a*) Flux pattern with magnetic object in the magnetic field. (*b*) Flux pattern with magnets and object removed. (*c*) Flux pattern with magnetic shield in the magnetic field. (*d*) Flux pattern with magnets, object, and shield removed.

Electric generator

Commutator

Brushes

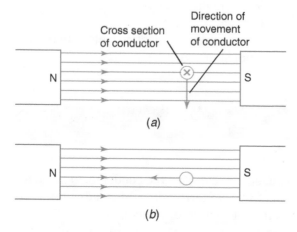

Fig. 7-25 (*a*) Voltage induced in a moving conductor. The conductor must cut the lines of flux. (*b*) No voltage induced in the conductor.

Generator Action

Induced voltage is the principle on which an *electric generator* operates. When the shaft of a generator is turned, a conductor loop is forced through a magnetic field. A simplified dc generator is shown in Fig. 7-26. As illustrated, the ends of the loop are connected to *commutator* bars and *brushes*. The commutator and brushes serve two purposes. First, they make an electric connection to the rotating loop of wire. The ends of the loop are permanently attached to the commutator bars. As the loop turns, the commutator turns with it. The commutator bars touch the brushes and slide against them as the loop rotates. Second, the commutator and brushes reverse connections to the rotating loop every time the polarity of the induced voltage changes. The polarity of the induced voltage changes because the relative direction of

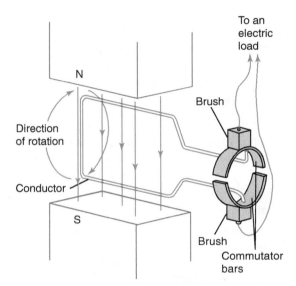

Fig. 7-26 Generator principle.

motion of the conductor reverses. The top part of the rotating loop in Fig. 7-26 is shown moving from left to right. Ninety degrees later it will start moving from right to left and its polarity will reverse. At the same time, the brushes will change commutator bars. Thus, one brush is always negative with respect to the other brush. This provides a *dc output* from the generator.

Transformer Action

Induced voltage is also created when a conductor is in the immediate vicinity of a *changing* magnetic *flux*. The conductor must be perpendicular to the lines of flux (Fig. 7-27). When the switch in Fig. 7-27 is closed and opened, the current in the circuit starts to flow and then stops. The changing current in the top coil produces a changing flux in the iron core. This changing flux in the core induces a voltage in the bottom coil. The magnitude of the induced voltage is a direct function of both the amount of flux change and the rate of flux change.

History of
Electronics

William Sturgeon (1783–1850)
William Sturgeon was self-educated in electrical science. In 1823 he built the first electromagnet. Sturgeon discovered that he could create a magnetic field by running an electric current through a coil of copper wire wrapped around a soft piece of iron. In 1832 he invented the commutator for electric motors, and in 1836 made the first moving-coil galvanometer.

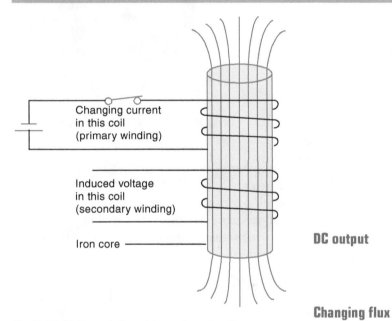

Fig. 7-27 Voltage induced by a changing flux.

Voltage induced by changing flux is the principle on which *transformers* and other devices operate. The ignition coil (which is a transformer) of an automobile operates similarly to the circuit shown in Fig. 7-27. In the ignition coil, the secondary coil has thousands of turns of fine wire. This provides the thousands of volts needed by the spark plugs.

 Self-Test

Answer the following questions.

33. List the factors that determine the amount of voltage induced in a conductor by transformer action.

34. What are the functions of the commutator and brushes in a generator?

35. What is the principle on which an automotive ignition coil operates?

History of
Electronics

Wilhelm Eduard Weber

Named for Wilhelm Eduard Weber, a weber is a unit of magnetic flux, representing one line of flux in the meter-kilogram-second (mks) or Système International (SI) system of units. The weber is equivalent to a volt-second.

7-13 Magnetic Quantities and Units

A number of magnetic quantities, such as flux, have already been discussed. However, we can develop a deeper understanding of magnetism and magnetic devices when we know the units in which these quantities are specified and measured.

Magnetomotive Force— The Ampere-Turn

You May Recall

. . . that in Sec. 7-6 we defined magnetomotive force (mmf) as the force that creates flux.

Weber (Wb)

Ampere-turn (A · t)

We will use the *ampere-turn* (A · t) for the base unit of mmf. One ampere-turn is the mmf created by 1 A flowing through one turn of a coil. Three ampere-turns of mmf are created by 1 A moving through three turns. Three ampere-turns are also equal to 3 A through one turn.

EXAMPLE 7-1

What is the mmf of the coil in Fig. 7-27 if 2.8 A of current is flowing in the circuit?

| **Given:** | Number of turns (N) = 3 t |
| | Current (I) = 2.8 A |

Find:	mmf
Known:	mmf = turns × current
Solution:	mmf = 3 t × 2.8 A = 8.4 A · t
Answer:	The mmf of the coil is 8.4 A · t.

Three ampere-turns can be created by any combination in which the product of amperes times turns equals 3.

Strictly speaking, the base unit of mmf in the metric system is just the ampere. At first, this may seem confusing because the ampere is also the base unit of current in the metric system. The reasonableness of using ampere as the unit of mmf is illustrated in Fig. 7-28. In Fig. 7-28(*a*), 4 A of current is moving around the iron core and creating flux. In Fig. 7-28(*b*), 4 A is again moving around the iron core. The fact that the 4 A is split into two paths does not change the total force exerted in creating flux. The same reasoning extends to Fig. 7-28(*c*). The four 1-A currents across the iron produce the same mmf as one 4-A current. Thus, it is the total current flowing around the iron that determines the mmf. The number of turns carrying the current really does not matter. However, the easiest way to determine the total current flowing by the iron is to multiply the coil current by the number of turns. Thus, the ampere-turn is a more descriptive unit than the ampere. Therefore, we will use ampere-turn as the unit of mmf.

Flux—The Weber

The base unit of magnetic flux (ϕ) is the weber. A *weber* (Wb) can be defined only in terms of a change in the flux in a magnetic circuit. One

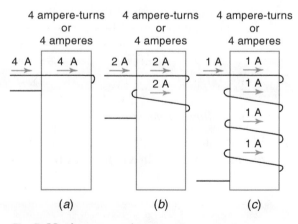

Fig. 7-28 Ampere and ampere-turn.

weber is the amount of flux change required in 1 s to induce 1 V in a single conductor. Refer to Fig. 7-27 for an example of the use of the weber. The bottom coil has a voltage induced in it when the flux changes. Note that this coil is a three-turn coil. When the flux in the core changes 1 Wb in 1 s, 3 V is induced in the coil.

Magnetic Field Strength— Ampere-Turn per Meter

Magnetizing force, field intensity, and *magnetic field strength* all mean the same thing. They refer to the amount of mmf available to create a magnetic field for each unit length of a magnetic circuit. The symbol for magnetic field strength is *H*. The base unit of magnetic field strength is the *ampere-turn per meter* (A · t/m). In the magnetic circuit of Fig. 7-29(*a*), a four-turn coil carries 3 A of current. Therefore, the mmf is 12 A · t (3 A × 4 t = 12 A · t). The average length of the magnetic circuit (iron core) of Fig. 7-29(*a*) is 0.25 m; therefore, the magnetic field strength is the mmf divided by the circuit length:

$$H = \frac{\text{mmf}}{\text{length}}$$
$$= \frac{12 \text{ A} \cdot \text{t}}{0.25 \text{ m}}$$
$$= 48 \text{ A} \cdot \text{t/m}$$

In Fig. 7-29(*b*) the mmf is 4.8 A × 6 t = 28.8 A · t, and the magnetic field strength is

$$H = \frac{\text{mmf}}{l} = \frac{28.8 \text{ A} \cdot \text{t}}{0.6 \text{ m}} = 48 \text{ A} \cdot \text{t/m}$$

where *l* is the length.

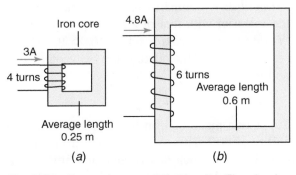

Fig. 7-29 Comparing magnetic circuits. The circuits in (*a*) and (*b*) have the same magnetic field strength (48 A · t/m).

History of
Electronics

Karl Friedrich Gauss (1777–1855)
Karl Gauss was a German mathematician who made important contributions to the field of electricity and magnetism. In fact, Gauss had such an impact on the new science of electricity and magnetism that the cgs unit of magnetic flux density, the gauss, is named for him.

Notice that the magnetic field strength is the same for both circuits in Fig. 7-29. Yet the two circuits have different currents, different numbers of turns, different mmfs, and different lengths. Specifying magnetic field strength rather than mmf makes it possible to compare magnetic circuits without specifying physical dimensions.

Magnetizing force [*H*]

Magnetic field strength

Ampere-turn per meter (A · t/m)

EXAMPLE 7-2

What is the magnetizing force of a magnetic circuit with 150 turns on a core with an average length of 0.3 meters if the current is 0.4 amperes?

Given: Number of turns (*N*) = 150 t
             Current = 0.4 A
             Length (*l*) = 0.3 m
Find: Magnetizing force (*H*)
Known: mmf = *IN*

$$H = \frac{\text{mmf}}{l}$$

Solution: mmf = 0.4 A × 150 t = 60 A · t

$$H = \frac{60 \text{ A} \cdot \text{t}}{0.3 \text{ m}} = 200 \text{ A} \cdot \text{t/m}$$

Answer: The magnetizing force is 200 ampere-turns per meter.

Flux Density—The Tesla

The amount of flux per unit cross-sectional area is called *flux density* (*B*). The base unit of flux density is the *tesla* (T). One tesla is equal to one weber per square meter. Specifying flux density

Flux density (*B*)

Tesla (T)

Permeability (μ)

$$\mu = \frac{B}{H}$$

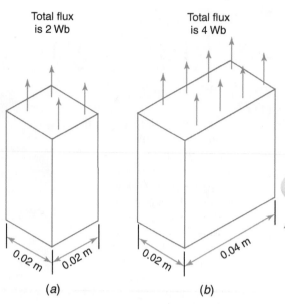

Total flux is 2 Wb

Total flux is 4 Wb

0.02 m   0.02 m

0.02 m   0.04 m

(a)

(b)

Fig. 7-30 Comparing flux density. Both (a) and (b) have a flux density of 5000 T.

rather than flux also makes it easier to compare magnetic circuits of unequal size. For example, the iron cores in Fig. 7-30 have equal flux density. Yet they have different cross-sectional areas. Flux density can be calculated by dividing the total flux by the cross-sectional area. In Fig. 7-30(a), the cross-sectional area is

Area = 0.02 m × 0.02 m = 0.0004 m²

The flux density therefore is

$$\text{Flux density} = \frac{\text{flux}}{\text{area}} = \frac{2 \text{ Wb}}{0.0004 \text{ m}^2}$$

$$= 5000 \text{ T}$$

If you calculate the flux density for Fig.7-30(b), you will find that it is also 5000 T.

Relative permeability μ_r

Permeability of air μ_0

$$\mu_r = \frac{\mu}{\mu_0}$$

EXAMPLE 7-3

How much flux will there be in a core that is 0.03 m × 0.05 m and has a flux density of 400 teslas?

Given: Dimensions = 0.03 m × 0.05 m
               Flux density (B) = 400 T

Find: Flux (ϕ)

Known: $B = \dfrac{\phi}{\text{area}}$; therefore, $\phi = B$ area

Solution: ϕ = 400 T × 0.03 m × 0.05 m
             = 0.6 Wb

Answer: The flux is 0.6 weber.

Permeability and Relative Permeability

Permeability refers to the ability of a material to conduct flux. Its symbol is μ. Permeability is defined as the ratio of flux density to magnetic field strength:

$$\mu = \frac{B}{H}$$

You May Recall

... that both flux density and field strength are independent of size.

Therefore, permeability is also independent of size. Substituting the base units for the flux density and magnetizing force into the above formula show that the base unit for permeability is the weber per ampere-turn-meter [Wb/(A · t · m)].

EXAMPLE 7-4

What is the permeability of a material when a magnetizing force of 100 A · t/m produces a flux density of 0.2 T?

Given: $H = 100$ A · t/m, $B = 0.2$ T

Find: μ

Known: $\mu = B/H$, T = Wb/m²

Solution: $\mu = \dfrac{0.20 \text{ Wb/m}^2}{100 \text{ A} \cdot \text{t/m}}$
              = 0.002 Wb/(A · t · m)

Answer: The permeability is 0.002 weber per ampere-turn-meter.

Often the permeability of a material is specified as relative permeability (μ_r). *Relative permeability* compares the permeability of the material with that of air. The *permeability of air* (μ_0) is approximately 1.26×10^{-6} Wb/(A · t · m). Thus the relationship between μ, μ_0, and μ_r is

$$\mu_r = \frac{\mu}{\mu_0}$$

Suppose a piece of iron has a relative permeability of 600. This means that the iron carries 600 times as much flux as an equal amount of air. Since relative permeability is a ratio of two permeabilities, the units cancel out. Thus, relative permeability is a unitless (pure) number.

EXAMPLE 7-5

A magnetic material has a permeability (μ) of 6.3×10^{-4} Wb/(A · t · m). Determine its relative permeability (μ_r).

Given: $\mu = 6.3 \times 10^{-4}$ Wb/(A · t · m)

Find: μ_r

Known: $\mu_o = 1.26 \times 10^{-6}$ Wb/(A · t · m)

$\mu_r = \dfrac{\mu}{\mu_o}$

Solution: $\mu_r = \dfrac{6.3 \times 10^{-4} \text{ Wb/(A · t · m)}}{1.26 \times 10^{-6} \text{ Wb/(A · t · m)}}$

Answer: $\mu_r = 500$

The relative permeability of all nonmagnetic materials is very close to 1. Relative permeabilities of magnetic materials range from about 30 to more than 6000.

The permeability of a magnetic material decreases as the flux density increases. As a material approaches saturation, its permeability

History of
Electronics

Nikola Tesla (1857–1943)
The tesla is the SI unit for magnetic flux density, equal to 1 weber per square meter. It was named for Nikola Tesla, who immigrated to the United States from Europe in 1884 and went to work for Thomas Edison. Tesla wanted to develop an ac induction motor, but Edison did not want alternating current to compete with his dc power plants. Tesla left and received many patents for ac uses. Backed by George Westinghouse, Tesla's ac power was adopted in many cities and became the worldwide power standard.

decreases rapidly. When the material is fully saturated, the permeability is only a small fraction of its value at lower flux densities.

Self-Test

Answer the following questions.

36. What is the mmf of a 200-t coil when the coil current is 0.4 A?
37. Magnetic field strength is also called _____ or _____.
38. The base unit for magnetic field strength is _____.
39. The base unit for flux is _____.
40. The base unit for flux density is _____.
41. The ratio of B to H is called _____.
42. What is the advantage of specifying flux density rather than flux?
43. Define relative permeability.
44. True or false. One tesla equals one weber per meter.

7-14 Electromagnets

You May Recall

. . . that we have already discussed many of the ideas of an electromagnet in Secs. 7-3 and 7-5.

A coil of wire with a current through it is an electromagnet. However, most useful electromagnets are wound on, and partially encased in, a material that can become a temporary magnet.

The *strength of an electromagnet* depends upon four factors:

1. The type of core material (the magnetic material)
2. The size and shape of the core material
3. The number of turns on the coil
4. The amount of current in the coil

In general, an electromagnet is strongest when

1. The core material has the highest permeability

Strength of an electromagnet

Commutator

Brushes

2. The core material has the largest cross-sectional area and the shortest length for flux lines. These attributes result in a low reluctance.
3. The number of turns and amount of current are greatest. This provides a large mmf.

Electromagnets are used in industry for a variety of jobs. They are used to hold steel while it is being machined, sort magnetic from nonmagnetic materials, lift and move heavy iron and steel products. Figure 7-31 shows a large electromagnet piling scrap metal.

7-15 DC Motors

DC motor

Physically and electrically, a *dc motor* resembles a dc generator. In fact, in some cases a single machine may be used as either a generator or a motor. Whereas the shaft of a generator turns because of some outside mechanical force, the shaft of a motor turns because of the interaction of two magnetic fields within the motor. The principle of a simple dc motor is illustrated in Fig. 7-32. Current from an external source flows through the brushes, commutator, and *armature coil*. This produces a magnetic field in the armature iron. The armature poles are attracted by the *field poles* of the permanent magnet. The result is a rotational force called *torque,* as shown in Fig. 7-32(a).

Armature coil

Field poles

Torque

Fig. 7-31 Large industrial electromagnet.

The *commutator* and *brushes* change the direction of current in the armature coil every 180°. This change reverses the magnetic polarity of the armature so that the direction of the torque (rotational force) is always the same. When the armature in Fig. 7-32(a) is rotated 90°, its south pole is in line with the field's north pole. However, the inertia (which is the tendency of a moving body to keep moving) of the moving armature carries it to just past the vertical position. At this time, the current in the armature coil changes and the armature field reverses. Now the north pole of the armature is repelled by the north pole of the field [Fig. 7-32(b)].

In general, motors have more than two commutator segments and many more armature coils. With more commutator segments, it is not necessary to rely on inertia as the armature polarity is reversed. As seen in Fig. 7-33, the segments change an instant before the poles of the fields are aligned. During the time that each brush is in contact with two segments of the commutator, four poles are created. However, as shown in Fig. 7-33(b), all four poles provide torque in the same direction.

The field poles in a dc motor can be either permanent magnets or electromagnets. When electromagnets are used, the current in the field never changes direction. Thus dc motors with

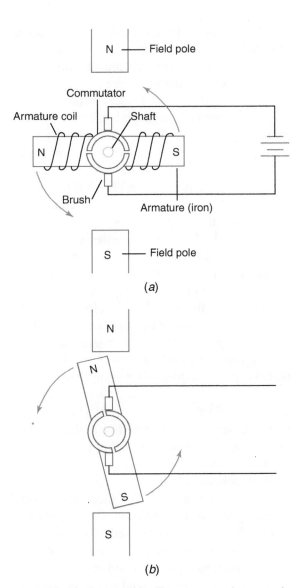

Fig. 7-32 Motor principle. The armature is rotated by alternate attraction (a) and repulsion (b) of magnetic poles.

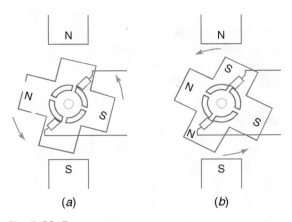

Fig. 7-33 Four-segment commutator.

electromagnetic fields operate in the same way as those with permanent-magnet fields.

7-16 Solenoids

A *solenoid* is an electromagnetic device that enables an electric current to control a mechanical mechanism. The movable *plunger* in Fig. 7-34 can be used to operate, for example, a mechanical brake, a clutch, or a valve to control the flow of a liquid.

When the solenoid is deenergized [Fig. 7-34(b)], there is no magnetic force to hold the plunger in the center of the coil. When current is supplied to the coil, the solenoid is energized and the plunger is pulled into the coil because of its high permeability. The plunger has much less reluctance than the air it replaces. As shown in Fig.7-34(c), the flux path is almost entirely through iron when the solenoid is energized. The pull of the plunger of a solenoid is dependent on its magnetic properties. In general, increasing the mmf or decreasing the reluctance increases the pull.

Most solenoids are made to operate from an ac supply. However, dc solenoids are also available. The two types (ac and dc) are not interchangeable.

Solenoid

Plunger

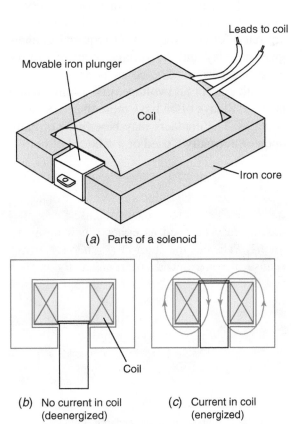

(a) Parts of a solenoid

(b) No current in coil (deenergized)

(c) Current in coil (energized)

Fig. 7-34 Movable plunger.

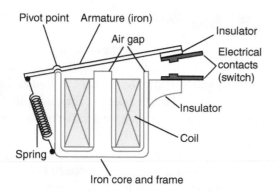

Fig. 7-35 Relay. The armature is attracted to the iron core when current flows through the coil.

7-17 Relays

Magnetic relay

Armature

A *magnetic relay* uses the attraction between an iron *armature* and an energized coil to operate a pair of electric contacts. Figure 7-35 shows the cross section of a relay in the deenergized position. When current flows through the coil, the armature is pulled down against the iron core. This closes the set of contacts that can complete another electric circuit. When current stops flowing in the coil, the spring pulls the armature up and opens the contacts.

Relays are manufactured in a wide variety of ratings for both the coil and the contacts. Coils are rated for both the current required to energize the relay and the voltage required to produce that current. The contacts are also rated for both current and voltage, just as any switch is rated. Relays often have more than one set of contacts. The contacts may be either normally open or normally closed or a mixture of both.

Relay contacts have a thick layer of special materials that minimize arcing, pitting, and wear as the relay turns circuits off and on. This surface layer can be seen in the picture in Fig. 7-36(*a*), which shows the normally closed contacts (two sets) that were unused in a relay. This particular relay used only the two sets of normally open contacts. The normally open contacts that turned on and off a 240-V circuit that powered an 11-kW load are pictured in Fig. 7-36(*b*). When the special materials layer on the contacts is worn away, the base metal of the contacts can arc and weld the contacts together to the extent that the contacts remain closed even when the solenoid coil is deenergized. This is what happened with the contacts shown in Fig. 7-36(*b*). As a result of these sticking (welded) contacts, the load was not turned off until other parts of the system were damaged to the extent that excess current finally tripped the system's main circuit breaker. If accessible, relays should be checked often and the contacts cleaned with a burnishing tool or fine-grit energy paper. However, the relay (or the contacts) should be replaced when the contacts are worn (or deeply pitted) nearly to the base metal.

Two schematic symbols used for relays are shown in Fig. 7-37. For electrical control devices, such as a motor controller, the symbol in Fig. 7-37(*b*) is usually used. The symbol in Fig. 7-37(*a*) is commonly used for electronic systems such as computers and transmitters.

A relay may be energized by either alternating or direct current. However, ac and dc relays are not interchangeable. A dc relay operated

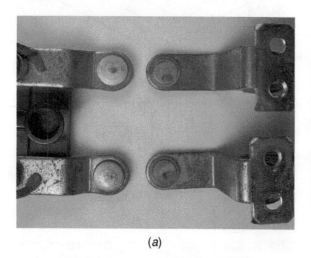

(a) (b)

Fig. 7-36 Condition of relay contacts. (*a*) Contacts that have not turned a circuit on and off. (*b*) After many on and off operations, these contacts show severe wear from arcing.

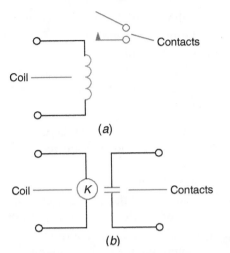

Fig. 7-37 Schematic symbols for a relay with one set of open contacts. (*a*) Symbol commonly used in schematics for electronic devices. (*b*) Symbol commonly used in schematics for electrical control devices.

on alternating current will chatter because the flux in the core builds up and collapses as the current builds up and decays back to zero. An ac relay has a *shading coil* (separate from the main coil) that delays the collapse of the flux in part of the core material until flux starts to rebuild in another part of the core. This eliminates the chatter of the relay.

One of the chief advantages of a relay over a simple switch is that it allows *remote operation*. A low-voltage, low-current supply can control the relay coil. Then the relay contacts can control a high-voltage and/or high-current circuit. The switch that operates the coil can be located in a remote place. Only a small two-conductor cable has to run between the switch and the relay.

Remote operation

Self-Test

Answer the following questions.

45. True or false. The core of an electromagnet should be made from a material with high reluctance.
46. True or false. Changing the core of an electromagnet from a low-permeability material to a high-permeability material will increase the strength of the magnet.
47. True or false. The field in a dc motor periodically changes magnetic polarity.
48. True or false. Most solenoids can be operated with either alternating current or direct current.
49. True or false. A single relay can control power to two circuits even if the circuits operate from different power sources.
50. What is the purpose of the commutator and brushes in a dc motor?
51. Why is the plunger of a solenoid pulled into the center of the coil?
52. List the major electrical ratings needed to specify a relay.

7-18 Hall-Effect Devices

Edwin H. Hall, an American physicist, discovered the *Hall effect* in 1879. He found that a voltage was produced on opposing surfaces of a current-carrying conductor when the conductor was in a magnetic field. This voltage is called the *Hall-effect voltage* or simply the *Hall voltage*. As illustrated in Fig. 7-38, the flux must be perpendicular to the direction of the current and the Hall voltage is produced transverse to the direction of the current flow. The magnitude of the Hall voltage is a function of three things:

1. The type of material carrying the current
2. The magnitude of the current
3. The amount of flux

Hall effect

Semiconductor materials tend to produce the largest amount of Hall voltage. Typical Hall voltages are in the millivolt range.

The small voltage generated by the Hall-effect plate can be amplified by electronic circuits integrated into the same semiconductor material used for the Hall-effect plate. A simple Hall-effect device, including the integrated electronic circuitry, is often packaged in a

Hall voltage

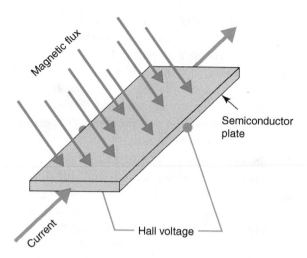

Magnetic flux

Semiconductor plate

Current

Hall voltage

Fig. 7-38 The Hall effect. The Hall-effect voltage is transverse to the direction of the current flow.

three-pin unit. One pin is the common negative for the supply voltage and the output voltage. Another pin is for the positive supply voltage, and the third pin provides the output voltage. If the device is a linear sensor, the output voltage is an analog voltage that varies as the flux density acting on the device varies. If the device is a Hall-effect switch, the output will be one of two discrete voltage levels. When the flux is below a certain critical density, the output voltage is at the lower level. When the flux increases to the specified critical density, the output immediately and rapidly switches to the

higher voltage level. These two voltage levels can be used to represent the 0 and 1 digits needed to control digital systems.

Hall-effect sensors and switches are used in a wide range of industrial and automotive systems. Several examples are:

1. A Hall-effect sensor can be used to measure the level of current in a conductor. The greater the current in the conductor, the denser the flux around the conductor and the greater the Hall-voltage output of the device subjected to this flux.
2. The brake lights on an automobile can be activated when a magnet attached to the brake pedal is pulled away from a Hall-effect switch.
3. Ferromagnetic parts can be counted as they pass between a Hall-effect switch and a permanent magnet acting on the switch. When the part is between the switch and the magnet, the reluctance between the switch and the magnet is greatly reduced and the flux increases to the critical level needed to activate the switch.
4. The position of the permanent-magnet rotor in a brushless dc motor can be detected by Hall-effect devices. The output of the Hall-effect devices determines which coils in the motor are to be energized (receive current) at which times.

 Self-Test

Answer the following questions.

53. What determines the magnitude of the Hall-effect voltage?

54. How does a Hall switch differ from a Hall sensor?

55. Can the Hall-effect voltage be increased by increasing the current through the device?

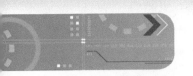

Chapter 7 Summary and Review

Summary

1. Magnetism is an invisible force field.
2. A magnetic field exists around a magnet.
3. The lines of force of a magnetic field are called magnetic flux.
4. The denser the flux, the stronger the magnetic field.
5. The magnetic field is strongest at the poles of a magnet.
6. The direction of external flux is from the north pole to the south pole.
7. Like magnetic poles repel; unlike poles attract.
8. When magnetic poles attract each other, their flux lines join together.
9. Lines of flux are continuous loops.
10. Magnetic fields exist around all current-carrying conductors.
11. The magnetic field around a straight conductor has no poles.
12. The left-hand rule is used to determine the direction of flux around a conductor.
13. The strength of the magnetic field around a conductor is directly proportional to the current in the conductor.
14. Two parallel conductors attract each other if their currents are in the same direction. They repel if the currents are in opposite directions.
15. A coil of wire carrying a current forms magnetic poles.
16. The north pole of a coil can be determined by use of the left-hand rule.
17. Decreasing the space between turns of a coil increases the flux at the poles of a coil.
18. Iron, nickel, and cobalt are all magnetic materials.
19. Iron is the most magnetic of the elements.
20. Nonmagnetic materials do not prohibit flux from passing through them.
21. The domains of a permanent magnet remain aligned after the magnet is removed from the magnetizing field.
22. Magnetic materials are magnetized when put in the fields of other magnets.
23. Magnetomotive force can be increased by increasing the current or the turns in a coil.
24. When a material is saturated, increasing the mmf does not increase the flux.
25. Residual magnetism refers to the flux that remains in a temporary magnetic material after it is removed from a magnetic field.
26. Reluctance is the opposition to magnetic flux.
27. Magnetic materials have less reluctance than nonmagnetic materials.
28. A changing magnetic flux can induce a voltage in a conductor.
29. Magnetic field strength is also known as field intensity and magnetizing force.
30. Flux density (B) is the flux per unit cross-sectional area of a magnetic material.
31. One tesla is one weber per square meter.
32. Permeability refers to the ability of a material to pass or carry magnetic flux. It is equal to flux density divided by magnetic field strength.
33. Relative permeability is the ratio of the permeability of a material to the permeability of air.
34. The strength of an electromagnet is determined by the reluctance and permeability of the core and by the mmf.
35. Direct current motors rotate because of attraction and repulsion between the magnetic fields of the field poles and the armature.
36. The plunger of a solenoid is made from a high-permeability material.
37. The pull of a solenoid depends on the mmf and on the reluctance of its magnetic circuit.
38. A relay is an electromagnetic switch.

Related Formulas and Tables

mmf = turns × current

$$H = \frac{\text{mmf}}{\text{length}}$$

$$B = \frac{\phi}{\text{area}}$$

$$\mu = \frac{B}{H}$$

$$\mu_r = \frac{\mu}{\mu_0}$$

Magnetic Quantities and Units

Quantity		Unit	
Name	Symbol	Name	Symbol
Magneto-motive force	mmf	Ampere-turn	A · t
Flux	ϕ	Weber	Wb
Magnetic field strength	H	Ampere-turn per meter	A·t/m
Flux density	B	Tesla	T
Permeability	μ	Weber per ampere-turn-meter	Wb/(A·t·m)

Chapter Review Questions

For questions 7-1 to 7-10, determine whether each statement is true or false.

7-1. In a magnetic circuit, flux leaves the north pole and enters the south pole. (7-2)

7-2. Breaking a magnet into two pieces demagnetizes the magnet. (7-2)

7-3. Unlike magnetic poles repel each other. (7-2)

7-4. Both a north and a south pole are produced by all magnetic fields. (7-2)

7-5. Lines of flux are continuous loops that follow the path of least opposition. (7-2 and 7-10)

7-6. Decreasing space between turns of a coil will decrease the pole flux. (7-3)

7-7. Permanent-magnet materials are usually used in motors and transformers. (7-4)

7-8. An ac coil can be used either to magnetize or to demagnetize a permanent-magnet material. (7-8)

7-9. Permanent-magnet materials have very little residual magnetism. (7-9)

7-10. When the current in a conductor is flowing away from you, the direction of the flux will be counterclockwise. (7-3)

For questions 7-11 to 7-21, supply the missing word or phrase in each statement.

7-11. Nonmagnetic materials do not have magnetic _____. (7-4)

7-12. Magnetic shields are made from _____ materials. (7-11)

7-13. The _____ is the rotating part of a dc motor. (7-15)

7-15. The _____ is the base unit of flux density. (7-13)

7-15. The _____ is the base unit of flux. (7-13)

7-16. The _____ is the base unit of mmf. (7-13)

7-17. The _____ is the base unit of magnetizing force. (7-13)

7-18. Inducing a voltage into a moving conductor is called _____ action. (7-12)

7-19. Inducing a voltage into a conductor by a changing magnetic field is called _____ action. (7-12)

7-20. The most magnetic element is _____. (7-4)

7-21. The invisible lines of force associated with a magnet are called _____. (7-2)

For questions 7-22 to 7-25, choose the letter that best completes each sentence.

7-22. If all other factors are equal, the material with the highest permeability will (7-13)
a. Have the least reluctance
b. Saturate most easily
c. Have the most residual magnetism
d. Have the lowest flux density

7-23. The pull of a solenoid is increased by (7-16)
a. Forcing more current through the coil
b. Using fewer turns of wire in the coil
c. Using a core material of lower permeability
d. Increasing the air gap around the plunger

7-24. A magnet is a device that (7-1)
   a. Attracts materials like paper and glass
   b. Always has two poles
   c. Produces a force field
   d. Creates reluctance in a material

7-25. Doubling the distance between two magnetic poles (7-2)
   a. Doubles the force between the poles
   b. Quadruples the force between the poles
   c. Makes the force between the poles one-half as great
   d. Makes the force between the poles one-fourth as great

Chapter Review Problems

7-1. What is the mmf of a 300-t coil that draws 3 A of current if the coil is 3 in. long? (7-13)

7-2. What is the flux density of a magnetic circuit with 0.1 base unit of flux and cross-sectional dimensions of 0.02 by 0.03 m? (7-13)

7-3. Determine the magnetic field strength of a circuit with an average core length of 0.4 m, a 300-turn coil, and a coil current of 0.6 A. (7-13)

7-4. Calculate the permeability of a material that requires 150 base units of magnetizing force to create 0.3 T in the material. (7-13)

7-5. A magnetic circuit has an average core length of 0.3 m and a coil current of 0.54 A. Determine the required number of turns in the coil to provide 360 base units of magnetic field strength. (7-13)

7-6. How much current is flowing in a 620-turn coil that provides 800 base units of mmf?

7-7. A magnetic circuit has a cross-sectional dimension of 0.0008 m^2 and a flux density of 400 base units. Determine the flux.

7-8. A 6-turn coil has 4 A flowing through it. How much current must flow through an 8-turn coil to provide the same amount of mmf as the 6-turn coil is providing?

7-9. A 20-turn coil is wound on a magnetic core with a length of 0.2 m. What is the magnetizing force when 250 mA flows through the coil?

7-10. What is the flux density when 200 base units of magnetic field strength is applied to a magnetic core made of material with a permeability of 0.3 base units?

7-11. Determine the permeability (μ) of a material that has a relative permeability (μ_r) of 800.

Critical Thinking Questions

7-1. Why do the domains in some magnetic materials return to a random pattern when the material is removed from a magnetic field?

7-2. Determine the rate of flux change in a circuit in which 15 V is induced in a 30-turn coil.

7-3. How much current is required in an 85-t coil to produce a magnetic field strength (H) of 50 A·t/m in a core with an average length of 0.45 m?

7-4. The permeability of air is 1.26×10^{-6} Wb/ (A · t · m). Determine the flux density in a core

if its relative permeability is 5000 when the magnetizing force is 50 A · t/m.

7-5. If you want to magnetize many pieces of permanent magnetic material, would you use ac or dc to power the coil? Why?

7-6. Discuss some limitations of using an electromagnet to pick up, move, and release metals in a scrap-metal yard.

1. Magnetism is a force field that exists around certain materials and devices.
2. A magnetic field is the area of force extending around a magnet.
3. flux
4. north pole
5. north, south
6. quadruple
7. south
8. yes
9. no
10. no
11. perpendicular
12. F
13. counterclockwise
14. coil
15. above
16. decreases
17. F
18. iron
19. Silicon
20. magnetic domain
21. silicon
22. T
23. The flux disappears.
24. The steel becomes a permanent magnet, and the flux remains as its field.
25. magnetomotive force
26. saturated
27. F
28. F
29. residual magnetism
30. permanent-magnet material
31. because the reluctance of the magnetic material is much less than that of the air which it replaces
32. F

33. the amount and the rate of flux change
34. to make connections to the rotating conductor and to provide dc output
35. A changing flux caused by one coil induces a voltage in another coil.
36. 80 A · t
37. magnetizing force, field intensity
38. ampere-turn per meter
39. weber
40. tesla (weber per square meter)
41. permeability
42. The dimensions of the circuit are unimportant when flux density is specified
43. Relative permeability is the ability of a material to pass flux compared with the ability of air to pass flux.
44. F
45. F
46. T
47. F
48. F
49. T
50. They periodically reverse the direction of current in the armature.
51. Because the plunger has a much higher permeability than air does. When the plunger is pulled into the coil, the reluctance of the circuit is greatly reduced.
52. current and voltage ratings for both the coil and the contacts as well as the switching arrangement of the contacts
53. the type of plate material, the current through the plate, and the flux density acting on the plate
54. The switch has only two discrete levels of output voltage, whereas the sensor has a continuously variable (analog) output voltage
55. yes

Alternating Current and Voltage

Learning Outcomes

This chapter will help you to:

8-1 *Understand* what is meant by terms like ac voltage and ac power.

8-2 *Remember* that electrical waveforms represent the variations of an electrical quantity during a period of time.

8-3 *List and describe* three common types of ac waveforms.

8-4 *Define* these ac terms: cycle, period, frequency, Hertz, I_p, I_{av}, rms value, and V_{p-p}.

8-5 *List* the three factors that determine the amount of voltage induced in a conductor.

8-6 *Describe* what determines the frequency of the output of an ac generator.

8-7 *Explain* two advantages of alternating current over direct current.

8-8 *Discuss:* (a) how three-phase ac is generated and distributed, and (b) the advantages of three-phase ac over single-phase ac.

Most electric energy is distributed in the form of alternating current and voltage. Our homes and our factories are both powered by ac electricity. These, of course, are the major consumers of electric energy.

8-1 AC Terminology

Alternating current periodically changes the direction in which it is flowing. It also changes magnitude (amount), either continuously or periodically. With most types of alternating current, the magnitude is changing continuously.

Of course, if there is an *alternating current,* there must also be an *alternating voltage* and power. Although it is not proper English, we often refer to alternating voltage as ac voltage. When written or spoken in the abbreviated form, the term *ac voltage* looks and sounds all right. However, when spelled out as *alternating current voltage,* the term can be confusing. An ac voltage is a voltage that produces an alternating current when used to power a circuit. Likewise, *ac power* refers to power that is produced by alternating current and alternating voltage.

Alternating current

Alternating voltage

AC power

8-2 Waveforms

An *electrical waveform* is represented by a line on a graph, as shown in Fig. 8-1(*a*). The line is produced by plotting points on a graph and then connecting the points. The points represent the value of some electrical quantity at different times. The magnitude and direction of the electrical quantity (voltage or current) are indicated on the vertical axis of the graph. Time is marked on the horizontal axis.

The waveform shown in Fig. 8-1(*a*) could represent the voltage across the resistor in Fig. 8-1(*b*). In this case the units on the vertical axis would be volts. The waveform shows that the voltage rapidly increases to its maximum

Electrical waveform

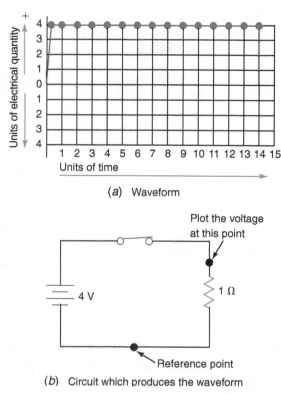

(a) Waveform

Plot the voltage at this point

4 V

1 Ω

Reference point

(b) Circuit which produces the waveform

Fig. 8-1 Plotting a waveform. An electrical quantity such as voltage is plotted against time.

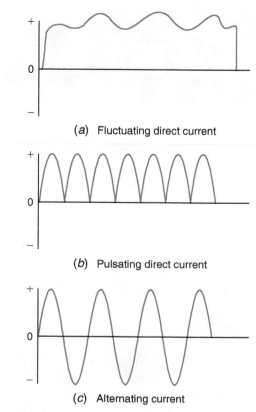

(a) Fluctuating direct current

(b) Pulsating direct current

(c) Alternating current

Fig. 8-2 Electrical waveforms.

value (4 V) when the circuit is energized. The voltage remains at its maximum value until the circuit is opened. When the circuit is opened, the voltage across the resistor suddenly drops to zero.

The waveform of Fig. 8-1(a) could just as well represent the current in Fig. 8-1(b). The only thing that would change would be the units on the vertical axis. For the current waveform, the units would be amperes.

Notice in Fig. 8-1(a) that the waveform is always above the zero reference line. This means that the polarity of the voltage is always positive with respect to some reference point. Or, in the case of a current waveform, it means that the direction of the current never reverses. In other words, Fig. 8-1(a) is the waveform of a pure (steady) direct current or voltage. The magnitude of the voltage or current changes only when the circuit is turned on or off.

Other common waveforms are shown in Fig. 8-2. The *fluctuating direct current* in Fig. 8-2(a) is the type of current produced in an amplifying transistor. The pulsating waveform of Fig. 8-2(b) represents the type of current (or voltage) produced by a battery charger. Notice that *pulsating direct current* periodically reduces to zero while fluctuating direct current

does not. Figure 8-2(c) shows one type of alternating current. Notice that the ac waveform goes below the zero reference line. This means that the polarity of the voltage reverses and that the direction of the current changes.

8-3 Types of AC Waveforms

The most common type of ac waveform is the *sine wave,* shown in Fig. 8-3(a). An alternating current with this type of waveform is referred to as *sinusoidal* alternating current. The alternating current (and voltage) supplied to homes and factories is sinusoidal. A pure *sine wave* has a very specific shape. Its shape can be precisely defined by mathematics. All the formulas we will be using to solve ac circuit problems are based on the sine wave. Therefore, these formulas are appropriate only when working with sinusoidal alternating current.

Figure 8-3(b) shows a *square wave.* This type of ac waveform is used extensively in computer circuits. With a square wave the magnitude of the current (or voltage) is not continuously varying. However, both amplitude and direction do periodically change.

Sine wave

Fluctuating direct current

Square wave

Pulsating direct current

The *sawtooth waveform* of Fig. 8-3(c) is used in television receivers, radar receivers, and other electronic devices. Sawtooth voltages and currents are used in the circuits that produce the picture on a television screen.

Wave shapes other than those shown in Fig. 8-3 are possible. In fact, alternating current can be electronically produced in an almost infinite variety of waveforms. Electronic music is created by producing and mixing together a wide variety of waveforms.

Sawtooth waveform

 Self-Test

Answer the following questions.

1. True or false. The vertical axis of a waveform graph is usually marked in units of time.
2. True or false. The horizontal axis of a waveform graph is usually marked in units of power.
3. True or false. A pulsating dc waveform periodically returns to the zero axis.
4. True or false. Fluctuating direct current periodically fluctuates above and below the zero reference line.
5. A(n) _____ is produced when values of voltage and time are plotted on a graph.
6. Three common types of alternating current are _____, _____, and _____.

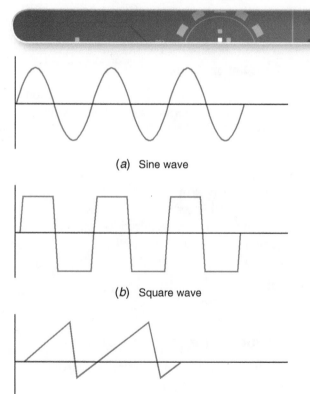

(a) Sine wave

(b) Square wave

(c) Sawtooth wave

Fig. 8-3 Common types of ac waveforms.

8-4 Quantifying Alternating Current

Fully describing alternating current (or ac voltage) requires the use of a number of terms.

Some of these terms have general meanings, but when used in electricity they have very specific meanings.

Cycle

The waveform in Fig. 8-4 shows four *cycles* of alternating current. A cycle is that part of a waveform that does not repeat or duplicate itself. Each cycle in Fig. 8-4 is a duplicate of every other cycle in the figure.

The part of the cycle above the horizontal line in Fig. 8-4 is called the positive half-cycle. A half-cycle is also called an *alternation*. Therefore, the positive half-cycle could be called the positive alternation. The negative half-cycle, of course, is that part below the horizontal reference line.

Cycles

Alternation

Period

The time required to complete one cycle is the *period* (T) of a waveform. In Fig. 8-4 it takes 0.25 s to complete one cycle. Therefore the period T of that waveform is 0.25 s.

Period (T)

Frequency

The rate at which cycles are produced is called the *frequency* (f) of an ac current or ac voltage. Frequency, then, refers to how rapidly

Frequency (f)

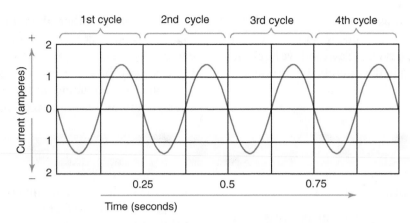

Fig. 8-4 Cycle, period, and frequency. The waveform has a period of 0.25 s and completes 4 cycles per second.

the current reverses or how often the voltage changes polarity.

Unit of Frequency—The Hertz

Hertz (Hz)

$$T = \frac{1}{f}$$

$$f = \frac{1}{T}$$

The base unit of frequency is the *hertz* (Hz). One hertz is equal to one cycle per second. The sine wave in Fig. 8-4 goes through four cycles in 1 s. Therefore, it has a frequency f of 4 Hz.

Commercial power in North America is distributed at a frequency of 60 Hz. It is sinusoidal alternating current. In many European countries a frequency of 50 Hz is used for electric power.

Electronic circuits use a wide range of frequencies. For example, an audio amplifier usually amplifies all frequencies between 20 Hz and 20 kHz. Frequencies between 540 and 1600 kHz are used for AM radio, and television uses frequencies up to 980 MHz.

Frequency and period are reciprocally related. That is

$$T = \frac{1}{f} \text{ and } f = \frac{1}{T}$$

You May Recall

. . . that 1 Hz is equal to one cycle per second. Therefore, the period is in seconds when the frequency is in hertz.

History of Electronics

Heinrich Hertz

In 1887 German physicist Heinrich Hertz demonstrated the effect of electromagnetic radiation through space. In his honor, the hertz (Hz) is now the standard unit for the measurement of frequency (1 Hz equals 1 complete cycle per second).

EXAMPLE 8-1

How much time is required to complete one cycle if the frequency is 60 Hz?

Given:	$f = 60$ Hz
Find:	T
Known:	$T = \frac{1}{f}$
Solution:	$T = \frac{1}{60 \text{ Hz}} = 0.0167$ s
Answer:	The time required is 0.0167 s, or 16.7 ms.

EXAMPLE 8-2

What is the period of a 2-MHz sine wave?

Given:	$f = 2$ MHz
Find:	T
Known:	$T = \frac{1}{f}$

Solution: $T = \dfrac{1}{2,000,000} = 0.0000005$ s

Answer: The period is 0.0000005 s, or 0.5 μs.

If we had expressed the data in example 8-2 in powers of 10, the solution would be

$$T = \frac{1}{2 \times 10^6} = 0.5 \times 10^{-6} \text{ s}$$

Notice how the use of powers of 10 can simplify working with large and small numbers.

EXAMPLE 8-3

What is the frequency of a waveform that requires 0.01 s to complete one cycle?

Given: $T = 0.01$ s
Find: f
Known: $f = \dfrac{1}{T}$
Solution: $f = \dfrac{1}{0.01} = 100$ Hz
Answer: The frequency is 100 Hz.

Specifying Amplitude

The *amplitude* of an ac waveform can be specified in several different ways, as shown in Fig. 8-5. These ways of specifying amplitude are appropriate for either voltage or current.

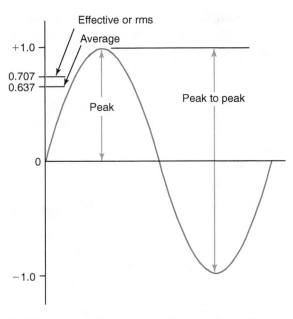

Fig. 8-5 Values of a sine wave. The amplitude of a sine wave can be specified in any one of four ways.

For symmetrical waveforms the *peak value* (V_p) (also called the maximum value) of the negative and of the positive half-cycle are equal. Therefore, the *peak-to-peak value* $(V_{p\text{-}p})$ is twice as great as the peak value. For a sinusoidal ac voltage we can write

$$V_{p\text{-}p} = 2\,V_p$$

Many of the waveforms in electronic circuits are not symmetrical. Such waveforms can be specified as either peak or peak-to-peak values. When specifying a peak value, one must indicate whether it is the positive or the negative peak.

The *average value* (V_{av}) of a waveform is the arithmetical mean. Finding the arithmetical mean of a waveform is illustrated in Fig. 8-6(*a*). The instantaneous value of the waveform is determined at a number of equally spaced intervals along the horizontal axis. Then, all the instantaneous values are added together and divided by the total number of intervals. The closer the intervals, the more accurate the estimate of the average value of the waveform. Figure 8-6(*b*) provides a

Peak value

Peak-to-peak value

$V_{p\text{-}p} = 2\,V_p$

Average value (V_{av})

Amplitude

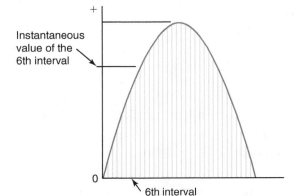

(*a*)   Method of determining the average value

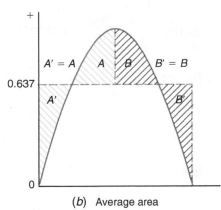

(*b*)   Average area

Fig. 8-6 Average value of a waveform. The combined area of A and B is equal to the area of A' plus B'.

further idea of the meaning of the average value of a waveform. The area of the waveform above the average value is equal to the "missing" areas below the average value. The area labeled A is equal to the area labeled A' in Fig. 8-6(b). Area B is equal to area B'.

For a sine wave, the relationship between peak value and average value is indicated in Fig. 8-5. For voltage, the relationship can be written as

$$V_{av} = 0.637\ V_p$$

$V_{av} = 0.637\ V_p$

Or, solving for V_p, the relationship is

$$V_p = 1.57\ V_{av}$$

$V_p = 1.57\ V_{av}$

The relationship between average and peak voltage is used when analyzing circuits that convert ac voltage to pulsating dc voltage.

The most common way of specifying the amount of alternating current is by stating its *effective,* or *rms, value.* The effective value of an alternating current is that value that produces the same heat in a resistive circuit as a direct current of the same value. Also, equal amounts of dc voltage and effective ac voltage produce equal power across resistors of equal value. In Fig. 8-7(a), 20 V of effective ac voltage forces

Effective, or rms, value

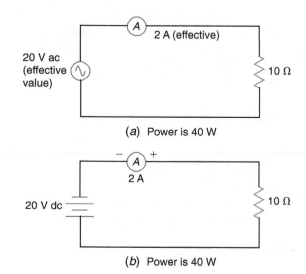

(*a*) Power is 40 W

(*b*) Power is 40 W

Fig. 8-7 Effective current and voltage. Direct current and effective alternating current produce the same amount of heat.

Now that you know what *effective value* means, let us examine why it is also the *rms value.* The effective value of a waveform can be determined by a mathematical process known as root mean square (rms). With this process the waveform is divided into many intervals, as was done in Fig. 8-6(a). The instantaneous value of each interval is squared. The mean (average) of the squared values is then determined. Finally, the square root of the mean is calculated. When many intervals are used, this process yields the effective value.

An example may help clarify how this mathematical process yields the effective value. Suppose a sine wave represents the current through a 1-Ω resistor. Because $P = I^2 R$, and $R = 1\ \Omega$, the squared value at any interval represents the instantaneous power at that interval. Therefore, the mean of all of these squared values yields

EXAMPLE 8-4

What is the average voltage of a sinusoidal waveform if its peak-to-peak voltage is 300 V?

Given:	$V_{p\text{-}p} = 300\ \text{V}$
Find:	V_{av}
Known:	$V_p = \dfrac{V_{p\text{-}p}}{2}$
	$V_{av} = 0.637\ V_p$
Solution:	$V_p = \dfrac{300\ \text{V}}{2} = 150\ \text{V}$
	$V_{av} = 0.637 \times 150\ \text{V} = 95.6\ \text{V}$
Answer:	The average voltage is 95.6 V.

2 A of effective alternating current through a 10-Ω resistor. This causes the 10-Ω resistor to dissipate 40 W:

$$P = IV$$
$$= 2\ \text{A} \times 20\ \text{V} = 40\ \text{W}$$

As shown in Fig. 8-7(b), 20 V of dc voltage across a 10-Ω resistor also dissipates 40 W.

EXAMPLE 8-5

What is the peak voltage of 120-V rms?

Given:	$V_{rms} = 120\ \text{V}$
Find:	V_p
Known:	$V_p = 1.414\ V_{rms}$
Solution:	$V_p = 1.414 \times 120\ \text{V}$
	$= 169.7\ \text{V}$
Answer:	The peak voltage is 169.7 V.

the mean power. Solving $P = I^2R$ for I, gives $I = \sqrt{P/R}$. And, since $R = 1\ \Omega$, the square root of P yields the effective value of I. Thus, we can use the terms *rms value* and *effective value* interchangeably.

For sinusoidal alternating current, the rms (effective) value and the peak value are related by the following formulas:

$$V_{rms} = 0.707\ V_p \text{ and } V_p = 1.414\ V_{rms}$$

$V_{rms} = 0.707\ V_p$

$V_p = 1.414\ V_{rms}$

Self-Test

Answer the following questions.

7. The base unit of frequency is the _____ and its abbreviation is _____ .

8. One cycle per second is equal to 1 _____ .

9. The part of a waveform that does not duplicate any part of itself is called a(n) _____ .

10. The segment of a waveform below the zero reference line is the negative _____ or the negative _____ .

11. How rapidly a waveform is produced is determined by the _____ of the waveform.

12. The _____ value of a waveform is also known as the effective value.

13. The symbol for frequency is _____ .

14. Determine the following for a 400-Hz, 28-V sine wave:
    a. Period
    b. V_{av}
    c. $V_{p\text{-}p}$

15. Determine the peak current when $I_{av} = 6$ A.

16. Determine the rms current when $I_p = 8$ A.

17. Determine the frequency of a waveform that requires 0.005 s to complete one alternation.

8-5 The Sine Wave

You May Recall

. . . that the principles of *induced voltage* were introduced in the previous chapter.

To understand how a sine wave is produced, however, we must know more about induced voltages.

Induced Voltage and Current

When a conductor cuts across magnetic flux, a voltage is induced in the conductor. This voltage, of course, causes a current to flow if there is a complete circuit. Therefore, one can also say that a current is induced in the conductor.

Alternating currents and ac voltages are usually specified as rms values. It is common practice to assume that alternating current is in rms units unless a subscript (p-p, p, or av)

is included. For example, the voltage at the outlets in your home may be specified as 120 V. This is the rms value of the ac voltage.

The amount of *induced voltage* is a function of the amount of flux cut by the conductor per unit of time. The amount of *flux cut per unit of time* is in turn determined by:

1. The speed of the conductor
2. The flux density
3. The angle at which the conductor crosses the magnetic flux

As shown in Fig. 8-8(*a*), no voltage (or current) is induced when the conductor moves parallel to the flux. This is because no lines of flux are crossed or cut. When the motion is 45° to the flux, as in Fig. 8-8(*b*), some voltage is induced because some flux is cut. For a given conductor speed, a conductor provides maximum *induced current* when the motion is 90° to the flux, as shown in Fig. 8-8(*c*).

Induced voltage

Flux cut per unit of time

Induced current

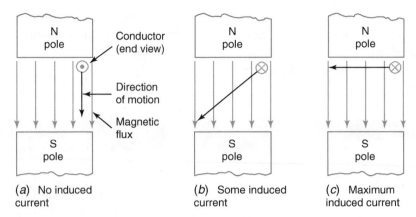

(a) No induced current

(b) Some induced current

(c) Maximum induced current

Fig. 8-8 Moving a conductor in a magnetic field. The amount of induced current depends on the angle at which the conductor cuts the flux.

You May Recall

. . . that the $\times$ and $\cdot$ in the center of the conductors in Figs. 8-8 and 8-9 indicate the direction of current, as explained in the previous chapter.

The direction of an induced current (or the polarity of an induced voltage) is determined by two factors:

1. The direction in which the conductor is moving
2. The polarity of the magnetic field or the direction of the flux

Figure 8-9 shows the four possible combinations of conductor movement and polarity.

Left-hand rule

The direction of induced current can be determined by using the *left-hand rule,* which is illustrated in Fig. 8-10. In this figure, the thumb indicates the direction of conductor movement. The index finger (pointing straight out) is aligned with the direction of the flux. The middle finger (bent at 90° to the palm) indicates the direction the current flows. Apply the left-hand rule to the diagrams in Fig. 8-9. Your middle finger should point into the page for those diagrams with an $\times$ in the conductor.

Producing the Sine Wave

A sine wave is produced when a conductor is rotated in a magnetic field. The conductor must rotate in a perfect circle and at a constant speed. The magnetic field has to have a uniform flux density. Production of a sine wave is illustrated in Fig. 8-11. At position 1, the conductor is

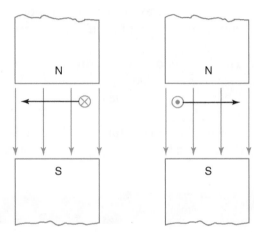

(a) Changing direction of conductor movement

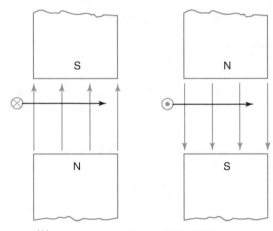

(b) Changing the polarity of the field

Fig. 8-9 Changing the direction of movement or the polarity of the field changes the direction of the induced current.

moving parallel to the flux and no voltage is induced in the conductor. (The instantaneous direction of the conductor is indicated by the arrow.) As the conductor rotates from position

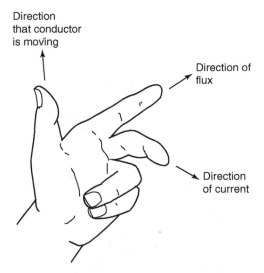

Direction
that conductor
is moving

Direction of
flux

Direction
of current

Fig. 8-10 Determining the direction of induced
current using the left-hand rule.

1 to position 2, it starts to cut the flux at a slight angle. Thus, a small voltage is induced in the conductor. The current produced by this voltage is flowing out of the page, as indicated by the dot in the conductor. As the conductor rotates on through positions 3 and 4, the angle at which the conductor cuts the flux increases. At position 4, the conductor is moving perpendicular to the flux. Maximum flux per unit of time is now being cut. Therefore, maximum voltage V_p is being produced, as shown in Fig. 8-11(*b*). The positive peak of the sine wave is produced. As the conductor moves on to positions 5 and 6, less flux is cut. This is because the conductor's direction of movement again starts to parallel the flux. (For a given length of movement of the conductor, less flux is being cut.) By the time the conductor reaches position 7, no voltage is being induced. The first alternation (positive half-cycle) of the sine wave has been produced. Notice in Fig. 8-11(*a*) that the direction in which the conductor cuts the flux reverses as the conductor leaves position 7. Thus, the polarity of the induced voltage is reversed and the negative half of the cycle is started. At position 10, the conductor is producing its peak negative voltage. Finally, as the conductor returns to position 1, the induced voltage drops to zero. The first cycle is completed; a sine wave has been produced. Each additional revolution of the conductor produces another cycle of a sine wave.

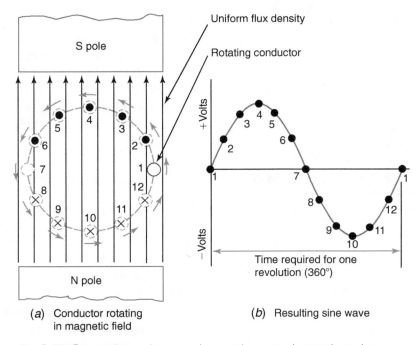

(*a*) Conductor rotating
in magnetic field

(*b*) Resulting sine wave

Fig. 8-11 Generating a sine wave by rotating a conductor through a
magnetic field.

Electrical and Mechanical Degrees

Notice in Fig. 8-11(a) that the conductor rotates 360° (one revolution) in producing one cycle of a sine wave. The horizontal axis of the waveform can therefore be marked off in degrees rather than units of time. This is done in Fig. 8-12 for each of the 12 positions of the conductor in Fig. 8-11. It can be seen from Fig. 8-12 that a sine wave reaches its peak values at 90° and 270°. Its value is zero at 0° (or 360°) and 180°. When degrees are used to mark off the horizontal axis of a waveform, they are referred to as *electrical degrees*. One alternation of a sine wave contains 180 electrical degrees; one cycle has 360 electrical degrees.

The advantage of marking the horizontal axis in degrees (rather than seconds) is that electrical degrees are independent of frequency. A cycle has 360° regardless of its frequency. However, the time required for a cycle (that is, the period) is totally dependent on the frequency. When the horizontal axis is marked off in seconds, the numbers are different for each frequency. Thus, you can see that the use of electrical degrees is very convenient and practical.

In Fig. 8-11 the number of electrical degrees is equal to the number of *mechanical degrees*. That is, the conductor rotates 360 mechanical degrees to produce the 360 electrical degrees of the sine wave. This one-to-one relationship of mechanical and electrical degrees is not essential. In fact, most ac generators produce more than one cycle per revolution.

Electrical degrees

Mechanical degrees

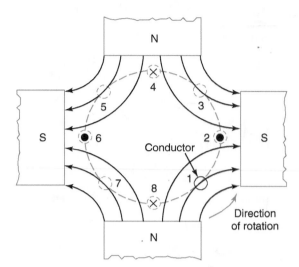

(a) Four-pole magnetic field

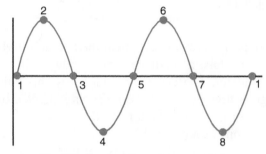

(b) 720 electrical degrees per revolution

Fig. 8-13 Electrical and mechanical degrees. The mechanical and electrical degrees of a generator need not be equal.

Figure 8-13 shows the magnetic field produced by two pairs of magnetic poles. This field produces two cycles per revolution of the conductor. Every 45° of rotation (position 1 to position 2, for example) produces 90 electrical degrees. The conductor in Fig. 8-13 cuts flux under a south pole and a north pole for each 180° of rotation. This produces one cycle (360 electrical degrees) for 180 mechanical degrees.

8-6 AC Generator

In the preceding section we saw how a conductor rotating in a magnetic field produces an ac voltage. Now, let us see how this voltage can be connected to a load.

The essential electric parts of an *ac generator* are shown in Fig. 8-14(a). In the drawing only one loop of wire (conductor) is shown. In a generator a coil consisting of many loops of wire would be used. This multilooped coil is called a *winding*. The winding is wound on a silicon-steel core

AC generator

Winding

Fig. 8-12 Degrees of a waveform. One cycle has 360 electrical degrees.

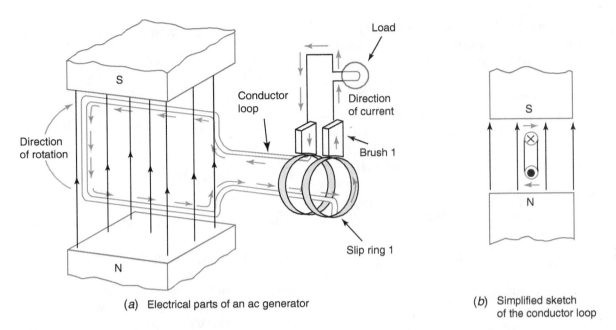

| (a) Electrical parts of an ac generator | (b) Simplified sketch of the conductor loop |

Fig. 8-14 An ac generator. The rotating conductor is connected to the load through the slip rings and brushes.

called an *armature*. The armature winding is composed of many coils (armature coils).

As seen from Fig. 8-14(*a*), both the top and the bottom of the loop move through the magnetic field. The voltage induced in the top of the loop aids the voltage induced in the bottom of the loop. Both voltages force current in the same direction through the load. Figure 8-14(*b*) shows the top and bottom conductors of the loop and uses the symbols developed in the previous section. This sketch again shows that the voltages (and currents) induced into the conductors of the rotating loop reinforce each other.

Notice in Fig. 8-14(*a*) that the bottom conductor always makes contact with *slip ring* 1. This ring always rotates on brush 1. Notice further that *brush* 1 is negative with respect to the other brush when the loop is in the position shown. It remains negative until the loop rotates another 90°. After 90° it becomes positive as the bottom loop starts to pass under the south pole. It remains positive for the next 180° while the conductor is moving under the south pole. Then it again becomes negative while the conductor moves across the face of the north pole. Thus, the polarity of the terminals of an ac generator reverses every 180 electrical degrees.

In Fig. 8-14 the magnetic field for the generator is created by a permanent magnet. In many generators the magnetic field is created by an electromagnet. The coils that create the

electromagnet are called *field coils*. The field coils are powered by a direct current so that the polarity of the field never changes. Also, in many large generators the conductors (loops) do not rotate. Instead, they are held stationary and the magnetic field is rotated around them. The end result is, of course, the same: voltage is induced in the conductors.

Armature

Field coils

Generator Voltage

You May Recall

. . . that the voltage induced in a rotating conductor is determined by the flux density and the rotational speed.

Slip ring

Brush

In a generator, the rotating winding consists of multiturn coils (many loops). The voltage induced in each turn of a coil aids the voltage induced in every other turn. Thus, the output voltage of a generator is dependent on

1. The number of turns in the rotating (armature) coils
2. The speed at which the coils rotate
3. The flux density of the magnetic field

Generator Frequency

The speed at which the generator coils rotate influences the frequency as well as the output

Pairs of magnetic poles

voltage of a generator. The frequency of the sine wave produced by a generator is determined by (1) the number of *pairs of magnetic poles* and (2) the rotational speed of the coils (the revolutions per minute of the generator shaft).

A two-pole (one-pair) generator, like the one in Fig. 8-14, produces one cycle for each revolution. If this generator is rotating at 60 revolutions per minute (r/min), which is 1 r/s, the frequency is 1 Hz (one cycle per second). With a four-pole generator, 60 r/min produces a frequency of 2 Hz. Stated as a formula, the frequency of a generator's output is

$$\text{Frequency} = \frac{\text{r/min}}{60} \times \text{pairs of poles}$$

Using this formula, the frequency is in base units of hertz.

EXAMPLE 8-6

What is the frequency of a six-pole generator rotating at 1200 r/min?

Given: Pairs of poles = 3
r/min = 1200

Find: f

Known: $f = \dfrac{\text{r/min}}{60} \times \text{pairs of poles}$

Solution: $f = \dfrac{1200}{60} \times 3 = 60 \text{ Hz}$

Answer: The frequency of the generator's output is 60 Hz.

8-7 Advantages of Alternating Current

It is easier to transform (change) alternating current from one voltage level to another than to transform direct current. That is why electric energy is distributed in the form of alternating current. In rural areas, power transmission lines can be operated at very high voltages (more than 400,000 V). When the power lines approach a city, the voltage can be reduced (by a transformer) to a few thousand volts. This lower voltage is then distributed in the city.

Another advantage of alternating current is that ac motors are less complex than dc motors. Alternating current motors can be built without brushes and commutators. This greatly reduces the amount of maintenance needed on an ac motor.

 Self-Test

Answer the following questions.

18. True or false. When operated at the same speed, an eight-pole generator produces a higher frequency than a six-pole generator.
19. True or false. The field coils of an ac generator are excited with alternating current.
20. True or false. No voltage is induced when a conductor moves parallel to lines of flux.
21. True or false. A conductor moving at 30° to the lines of flux produces more voltage than one moving at 60°.
22. Two revolutions of a conductor rotating in a four-pole magnetic field produce _____ electrical degrees.
23. Voltage is induced in the _____ coils of an ac generator.
24. The brushes in an ac generator make contact with the _____ .

25. A six-pole generator rotating at 1000 r/min produces a frequency of _____ .
26. What determines the magnitude of the voltage induced in a conductor?
27. Determine the speed (r/min) of an 8-pole generator that produces a 400-Hz, 80-V sine wave.
28. List the factors that determine the polarity of an induced voltage.
29. List the necessary conditions for inducing a sine wave in a conductor.
30. List the factors that determine the output voltage of a generator.
31. List the factors that determine the frequency of a generator's output.
32. List two advantages of alternating over direct current.

8-8 Three-Phase Alternating Current

Electricity is produced at an electric power plant in the form of *three-phase alternating current. Phase* is often abbreviated with the symbol ϕ. Therefore, *three-phase* is sometimes written as 3-ϕ. In three-phase alternating current (Fig. 8-15), each phase is a sine wave. Each phase is displaced from the other two phases by *120 electrical degrees.* Phase 2 in Fig. 8-15 starts its positive alternation 120° after phase 1 but 120° before phase 3.

Figure 8-15 shows that the algebraic sum of the three phases at any instant is zero. For instance, at 60°, phase 1 is +8.66 V, phase 2 is −8.66 V, and phase 3 is zero; the sum of the three is zero. At 150°, phase 1 and 2 are both at +5 V while phase 3 is at −10 V. Again the algebraic sum is zero. At 260° the positive voltages of phases 2 and 3 are exactly canceled by the negative voltage of phase 1. Pick any instant of time you desire to compare the three phases. You will find that the sum of two of the phases is always equal in magnitude to the remaining phase and opposite it in sign.

Generation of Three-Phase Alternating Current

Suppose three loops of wire (as in Fig. 8-16) are spaced 120° apart and rotating in a magnetic field. If each loop is electrically isolated from the others and has its own slip rings and brushes, each produces a sine wave. Each sine wave is displaced 120 electrical degrees from the other two sine waves. In other words, we have a *three-phase generator.*

Figure 8-16 is drawn to represent the beginning (zero electrical degrees of phase 1) of the waveforms in Fig. 8-15. In Fig. 8-16 the colored end of a loop is considered to be the reference point for that phase. Note that each phase has its own reference point. Thus, phase 2 is now negative and will become more negative as the generator turns a few more degrees. Phase 1 is now zero and will become positive, and phase 3 is positive but will decrease in value. Notice how this agrees with Fig. 8-15. When phase 1 in Fig. 8-15 becomes positive, phase 3 becomes less positive and phase 2 becomes more negative. After the generator in Fig. 8-16 has rotated 90°, phase 1 will be at peak positive

Three-phase alternating current

Three-phase generator

120 electrical degrees

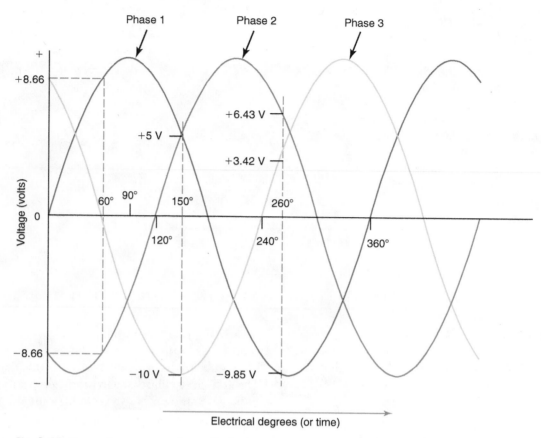

Fig. 8-15 Three-phase ac waveform. Each phase is separated by 120 electrical degrees.

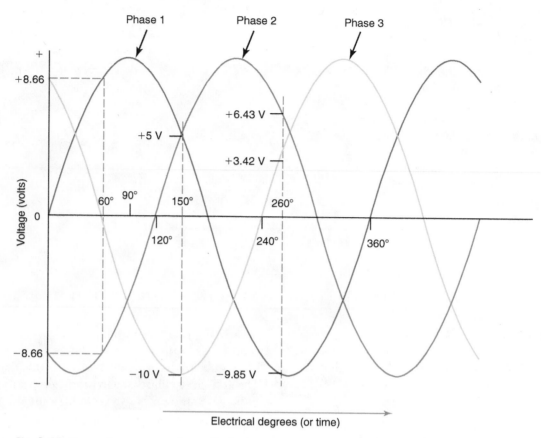

Fig. 8-16 Three-phase generator. The rotating conductors are mechanically locked together but electrically isolated.

voltage, phase 3 will be negative (and increasing in value), and phase 2 will be less negative.

The three phases of a generator can be connected so that the load can be carried on only three conductors. Thus, only three conductors

are needed to carry three-phase power from a power plant to its place of use. The three phases of a generator can be connected in either a *delta connection* or a *wye connection*.

Delta Connection

A delta connection is shown in Fig. 8-17. In this figure, each of the generator symbols represents one of the phases of a three-phase generator. The dot on one end of the generator symbol indicates the *reference end* of the *phase winding*. Notice that the reference ends are not connected to each other. The polarities and voltages indicated for the generators are momentary polarities and voltages. They are the ones that occur at 150° in Fig. 8-15. The following discussion assumes a balanced system. That is, all *phase voltages* are equal and all loads connected across them are equal.

The three phases are connected in one continuous loop for a delta connection. Yet no current flows in the loop until a load is applied to the generator. This is because the sum of the voltages of any two of the phases is equal in magnitude, but opposite in sign, to the other

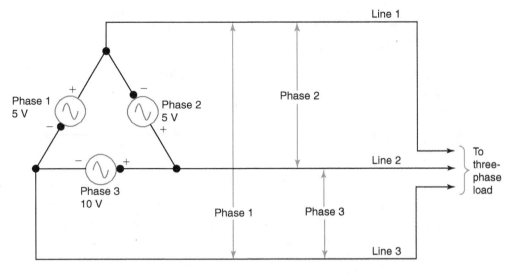

Fig. 8-17 Delta connection of generator. The line-to-line voltage is equal to the phase voltage.

phase. For instance, the voltages of phases 1 and 2 are aiding each other. The sum of these two voltages is 10 V (Fig. 8-17), which exactly cancels the 10 V of phase 3.

EXAMPLE 8-7

Determine the instantaneous voltage of the phase-2 coil in a delta-connected, three-phase generator when the instantaneous voltage of phase 1 is +50 V and of phase 3 is –30 V.

Given: $V_{phase1} = +50\,V$
$V_{phase3} = -30\,V$
Three-phase delta connection

Find: V_{phase2}

Known: For a delta connection, $V_{phase1} + V_{phase2} + V_{phase3} = 0$

Solution: $V_{phase2} = 0 - (V_{phase3}) - (V_{phase1})$
$= 0 - (-30\,V) - (+50\,V) = -20\,V$

Answer: $V_{phase2} = -20\,V$

The answer to example 8-7 can be cross-checked by entering the calculated and given voltages into the formula listed for "Known":

$$V_{phase1} + V_{phase2} + V_{phase3} = 0$$
$$(+50\,V) + (-20\,V) + (-30\,V) = 0$$

It is a good idea to develop the habit of cross-checking the answer to a problem you have solved!

Although the three phase voltages cancel each other within the continuous loop, each individual phase voltage is readily available. The full sinusoidal voltage produced by phase 1 of the generator is available between lines 1 and 3 in Fig. 8-17. Phase 2 is available between lines 1 and 2, and phase 3 exists between lines 2 and 3. Thus, the three phases can serve a load with a total of three lines.

Notice in Fig. 8-17 that the voltage between any two lines is exactly equal to one of the phase voltages. In other words, the *line voltages* and the phase voltages are the same. For a delta connection

$$V_{line} = V_{phase}$$

When a delta-connected generator is loaded (Fig. 8-18), currents flow in the loads, lines, and phase windings. However, the *line currents* and the *phase currents* are not equal. With equal loads connected to each phase, the line current is 1.732 times as great as the phase current. For a delta connection

$$I_{line} = 1.732\,I_{phase}$$

The generator's phase currents and the load currents are equal. Therefore, the line currents are also 1.732 times as great as the load current. Line currents are not equal to phase currents because each line carries current from two phases. The currents from any two phases are 120° out of phase.

Notice in Fig. 8-18 that neither voltage polarity nor current direction is indicated. This is because it is an ac system in which polarity

Line voltages

Line currents

Phase currents

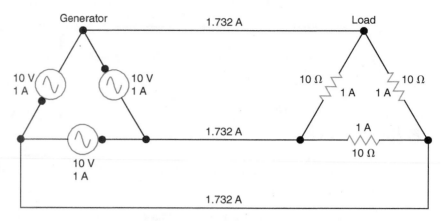

Fig. 8-18 Currents in a three-phase delta system. The line current is 1.732 times as great as the phase current.

and direction reverse periodically. The current and voltage are assumed to be in rms values because they are not specified otherwise.

Remember the following points about a *balanced delta* connection:

1. No current flows in the phase windings until a load is connected.
2. The line voltage and the phase voltage are equal.
3. The line current is 1.732 times as great as the phase current or the load current.

Wye Connection

Figure 8-19 shows a *wye-connected* three-phase generator and load. Again in this diagram, each generator symbol represents one phase of the three-phase generator. The voltages and currents are rms (effective) values. With the wye

connection and balanced (equal) loads, all currents are the same. For a wye connection

$$I_{line} = I_{phase}$$

However, notice in Fig. 8-19 that the line voltages are 1.732 times as great as the phase voltages. For a wye connection.

$$V_{line} = 1.732 \, V_{phase}$$

The *voltage* between any *two lines* is the result of two phase voltages. Since the two phase voltages are separated by 120 electrical degrees, their values cannot be simply added together.

Adding phase voltages to obtain line voltages is further complicated by the different reference points used for these voltages. Referring to Fig. 8-20 will help you understand reference points and addition of phase voltages. Figure 8-20(*b*) shows that phase 1 is

Balanced delta

Line-to-line voltage

Wye-connected

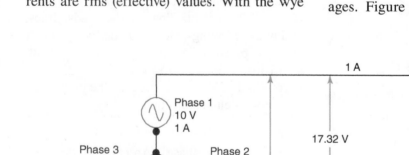

Fig. 8-19 Three-phase wye system. The line voltage is 1.732 times as great as the phase voltage.

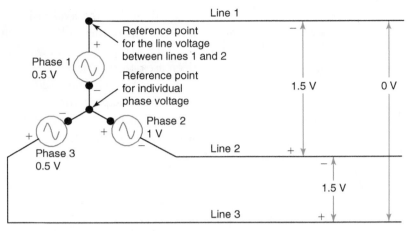

(a) Instantaneous (30°) phase and line voltage and polarity

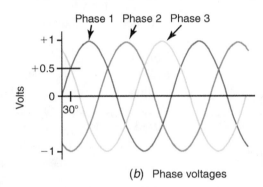

(b) Phase voltages

Fig. 8-20 Instantaneous voltages in a wye system.

+0.5 V at 30 electrical degrees. What does "+0.5 V" mean? Nothing, unless we have a reference point.

The *reference point* for all three phase voltages is the center of the wye in Fig. 8-20(a). With this reference point established, the voltages read from the waveform graph in Fig. 8-20(b) have meaning. At 30° both phases 1 and 3 are 0.5 V positive with respect to the center of the wye. Phase 2 is 1 V negative with respect to the center of the wye. These voltages and polarities are marked on the diagram in Fig. 8-20(a). Notice that the *instantaneous voltages* of phases 1 and 2 are in series between lines 1 and 2. These two voltages (phases 1 and 2) are aiding each other at this instant, and so they can be added together. Together, they produce 1.5 V between lines 1 and 2. Referring to

Fig. 8-20(b), notice that phases 1 and 2 at 30° are of opposite polarities. When connected in a wye configuration, however, the two voltages aid each other. Also notice in Fig. 8-20(a) that phases 1 and 3 are series-opposing. Their momentary voltages cancel each other to provide a net voltage of zero between lines 1 and 3.

In Fig. 8-20 we are looking only at instantaneous values taken at 30° on the three-phase waveform. Of course, these instantaneous values are different for each degree (or part of a degree) of the waveform. If all possible instantaneous values between two lines are plotted on a graph, the result is a sine wave that is 1.732 times as large as the sine wave of the phases which produced it. In plotting a waveform of this line voltage, a reference point is needed. The selection of this point is arbitrary. Let us select line 1 as the reference point. Then we can plot a waveform of the line voltage between lines 1 and 2. This waveform is shown in Fig. 8-21(a) superimposed on the waveforms of phases 1 and 2. Figure 8-21(b) lists some of the instantaneous values used to plot the waveform of the line voltage. Notice in Fig. 8-21

Reference point

Instantaneous voltage

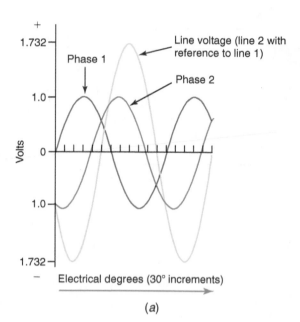

Electrical degrees	Instantaneous values (volts)		
	Phase 1	Phase 2	Line
0	0	−0.866	−0.866
30	+0.5	−1	−1.5
60	+0.866	−0.866	−1.732
90	+1	−0.5	−1.5
120	+0.866	0	−0.866
150	+0.5	+0.5	0
180	0	+0.866	+0.866
210	−0.5	+1	+1.5
240	−0.866	+0.866	+1.732

(b)

Fig. 8-21 Adding wye-connected phase voltages to obtain the line voltages.

that the line voltage is displaced 30° from the phase-2 voltage. The line voltage lags the phase-2 voltage by 30°. That is, the negative peak of the line voltage occurs 30° after the negative peak of the phase voltage.

Suppose line 2 had been used as the reference point for the line voltage in Fig. 8-20. Then the polarity of the line voltage would have started out positive rather than negative. The line voltage waveform in Fig. 8-21 would be completely inverted. It would start at +0.866 V rather than −0.866 V. The *line voltage* would then lead the phase-1 voltage by 30°.

In summary, the major characteristics of the *balanced* wye-connected three-phase system are:

1. The phase and line currents are equal.
2. The line voltage is 1.732 times greater than the phase voltage.
3. The line voltage and the phase voltage are separated by 30°.

Line voltage

Balanced wye

If any one of the load resistances in Fig. 8-19 changes value, then all of the line currents and the load voltage will change. If a load resistance decreases, then all line currents increase and the voltage across the decreased resistance decreases while the voltage across the two unchanged resistances increases. The effects of one resistance increasing are exactly the reverse of those stated for a resistance decrease.

EXAMPLE 8-8

The three coils of a three-phase generator are to be wye-connected to power a wye-connected, balanced, resistance load. The output voltage and maximum current rating of each coil is 139 V and 10 A. Determine (1) the line-to-line voltage, (2) the minimum value of commonly available resistors with ±5% tolerance (see Fig. 4-23) that could be used for a load, and (3) the minimum power rating for each resistor.

Given: $V_{phase} = 139$ V
I_{phaze} maximum $= 10$ A

Wye-connected phase coils and load resistors

Find: Line-to-line voltage

Minimum resistor nominal value for ±5% tolerance resistors

Minimum power rating for resistors

Known: $V_{line} = 1.732 \times V_{coil}$ and
$I_{coil} = I_{line} = I_{load}$
$R = V/I$
$P = I \times V$
$R_{actual} = R_{nominal} \pm 5\%$

Solution: Line-to-line voltage $= 1.732 \times 139$ V $= 240.75$ V

Resistor nominal-value $= (240.75$ V $\div 10$ A$) + 0.05(240.75$ V $\div 10$ A$) = 24.075$ Ω $+ 1.204$ Ω $= 25.279$ Ω

According to Fig. 4-23, the 27-Ω ±5% resistor is the smallest usable value.

$I_{max} = 240.75$ V $\div [27$ Ω $- (0.05 \times 27$ Ω$)] = 240.75$ V $\div (27$ Ω $- 1.35$ Ω$) = 9.386$ A

Minimum power rating = 9.386 A
$\times$ 240.75 V = 2259.7 W

Answer: Line-to-line voltage = 240.75 V
Minimum value of commonly
available resistor = 27 Ω
$P_{\text{min rating}}$ = 2259.7 W

Four-Wire Wye System

Figure 8-22 illustrates a *four-wire wye* system. In this figure, the generator symbol for each phase has been replaced by a coil symbol. The coil symbol represents one of the phases in a three-phase generator. This is the more traditional way of representing a three-phase generator.

The fourth wire of the four-wire system comes from the common center connection of the phase windings. (This common connection is sometimes referred to as the *star point* or connection.) This fourth wire is often called the *neutral wire* because it is electrically connected to ground (earth). Thus, the fourth wire is neutral (has no voltage) *with respect to ground.*

The neutral wire carries current under two conditions. One is when a single-phase load is connected between neutral and a line. The other is when an unbalanced wye-connected, three-phase load has its star point connected to the neutral.

Inspection of Fig. 8-22 reveals the main feature of the four-wire wye system. This system can supply (with only four wires) all the voltage requirements of a small manufacturing facility. Single-phase, 120-V circuits are connected between the neutral and any line; single-phase, 208-V circuits are connected between any two of the three lines; and three-phase, 208-V devices are connected to the three lines. Common voltages in four-wire wye systems are 120/208 V and 277/480 V. The first value in each case is the phase, or single-phase, voltage. The second value in each case is the three-phase voltage. Notice that the three-phase voltage is 1.732 times the single-phase voltage.

Any combination of single-phase and three-phase loads can be connected to the four-wire wye system without causing a change in the system voltages. The three-phase loads can be delta-connected and/or wye-connected. These loads can be either balanced or unbalanced. Unbalanced loads change system line currents but not system voltages.

Advantages of Three-Phase Systems

The three-phase system has advantages over the single-phase system. It

1. Makes more efficient use of copper
2. Provides a more constant load on the generator
3. Operates a motor that is less complex and provides more constant torque
4. Produces a smoother dc voltage and current when rectified

Suppose a three-phase system and a single-phase system are delivering the same amount

Four-wire wye

Star point

Neutral wire

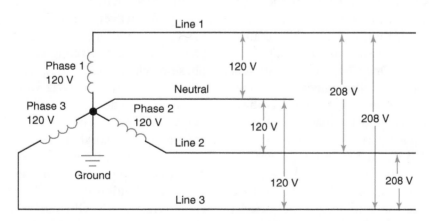

Fig. 8-22 Four-wire wye system. Four wires provide 120-V single phase, 208-V single phase, and 208-V three phase.

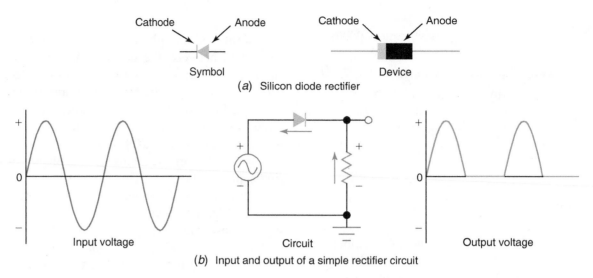

(a) Silicon diode rectifier

(b) Input and output of a simple rectifier circuit

Fig. 8-23 A single-phase, half-wave rectifier circuit produces one output pulse of dc for each cycle of ac input.

of power over the same length of lines. The single-phase system requires about 1.15 times as much copper for power lines as the three-phase system does.

The power required by a resistive load is equal to I^2R. With single-phase current, the power required from the generator follows the current variations. Thus, the load on the generator goes from zero to maximum power and back to zero with each alternation. With three-phase current at least two of the phases are providing current (and thus power) at any instant. The load on the three-phase generator never reduces to zero. The more uniform load provides for smoother mechanical operation of the generator.

Most ac motors can start only if the magnetic field behaves as if it were rotating. With single-phase alternating current, this rotating magnetic field is created through an auxiliary circuit in the motor. Once the motor has started, this auxiliary circuit is disconnected by a mechanically operated switch. In a three-phase motor, the rotating field is created by having each phase produce part of the magnetic field of each pole. No internal switch (or mechanical mechanism to operate it) is needed with the three-phase motor. Of course, the three-phase motor has more uniform torque because at least two phases are always producing magnetic fields.

Alternating current can be converted to direct current by a process known as *rectification*. The *silicon diode* rectifier, shown in Fig. 8-23(a), is a popular device used to rectify ac. A rectifier allows current to flow in only one direction. It has almost infinite resistance when the anode is negative with respect to the cathode and practically no resistance when the polarity is reversed. Thus, current flows only in the direction shown in the single-phase, *half-wave* rectifier circuit in Fig. 8-23(b). Notice that only half of the input cycle (the positive alternation) produces an output pulse.

The output waveform for a three-phase, half-wave rectifier is compared to that of the single-phase, half-wave rectifier in Fig. 8-24. The output waveforms are superimposed over the input waveforms. Notice that the single-phase produces pulsating dc while the three-phase produces fluctuating dc that reduces to only 50 percent of its peak value.

Ac can also be rectified with full-wave rectifier circuits to provide smoother dc output. A comparison of single-phase and three-phase, full-wave rectification is provided in Fig. 8-25. Notice that the output of three-phase, full-wave rectification never reduces below 0.866 of its peak value. Obviously the three-phase system produces a dc output with much less fluctuation.

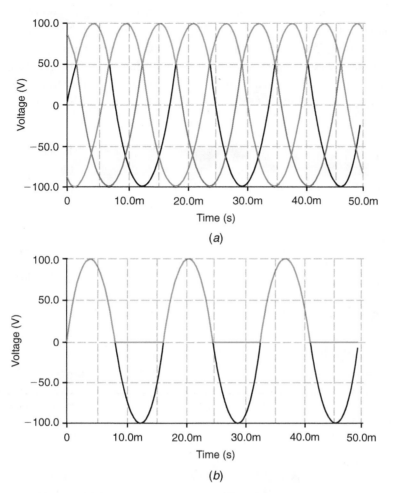

Fig. 8-24 A comparison of half-wave rectification of single-phase and three-phase ac. (*a*) Output waveform (red) for three cycles of a 200 $V_{p\text{-}p}$ input to a three-phase, half-wave rectifier. (*b*) Output waveform (red) for three cycles of a 200 $V_{p\text{-}p}$ input to a single-phase, half-wave rectifier.

Self-Test

Answer the following questions.

33. The sine waves of a three-phase system are separated by _____ electrical degrees.
34. The instantaneous voltage of phase 2 is _____ when phase 1 is −67 V and phase 3 is +18 V.
35. The windings of a three-phase generator are connected in either the _____ or the _____ configuration.
36. Transmission of three-phase power requires _____ conductors.

37. For a(n) _____ connection, the three-phase windings are connected in a continuous loop.
38. In the _____ connection, the phase current equals the line current.
39. In the _____ connection, the line current is _____ times as large as the phase current.
40. In the _____ connection, the three phases are connected together at the _____ point.
41. In the _____ connection, the line voltage is _____ times the phase voltage.

42. In a four-wire, three-phase system, the fourth wire is called the_____ wire.

43. Under what conditions does the fourth wire in a three-phase system carry current?

44. List the three voltages that are available from a four-wire, three-phase, 208-V system.

45. A three-phase generator produces a phase voltage of 277 V. What is the largest line voltage it can produce?

46. What type of dc output does half-wave, single-phase rectification provide?

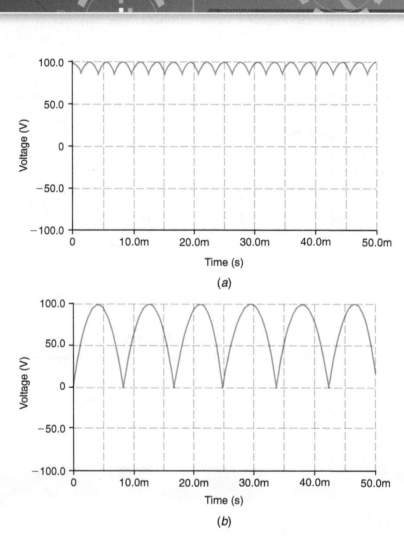

Fig. 8-25 Comparison of full-wave rectification of three-phase and single-phase ac. (a) Output waveform of a full-wave, three-phase rectifier with an input of 200 V_{p-p}. (b) Output waveform of a full-wave, single-phase rectifier with an input of 200 V_{p-p}.

Chapter 8 Summary and Review

Summary

1. The great majority of electric energy is used in the form of alternating current and ac voltage. A sine wave is the most common waveform for alternating current.

2. Alternating current periodically reverses the direction in which it flows.

3. Alternating current periodically or continuously changes magnitude.

4. A dc waveform never crosses the zero reference line.

5. A cycle is that part of a periodic waveform which occurs without repeating itself. It is composed of two alternations: a negative alternation and a positive alternation.

6. The period (T) is the time required to complete one cycle.

7. The frequency (f) is the rate at which cycles are produced. Its base unit is the hertz, abbreviated Hz.

8. One hertz is equal to one cycle per second.

9. Power in North America is distributed at a frequency of 60 Hz.

10. Unless otherwise indicated, ac quantities are assumed to be in rms (effective) values.

11. Effective, or rms, values of alternating current produce the same heating effect as the same value of direct current.

12. A conductor rotating in a perfect circle at constant speed in a uniform magnetic field produces a sine wave.

13. There are 180 electrical degrees per alternation and 360 electrical degrees per cycle.

14. Generator voltage is determined by (1) speed of rotation, (2) number of turns in the coils, and (3) flux density.

15. Generator frequency is determined by (1) speed of rotation and (2) number of pairs of magnetic poles.

16. The magnitude of ac voltage is easier to change than the magnitude of dc voltage.

17. Electric power is usually transmitted from a power plant as three-phase alternating current.

18. The algebraic sum of the instantaneous phase values of three-phase voltage or current is always zero.

19. The voltages in a three-phase system are separated by 120 electrical degrees.

20. The fourth wire in a four-wire system is the neutral wire. When the loads on all three phases are equal (balanced), the fourth wire carries no current.

21. The advantages of three-phase alternating current over single-phase alternating current are (1) more efficient transfer of power, (2) more constant load on the generator, (3) more constant torque from motors, and (4) less fluctuation when rectified to direct current.

22. A common four-wire system provides 120/208 V.

Related Formulas

$$T = \frac{1}{f}$$

$$f = \frac{1}{T}$$

$$f = \frac{r/min}{60} \times \text{pairs of poles}$$

Peak = 1.414 rms
Av = 0.637 peak
rms = 0.707 peak

In a delta connection:

$$V_{line} = V_{phase}$$

$$I_{line} = 1.732\, I_{phase}$$

In a wye connection:

$$I_{line} = I_{phase}$$

$$V_{line} = 1.732\, V_{phase}$$

For questions 8-1 to 8-16, determine whether each statement is true or false.

8-1. One half-cycle is also known as an alternation. (8-4)

8-2. One alternation per second is equal to 2 Hz. (8-4)

8-3. Fluctuating direct current never crosses the zero line of a waveform graph. (8-3)

8-4. The average value of a sine wave is smaller than the rms value. (8-4)

8-5. The mechanical degrees and the electrical degrees are equal for a four-pole generator. (8-5)

8-6. Both frequency and output voltage are affected by the rotational speed of a generator. (8-6)

8-7. DC motors require more maintenance than ac motors. (8-7)

8-8. The algebraic sum of the phase voltages in a three-phase system is always zero at any given instant. (8-8)

8-9. A three-phase system requires six conductors to transfer power between two locations. (8-8)

8-10. The line voltages and phase voltages are equal in a wye-connected system. (8-8)

8-11. The line currents and phase currents are equal in a wye-connected system. (8-8)

8-12. When line currents and load currents are not equal in a balanced three-phase system, they are related by a factor of 1.732. (8-8)

8-13. The line voltage in a wye-connected system is always displaced 30° from one of the phase voltages. (8-8)

8-14. In a 208-V, four-wire system, 120 V is available between the neutral and any line. (8-8)

8-15. In a delta connection, the reference ends of the phase windings of a generator are all connected together. (8-8)

8-16. In some ac generators, the magnetic field is rotated. (8-6)

For questions 8-17 to 8-23, supply the missing word or phrase in each statement.

8-17. The waveform used to distribute electric energy is the _____ . (8-7)

8-18. The base unit of frequency is the _____ , which is abbreviated _____ . (8-4)

8-19. The time required to complete a cycle is called the _____, which is abbreviated _____. (8-4)

8-20. In a(n) _____ four-wire, three-phase system, the _____ wire carries no current. (8-8)

8-21. V_p = 20 V is meaningful for a(n) _____ wave. (8-4)

8-22. One hertz can be defined as a(n) _____. (8-4)

8-23. The _____ wire is connected to the star point of a four-wire, three-phase system. (8-8)

Answer the following questions:

8-24. What determines the amount of voltage induced in a conductor moving through a magnetic field? (8-5)

8-25. List three factors that determine the output voltage of a generator. (8-6)

8-26. List four advantages of a three-phase system over a single-phase system. (8-8)

8-27. What type of dc output does three-phase rectification provide? (8-8)

8-1. Determine the frequency and the period of an electrical waveform that goes from zero to maximum positive and back to zero in 0.005 s. (8-4)

8-2. Determine the period of a 200-Hz sine wave. (8-4)

8-3. What is the rms voltage of a sine wave with a peak-to-peak value of 280 V? (8-4)

8-4. Determine the average value of a 120-V electric outlet. (8-4)

8-5. What is the frequency of an eight-pole generator turning at 900 r/min? (8-6)

8-6. What value of I_{p-p} is equivalent to 4.24 A of direct current? (8-4)

8-7. What is the rms value of the input voltage in Fig. 8-24(b)? (8-4)

8-8. What is the frequency of the input voltage in Fig. 8-24(a)? (8-4)

8-9. What is the value of the line voltage of a three-phase, wye-connected generator when the phase voltage is 30 V? (8-8)

8-10. What is the peak-to-peak value of the line current in a three-phase, delta-connected

system when the generator phase current is 2 A? (8-8)

8-11. If the 2 A of phase current in problem 8-10 is caused by three 15-Ω, delta-connected resistors, what is the phase voltage of the generator? (8-8)

8-12. If the generator in problem 8-9 is connected to a load of three 10-Ω, wye-connected resistors, what is the phase current of the generator? (8-8)

Critical Thinking Questions

8-1. Draw a schematic diagram that shows how to connect an automatic-reset circuit breaker, two cells, a relay, and a resistor so that square-wave current flows in the resistor. The symbol for a relay is shown below.

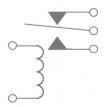

8-2. Using only the symbols presented in previous chapters, draw a schematic diagram for a circuit that will produce a pulsating direct current in a resistor. Also, draw the waveform for the voltage across the resistor.

8-3. How could you double the frequency of a generator and not change the output voltage?

8-4. The neutral wire in a four-wire wye system opens at the star point of the generator. How will this affect the 208-V loads? How will this affect the 120-V loads?

8-5. A phase coil of a delta-connected generator opens. How will this affect the operation of single-phase motors and three-phase delta-connected motors?

8-6. A phase coil of a wye-connected generator opens. How will this affect the operation of

single-phase motors and three-phase, wye-connected motors?

8-7. What is the magnitude relationship between the phase current of the source and the phase current of the load when a balanced wye load is connected to a delta source?

8-8. What is the magnitude relationship between the phase current of the source and the phase current of the load when a balanced delta load is connected to a wye source?

8-9. Why is frequency expressed in hertz when solving this practical working formula?

$$f = \frac{\text{r/min}}{60} \times \text{pairs of poles}$$

8-10. The seven positions on the rotary switch in Fig. 8-26 are spaced 45° apart. From the time the pole contacts a position until it breaks contact, it uses 30° of rotation. Graph the output voltage waveform for 215° of counterclockwise rotation from the location shown in Fig. 8-26 where contact has just been made with a position.

a. Make one graph using degrees of rotation on the horizontal axis.

b. Make a second graph using seconds of time on the horizontal axis.

Assume the rotational speed of the pole is 25 revolutions per minute.

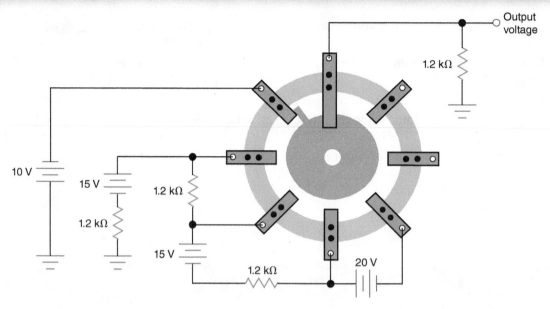

Fig. 8-26 Circuit for critical thinking question 8-10.

Answers to Self-Tests

1. F
2. F
3. T
4. F
5. waveform
6. square wave, sine wave, sawtooth wave
7. hertz, Hz
8. hertz
9. cycle
10. alternation, half-cycle
11. frequency
12. Root mean square (rms)
13. f
14. 2.5 ms, 25.2 V_{av}, 79.2 V_{p-p}
15. 9.42 $A_{(peak)}$
16. 5.656 A
17. 100 Hz
18. T
19. F
20. T
21. F
22. 1440
23. armature
24. slip rings
25. 50 Hz
26. the flux cut per unit of time

27. 6000 r/min
28. the polarity of the magnetic field and the direction of conductor movement
29. uniform magnetic field, constant speed of rotation of the conductor, and rotation in a perfect circle
30. number of turns in the armature coils, strength of the poles, and speed at which the armature rotates
31. number of magnetic poles and revolutions per minute of the armature
32. It is easier to transform voltage, and ac motors are easier to construct and maintain.
33. 120
34. +49 V
35. delta, wye
36. three
37. delta
38. wye
39. delta, 1.732
40. wye, star
41. wye, 1.732
42. neutral
43. when the phases are unequally loaded
44. 120-V, single-phase; 208-V, single-phase; 208-V, three-phase
45. 480 V
46. pulsating

CHAPTER 9

Power in AC Circuits

Learning Outcomes

This chapter will help you to:

9-1 *Realize* that in ac circuits that have only a resistance load, power is calculated using the same formulas used for dc circuits.

9-2 *Appreciate* why the functions of a right triangle are important in working with ac circuits that contain both resistance and reactance.

9-3 *Differentiate* between true power and apparent power, and learn to measure and/or calculate each type of power.

9-4 *Understand* (a) the relationships between $\cos \theta$, P_{app}, P, and PF and (b) the importance of PF to electric energy companies and manufacturers that use large amounts of electric energy.

9-5 *Determine* total power in three-phase systems with either balanced or unbalanced loads.

The most common form of electric power is the power produced by alternating current and ac voltage. In many ac circuits, the current and the voltage do not rise and fall in step with each other. Thus, we need to learn some new techniques to determine the power in ac circuits.

9-1 Power in Resistive AC Circuits

When the load on an ac source contains only resistance, the current and voltage are *in phase*. *In phase* means that two waveforms arrive at their peak positive values, zero values, and peak negative values simultaneously. The current in Fig. 9-1 is in phase with the voltage. The *resistive load* in Fig. 9-1(a) might represent an electric heater, a clothes iron, or a stove. As with direct current, power is the product of the current and voltage. With alternating current, the values of current and voltage vary with time. At any instant, the power is equal to the current at that instant multiplied by the voltage at that instant. As shown in Fig. 9-1(b), the product of the instantaneous current and voltage is the instantaneous power. Plotting all the instantaneous powers produces a power waveform. However, notice that the power waveform is always positive for in-phase currents and voltages. A negative current multiplied by a negative voltage yields a positive power. This means that the resistive load is converting electric energy into heat energy during the complete cycle. For in-phase circuits, that is, resistive circuits, we can state

In phase

Resistive load

$$P = IV = I^2R = \frac{V^2}{R}$$

where I and V are the rms current and voltage.

Suppose a technician measures the sinusoidal ac voltage across a resistor so that he can

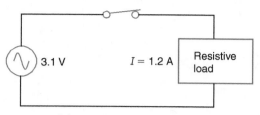

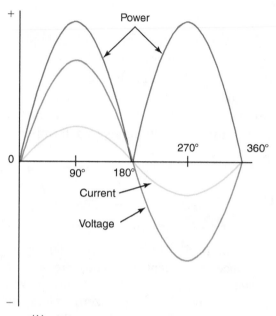

(a) Resistive ac circuit

(b) Current, voltage, and power waveforms

Fig. 9-1 Power in a resistive ac circuit.

Notice that power calculations in resistive ac circuits are no different from those in dc circuits.

determine the power dissipated by the resistor. If the technician measures the voltage with a VOM or DMM and uses the indicated value in the formula $P = \dfrac{V^2}{R}$, the calculated power will be correct because either of these meters indicates the rms value. However, if the voltage is measured with an oscilloscope (discussed in Appendix I), which displays peak-to-peak voltage, and the technician forgets to convert from $V_{p\text{-}p}$ to V_{rms}, the calculated power will be off by a factor of 2.8 $\times$ 2.8 $= 7.84$.

Reactance

Capacitive reactance

9-2 Power in Out-of-Phase Circuits

An ac circuit may contain a load that is not entirely composed of resistance. The load may also contain reactance. *Reactance* is the name given to the opposition to current caused by capacitors and inductors. How and why capacitors and inductors provide reactance is discussed in later chapters. Here, we will be studying only the effect of reactance (and thus capacitors and inductors) on power in ac circuits.

Although reactance opposes current in ac circuits, it is different from resistance. It causes current and voltage to be 90° out of phase with one another. That is, reactance causes a phase shift to occur. The reactance of a capacitor, which is called *capacitive reactance*, causes the current to lead the voltage. In Fig. 9-2(a) the voltage is zero at time zero. At this time, however, the current has already reached its maximum value. Thus, the current leads the voltage by 90°.

ABOUT ELECTRONICS

Flash of Brilliance Flashes far above storm clouds are caused by nitrogen molecules becoming agitated by electron collisions. During the process, lightning sends pulses of energy upward and causes slow changes in the atmosphere's electric field.

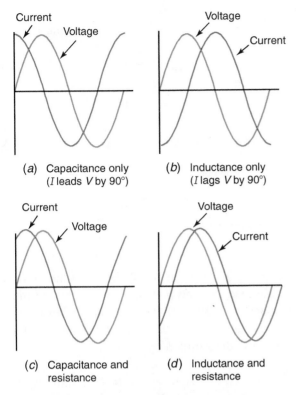

(a) Capacitance only
(I leads V by 90°)

(b) Inductance only
(I lags V by 90°)

(c) Capacitance and resistance

(d) Inductance and resistance

Fig. 9-2 Phase shift in ac circuits.

Inductors produce *inductive reactance*. The effect of inductive reactance is to make the current lag the voltage by 90°. In Fig. 9-2(b) the voltage has already reached its peak positive value when the current is only beginning to go positive.

Loads containing only capacitance or only inductance cause the current and voltage to be exactly 90° out of phase. In other words, such loads cause a 90° phase shift between current and voltage. As shown in Fig. 9-2(c) and (d), loads containing a combination of resistance and reactance cause a phase shift of less than 90°. The exact amount of phase shift is determined by the relative amounts of resistance and reactance.

A good example of a combination load is an electric motor. It contains both inductance and resistance. When a small ac motor has no load on its shaft, the phase shift is about 70°. When the motor is loaded, it acts more like a resistance. When the motor is loaded, the phase shift reduces to about 30°. Of course, in a purely resistive circuit, the phase shift is 0°.

For circuits in which *phase shift* occurs, calculating power is no longer a simple matter of multiplying current and voltage. In Fig. 9-3, three examples of phase shift and the resulting power are shown. Notice that in some cases the current and voltage are of opposite polarities. This yields a negative power for that part of the cycle. What does it mean to have negative power? It means that the load is not taking power from the source during that period of time. In fact, it means that the load is returning power to the source for

Phase shift

Inductive reactance

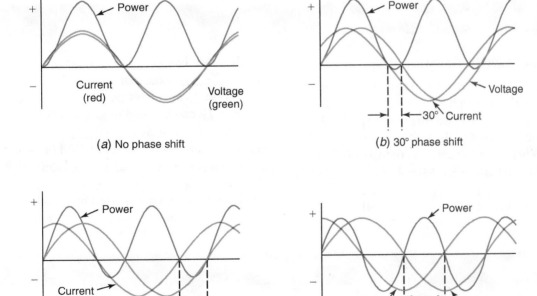

(a) No phase shift

(b) 30° phase shift

(c) 60° phase shift

(d) 90° phase shift

Fig. 9-3 Power in phase-shifted circuits. At 90° of phase shift, the power is zero.

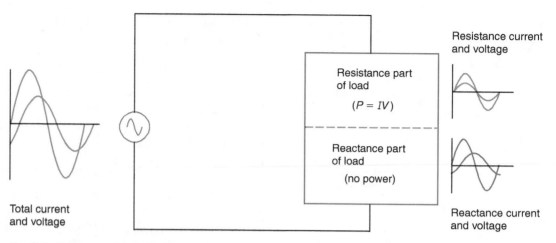

Fig. 9-4 Current and voltage in a combination load.

that part of the cycle. Therefore, the net power used by the load is the difference between the positive power and the negative power. Notice also in Fig. 9-3 that the greater the phase shift, the smaller the net power taken from the source. At 90° phase shift, no net power (thus no energy) is taken from the source.

Since capacitive and inductive reactances cause a 90° phase shift, we can conclude that reactance uses no power. Only resistance uses power. Therefore, in circuits with combination loads we can conclude several things (Fig. 9-4). First, the resistance part of the load uses power. In the resistance part, the current and the voltage are in phase and the power

is simply $P = IV$. Second, the reactance part of the load uses no power. The current and the voltage are a full 90° out of phase. Third, the phase relationship between the total voltage and current is determined by the relative amount of resistance and reactance.

We need a way in which to determine what part of the total current and voltage is associated with the resistance of the load. Then we can determine the net (or real) power used by any combination load with any amount of phase shift. Therefore, let us next develop a way of breaking a current or voltage into a resistive part and a reactive part. We will do this through the use of *phasors* and *right triangle* relationships.

Self-Test

Answer the following questions.

1. _____ causes current and voltage to be 90° out of phase.
2. _____ causes no phase shift between the current and the voltage.
3. When the phase shift is between 0 and 90°, the circuit contains both _____ and _____.
4. The current in an ac motor will _____ the voltage.
5. A motor has _____ reactance.
6. _____ does not convert electric energy to heat energy.
7. $P = IV$ can be used in ac circuits when the load is purely _____.
8. How much power does a 400-Ω resistor use when it carries 25 mA of alternating current?
9. How much power is used by a 500-Ω resistor connected to a 100-V_{p-p} source?

Phasors

Drawing sinusoidal waveforms to show amplitude and phase shift is slow and tedious work.

A much simpler technique of showing amplitude and phase shift is the use of phasors.

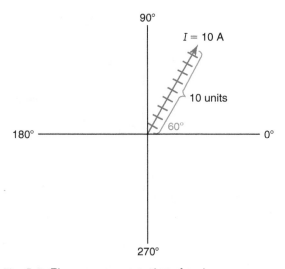

Fig. 9-5 Phasor representation of a sine wave.

A *phasor* is a line whose *direction* represents the phase angle in electrical degrees and whose length represents the *magnitude* of the electrical quantity. You may have studied *vectors* in mathematics or physics classes. From the definition of a phasor, you can see that phasors and vectors are very similar. In electricity and electronics, the term *phasor* is used to emphasize the fact that the position of the line changes with time as the wave moves through its cycle from 0 to 360°. The phasor in Fig. 9-5 represents 10 A of current at 60°. By convention, 0 electrical degrees is on the right-hand side of the horizontal axis. Figure 9-6 shows the phasors of two voltages that are 45° out of phase. In this figure the two phasors have to use the same scale since they both represent voltage. For convenience, one phasor of a group is typically drawn on the 0° line. This phasor,

then, becomes the reference to which the other phasors are related.

Remember that phasors are actually moving with time. By convention, phasors are assumed to rotate in a counterclockwise direction. By showing V_1 on the 0° line, we stopped its motion at that instant. Of course, the motion of V_2 is also stopped at the same instant. Thus, the phase relationship between V_1 and V_2 does not change. Since the motion of phasors is counterclockwise, V_2 in Fig. 9-6 is lagging behind V_1, which is the reference phasor. Since V_2 is out of phase with V_1 by 45°, we can conclude from Fig. 9-6 that V_2 lags V_1 by 45°.

A *voltage phasor* and a *current phasor* are shown in Fig. 9-7. Since current and voltage are different quantities, these two phasors can use different scales for their magnitude. The phasor diagram shows that the current leads the voltage by 45°. The power resulting from this current and voltage can be found by breaking the current phasor into two smaller phasors. One of the smaller phasors represents the current in the resistive portion of the load. It is in phase with the voltage phasor. The other smaller phasor represents the current in the reactive portion of the load. It, of course, is 90° out of phase with the voltage. The two smaller phasors are separated by 90°.

The idea of breaking a phasor into two smaller phasors is illustrated in Fig. 9-8. In this figure, we are using vectors because the positions and magnitudes do not change with time. One vector of 500 newtons (N), approximately 112 lb, of force is exerted on the spring in the northeast direction. In the first case

Voltage phasor

Current phasor

Phasors

Direction

Magnitude

Vectors

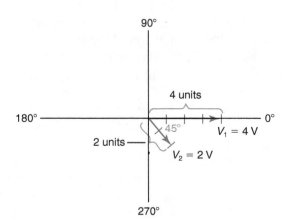

Fig. 9-6 Out-of-phase voltages. Amplitude and phase shift can be shown by phasors.

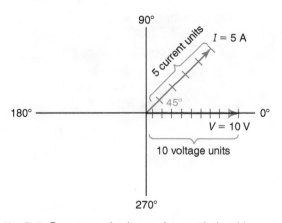

Fig. 9-7 Current and voltage phase relationship represented by phasors. *I* leads *V* by 45°.

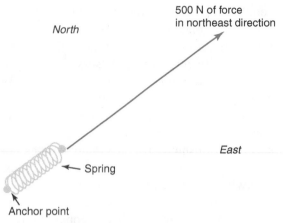

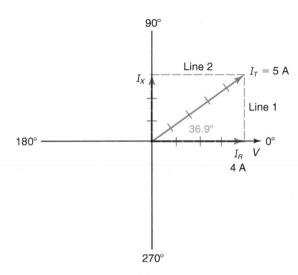

Fig. 9-9 Graphical analysis of a phasor.

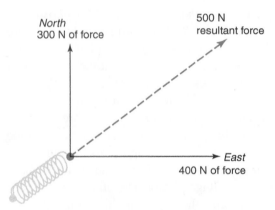

Fig. 9-8 Equivalency of vectors. A single vector can be represented by two smaller vectors separated by 90°.

[Fig. 9-8(*a*)], the force is created by one person pulling in a northeast direction with a force of 500 N. In the second case [Fig. 9-8(*b*)], one person pulls north with 300 N of force and

another person pulls east with 400 N. The net result is 500 N of force in the northeast direction. The effect on the spring is exactly the same in both cases.

The vectors in Fig. 9-8(*b*) represent two forces at a right angle (90°) to each other. However, they could equally well represent any other quantity that has magnitude and direction. Thus, they could be phasors representing resistive current and reactive current in an electric circuit.

The magnitude of resistive and reactive current can be determined graphically. The process is shown in Fig. 9-9. In this diagram, V is at the 0° reference line. The total current I_T is 5 A and is leading the voltage by 36.9°. The first step is to draw the original phasor I_T to scale at the correct angle. Then draw a vertical line (line 1) from the tip of the I_T phasor through the horizontal axis. Line 1 must be parallel to the vertical axis. The intersection of line 1 and the horizontal axis locates the end of the resistive current phasor I_R. The end of the reactive current phasor I_X is found by drawing line 2 from the tip of I_T to the intersection of the vertical axis. Again, line 2 must be parallel to the horizontal axis. Finally, the magnitude of the resistive and reactive currents are measured with the scale used to measure I_T. The 4-A resistive current in Fig. 9-9 is the current that is in phase with the voltage. It is the current that can be used to calculate the power when 5 A of total current leads the total voltage by 36.9°.

Answer the following questions.

10. True or false. A phasor represents the magnitude and the phase angle of an electrical quantity.

11. True or false. The same scale must be used for all voltage phasors in a phasor diagram.

12. True or false. Current and voltage phasors must use the same scale in a phasor diagram.

13. Draw a phasor for 10 V at 280°.

14. Draw a phasor for 8 A at 45°.

15. A current of 6 A lags a voltage of 10 V by 90°. Draw a phasor diagram of this current and voltage, using current as the reference phasor.

16. Draw a sketch that shows the resistive part of the current phasor in Fig. 9-5.

Functions of a Right Triangle

The currents determined by the *graphical method* illustrated in Fig. 9-9 are as accurate as the drawing itself. For most applications the graphical method is accurate enough. However, it is also a slow and tedious method. Let us develop a faster, easier way of determining these currents. This easier way is through the use of the properties of a right triangle. Notice in Fig. 9-9 that the reactive current phasor is exactly the same length as line 1. Also, notice that it is parallel to line 1. Therefore, line 1 can be replaced by the reactive current phasor. This has been done in Fig. 9-10. It can be clearly seen in Fig. 9-10 that the three current phasors form a right triangle. Now can you see why we need to know how to work with right triangles?

The right triangle of Fig. 9-10 is redrawn and relabeled in Fig. 9-11. The new labels are terms used to describe any right triangle. The symbol θ (which is the Greek letter *theta*) is used to indicate the angle that will be used to analyze the triangle. In the phasor diagram of Fig. 9-9, θ is the angle by which the total current leads the voltage. The *adjacent side* is always the shorter side that makes up angle θ. The side across the triangle from angle θ is always called the *opposite side*. The longest side of a right triangle is always called the *hypotenuse*.

Graphical method

The three angles of any triangle always contain a total of 180°. This means that specifying θ for a right triangle also determines the other angle. Thus, the only possible difference between right triangles with the same angle θ is their size. This is illustrated in Fig. 9-12. Triangles *ABG*, *ACF*, and *ADE* all have the same angle θ. Only their sizes differ.

Notice in Fig. 9-12 that the sides and the hypotenuse increase proportionately. That is, when the hypotenuse is doubled in length, the opposite and adjacent sides must also be

Theta

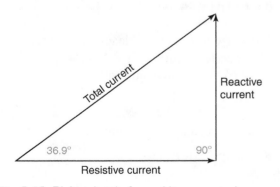

Fig. 9-10 Right triangle formed by current phasors.

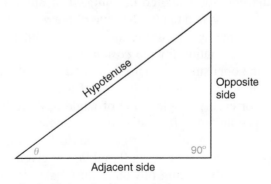

Fig. 9-11 Right triangle terminology.

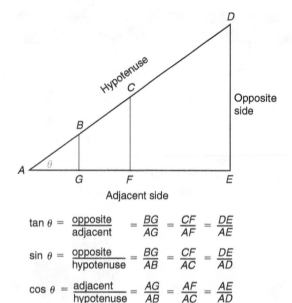

$$\tan \theta = \frac{\text{opposite}}{\text{adjacent}} = \frac{BG}{AG} = \frac{CF}{AF} = \frac{DE}{AE}$$

$$\sin \theta = \frac{\text{opposite}}{\text{hypotenuse}} = \frac{BG}{AB} = \frac{CF}{AC} = \frac{DE}{AD}$$

$$\cos \theta = \frac{\text{adjacent}}{\text{hypotenuse}} = \frac{AG}{AB} = \frac{AF}{AC} = \frac{AE}{AD}$$

Fig. 9-12 Trigonometric functions of a right triangle.

doubled. For example, all the sides of triangle *ACF* are twice as great as the corresponding sides of triangle *ABG*. The sides of triangle *ADE* are four times as large as the corresponding sides of triangle *ABG*.

Since the sides of a right triangle increase proportionately, the ratio of any two sides remains constant. In Fig. 9-12, side *BG* divided by side *AG* is exactly equal to side *CF* divided by side *AF*. The opposite side divided by the adjacent side is a constant for all right triangles that have the same angle θ. This same statement can be made for any other combination of sides of a right triangle. Remember that the hypotenuse is also a side.

The ratios of the sides of a right triangle are called the *trigonometric functions*. Three of the common ones used in electricity are shown in Fig. 9-12. The ratio of the *opposite side* to the adjacent side is called the *tangent* of angle θ and abbreviated tan θ. The ratio of the *adjacent side* to the hypotenuse is called cos θ, which is the abbreviation for the *cosine* of θ. The ratio of the opposite side to the *hypotenuse* is called sin θ, which stands for the *sine* of θ.

For each possible value of angle θ, there is a specific value for the sine, the tangent, and the cosine. A table of trigonometric functions for even angles between 0 and 90° is given in Appendix G. Using this table (or a calculator) and the formulas in Fig. 9-12, we can easily

solve right triangle problems. For example, if we know the angle and the length of the hypotenuse, we can solve for the length of the adjacent side.

EXAMPLE 9-3

How long is the adjacent side of a right triangle if the hypotenuse is 7 cm and the angle θ is 40°?

Given:	Hypotenuse = 7 cm
	θ = 40°
Find:	Adjacent side
Known:	$\cos \theta = \dfrac{\text{adjacent}}{\text{hypotenuse}}$
Solution:	From the cosine formula we can solve for the adjacent side:
	Adjacent = hypotenuse × cos θ
	With a calculator or a table of trigonometric functions (Appendix G) we can find the value of cos 40°:
	cos 40° = 0.766
	Then,
	Adjacent = 7 cm × 0.766
	= 5.362 cm
Answer:	The adjacent side is 5.362 cm long.

In an electric circuit the hypotenuse might represent the total current, and the adjacent side the resistive current (Fig. 9-10). In that case,

$$\cos \theta = \frac{\text{resistive current}}{\text{total current}}$$

Let us now use the functions of a right triangle to solve an electrical problem.

EXAMPLE 9-4

The current in a circuit leads the voltage by 25°. The total current is 8.4 A. What is the value of the resistive current?

Given:	θ = 25°
	I_T = 8.4 A
Find:	Resistive current
Known:	$\cos \theta = \dfrac{I_R}{I_T}$

Trigonometric functions

Opposite side

Tangent

Adjacent side

Cosine

Hypotenuse

Sine

Solution: From the cosine formula:

$$I_R = I_T \times \cos \theta$$

Determine the cosine:

$$\cos 25° = 0.906$$

Then,

$$I_R = 8.4 \text{ A} \times 0.906$$
$$= 7.610 \text{ A}$$

Answer: The resistive current is 7.61 A.

Notice in the trigonometric table of Appendix G that the cosine varies from 1 to 0 as the angle changes from 0° to 90°. In other words, at 0° the adjacent side and the hypotenuse are the same length and $\cos \theta$ is 1. At 90° the adjacent side reduces to 0 and $\cos \theta$ is 0. Referring to Fig. 9-10, we can see that there is no resistive current when the angle is 90°. When θ is 0°, there is no reactive current. Notice, then, that $\cos \theta$ tells what part, or percentage, of the total current is in phase with the voltage. That is, $\cos \theta$ is the portion of the total current that is resistive. For example, when θ is 45°, $\cos \theta$ is 0.707. Therefore, 0.707 of the total current is resistive current. That portion (0.707) of the current uses power.

Because $\cos \theta$ determines the amount of resistive current, it is included in the formula for calculating power. The power formula is

$$P = IV \cos \theta$$

This is the general formula for power. It is the correct formula for all types of circuits. However, when a circuit contains only resistance, $\cos \theta$ is 1. Therefore, $\cos \theta$ can be dropped when working only with resistance.

EXAMPLE 9-5

A 208-V motor draws 12 A. The current lags the voltage by 35°. What is the power input to the motor?

Given: $V = 208$ V
$I = 12$ A
$\theta = 35°$

Find: P

Known: $P = IV \cos \theta$

Solution: From Appendix G:
$\cos 35° = 0.819$

Then,
$$P = IV \cos \theta$$
$$= 12 \text{ A} \times 208 \text{ V} \times 0.819$$
$$= 2044 \text{ W}$$

Answer: The power required by the motor is 2044 W.

So far in developing our power equation, we have always shown the current out of phase with the voltage, with the voltage on the 0° reference line. Thus, we broke the total current into two parts: the resistive and the reactive. It is also possible (and common) for the current to be shown on the 0° line and the voltage to be shown at some other angle. In that case, the total voltage would be separated into a resistive voltage and a reactive voltage. The phasor diagram for such a case is illustrated in Fig. 9-13. In this case, we still use $\cos \theta$. Only now we use it to determine the part of the total voltage that is resistive. Then we will know the resistive voltage and current, and from that we can find the power.

Whether it is the total voltage that splits into a reactive and resistive voltage or the total current that splits into a reactive and resistive current depends on the circuit configuration. When the reactance and resistance are in series,

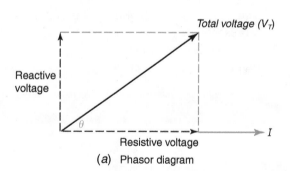

(a) Phasor diagram

$$P = IV \cos \theta$$

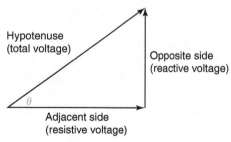

(b) Voltage triangle

Fig. 9-13 Determining the resistive voltage. The total voltage phasor is represented by its two equivalent phasors.

the voltage is split because in a series circuit the same current flows through all of the series components. Likewise, when the resistance and reactance are in parallel, there is only one common voltage and the total current splits.

EXAMPLE 9-6

In an electric circuit the voltage leads the current by 73°. The voltage is 80 V, and the current is 4 A. What is the power used?

Given: $\theta = 73°$
$V = 80$ V
$I = 4$ A

Find: P
Known: $P = IV \cos \theta$
Solution: $\cos 73° = 0.292$
$P = 4$ A $\times$ 80 V $\times$ 0.292
$= 93.4$ W
Answer: The power used by the circuit is 93.4 W.

In the above example the voltage was leading the current by 73°. However, the power would be the same if the voltage were lagging the current.

Self-Test

Answer the following questions.

17. Draw a right triangle which represents the force vectors shown in Fig. 9-8.
18. Draw a right triangle that has a 30° angle θ. Label all three sides of the triangle and mark the degrees in each angle.
19. The symbol that represents the angle between two phasors is _____.
20. The resistive current is represented by the _____ side of a right triangle.
21. The _____ side of a right triangle represents the reactive voltage.
22. The sin θ is equal to the _____ divided by the hypotenuse.

23. The resistive current divided by the total current yields the _____ of angle theta (θ).
24. How much resistive voltage is there when the total voltage is 80 V at 30°?
25. How much resistive current is there when 5 A of total current leads the source voltage by 70°?
26. Determine the power of a 120-V circuit in which the 7 A of current leads the voltage by 300°.
27. Determine the power used by a 440-V motor that draws 8 A and causes the current to lag the voltage by 35°.

9-3 True Power and Apparent Power

Apparent power

True power

So far we have been calculating the *true power* of ac circuits. That is, we have determined the *actual* power used by the circuit. We have done this by taking into account the phase shift between the total current and the total voltage. We can also determine true power in a circuit by measuring it with a *wattmeter*. A wattmeter is constructed so that it takes into account any phase difference between current and voltage.

Wattmeter

Sometimes it is as important to know the apparent power in a circuit as it is to know the true power. The *apparent power* is the power that appears to be present when the voltage and the current in a circuit are measured separately. The apparent power, then, is the product of the voltage and the current regardless of the phase angle θ. In a circuit containing both resistance and reactance, the apparent power is always greater than the true power.

Apparent power is calculated by the formula

$$P_{app} = IV$$

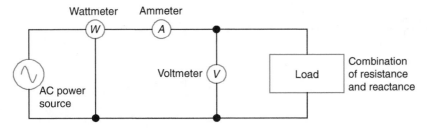

Fig. 9-14 Measuring power and apparent power. The apparent power is found by multiplying the current by the voltage. The true power is indicated by the wattmeter.

It has a base unit of *voltampere* (VA) to indicate that it is not true power, which has units of watts.

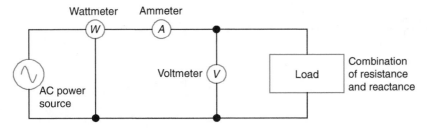

You May Recall

. . . that one watt of true power means that electric energy is converted into another form of energy at a rate of 1 J/s.

One voltampere of apparent power converts *less than* 1 J/s. How much less depends on how far the voltage and current are out of phase.

EXAMPLE 9-7

Determine the apparent power for the circuit specified in example 9-6.

Given:	$V = 80\text{ V}$
	$I = 4\text{ A}$
Find:	P_{app}
Known:	$P_{app} = IV$
Solution:	$P_{app} = 4\text{ A} \times 80\text{ V} = 320\text{ VA}$
Answer:	The apparent power of the circuit is 320 VA.

Look closely at the formulas for apparent power and true power:

$$P = IV \cos \theta$$
$$P_{app} = IV$$

Notice that the only difference between the two is that the true power includes a cos θ term. Combining these two formulas yields

$$\cos \theta = \frac{P}{P_{app}}$$

This relationship makes it relatively easy to determine cos θ and thus the phase relationship between current and voltage. All we need to know is the current, the voltage, and the true power. All three of these are easily measured, as shown in Fig. 9-14. With the measured current and voltage, we can calculate the apparent power. Then, with the measured power and the calculated apparent power, we can calculate cos θ. After calculating the value of cos θ, we can use a calculator, or a table of trigonometric functions, to find θ.

EXAMPLE 9-8

Refer to Fig. 9-14. Suppose the wattmeter reads 813 W, the voltmeter reads 220 V, and the ammeter reads 6 A. What is the phase angle (θ) between the current and the voltage?

Given:	$P = 813\text{ W}$
	$V = 220\text{ V}$
	$I = 6\text{ A}$
Find:	θ
Known:	$P_{app} = IV,\ \cos \theta = \dfrac{P}{P_{app}}$
Solution:	$P_{app} = 6\text{ A} \times 220\text{ V}$
	$\qquad = 1320\text{ VA}$
	$\cos \theta = \dfrac{813\text{ W}}{1320\text{ VA}} = 0.616$
	Using a calculator or Appendix G, we determine that the angle whose cosine is 0.616 is 52°.
Answer:	The current and voltage are 52° out of phase ($\theta = 52°$).

If the reactive part of the load in the above example were inductance, the current would lag the voltage by 52°. If it were capacitance, the current would lead the voltage by 52°. Phasor diagrams for both cases are drawn in Fig. 9-15.

Voltamperes

Cos $\theta = \dfrac{P}{P_{app}}$

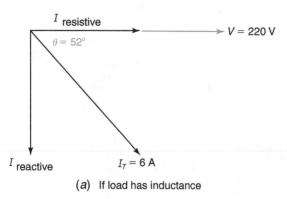

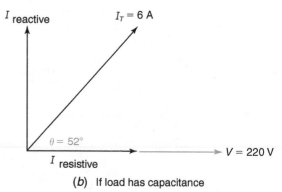

(a) If load has inductance   (b) If load has capacitance

Fig. 9-15 Phasor diagrams for example 9-8.

Self-Test

Answer the following questions.

28. _____ power does not account for any phase difference between current and voltage.
29. Voltampere is the base unit of _____.
30. _____ power can never be greater than _____ power.
31. The measured power, current, and voltage in a circuit are 470 W, 3.6 A, and 240 V respectively. Determine θ, $\cos \theta$, and P_{app}.

32. Determine P_{app}, PF, and I_T for a circuit in which I_T leads V_T by 40°, $P = 780$ W, and $V_T = 240$ V.
33. True or False. A wattmeter measures apparent power.
34. True or False. A wattmeter compensates for phase shift when measuring power.
35. True or False. One voltampere converts one joule per second.

9-4 Power Factor

Power factor (PF)

The ratio of the true power to the apparent power in a circuit is known as the *power factor* (PF). Since this ratio also yields cos θ, power factor is just another way of specifying cos θ. Mathematically we can write

$$PF = \cos \theta = \frac{P}{P_{app}}$$

Power factor, therefore, is a way of indicating that portion of the total current and voltage that is producing power. When current and voltage are in phase, the power factor is 1; the true power is the same as the apparent power. When current and voltage are 90° out of phase, the power factor is 0. Thus, the power factor of a circuit can be any value between 0 and 1. Sometimes power factor is expressed as a percentage. Then the power factor can vary from 0 to 100 percent.

EXAMPLE 9-9

An electric motor draws 18 A of current from a 240-V source. A wattmeter connected to the circuit indicates 3024 W. What is the power factor of the circuit?

Given: $I = 18$ A
               $V = 240$ V
               $P = 3024$ W

Find: Power factor

Known: $PF = \dfrac{P}{P_{app}}$, $P_{app} = IV$

Solution: $P_{app} = 18$ A $\times$ 240 V
                $= 4320$ VA
            $PF = \dfrac{3024 \text{ W}}{4320 \text{ VA}} = 0.7$

Answer: The power factor is 0.7, or 70 percent.

Power factor is very important to the utility company that sells electric energy. Its

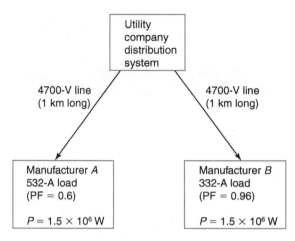

Fig. 9-16 Importance of power factor. It costs more to deliver a unit of power when the power factor is low.

importance can be best shown by an example. Suppose a utility company is selling energy to two different manufacturers. The two manufacturers are located the same distance from the utility's distribution center (Fig. 9-16). Both manufacturers receive power at the same voltage (4700 V), and both require the same true power (1.5 MW). However, manufacturer A uses a lot of reactive loads (motors) and operates with a power factor of 60 percent. Manufacturer B uses mostly resistive loads (heaters) and operates with a power factor of 96 percent. In order to serve manufacturer A, the utility company must provide the following number of voltamperes:

$$P_{app} = \frac{P}{PF} = \frac{1,500,000 \text{ W}}{0.6}$$
$$= 2,500,000 \text{ VA} = 2.5 \text{ MVA}$$

In order to serve this load, the utility company feeders (conductors) must be able to carry the following current:

$$I = \frac{P_{app}}{V} = \frac{2,500,000 \text{ VA}}{4700 \text{ V}}$$
$$= 532 \text{ A}$$

Manufacturer B uses the same true power as manufacturer A and thus pays the same amount to the utility company. Manufacturer B, however, requires the following apparent power:

$$P_{app} = \frac{P}{PF} = \frac{1,500,000}{0.96}$$
$$= 1,562,500 \text{ VA}$$

or about 1.56 MVA. This is almost 1 MVA less than required by manufacturer A. In addition, the current drawn by manufacturer B would be

$$I = \frac{P_{app}}{V} = \frac{1,560,000 \text{ VA}}{4700 \text{ V}}$$
$$= 332 \text{ A}$$

Notice that manufacturer B draws less current than manufacturer A (332 A versus 532 A) to obtain *exactly the same true power.*

The utility company (as well as the customer) designs its transmission and distribution systems according to the *apparent power* and current it must deliver. Since the customer is billed for the *true power* used, the utility company encourages the use of high-power-factor systems. Power factor can be improved (or corrected) by inserting a reactance opposite the one causing the low power factor. Thus, a lagging power factor can be improved by inserting a leading-power-factor device, such as a capacitor, in the system.

 Self-Test

Answer the following questions.

36. PF is numerically equal to _____.
37. Power divided by apparent power equals

    _____.

38. PF can be no larger than _____.
39. What is the power factor of a circuit in which the current and the voltage are 70° out of phase?

40. The apparent power in a circuit is 360 VA. The power is 295 W. What is the power factor expressed as a percentage?
41. Two manufacturing plants use equal amounts of true power for the same length of time. Plant X operates with a power factor of 0.75, and plant Y operates with a power factor of 0.89. If both plants operate at the same voltage, which plant draws more current?

42. Which plant in question 41 is utilizing the distribution equipment more efficiently?

43. How can the PF be improved in an industrial plant that primarily uses motors?

9-5 Power in Three-Phase Circuits

The power of any three-phase load, either balanced or unbalanced, is simply the sum of the powers dissipated by the load on each phase. When the loads are pure resistance, the individual phase voltages and currents are all that is needed to calculate the power. When the phase loads are a combination of resistance and reactance, angle theta must be known for each phase or the power must be determined with a wattmeter. When only balanced loads are involved, determining the power for one phase is all that is necessary. The total power is merely three times the individual phase power.

You May Recall

… that in Chapter 8 the relationships between line current and load current and between line voltage and load voltage were determined for balanced wye and balanced delta loads. Thus, with balanced resistive loads we can determine the power of a three-phase system if we know any line current and line voltage and the type of connection (delta or wye).

EXAMPLE 9-10

Determine the power of a three-phase, delta-connected, balanced-resistive load system with a line voltage of 480 V and a line current of 30 A.

Given: $V_{line} = 480$ V
$I_{line} = 30$ A
Balanced delta system
Resistive load

Find: Total power

Known: $V_{line} = V_{phase}$
$I_{line} = 1.732\, I_{phase}$
$\cos \theta = 1$
$P_{phase} = V_{line} \times \dfrac{I_{line}}{1.732}$
$P_T = 3\, P_{phase}$

Solution: $P_{phase} = 480$ V $\times \dfrac{30\ A}{1.732}$
$= 8314$ W

$P_T = 3 \times 8314$ W $= 24{,}942$ W.

Answer: The total power is 24.942 kW.

EXAMPLE 9-11

Determine the phase voltage of a wye-connected, three-phase system connected to a balanced-resistive load that draws 4 A of line current and dissipates 3 kW of total power.

Given: $P_T = 3$ kW
$I_{line} = 4$ A
Wye system
Balanced-resistive load

Find: V_{phase}

Known: $V_{line} = 1.732\, V_{phase}$
$I_{line} = I_{phase}$
$P_T = 3\, P_{phase}$
$\cos \theta = 1$
$P_{phase} = \dfrac{V_{line}}{1.732} \times I_{line}$

Solution: $P_{phase} = \dfrac{P_T}{3} = \dfrac{3\text{ kW}}{3} = 1$ kW
Solving the last "known" formula for V_{line} yields
$V_{line} = \dfrac{1.732\, P_{phase}}{I_{line}}$
$= \dfrac{1.732 \times 1000\text{ W}}{4\text{A}} = 433$ V
$V_{phase} = \dfrac{V_{line}}{1.732} = 250$ V

Answer: The phase voltage is 250 V.

Answer the following questions.

44. Under what condition does $P_T = 3P_{phase}$?

45. What is the value of the load resistors in example 9-11?

46. When is it appropriate to use formula

$$P_{phase} = \frac{V_{line}\,I_{line}}{1.732}?$$

47. Write a formula for total power that is appropriate for any 3-ϕ system with any type of load.

48. Why is $P_{phase} = I_{line}\,V_{line}$ not a valid formula?

49. Which formula(s) used in example 9-10 would not be valid if the balanced load were part resistive and part reactive?

Chapter 9 Summary and Review

Summary

1. In a pure resistance circuit, current and voltage are in phase.
2. When I and V are in phase, $P = IV$.
3. Reactance causes current and voltage to be 90° out of phase.
4. Circuits with inductive reactance cause current to lag voltage.
5. Circuits with capacitive reactance cause current to lead voltage.
6. A phasor represents the magnitude and direction of a quantity.
7. A single phasor can be broken into two smaller phasors which are at right angles to each other.
8. Only the resistive part of a current or a voltage can use power (convert electric energy).
9. The cosine of angle θ is the ratio of the adjacent side to the hypotenuse.
10. The cosine of θ can be the ratio of the resistive current to the total current.
11. The cosine of θ can be the ratio of the resistive voltage to the total voltage.
12. In any electric circuit, power can be calculated with the formula $P = IV \cos \theta$.
13. Apparent power does not take into account the phase relationship between current and voltage.
14. The unit for apparent power is the voltampere (VA).
15. The ratio of power to apparent power is equal to $\cos \theta$. It is also equal to the power factor.
16. Power factor varies from 0 to 1 or from 0 to 100 percent.
17. Energy used at a low power factor is more costly than the same energy used at a high power factor.

Related Formulas

$$\cos \theta = \frac{\text{adjacent}}{\text{hypotenuse}} = \frac{P}{P_{\text{app}}} = \text{PF}$$

$$P = IV \cos \theta$$

$$P_{\text{app}} = IV$$

Chapter Review Questions

For questions 9-1 to 9-9, determine whether each statement is true or false.

9-1. Resistance causes electric energy to be converted into heat energy. (9-1)

9-2. In a circuit containing only resistance, the current and the voltage are 90° out of phase. (9-1)

9-3. All voltage phasors in a phasor diagram must be drawn to the same scale. (9-2)

9-4. The sine of an angle is equal to the adjacent side divided by the hypotenuse. (9-2)

9-5. A current that leads its voltage source by 30° uses more power than a current that lags its voltage source by 30°. (9-2)

9-6. The apparent power and the true power are equal in a circuit containing both resistance and inductance. (9-3)

9-7. The true power in a circuit is often greater than the apparent power. (9-3)

9-8. Power factor and $\cos \theta$ have the same numerical value. (9-4)

9-9. Industry pays more for a unit of energy used at a high power factor than one used at a low power factor. (9-4)

For questions 9-10 to 9-22, supply the missing word or phrase in each statement.

9-10. Phasors are like vectors except that they are _____. (9-2)

9-11. In a phasor diagram, the direction of a phasor represents _____. (9-2)

9-12. Most motors cause the current to _____ the voltage. (9-2)

9-13. Capacitance causes current to _____ voltage by _____ electrical degrees. (9-2)

9-14. _____ is a form of opposition to current that uses no power. (9-2)

9-15. The base unit of apparent power is the _____. (9-3)

9-16. Any phasor can be represented by two other phasors that are separated by _____ degrees. (9-2)

9-17. θ is the symbol for _____. (9-2)

9-18. Power factor can be determined by dividing the _____ power by the _____ power. (9-4)

9-19. In some circuits, the total current can be divided into _____ current and _____ current. (9-2)

9-20. A wattmeter measures _____ power. (9-3)

9-21. Unless theta = 0°, _____ power is _____ than true power. (9-3)

9-22. A current that leads the voltage by 340° also lags the voltage by _____ degrees. (9-2)

Answer the following questions:

9-23. Does a load with a PF of 0.6 or a PF of 0.9 make the most efficient use of the power distribution equipment? Why? (9-4)

9-24. If voltage and current are in units of Vp and Ip, respectively, will calculated power be in units of MW? Why? (9-1)

9-25. Which trigonometric function is equal to the adjacent side divided by the hypotenuse side? (9-2)

9-26. List the power formula that is appropriate for all types of circuits. (9-2)

Chapter Review Problems

9-1. A pure capacitance load draws 6 A from an 80-V ac source. Determine the power, apparent power, and θ. (9-2)

9-2. Determine the reactive current when the total current is 7 A at 40° and the total voltage is 240 V at 0°. (9-2)

9-3. Determine the power, apparent power, and power factor for problem 9-2. (9-4)

9-4. Determine θ when the measured current, voltage, and power in a circuit are 5 A, 120 V, and 400 W, respectively. (9-3)

9-5. Determine V and P_{app} when I leads V by 30°, P = 500 W, and I = 3 A. (9-3)

9-6. How much power is dissipated by a 2.2 kΩ resistor that drops 16 V_{p-p}? (9-1)

9-7. How much power is dissipated by a 3.9 kΩ resistor when the current is 2.8 mA? (9-1)

9-8. Determine the reactive current in a 240-V circuit when P = 800 W and P_{app} = 1000 VA. (9-3)

9-9. Determine the resistance in the circuit in problem 9-8. (9-1)

9-10. Determine R_{phase} and P_T for a wye-connected, three-phase system with a line voltage of 240 V and a balanced resistive load where each phase load draws 6 A. (9-5)

9-11. A 240-V, 5-hp motor is driving a 4.5-hp load. Instruments (voltmeter, ammeter, and wattmeter) connected to the motor indicate the motor inputs are 240 V, 17.5 A, and 3850 W. Determine the P_{app}, PF, θ, and percent efficiency. (9-4)

9-1. Could the force of attraction (or repulsion) between two electromagnets be used to measure power in an ac circuit? How?

9-2. To optimize the power factor, should electric motors be loaded to rated value or lightly loaded? Why?

9-3. Draw a phasor diagram for the line voltages of a wye-connected three-phase generator that has a phase voltage of 100 V.

9-4. A 500-Ω resistor (R_1) in series with a 1000-Ω resistor (R_2) is connected to a 30-V ac source. Draw the phasor diagram for the circuit.

9-5. What is the efficiency of a motor that provides 5/8 hp of shaft power when it draws 4 A at 240 V and has a power factor of 0.72?

9-6. Does it matter whether we say "the current leads the voltage by 290 degrees" or "the current lags the voltage by 70 degrees"? Why?

9-7. When a resistance load and a reactance load are in series, which phasor (voltage or current) is used as the reference? Why?

9-8. When a resistance load and a reactance load are in parallel, which phasor (voltage or current) is used as the reference? Why?

9-9. When power is calculated, does it matter whether the current or the voltage is the reference? Why?

Answers to Self-Tests

1. reactance
2. resistance
3. reactance, resistance
4. lag
5. inductive
6. Reactance
7. resistive
8. 0.25 W
9. 2.5 W
10. T
11. T
12. F
13.

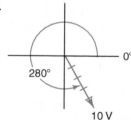

14.

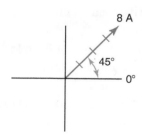

15.

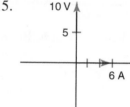

16.

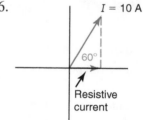

17.

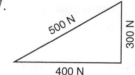

18.

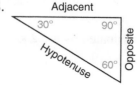

19. (θ) (theta)
20. adjacent
21. opposite
22. opposite
23. cosine
24. 69 V
25. 1.7 A
26. 420 W
27. 2883 W
28. apparent
29. apparent power
30. true, apparent
31. $\theta = 57°$
    $\cos \theta = 0.544$
    $P_{app} = 864$ VA
32. $P_{app} = 1018.2$ VA
    PF = 76.6 percent
    $I_T = 4.24$ A
33. F
34. T

35. F
36. $\cos \theta$
37. $\cos \theta$ or PF
38. 1 or 100 percent
39. 0.342 or 34.2 percent
40. 81.9 percent
41. plant X
42. plant Y
43. Improve the PF by inserting a capacitor in the system.
44. When the 3-ϕ load is balanced.
45. 62.5 Ω
46. When the three-phase system (either delta or wye) is connected to a balanced-resistive load.
47. $P_T = P_{phase1} + P_{phase2} + P_{phase3}$
48. Because with a delta-connection, I_{line} is 1.732 times larger than I_{phase}, and with a wye connection, V_{line} is 1.732 times larger than V_{phase}.
49. $P_{phase} = V_{line} \times \dfrac{I_{line}}{1.732}$

CHAPTER 10

Capacitance

Learning Outcomes

This chapter will help you to:

10-1 *Understand* the terminology used in working with capacitors and capacitance.

10-2 *Visualize* how a capacitor stores energy as it charges.

10-3 *Explain* why a capacitor has a voltage rating usually called the direct current working voltage (DCWV).

10-4 *Remember* that the farad (F) is the base unit of capacitance (*C*) and a farad is equal to a coulomb/volt.

10-5 *List and discuss* the four factors that determine the capacitance of a capacitor.

10-6 *Realize* that capacitors may be classified (named) for their intended use, their type of enclosure, their type of dielectric material, or the construction process used to make them.

10-7 *Use* the correct schematic symbol for a specific capacitor and remember there is a symbol for fixed nonpolarized, fixed polarized, variable, and ganged variable.

10-8 *Define* the term "time constant" for an *RC* circuit. Also discuss voltage distribution between series capacitors in a dc circuit.

10-9 *Explain* how a capacitor causes *I* and *V* to be 90° out-of-phase in an ac circuit. Also, discuss the importance of the "quality" of a capacitor.

10-10 *Discuss* why connecting capacitors in series (a) decreases the total capacitance, and (b) increases total reactance (X_{CT}).

10-11 *Discuss* why connecting capacitors in parallel (a) increases the total capacitance, and (b) decreases total reactance.

10-12 *Discuss* the limitations of testing capacitors with an ohmmeter.

10-13 *Define* "stray capacitance" and explain when it is most likely to cause problems.

10-14 *List* five specifications to be considered when ordering capacitors.

10-15 *Explain* what a coupling capacitor does and how it does it.

Capacitors, devices that possess capacitance, are almost as common as resistors in electric and electronic circuits. They are used in electric motors, automobile ignition systems, electronic photoflashes, and fluorescent lamp starters, to name just a few applications.

Large capacitors are used by electric utility companies and manufacturers to improve the power factor of their system. Operating with a high (close to one) power factor is beneficial to both the utility company and the manufacturer.

10-1 Terminology

The ability to store energy in the form of electric charge is called *capacitance*. The symbol used to represent capacitance is *C*. A device designed to possess capacitance is called a *capacitor.* In its simplest form a capacitor is nothing more than two conductors separated by an insulator. Figure 10-1 illustrates a capacitor consisting of two pieces of aluminum (a conductor) separated by a sheet of polyester plastic (an insulator). In capacitors the conductors are called *plates* and the insulator is called a *dielectric.*

Capacitance (C)

Plates

Dielectric

10-2 Basic Capacitor Action

A capacitor stores energy when an electric charge is forced onto its plates by some other energy source, such as a battery. When a capacitor is connected to a battery, a current flows until the capacitor becomes charged (Fig. 10-2). Electrons from the negative terminal of the battery move through the connecting lead and pile up on one of the plates. At the same instant, electrons from the other plate move through the connecting lead to the positive terminal of

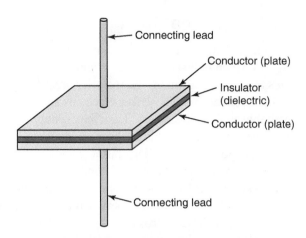

Fig. 10-1 Simple capacitor. The plates of a capacitor are always insulated from each other.

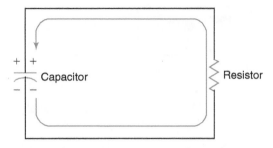

Fig. 10-3 A capacitor is an energy source. A capacitor forces current through a resistor until the capacitor is discharged.

the battery. The net result is that one plate of the capacitor ends up with an excess of electrons (negative charge). The other plate ends up with a deficiency of electrons (positive charge). These charges on the plates of the capacitor represent a voltage source just as the charges on the plates of a cell do. Note that in the process of charging the capacitor, no electrons move from one plate through the dielectric to the other plate.

Because of opposing voltages, the current stops once the capacitor is charged. The opposite charges on the plates of the capacitor create a new energy source. The energy stored in the capacitor produces a voltage equal to that of the battery (the original energy source). Since the capacitor's voltage is equal to, and in opposition to, the battery voltage, a state of equilibrium exists. No current can flow in either direction.

A *charged capacitor* can be disconnected from the original energy source (the battery) and used as a new energy source. If a voltmeter is connected to the capacitor, it will register a

voltage. If a resistor is connected to the capacitor (Fig. 10-3), current will flow through the resistor. However, a capacitor has limited use as a primary energy source for two reasons:

1. For its weight and size, the amount of energy it can store is small compared with what a battery can store.
2. The voltage available from the capacitor rapidly diminishes as energy is removed from the capacitor.

Although the amount of energy stored in a capacitor may be small, a capacitor can deliver a shock. The shock can be very severe (even fatal) if the capacitor is large and charged to a high voltage. Treat charged capacitors as you would any other electric energy source.

When a capacitor is charged, an electric field is established between the two charged plates. The lines in Fig. 10-4 represent an electric field in the same way that lines between magnetic poles represent a magnetic field. The electric field exerts its force on any charge within the field. A negatively charged particle, such as an electron, is forced toward the positive plate of the capacitor. The dielectric material between

Charged capacitor

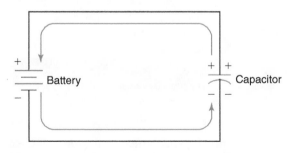

Fig. 10-2 Capacitor being charged. The current stops when the capacitor is fully charged.

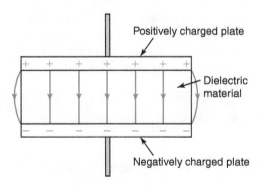

Fig. 10-4 Electric field of a charged capacitor. The field stresses the dielectric material.

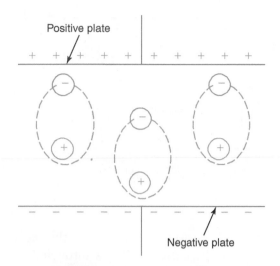

Positive plate

Negative plate

Fig. 10-5 Stress on the dielectric material. Energy is stored in the stressed dielectric material.

Direct current working voltage (DCWV)

the plates of a capacitor is under stress because of the force of the electric field. One way to visualize this stress is illustrated in Fig. 10-5. In Fig. 10-5 the electric field modifies the orbital path of an electron by forcing it toward the positive plate and away from the negative plate. The modified orbital path of the electron raises the energy level of the electron. This is the way the capacitor stores energy.

10-3 Voltage Rating

A dielectric material, such as that illustrated in Fig. 10-5, can withstand only a certain amount of stress before electrons in the material break free of the parent atoms. These free electrons are then available as current carriers. Then the capacitor's negative plate can discharge through the dielectric to the positive plate. This burns a hole in the dielectric material and ruins the capacitor.

The strength of the electric field in a capacitor is related to the distance between the plates and to the voltage across the plates. Therefore, a specific capacitor can withstand being charged only to a given voltage before its dielectric breaks down. The maximum voltage to which a capacitor should be charged is determined by the manufacturer. It is listed as part of the specifications of the capacitor. The manufacturer's voltage rating is usually listed as the *direct current working voltage* (DCWV). When a capacitor is rated at 100 DCWV, it can be operated at 100 V for long periods of time with no deterioration of the dielectric. If the capacitor is subjected to 125 or 150 V dc, the dielectric may not immediately break down. However, its life expectancy will be greatly shortened.

Self-Test

Answer the following questions.

1. A(n) _____ stores energy in the form of electric charges.
2. The symbol, or abbreviation, for capacitance is _____.
3. A capacitor has two plates and a(n) _____.
4. The _____ in a charged capacitor is stressed by the electric field.
5. True or false. A capacitor rapidly discharges once it is disconnected from the battery that charged it.

6. True or false. Electrons travel from one plate through the dielectric to the other plate when a capacitor is charged.
7. Where is the energy in a capacitor stored?
8. What abbreviation is used to specify the voltage rating of a capacitor?
9. Why do capacitors have voltage ratings?
10. What two factors determine the electric field strength in a capacitor?

10-4 Unit of Capacitance

Farad (F)

The base unit of capacitance is the *farad.* The abbreviation for farad is F. One farad is

that amount of capacitance that stores 1 C of charge when the capacitor is charged to 1 V. Thus, a farad is a coulomb per volt (C/V).

. . . from an earlier chapter that one letter is often used to denote two different quantities. In that chapter it was pointed out that the letter "W" is used as a symbol for energy and as the abbreviation for the base unit for power, the watt.

Now, notice that the letter "C" is used as the symbol for capacitance and as the abbreviation for the base unit of charge, the coulomb.

Farad is the unit of capacitance (C), coulomb is the unit of charge (Q), and volt is the unit of voltage (V). Therefore, capacitance can be mathematically expressed as

$$C = \frac{Q}{V}$$

EXAMPLE 10-1

What is the capacitance of a capacitor that requires 0.5 C to charge it to 25 V?

Given:	Charge (Q) = 0.5 C
	Voltage (V) = 25 V
Find:	Capacitance (C)
Known:	$C = \dfrac{Q}{V}$
Solution:	$C = \dfrac{0.5 \text{ C}}{25 \text{ V}} = 0.02 \text{ F}$
Answer:	The capacitance is 0.02 F.

This relationship is very useful in understanding voltage distribution in series capacitor circuits.

The 0.02 F just calculated is a very high value of capacitance. In most circuits the capacitances used are much lower. In fact, they are usually so much lower that the base unit of farad is too high to conveniently express their values. The microfarad (μF), which is 1/1,000,000 or 1×10^{-6} farad, and the picofarad (pF), which is 1/1,000,000,000,000 or 1×10^{-12} farad, are more convenient units. Both are used extensively in specifying values of capacitors in electronic circuits. Often it is necessary to convert from one unit to another. For example, a manufacturer may specify a 1000-pF capacitor in a parts list. The technician repairing the equipment may not have a capacitor marked 1000 pF. However, an equivalent

History of
Electronics

Michael Faraday
The unit of measure for capacitance, the farad, was named for Michael Faraday, an English chemist and physicist who discovered the principle of induction.

capacitor marked in microfarads may be available. The technician must convert picofarads into microfarads by multiplying picofarads by 10^{-6}. The net result is

$$1000 \text{ pF} = 1000 \times 10^{-6} \text{ } \mu\text{F} = 0.001 \text{ } \mu\text{F}$$

To convert from microfarads to picofarads, the reverse process is used. Divide the microfarads value by 10^{-6} to obtain the equivalent picofarad value. Since dividing by 10^{-6} is the same as multiplying by 10^6, the conversion proceeds as follows:

$$0.001 \text{ } \mu\text{F} = 0.001 \times 10^6 \text{ pF}$$
$$= 1000 \text{ pF}$$

which agrees with the results above.

$$C = \frac{Q}{V}$$

EXAMPLE 10-2

How much voltage is required to force 4000 μC of charge on a 20 μF capacitor?

Given:	$Q = 4000 \text{ } \mu\text{C}$
	$C = 20 \text{ } \mu\text{F}$
Find:	V
Known:	$C = Q \div V$ rearranged
	$V = Q \div C$
Solution:	$V = Q \div C = 4000 \times 10^{-6} \text{ C}$
	$\div 20 \times 10^{-6} \text{ F} = 200 \text{ V}$
Answer:	The voltage is 200 volts.

Occasionally, capacitance will be specified in nanofarads (nF).

You May Recall

. . . from an earlier discussion on multiple and submultiple units that *nano* means 10^{-9}. Thus, 1 nF = 1000 pF = 1/1000 μF. Therefore, a 1000-pF capacitor will occasionally be called a 1-nF capacitor.

EXAMPLE 10-3

Determine the charge on a 20 nF capacitor that is charged to 50 V.

Given: Capacitance (C) = 20 nF
Voltage (V) = 50 V

Find: Charge (Q)

Known: $C = \dfrac{Q}{V}$ and rearranged

$Q = CV$

Solution: $Q = 20 \times 10^{-9}$ F $\times$ 50 V
$= 1 \times 10^{-6}$ C

Answer: The charge is 1×10^{-6} C or 1 μC.

10-5 Determining Capacitance

The capacitance of a capacitor is determined by four factors:

1. Area of the plates
2. Distance between the plates
3. Type of dielectric material
4. Temperature

Temperature of the capacitor is the least significant of the four factors. It need not be considered for many general applications of capacitors. However, in more critical applications (such as oscillator circuits), the temperature characteristic of a capacitor is very important. The temperature characteristic of a capacitor is determined primarily by the type of dielectric material used in constructing the capacitor. Some dielectric materials cause an increase in capacitance as the temperature increases. This is referred to as a positive *temperature coefficient*. Other dielectric materials have negative temperature coefficients; that is, the capacitance increases as the temperature decreases. Still other dielectric materials have zero temperature coefficients. Their capacitance is independent of temperature. The temperature coefficient of a capacitor is specified by the manufacturer in *parts per million per degree Celsius*. Negative temperature coefficient is abbreviated N, positive temperature coefficient is P, and zero temperature coefficient is NPO. The reference temperature at which the capacitor is rated is 25°C. (Note that the letter "C" is used here as the abbreviation for Celsius. With only 26 letters in the alphabet and thousands of abbreviations and symbols, it is no wonder that some letters are used more than once.)

The capacitance of a capacitor is directly proportional to the area of its plates (or the area of its dielectric). All other factors remaining the same, doubling the plate area doubles the capacitance [Fig. 10-6(*a*)]. This is so because doubling the *plate area* also doubles the area of dielectric material. For a given voltage across the capacitor, the strength of the electric field is independent of the plate area. Thus, when the dielectric area is increased, the amount of energy stored in the dielectric is increased and the capacitance is also increased. (Remember, capacitance is defined as *the ability to store energy.*)

Other factors being equal, the amount of capacitance is inversely proportional to the distance between the plates [Fig. 10-6(*b*)]. As the *distance between the plates* is doubled, the amount of dielectric material also doubles. However, doubling the distance causes the strength of the electric field to decrease by a factor of 4. The net effect is that the amount of energy stored in the capacitor, for a given voltage applied to the capacitor, decreases. Thus, the capacitance decreases.

The capacitance of a capacitor is also controlled by the type of dielectric material used [Fig. 10-6(*c*)]. When subjected to the same electric field, some materials undergo greater molecular distortion than others. In general, those materials which undergo the greatest distortion store the most energy. The ability of a dielectric material to distort and store energy is indicated by its *dielectric constant K*. The dielectric constant of a material is a pure number (that is, it has no units). It compares the material's ability to distort and store energy when in an electric field with the ability of air to do the same. Since air is used as the reference, it has been given a K equal to 1. Mica, often

Plate area

Distance between the plates

Dielectric constant (K)

Temperature coefficient

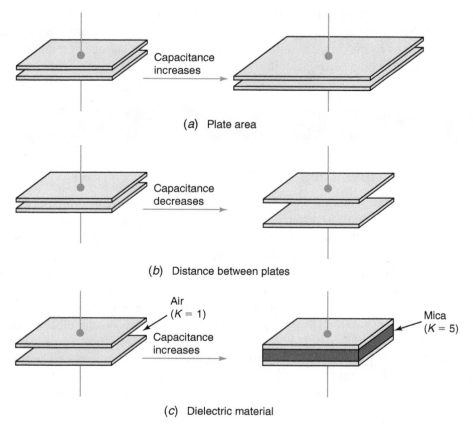

(a) Plate area

(b) Distance between plates

Air
(K = 1)

Capacitance
increases

Mica
(K = 5)

(c) Dielectric material

Fig. 10-6 Major factors affecting capacitance.

used as a dielectric, has a dielectric constant approximately five times that of air. Therefore, for mica, $K = 5$ (approximately). Suppose all other factors (plate area, distance between plates, and temperature) are the same. Then a capacitor with a mica dielectric will have five times as much capacitance as one using air as its dielectric. Dielectric constants for materials commonly used for dielectrics range from 1 for air to more than 4000 for some types of ceramics.

Self-Test

Answer the following questions.

11. The base unit for capacitance is the
    _____.

12. The abbreviation for the base unit of capacitance is _____.

13. A 4700-pF capacitor is equal to a(n) _____-μF capacitor.

14. The pF rating of a 0.003-μF capacitor would be _____.

15. A farad is equal to a(n) _____ divided by a(n) _____.

16. List four factors that determine the capacitance of a capacitor.

17. Which of the four factors in question 16 has the least effect on capacitance?

18. What happens to the capacitance of a capacitor when the plate area is doubled?

19. Suppose the distance between the plates of a capacitor is reduced by half. What happens to the capacitance?

20. What does the dielectric constant of a material indicate?

21. Does air have a smaller or larger dielectric constant than other common dielectric materials used in making capacitors?

22. How much charge is on a 6-μF capacitor that is charged 150 V?

23. How much capacitance is needed to store 0.5 C when the capacitor is charged to 30 V?

24. How much voltage is required to store 0.8 C of charge on a 2000-μF capacitor?

10-6   Types of Capacitors

Many different types and styles of capacitors (Fig. 10-7) are manufactured to satisfy the needs of the electronics industry. Capacitors may be named to indicate their dielectric material, their enclosure, the process used in their construction, or their intended use.

Look closely at the capacitance ratings on the capacitors in Fig. 10-7(c) and (d). The 1000UF in Fig. 10-7(c) stands for 1000 μF. It is quite common to use a "U" in place of a "μ" when the μ symbol is not readily available. The MFD in Fig. 10-7(f) also stands for μF, but MFD is rarely seen on newer capacitors.

Electrolytic Capacitors

Electrolytic capacitors [Fig. 10-7(c) and (e)] provide more capacitance for their size and weight than do any of the other types of capacitors. This is their primary advantage over other capacitors.

A common electrolytic capacitor consists of two aluminum plates separated by a layer of fine gauze or other absorbent material. The plates and separators are long, narrow strips. These strips are rolled up and inserted into an aluminum container (Fig. 10-8). One plate (the negative plate) is usually electrically connected to the aluminum can (container). Electric terminals from the plates are brought out through one end of the container. Then the container is sealed.

One of the aluminum plates in Fig. 10-8 is oxidized. Since aluminum oxide is an insulator, it is used for the dielectric. The separators are saturated with a chemical solution (such as borax) which is called the *electrolyte*. The electrolyte, having relatively good conductivity, serves as part of one of the plates of the capacitor (Fig. 10-9). It is in direct contact with both the dielectric and the pure aluminum plate. The electrolyte is also necessary in forming and maintaining the oxide on one of the plates.

The method used to produce electrolytic capacitors results in the plates being *polarized*. When using these capacitors, always keep the same voltage polarity as was used in manufacturing them. This polarity is always marked on the body of the capacitor. Again, the aluminum container (which may be encased in an insulating tube) is usually connected to the negative plate. It is never connected to the positive plate. Reverse polarity on an electrolytic capacitor causes excessively high current in the capacitor. It causes the capacitor to heat up and possibly to explode. Thus, the common electrolytic capacitor is limited to use in dc circuits.

The plates in an electrolytic capacitor are given a rough-textured surface to increase their effective area. The aluminum-oxide dielectric is very thin (approximately 0.25 μm thick), so the plates are very close together. These two factors, large plate area and thin dielectric, make the capacitance of an electrolytic capacitor very high for its weight and volume.

An oxide dielectric is far from a perfect insulator. Compared to other dielectric materials, it has a rather low dielectric resistance. Thus, a significant number of electrons flow through the dielectric from one plate to the other plate. This small current through the dielectric is called *leakage current*. An ideal capacitor would have zero leakage current.

Special electrolytic capacitors are manufactured for use in ac circuits. These capacitors are usually listed in parts catalogs as *nonpolarized* or *ac* electrolytic capacitors. An ac electrolytic capacitor is really two electrolytic capacitors packaged in a single container (Fig. 10-10). The two internal capacitors are in series, with their positive ends connected together. Regardless of the polarity on the leads of the ac electrolytic capacitor, one of the two internal capacitors will be correctly polarized. The correctly polarized capacitor will limit the current flowing through the reverse-polarized capacitor.

Polarized

Electrolytic capacitor

Leakage current

Electrolyte

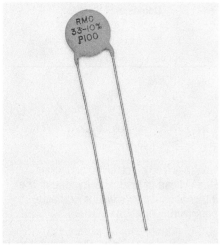

(a) Disk ceramic

(b) Tuning (variable)

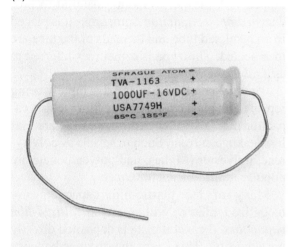

(c) Electrolytic (axial lead)

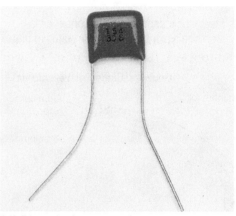

(d) Dipped mica

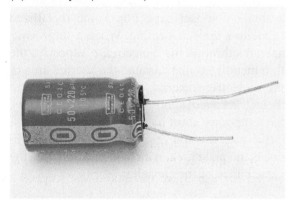

(e) Electrolytic (radial lead)

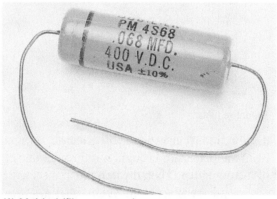

(f) Molded (film or paper)

Fig. 10-7 Types of capacitors.

Many nonpolarized electrolytic capacitors are designed for intermittent use only. These capacitors overheat (and often rupture) if used for extended periods of time.

Tantalum, which is a conductive metallic element, is also used for the plates of electrolytic capacitors. Tantalum electrolytic capacitors are smaller, more stable, and more reliable than aluminum electrolytic capacitors. Tantalum electrolytic capacitors have less leakage current (higher dielectric resistance) than aluminum electrolytic capacitors do. However, tantalum electrolytic capacitors are more expensive than their aluminum counterparts.

Tantalum

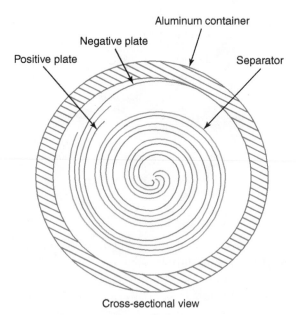

Fig. 10-8 Electrolytic capacitor. The container is electrically connected to the negative plate.

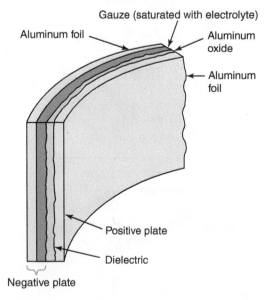

Fig. 10-9 Section of an electrolytic capacitor.

Film and Paper Capacitors

Paper capacitors and *film capacitors* are constructed using a rolled-foil technique, illustrated in Fig. 10-11. Notice in this figure that the two foil plates are offset so that one plate extends out each end. The leads for the capacitor are then connected to the ends of the completed roll. After the leads are attached, the complete assembly (plates, lead connections, and dielectric) is covered with a protective coating of insulation. Sometimes the insulation is molded around the assembly to produce a

Molded capacitor

Dipped capacitor

Tubular capacitor

Metallized-film capacitor

Paper capacitor

Film capacitor

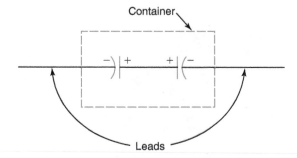

Fig. 10-10 Nonpolarized electrolytic capacitor. Regardless of lead polarity, one of the internal capacitors will be correctly polarized.

molded capacitor. Sometimes the capacitor is dipped in a plastic insulating material to produce a *dipped capacitor.* Sometimes it is placed in an insulated tube and the ends of the tube are then sealed. This type is sometimes referred to as a *tubular capacitor.*

Paper as a dielectric material is rapidly being replaced by plastic film. Plastic film yields a capacitor that is smaller and more stable and has less leakage current. Such materials as polystyrene, polyester (Mylar), and polycarbonate are popular replacements for paper.

Many of the plastic-film capacitors use metallized plates. With these *metallized-film capacitors,* the metal plate is deposited directly onto the film. This keeps the distance between the plates as small as possible and produces a small, compact capacitor. Some metallized capacitors are self-healing. When a surge voltage arcs through the dielectric, it vaporizes the thin metallized plate around the charred area of the dielectric material. Therefore, the charred dielectric cannot short the plates together. Thus, the temporary short has been "healed."

Of course vaporizing part of the plate reduces the plate area, which, in turn, reduces the capacitance of the capacitor.

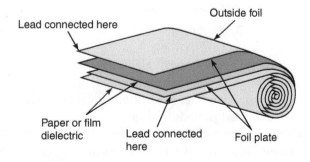

Fig. 10-11 Paper or film capacitor construction.

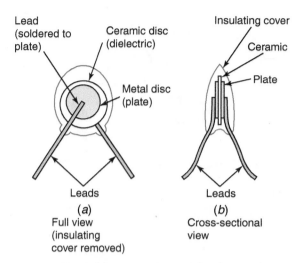

Leads

(a)
Full view
(insulating
cover removed)

(b)
Cross-sectional
view

Lead
(soldered to
plate)

Ceramic disc
(dielectric)

Metal disc
(plate)

Insulating cover

Ceramic

Plate

Leads

Fig. 10-12 Ceramic capacitor construction.

Film and paper capacitors used in industrial applications can be as large as several hundred microfarads. They are often rated in terms of apparent power (VA) as well as DCWV. These high-value capacitors are usually enclosed in a metal container filled with a special insulation oil. However, in most electronic circuits, paper and film capacitors have a value of less than 1 μF.

Ceramic Capacitors

For low-value capacitors (less than 0.1 μF), ceramic is a popular dielectric material. The most common style of *ceramic capacitor* is the disc ceramic [Fig. 10-7(a)].

The structure of a typical ceramic capacitor is detailed in Fig. 10-12. As indicated in this figure, these capacitors have a simple, strong structure. This makes them a tough, reliable, general-purpose capacitor. They are used in a wide variety of applications where low values of capacitance are needed.

Mica Capacitors

Like ceramic capacitors, *mica capacitors* are made with two plates (with attached leads) that are separated by dielectric material. However, the dielectric is a thin piece of mica instead of a ceramic material. The assembled plates, leads, and dielectric are then dipped in an insulating material to form a dipped mica capacitor like the one shown in Fig. 10-7(d). Mica capacitors are limited to even lower values than ceramic capacitors. This is because mica has a lower dielectric constant than ceramic. However, it is easier to control production tolerances with mica dielectrics, and mica has good high-temperature characteristics.

It takes less space to provide a given value of capacitance with ceramic than with mica. Also, it is easier to construct a ceramic capacitor. Therefore, mica capacitors are not as common as ceramic capacitors.

Specific-Use Capacitors

So far we have classified capacitors by the type of dielectric materials used in their manufacture. Capacitors can also be classified by their function or by the way they are connected in a circuit. Some of the more common names in this classification are feed-through, filter, padder, trimmer, tuning, and stand-off.

Padders, trimmers, and tuning capacitors are all variable capacitors. They are used in circuits that tune radio and television sets to a particular station. An example of a miniature *trimmer* capacitor is pictured in Fig. 10-13.

One of the plates of a *feed-through capacitor* is the tubular metal case which encloses the capacitor (Fig. 10-14). The two leads extending from the ends of the capacitor are both connected to the other plate. In use, the capacitor protrudes through a hole in the electrical (metal) shield that separates two parts of an electronic system. The case (one plate) is electrically connected to the shield through which it protrudes. The case may be soldered as in Fig. 10-14, or it may be threaded and held in place with a nut. This arrangement allows direct current to feed through the grounded shield, while radio frequencies are routed to ground by the capacitor. The radio frequencies are bypassed from either lead.

Stand-off capacitors (Fig. 10-15) are also used to bypass high frequencies to the ground of

Trimmers

Feed-through capacitor

Ceramic capacitor

Mica capacitor

Stand-off capacitor

Fig. 10-13 Trimmer capacitor.

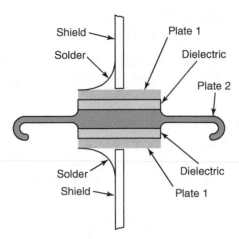

Fig. 10-14 Feed-through capacitor. The two leads connect to the same plate.

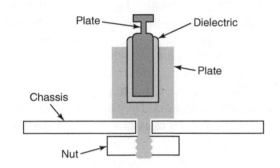

Fig. 10-15 Stand-off capacitor.

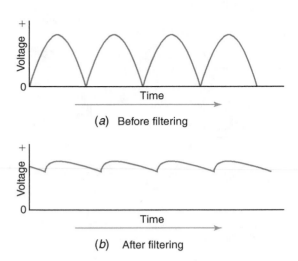

Fig. 10-16 Filtering pulsating direct current.

Energy-storage capacitor

an electronic system. These capacitors have one plate connected to a threaded terminal that bolts the capacitor to the grounded chassis. The other plate is connected to a terminal that serves as a tie point for other electronic components. Radio frequency going through the components connected to the standoff capacitors is bypassed to the chassis and effectively eliminated.

Filter capacitor

Filter capacitors are high-value capacitors used to filter (flatten) out pulsations in dc voltage. For example, a filter capacitor can change the pulsating direct current in Fig. 10-16(*a*) to that shown in Fig. 10-16(*b*). Many filter capacitors are electrolytic capacitors that are polarized. They must be used in circuits in which the polarity of the voltage applied to them will never be reversed.

$W = 0.5CV^2$

Most filter capacitors have a value somewhere between 1 and 5000 μF. The voltage rating may be anywhere from a few volts to thousands of volts.

Energy-Storage Capacitors

Capacitors designed to release their stored energy (discharge) during a short time interval

are sometimes called *energy-storage capacitors.* They may be either electrolytic or nonelectrolytic capacitors. During the rapid discharge of an energy-storage capacitor, the terminal leads and plates must carry a very high current. The dielectric material, plates, and enclosure must be able to withstand the extreme stress created by such a discharge. Therefore, energy-storage capacitors often have a peak current rating. This rating allows one to calculate the minimum resistance through which the fully charged capacitor can be discharged. Exceeding the peak current rating one time will not immediately destroy the capacitor. However, repeated discharging at excess current levels will cause premature failure in a capacitor.

In addition to the standard voltage and capacitance ratings, most energy-storage capacitors also have an energy rating. The energy rating specifies how many joules of energy the capacitor stores when charged to its rated voltage. The amount of energy a capacitor can store depends upon its capacitance and voltage rating. We can calculate the amount of stored energy with the formula

$$W = 0.5CV^2$$

The calculated energy is in joules when capacitance is in farads and voltage is in volts.

EXAMPLE 10-4

An energy-storage capacitor is rated at 300 μF and 450 V. How much energy can it store?

Given: $\qquad C = 300\ \mu$F
$\qquad\qquad\quad V = 450$ V

Find:	W
Known:	$W = 0.5CV^2$
Solution:	$W = 0.5 \times (300 \times 10^{-6} \text{ F})$
	$\times\ 450 \text{ V} \times 450 \text{ V}$
	$= 30.4 \text{ J}$
Answer:	The capacitor stores 30.4 J of energy.

The amount of energy in the above example is not very great. A 40-W lamp uses more energy every second than this capacitor stores. However, the capacitor can provide high power if it is discharged rapidly. For instance, if the capacitor in example 10-4 is discharged in 2 ms, it will produce the following power:

$$P = \frac{W}{t} = \frac{30.4 \text{ J}}{0.002 \text{ s}} = 15{,}200 \text{ W}$$

Large, high-voltage capacitors like the one shown in Fig. 10-17 can be charged to many thousands of volts. They can store several thousand joules of energy. These capacitors are used in such industrial applications as capacitor discharge welding. They can also be used for less demanding applications, such as filtering.

Fig. 10-17 Energy-storage capacitor.

SMD Capacitors

SMD capacitor

Chip capacitor

Surface-mount capacitors, commonly called *chip capacitors,* are about the same size and shape as the chip resistors shown in the chapter on circuit components. Depending on their specific characteristics, typical chip capacitors range from 0.050 to 0.180 in. in length, 0.040 to 0.169 in. in width, and 0.035 to 0.110 in. in thickness.

Ceramic chip capacitors are commonly available with capacitance values ranging from 1 pF to 0.27 μF and with voltage ratings of 25, 50, and 100 V. However, they are available with voltage ratings in the thousands of volts.

Solid-tantalum chip capacitors are also available. These surface-mount devices are, of course, polarized. They typically have voltage ratings in the 6- to 25-DCWV range. Capacitances up to 100 μF are common.

Surface-mount polymer aluminum electrolytic capacitors are also available in the 4.7- to 270-μF range, with voltage ratings as high as 16 V for the 4.7-μF size. These devices are about 7.5 mm long, 4.5 mm wide, and 1.5 to 4.5 mm high.

Supercapacitors

Technical advances in materials and manufacturing techniques have resulted in the development of ultra-high-capacitance capacitors. Capacitors with nominal values as large as 50 F are now available. However, these large-value capacitors have a very limited voltage rating. Capacitors up to 1 F may have a working voltage of 5 V, while those up to 50 F may be limited to 2.5 V.

Capacitor Codes

Many different alphanumeric codes and color codes are used to identify the capacitance and other ratings of capacitors—especially physically small capacitors like those in Fig. 10-7(*a*) and (*d*). For physically larger capacitors, like those in Fig. 10-7(*c*), (*e*), and (*f*), the value and ratings are printed on their cases. The most commonly used codes and color codes are listed and explained in Appendix H.

Answer the following questions.

25. True or false. For a given value of capacitance, a metallized-film capacitor is smaller than a foil-and-film capacitor.
26. True or false. The electrolyte is the dielectric material in an electrolytic capacitor.
27. True or false. The negative plate of an electrolytic capacitor is often connected to the aluminum container.
28. True or false. The plates in an electrolytic capacitor are oxidized to increase their effective plate area.
29. True or false. Single electrolytic capacitors are polarized.
30. True or false. Mica capacitors are more common than ceramic capacitors are.

31. _____ capacitors are sometimes self-healing.
32. A trimmer capacitor is classified as a(n) _____ capacitor.

33. A(n) _____ capacitor has both leads connected to the same plate.
34. Electrolytic capacitors use either _____ or _____ for plate material.
35. The dielectric material in an electrolytic capacitor is a metal _____.
36. A(n) _____ chip capacitor is polarized.
37. What happens when an electrolytic capacitor has reverse-polarity voltage connected to it?
38. List four types of capacitors named after the dielectric material used in them.
39. What are the advantages of film capacitors over paper capacitors?
40. What name is given to a capacitor that is used to smooth out pulsating dc voltage?
41. Where are energy-storage capacitors used?
42. How much energy is stored in a 150-μF capacitor that is charged to 300 V?

10-7 Schematic Symbols

Schematic symbols for capacitors

Capacitors can be broadly classified as either fixed or variable and as either polarized or nonpolarized. All variable capacitors are nonpolarized. *Schematic symbols for capacitors* are shown in Fig. 10-18. The curved line representing one of the plates in the symbols of Fig. 10-18 has no particular significance for some types of capacitors. For other types it indicates a specific lead of the capacitor.

For a polarized capacitor, the curved plate of the symbol indicates the negative terminal of the capacitor. With variable capacitors, the curved line identifies the plate that is electrically connected to the frame or adjusting mechanism. The curved line identifies the outside foil (Fig. 10-11) on a rolled-foil (film or paper) capacitor. On the capacitor itself, the lead connecting to the outside foil has to be identified. This is done with a color band around the end of the capacitor from which this lead protrudes.

Surge of current

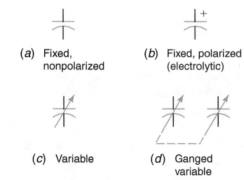

(*a*) Fixed, nonpolarized

(*b*) Fixed, polarized (electrolytic)

(*c*) Variable

(*d*) Ganged variable

Fig. 10-18 Schematic symbols for capacitors.

10-8 Capacitors in DC Circuits

When the switch of a dc circuit containing only capacitance is closed, a *surge of current* charges the capacitor. Once this surge is over, no more current flows (Fig. 10-19). The capacitor, once charged, appears to the battery just like an open

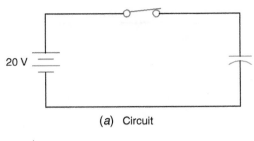

(a) Circuit

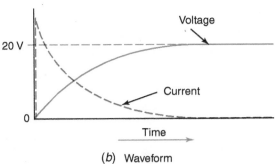

(b) Waveform

Fig. 10-19 Capacitor charging in dc circuit. The current reaches its maximum value almost the instant the switch is closed. The voltage developed across the capacitor increases more slowly.

switch. For practical purposes it has infinite opposition.

The magnitude of the surge of current in Fig. 10-19 is controlled by the resistance in the circuit. In this circuit the resistance is composed of the lead resistance and the internal resistances of the battery and the capacitor. These resistances are usually very low; therefore, the surge current can be very high. However, this surge current lasts only for an extremely short period of time.

Capacitor Time Constants

The time required for a capacitor to charge or discharge is determined by the amount of resistance and capacitance in the circuit. Substituting appropriate and equivalent units for resistance and capacitance, as has been done below, shows that multiplying resistance by capacitance yields time:

$$RC = \text{ohms} \times \text{farads}$$
$$= \frac{\text{volts}}{\text{amperes}} \times \frac{\text{coulombs}}{\text{volts}}$$
$$= \frac{\text{coulombs}}{\text{amperes}} = \frac{\text{coulombs}}{\text{coulombs/second}}$$
$$= \text{seconds}$$

The time obtained by multiplying resistance by capacitance is called a *time constant* and is abbreviated *T*. Therefore, we can write

Time constant (T)

Time constant (T) = resistance (R) $\times$ capacitance (C)

$$T = RC$$

$$\boldsymbol{T = RC}$$

If R and C are in base units of ohms and farads, T is in base units of seconds.

When a capacitor is charging, the time constant represents the time required for the capacitor to charge to 63.2 percent of the available voltage. When a charged capacitor is discharging, the time constant represents the time required for the capacitor to lose 63.2 percent of its available voltage. The available voltage when the capacitor is charging is the difference between the capacitor's voltage and the source voltage. When the capacitor is discharging, the available voltage is the voltage remaining across the capacitor. The curves in Fig. 10-20 demonstrate the meaning of a time constant. Figure 10-20(a) shows that a capacitor charges from 0 to 63.2 percent of the source voltage in one time constant. During the next time constant, it charges another 63.2 percent of the remaining available voltage. Thus, at the end of two time constants the capacitor is 86.5 percent charged [63.2 + (0.632 × 36.8) = 86.5]. The percentage of charge at the end of additional time constants is tabulated in Fig. 10-20(a).

The amount of time required to charge to 86.5 percent of the source voltage is strictly a function of resistance and capacitance. We can determine the exact time by calculating the value of a time constant. For example, the time constant for the resistance-capacitance circuit in Fig. 10-21 is

$$T = RC = 2 \times 10^6 \ \Omega \times 4 \times 10^{-6} \ \text{F}$$
$$= 8 \ \text{s}$$

The capacitor requires two time constants to charge to 86.5 percent of the source voltage. Therefore, it takes 16 s for the capacitor in Fig. 10-21 to charge to 86.5 percent of the battery voltage.

The exact amount of voltage on the capacitor after two time constants is dependent on the value of the source voltage. If the source voltage in Fig. 10-21 is 10 V, the voltage on the capacitor will be 8.65 V. If the source is 100 V, it will be 86.5 V.

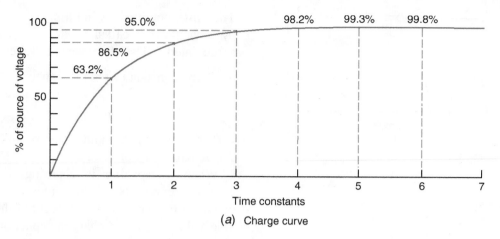

(a) Charge curve

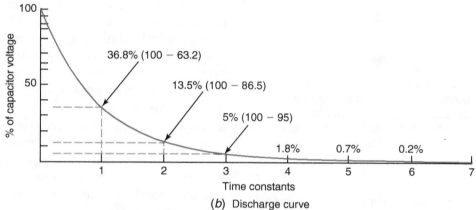

(b) Discharge curve

Charge-discharge curve

Fig. 10-20 Universal *charge-discharge curves*. After five or six time constants, the capacitor is considered to be fully charged or discharged.

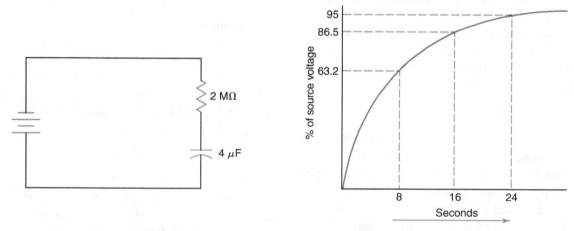

Fig. 10-21 Charging time for a capacitor. After 24 s (three time constants) the capacitor is 95 percent charged.

The discharge of a capacitor is illustrated in Fig. 10-22. Notice the values of the resistance and the capacitance. They are the same as used in Fig. 10-21; therefore, the time constant is still 8 s. When the switch is closed in Fig. 10-22, the capacitor will be charged to 200 V. When the switch is opened, the capacitor will discharge through the resistor. After 8 s (one time constant) the capacitor will have discharged 63.2 percent of its voltage. The voltage

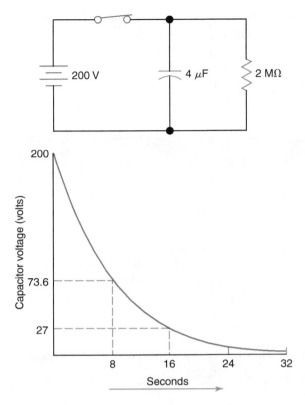

Fig. 10-22 Discharging time for a capacitor.

Given: $R = 270 \text{ k}\Omega = 0.27 \text{ M}\Omega$
$C = 3.2 \ \mu\text{F}$

Find: Time (t) needed to charge to 7.785 V

Known: $T = RC$

Number of time constants for a given % of charge (Fig. 10-20)

Desired voltage on the capacitor

Solution: % of $V_{available} = (V_{capacitor} \div V_{available}) \times 100 = (7.785 \text{ V} \div 9 \text{ V}) \times 100 = 86.5\%$

Fig. 10-20 shows that two time constants charge the capacitor to 86.5% of $V_{available}$

$T = RC = 0.27 \text{ M}\Omega \times 3.2 \ \mu\text{F}$
$\quad = 0.864 \text{ s}$

Required time (t) $= 2T = 2 \times 0.864 = 1.728 \text{ s}$

Answer: The time is 1.728 seconds.

---

across the capacitance will therefore decrease by 126.4 V. It will still have 73.6 V left. During the next time constant, it will discharge 63.2 percent of the remaining 73.6 V. Thus, it will discharge 46.5 V ($0.632 \times 73.6 = 46.5$). As shown in Fig. 10-22, there will be about 27 V left on the capacitor at the end of two time constants (16 s).

Inspect the curves in Fig. 10-20 again. Notice that a capacitor is essentially charged or discharged by the end of six time constants. For analyzing and designing circuits, many technicians consider a capacitor to be charged (or discharged) after only five time constants.

Resistance-capacitance circuits and their time constants are used extensively in electronic circuits. For example, they are used to control the frequency of many nonsinusoidal waveforms.

EXAMPLE 10-5

A 270-kΩ resistor and a 3.2-μF capacitor are connected in series. How long will it take the capacitor to charge to 7.785 V after this series combination is connected to a 9-V battery?

Voltage Distribution

A dc source voltage divides among series capacitors in inverse proportion to their capacitance (Fig. 10-23). The voltmeter across the 1-μF capacitor indicates twice as much voltage as the one across the 2-μF capacitor. This happens because capacitors in series must receive the same quantity of charge. (There is only one path for the charging current.) The same amount of charge produces less voltage in a large capacitor than it does in a small capacitor. Remember that $C = Q/V$; therefore,

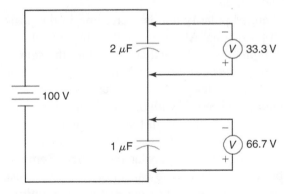

Fig. 10-23 The dc voltage distribution within series capacitors. The smallest capacitor develops the most voltage.

$V = Q/C$. If Q is the same for both capacitors, then the higher the capacitance, the lower the voltage.

Sometimes capacitors are connected in series because the voltage rating of either capacitor is lower than the source voltage. In this case, the capacitance of the two capacitors should be as close to equal as possible. Then the source voltage will divide equally between the two capacitors.

Answer the following questions.

43. The curved line of the capacitor symbol represents the _____ plate of an electrolytic capacitor.
44. Technicians figure it takes _____ time constants to charge a capacitor.
45. The outside foil of a film capacitor is identified by a(n) _____.
46. Define *time constant*.
47. What is the time constant of a 50-kΩ resistor and a 2.2-μF capacitor?
48. A dc source is connected to an 8-μF and a 5-μF capacitor connected in series. Which capacitor develops the larger voltage? Why?

10-9 Capacitors in AC Circuits

A capacitor controls the current in an ac circuit. It controls the current by storing energy, which produces a voltage in the capacitor. The voltage produced by the capacitor's stored energy is always in opposition to the source voltage.

Careful study of Fig. 10-24 shows exactly how a capacitor controls circuit current. The voltage waveform represents the source (generator) voltage applied across the capacitor. The current waveform represents the current flowing through the ammeter. Assume that the source voltage in Fig. 10-24(*a*) has just changed polarity and is increasing in the positive direction. Also assume that the capacitor is completely discharged at the instant the source voltage changes polarity. Now, as the generator (source) voltage builds up, the capacitor *must charge* to keep up with the increasing generator voltage. The current in the circuit is maximum at this time because the generator's voltage is changing most rapidly at this time. (Remember that the faster the capacitor's voltage must rise, the more charge per second it requires. And more charge per second means more current.) As the first quarter-cycle progresses in

Current leads the voltage by 90°

Fig. 10-24(*a*), the generator's voltage increases less rapidly. Consequently, the current in the circuit decreases. At the end of the first quarter-cycle there is an instant when the generator's voltage is stable (neither increasing nor decreasing). At that instant the circuit current is zero. The capacitor's voltage is equal to and opposite the source voltage.

Let us note four things from observing the action during this first quarter-cycle:

1. The generator is charging the capacitor. Thus electrons (current) leave the negative terminal of the generator and enter the positive terminal of the generator.
2. The generator is transferring energy to the capacitor as it charges the capacitor.
3. The capacitor allows just enough current to charge it to the value of the source voltage.
4. The current and voltage are 90° out of phase with each other. Specifically, the *current leads the voltage by 90°*.

Now let's look at the second quarter-cycle [Fig. 10-24(*b*)]. During this part of the cycle, the generator's voltage is decreasing. Since

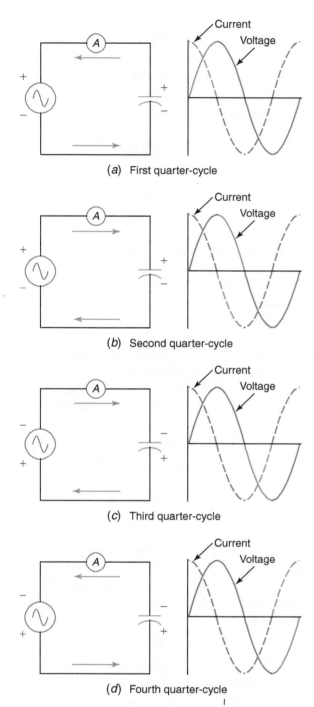

(a) First quarter-cycle

(b) Second quarter-cycle

(c) Third quarter-cycle

(d) Fourth quarter-cycle

Fig. 10-24 Energy transfer in a capacitor circuit. Every other quarter-cycle the capacitor becomes the energy source.

now returning its stored energy to the generator. Figure 10-24(b) shows that the decrease in the generator's voltage causes an increase in the current in the circuit in the opposite direction. The current is maximum when the generator voltage is zero. The current in this case, however, is a result of the capacitor's discharging. Again, let us summarize the major action during the second quarter-cycle:

1. The capacitor is discharging through the generator.
2. The capacitor is *transferring energy* back to the source (generator) as it discharges.
3. The capacitor furnishes just enough current to discharge it to the value of the source voltage.
4. The current still leads the voltage by 90°.

Transferring energy

Inspection of the third quarter-cycle [Fig. 10-24(c)] shows that it is essentially the same as the first quarter-cycle. The only difference is that the voltage and current have reversed. The generator again takes control and furnishes the energy needed to charge the capacitor. The capacitor stores energy during this third quarter of the cycle.

The fourth quarter-cycle [Fig. 10-24(d)] is like the second quarter-cycle. The capacitor now discharges through the generator and returns its stored energy to the source. As in the previous three quarters of the cycle, the current and voltage remain 90° out of phase.

Capacitive Reactance

A capacitor's opposition to alternating current is known as *reactance*. The symbol for reactance is X. Since inductors also provide reactance, we use a subscript with the symbol X. Thus X_C is used to signify capacitive reactance. That is, X_C means the reactance of a capacitor. Reactance is like resistance in that it is an opposition that can control the current in a circuit. Therefore, the base unit of reactance is also

Reactance (X)

the capacitor is connected directly across the generator, its voltage must also decrease. For the capacitor to lose voltage, it must discharge. This means that electrons must leave its negative terminal and enter its positive terminal. To discharge, the capacitor has to force electrons through the generator against (in opposition to) the generator's voltage. Thus, the capacitor is

the ohm (Ω). Reactance is unlike resistance in that it does not convert electric energy into heat energy. It is incorrect to interchange the terms *resistance* and *reactance* even though both are expressed in ohms.

Capacitive reactance is controlled by two factors: the frequency of the current (and voltage) and the amount of capacitance. Further, the reactance is inversely proportional to both these factors. If either frequency or capacitance is doubled, the reactance is halved. If either frequency or capacitance is halved, the reactance is doubled. Why reactance is inversely proportional to both frequency and capacitance is explained in the following discussion.

Figure 10-25 shows a capacitor circuit in which frequency, capacitance, and voltage are indicated. We know that the current in the circuit is controlled by the capacitive reactance. If the reactance increases, the current decreases and vice versa. Therefore, if we determine what happens to the current in the circuit, we will know what happens to the reactance. First let us determine the circuit current for the frequency, voltage, and capacitance given in Fig. 10-25. Then we can change either frequency or capacitance and see what happens to the current. To determine the current, we first determine the amount of charge needed to charge the capacitor to the source voltage. This can be done with the basic formula for capacitance:

$$C = \frac{Q}{V}$$

Rearranging yields

$$Q = CV$$

In Fig. 10-25 then, the charge is

$$Q = 1\,\text{F} \times 1\,\text{V} = 1\,\text{C}$$

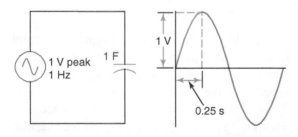

Fig. 10-25 Determining charge and average current. In 0.25 s the capacitor must receive 1 C of charge.

Next we determine the amount of time available to transfer the charge from the source to the capacitor. The time available is the time required for one-quarter of a cycle. The period of 1 Hz is 1 s. Thus the capacitor has 0.25 s in which to charge. Finally, we know that

$$\text{Current}\,(I) = \frac{\text{charge}\,(Q)}{\text{time}\,(t)}$$

$$= \frac{1\,\text{C}}{0.125\,\text{s}}$$

$$= 4\,\text{C/s} = 4\,\text{A}$$

A coulomb per second is an ampere, and so the average current for the first quarter-cycle in Fig. 10-25 is 4 A. The remaining quarters of the cycle handle the same charge in the same amount of time. Therefore, the average current in the circuit of Fig. 10-25 is 4 A.

Now suppose we double the frequency to 2 Hz. The new period is 0.5 s, and the time for one quarter-cycle is 0.125 s. Since the voltage and the capacitance are the same, the charge is the same. The current now increases to

$$I = \frac{Q}{t}$$

$$= \frac{1\,\text{C}}{0.125\,\text{s}}$$

$$= 8\,\text{C/s} = 8\,\text{A}$$

Since doubling the frequency doubles the current, the reactance of the circuit must be halved.

Now suppose we leave the frequency at 1 Hz and double the capacitance to 2 F. The new charge is

$$Q = CV = 2\,\text{F} \times 1\,\text{V} = 2\,\text{C}$$

The time for one quarter-cycle at 1 Hz is still 0.25 s. Therefore, the new current is

$$I = \frac{Q}{t}$$

$$= \frac{2\,\text{C}}{0.25\,\text{s}}$$

$$= 8\,\text{C/s} = 8\,\text{A}$$

Again, the current doubles and so the reactance must be half as great as it was with 1 F.

The exact value of capacitive reactance can be calculated with the formula

$$X_c = \frac{1}{2\pi f C}$$

where the Greek letter "pi" (π) is a unitless constant equal to approximately 3.14. Using this value for pi, the formula becomes

$$X_C = \frac{1}{6.28fC}$$

Capacitive reactance is in ohms when frequency is in hertz and capacitance is in farads.

EXAMPLE 10-6

What is the reactance of a $0.01\text{-}\mu$F capacitor at 400 Hz?

Given:	$C = 0.01 \ \mu$F
	$f = 400$ Hz
Find:	X_C
Known:	$X_C = \dfrac{1}{6.28fC}$
Solution:	$X_C =$
	$\dfrac{1}{6.28 \times 400 \times 0.01 \times 10^{-6}}$
	$= 39{,}800 \ \Omega$
Answer:	The capacitive reactance therefore equals 39,800 Ω, or about 40 kΩ.

From the above example we can quickly estimate the reactance of a $0.005\text{-}\mu$F capacitor at 400 Hz. Since $0.005 \ \mu$F is half as great as 0.01, the reactance will be twice as great. Thus, the $0.005\text{-}\mu$F capacitor at 400 Hz has approximately 80 kΩ of capacitive reactance.

Reactance cannot be measured with an ohmmeter. It must be calculated using either the reactance formula or Ohm's law. Since reactance is a form of opposition, Ohm's law can be used by replacing R with X_C. Thus, we can use any of the following formulas:

$$X_C = \frac{V_C}{I_C} \qquad I_C = \frac{V_C}{X_C} \qquad V_C = I_C X_C$$

EXAMPLE 10-7

How much current flows in the circuit shown in Fig. 10-26?

Given:	$f = 2$ MHz
	$C = 500$ pF
	$V = 20$ V

Find:	I
Known:	$I = \dfrac{V}{X_C}$ and $X_C = \dfrac{1}{6.28fC}$
Solution:	$X_C =$
	$\dfrac{1}{6.28 \times 2 \times 10^6 \times 500 \times 10^{-12}}$
	$= 159 \ \Omega$
	$I = \dfrac{20 \text{ V}}{159 \ \Omega} = 0.126$ A
	$= 126$ mA
Answer:	The current in the circuit is 126 mA.

$X_C = \dfrac{1}{6.28fC}$

Quality and Power of Capacitors

Capacitors (and inductors) are built to provide reactance so that current can be controlled without converting electric energy into heat energy. However, it is impossible to build capacitors and inductors (especially inductors) that do not also have some resistance. *Quality (Q)* is the term used to rate a capacitor (or inductor) on its ability to produce reactance with as little resistance as possible.

Quality *(Q)*

Quality (*Q*) is a ratio of reactance to resistance. Since it can be used to describe either capacitors or inductors, the formula for quality should include a subscript for the reactance. For capacitors, the formula is

$$Q = \frac{X_C}{R}$$

Since both X and R have units of ohms, Q has no units (it is a pure number). Because X (and also R to a lesser extent) changes with frequency, Q is frequency-dependent. That is, its value is always given for some specified frequency.

Pure capacitance can use no energy. It merely stores energy for one quarter-cycle and returns it the next quarter-cycle. Since capacitance

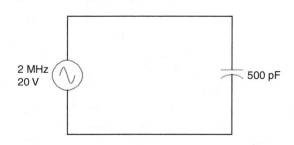

Fig. 10-26 Circuit for example 10-7

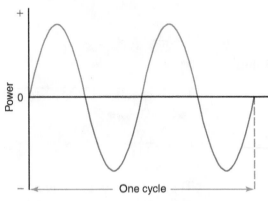

Fig. 10-27 Power waveform for capacitance. Capacitance uses no power.

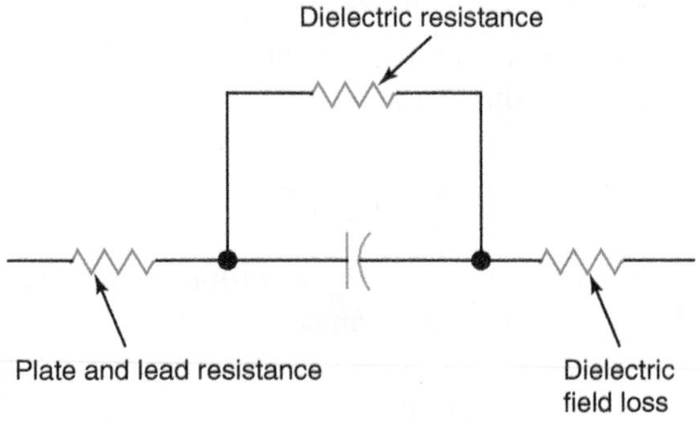

Dielectric resistance

Plate and lead resistance

Dielectric field loss

Fig. 10-28 Equivalent resistances in a capacitor. These resistances convert electric energy into heat energy.

uses no energy, it cannot consume power either (power = energy/time). This idea is illustrated in Fig. 10-27. This figure shows power (or energy) taken from the source as positive power, and power (or energy) returned to the source as negative power. The negative power exactly cancels the positive power.

You May Recall

. . . that you have seen the waveform of Fig. 10-27 before. It is the power waveform obtained in Chap. 9 for a current and voltage that are 90° out of phase.

Therefore, we can conclude two things about a circuit that contains only capacitance:

1. Current and voltage are 90° out of phase.
2. The circuit uses no net energy or power.

Notice that the term used is *capacitance,* not *capacitor. Capacitance* is used because capacitors do lose some energy.

Energy losses in capacitors can be classified as *resistance losses* and *dielectric losses.* The resistance loss is caused primarily by current flowing through the resistance of the capacitor leads and plates. There is also a very minute current flowing through the high resistance of the dielectric material. The dielectric resistance appears as a parallel resistance, whereas the plate and lead resistances appear as a series resistance (Fig. 10-28). Dielectric loss results from continually reversing the polarity of the electric field through the dielectric. Each time

the molecular structure of the dielectric has to adjust to a new polarity, a little electric energy is converted into heat energy. This dielectric loss appears as a small resistance in series with the capacitor. The effects of all these resistances are often combined as the *equivalent series resistance* (ESR). A capacitor with a small ESR has a small energy loss.

The relative amount of energy lost in a capacitor can be indicated by any of three terms. The terms are *dissipation factor* (DF), *power factor* (PF), and *quality* (Q). All these terms give about the same information. *Power factor* is the ratio of power (P) to apparent power P_{app}. In a capacitor, power represents the energy per second lost, and apparent power represents the energy per second stored. If a capacitor has very little energy loss, its power factor is low (less than 0.01, or 1 percent). *Quality* is the ratio of reactance to resistance. If a capacitor has very little energy loss, R is low and Q is high (more than 100). *Dissipation factor* is the reciprocal of quality. In other words, dissipation factor is the ratio of resistance to reactance. Obviously, if Q is typically a large number, dissipation factor will be a small number or percentage. For all but the lowest-quality capacitors, power factor and dissipation factor are essentially equal to each other. In summary, remember that capacitors with small energy losses have a high Q and low dissipation and power factors.

With lower-quality electrolytic capacitors, energy losses can be considerable (1 to 10 percent of the stored energy). However, with high-quality nonelectrolytic capacitors the losses are so small that they can be ignored in most applications. In this text, we will assume that capacitors have negligible energy loss. In terms of phase shift, this means that we will treat nonelectrolytic capacitors as ideal capacitors. We will assume that the current and voltage are a full 90° out of phase.

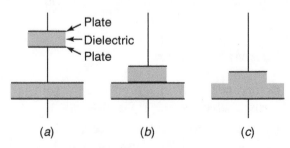

Self-Test

Answer the following questions.

49. How does a capacitor control current in an ac circuit?
50. Why doesn't capacitance consume power?
51. What is the phase relationship between current and voltage in an ideal capacitor?
52. Define quality.
53. List three terms used to indicate relative energy loss in a capacitor.
54. The opposition of a capacitor to ac current is called _____.
55. The symbol for a capacitor's opposition to ac current is _____.

56. ESR is the abbreviation for _____.
57. True or false. Capacitive reactance is inversely proportional to frequency.
58. True or false. Capacitive reactance is directly proportional to capacitance.
59. Calculate the reactance of a 2-μF capacitor at 150 Hz.
60. At what frequency will a 0.02-μF capacitor have a reactance of 400 Ω?
61. A 50-V, 1-kHz source is applied to a 0.1-μF capacitor. Determine the average current for the circuit.

10-10 Capacitors in Series

When capacitors are connected in series, the total capacitance decreases. In fact, the total capacitance of series capacitors is always less than the capacitance of the smallest capacitor. The logic of this statement can be seen by referring to Fig. 10-29, which shows two capacitors connected in series. Assume that these capacitors are made from flat metal plates separated by an insulator. Suppose the lead connecting the two capacitors is shortened until the plates of the capacitors touch each other. The result would be as indicated in Fig. 10-29(*b*), where the two plates combine to make one plate. The combined plates, isolated between two insulators, serve no useful purpose. If they are removed, as in Fig. 10-29(*c*), the result is a single equivalent capacitor. Look carefully at the plates of the equivalent capacitor. They are spaced farther apart than the plates of either of the original capacitors were. Also, the effective plate area is limited by the smaller plate. Therefore, the equivalent capacitor has less capacitance than either of the two series capacitors.

The total (or equivalent) capacitance of series capacitors can be calculated with any of these formulas:

General formula for series capacitors:

$$C_T = \cfrac{1}{\dfrac{1}{C_1} + \dfrac{1}{C_2} + \dfrac{1}{C_3} + \text{etc.}}$$

Fig. 10-29 Capacitors in series. The equivalent capacitance in (c) is less than the lower of the two capacitances in (a).

(a) (b) (c)

Plate
Dielectric
Plate

Two capacitors in series:

$$C_T = \frac{C_1 \times C_2}{C_1 + C_2}$$

n equal capacitors in series:

$$C_T = \frac{C}{n}$$

Notice that the general form of these formulas is the same as the form of the formulas used for parallel resistances. The formulas are indeed the same except that C is replaced by R.

EXAMPLE 10-8

What is the total capacitance of a 0.001-μF capacitor in series with a 2000-pF capacitor?

Given:	$C_1 = 0.001\ \mu F$
	$C_2 = 2000\ pF$
Find:	C_T
Known:	$C_T = \dfrac{C_1 \times C_2}{C_1 + C_2}$
Solution:	$2000\ pF = 0.002\ \mu F$
	$C_T = \dfrac{0.001\ \mu F \times 0.002\ \mu F}{0.001\ \mu F + 0.002\ \mu F}$
	$= \dfrac{0.000002}{0.003}$
	$C_T = 0.00067\ \mu F$
Answer:	The total series capacitance is 0.00067 μF.

Notice in the above example that the capacitances were converted into the same units (μF) before the formula was used. This way the total (equivalent) capacitance comes out in the same unit as the two original capacitances.

The reactances of series capacitors are additive. The total reactance can be computed by the formula

$$X_{C_T} = X_{C_1} + X_{C_2} + X_{C_3}$$

EXAMPLE 10-9

Determine the total reactance of the circuit shown in Fig. 10-30.

Given:	$C_1 = 2\ \mu F$

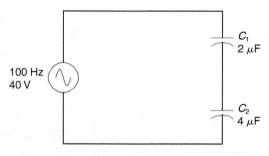

Fig. 10-30 Circuit for examples 10-9 and 10-10.

	$C_2 = 4\ \mu F$
	$f = 100\ Hz$
Find:	X_{C_T}
Known:	$X_{C_T} = X_{C_1} + X_{C_2}$
	$X_C = \dfrac{1}{6.28 fC}$
Solution:	$X_{C_1} = \dfrac{1}{6.28 \times 100 \times 2 \times 10^{-6}}$
	$= 796\ \Omega$
	$X_{C_2} = \dfrac{1}{6.28 \times 100 \times 4 \times 10^{-6}}$
	$= 398\ \Omega$
	$X_{C_T} = 796\ \Omega + 398\ \Omega$
	$= 1194\ \Omega$
Answer:	The total reactance of the circuit is 1194 Ω.

The answer to example 10-9 could also have been found by first determining the total capacitance and then finding the total reactance:

$$C_T = \frac{C_1 \times C_2}{C_1 + C_2} = \frac{2 \times 4}{2 + 4} = \frac{8}{6} = 1.33\ \mu F$$

$$X_{C_T} = \frac{1}{6.28\ fC_T}$$

$$= \frac{1}{6.28 \times 100 \times 1.33 \times 10^{-6}}$$

$$= 1194\ \Omega$$

Ohm's law is also applicable in capacitor circuits. Just replace R with X_C.

EXAMPLE 10-10

Determine the circuit current and the voltage across each capacitor in Fig. 10-30.

Given:	From the previous example:
	$X_{C_1} = 796\ \Omega,\ X_{C_2} = 398\ \Omega,$
	$X_{C_T} = 1194\ \Omega.$

Find: From Fig. 10-30:
$V_T = 40$ V
I_T, V_{C_1}, V_{C_2}

Known: $I_T = \dfrac{V_T}{X_{C_T}}$

$I_T = I_{C_1} = I_{C_2}$
$V = IX_C$

Solution: $I_T = \dfrac{40 \text{ V}}{1194 \; \Omega} = 0.0335$ A
$= 33.5$ mA
$V_{C_1} = 0.0335 \text{ A} \times 796 \; \Omega$
$= 26.7$ V
$V_{C_2} = 0.0335 \text{ A} \times 398 \; \Omega$
$= 13.3$ V

Answer: The total current is 33.5 mA. The voltage across C_1 is 26.7 V, and that across C_2 is 13.3 V.

As a check on the computations in the above example, we can apply Kirchhoff's voltage law:

$V_T = V_{C_1} + V_{C_2} = 26.7 \text{ V} + 13.3 \text{ V} = 40$ V

Notice in the above example that the largest capacitor dropped the least amount of voltage. This is because capacitance and capacitive reactance are inversely proportional.

In electronic circuits, capacitors can be used as ac voltage dividers. The advantage of using capacitors rather than resistors as ac voltage dividers is that capacitors use no power.

10-11 Capacitors in Parallel

The total capacitance of parallel capacitors is found by adding the individual capacitances:

$$C_T = C_1 + C_2 + C_3 + \text{ etc.}$$

Figure 10-31 shows that connecting capacitors in parallel effectively increases the plate area. Suppose the two capacitors shown in cross section in Fig. 10-31(a) are moved together until the ends of their plates touch. Then the two connecting conductors can be removed. The result is an equivalent capacitor [Fig. 10-31(b)] with a larger plate area. A larger plate area, of course, means more capacitance.

The reactances of capacitors in parallel can be treated like resistances in parallel. The formulas are

General formula for parallel reactances:

$$X_{C_T} = \frac{1}{\dfrac{1}{X_{C_1}} + \dfrac{1}{X_{C_2}} + \dfrac{1}{X_{C_3}} + \text{ etc.}}$$

Two reactances in parallel:

$$X_{C_T} = \frac{X_{C_1} \times X_{C_2}}{X_{C_1} + X_{C_2}}$$

n equal reactances in parallel:

$$X_{C_T} = \frac{X_C}{n}$$

Total reactance

The *total reactance* of parallel capacitors can also be obtained by either the reactance formula or Ohm's law. The formulas are

$$X_{C_T} = \frac{1}{6.28 \, fC_T} \quad \text{and} \quad X_{C_T} = \frac{V_T}{I_T}$$

Of course, total capacitance must be determined before the reactance formula can be applied.

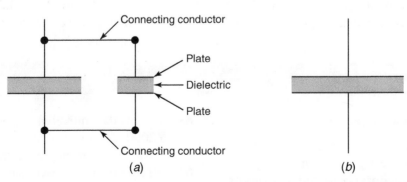

(a)   (b)

Fig. 10-31 Capacitors in parallel. The equivalent capacitance in (b) is equal to the sum of the capacitances in (a).

EXAMPLE 10-11

Refer to Fig. 10-32. Determine the total current, the current through C_1, the total capacitance, and the total reactance.

Given: $C_1 = 0.2\ \mu F$
$C_2 = 0.3\ \mu F$
$f = 1000\ Hz$
$V = 50\ V$

Find: I_T, I_{C_1}, C_T, and X_{C_T}

Known: $I_T = \dfrac{V_T}{X_{C_T}}$

$I_{C_1} = \dfrac{V_{C_1}}{X_{C_1}}$

$C_T = C_1 + C_2$

$X_{C_T} = \dfrac{1}{6.28\,fC_T}$

$V_T = V_{C_1} = V_{C_2}$

Solution: $C_T = 0.2\ \mu F + 0.3\ \mu F = 0.5\ \mu F$

$X_{C_T} =$

$\dfrac{1}{6.28 \times 1 \times 10^3 \times 0.5 \times 10^{-6}}$

$= 318\ \Omega$

$I_T = \dfrac{50\ V}{318\ \Omega} = 0.157\ A$

$= 157\ mA$

$X_{C_1} =$

$\dfrac{1}{6.28 \times 1 \times 10^3 \times 0.2 \times 10^{-6}}$

$= 796\ \Omega$

$I_{C_1} = \dfrac{50\ V}{796\ \Omega} = 0.0628\ A$

$= 62.8\ mA$

Answer: The total current is 157 mA, and I_{C_1} is 62.8 mA. The total capacitance is 0.5 μF, and the total reactance is 318 Ω.

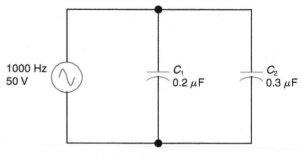

Fig. 10-32 Circuit for example 10-11.

In example 10-11, total reactance was determined without using a parallel-reactance formula or determining the value of X_{C_2}. Let us check the answers to this example by calculating X_{C_2} and then recalculating X_{C_T}:

$X_{C_2} = \dfrac{1}{6.28 \times 1 \times 10^3 \times 0.3 \times 10^{-6}}$

$= 531\ \Omega$

$X_{C_T} = \dfrac{796 \times 531}{796 + 531}$

$= 318\ \Omega$

Since the value obtained for X_{C_T} is the same in both cases, we have made no mathematical errors. If we wanted to know the current through C_2 in Fig. 10-32, we could use Kirchhoff's current law (we have already determined the value of I_T and I_{C_1}):

$I_T = I_{C_1} + I_{C_2}$

$I_{C_2} = I_T - I_{C_1}$

$= 157\ mA - 62.8\ mA$

$= 94.2\ mA$

We could also determine I_{C_2} by using Ohm's law because we know X_{C_2} and V_{C_2}:

$I_{C_2} = \dfrac{V_{C_2}}{X_{C_2}} = \dfrac{50\ V}{531\ \Omega}$

$= 0.0942\ A = 94.2\ mA$

᪥ Self-Test

Answer the following questions.

62. The equivalent capacitance of a 2-μF capacitor and a 1-μF capacitor is 3 μF. Are the capacitors connected in series or in parallel?

63. What is the equivalent capacitance of a 2000-pF capacitor and a 0.002-μF capacitor connected in series?

64. What is the total reactance at 600 Hz of a 0.2-μF and a 0.4-μF capacitor connected in parallel?

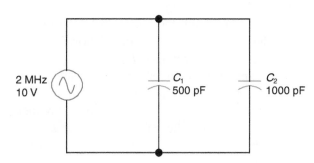

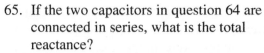

Fig. 10-33 Circuit for test question 67.

Fig. 10-34 Circuit for test questions 68 and 69.

65. If the two capacitors in question 64 are connected in series, what is the total reactance?
66. How much voltage is required to produce a current of 0.03 A in a capacitor that has 4500 Ω of reactance?
67. For the circuit of Fig. 10-33, find the following:
   a. X_{C_2}
   b. C_T

c. V_{C_1}
d. I_T
68. For the circuit of Fig. 10-34, find the following:
   a. X_{C_T}
   b. I_T
   c. C_T
69. In Fig. 10-34, which carries more current, C_1 or C_2?

10-12 Detecting Faulty Capacitors

An ohmmeter can be used to partially check the condition of a capacitor. When an ohmmeter is connected across a completely discharged capacitor, the capacitor will charge to a voltage equal to that of the battery or dc power supply used in the ohmmeter. How long it takes the capacitor to charge depends on two things: the size of the capacitor and the internal resistance of the ohmmeter. In other words, it depends on the time constant. When capacitors are being checked, the ohmmeter should be on a high range to provide a large time constant so that you can "see" the capacitor charge.

With *high values of capacitance,* the ohmmeter pointer of the VOM will deflect toward 0 Ω and then return to a high reading. If the capacitor is nonelectrolytic, the final indication on the VOM should be as high as the meter can go. With the DMM, each new reading will be higher than the previous reading until the DMM indicates overrange. (After the capacitor has charged, the ohmmeter is just measuring the resistance of the dielectric material.)

With low-quality electrolytic capacitors, the final reading may be as low as 100 kΩ. This is because the dielectric material is an oxide, which is not as good an insulator as plastics or ceramics.

With *low-value capacitors,* the charging current may not last long enough to cause a deflection on the ohmmeter. In this case, the meter just indicates a very high resistance. How large a capacitor must be to show a deflection depends on the type of meter. For a particular meter this can be determined by experiment.

So far we have dealt with ohmmeter action when a capacitor is not defective. How does it react to a defective capacitor? If the capacitor is open (a lead is not making contact with a plate), the meter won't deflect. If a capacitor is short-circuited, the pointer will never return to a high resistance reading. This means that current is flowing through the dielectric or that the plates are touching each other. If the ohmmeter indicates 0 Ω, we say that the capacitor is *dead-shorted* or completely shorted. This usually means that the plates are making direct contact. If the dielectric material has become contaminated or deteriorated, the ohmmeter

Low-value capacitors

High values of capacitance

may indicate a few thousand ohms. We would say that the capacitor is partially shorted, has a high-resistance short, or is *leaky*.

Leaky capacitor

A capacitor can pass the ohmmeter check and still not function properly in a circuit. The ohmmeter check does not determine whether the capacitor has the correct value or a high enough Q. Both value and Q can deteriorate as a capacitor ages or is subjected to excessive stress. The true condition of a capacitor can be determined only with a good-quality capacitor tester that tests these factors.

10-13 Undesired, or Stray, Capacitance

Any two conductive surfaces separated by insulation possess capacitance. Thus, every electric and electronic circuit has some undesired capacitance. For example, capacitance exists between

1. An insulated conductor and a metal chassis
2. Two conductors in an electric cable
3. The turns of a coil of wire
4. The input and output leads of a transistor

Stray capacitance

A complete list of places where *stray capacitance* is found would be almost endless.

Whether or not undesired capacitance affects a circuit depends on the amount of capacitance and the frequency of the circuit. For example, the capacitance between two parallel insulated conductors 5 cm long is very low. In dc and most ac power circuits, this amount of capacitance would have no effect. However, at the frequency used in a television tuner, even such a low capacitance would have a very pronounced effect. In fact, if the conductors were 6 cm rather than 5 cm long, the tuner might not function properly.

Relaxation oscillator

Sawtooth voltage

10-14 Capacitor Specifications

Specifications

When ordering capacitors, one has to include the *specifications* needed to ensure receiving the desired capacitors. The minimum specifications for ordering capacitors for general use are

1. Capacitance
2. Tolerance
3. Direct current working voltage
4. Type (ceramic disc, polystyrene, variable, etc.)
5. Lead configuration (axial or radial)

If the capacitors are for circuits in which the capacitance value is critical, then the temperature coefficient should also be specified. For extreme environmental conditions, the operating temperature range should be included in the specifications. When available space is limited, capacitor size and shape should also be specified.

10-15 Uses of Capacitors

Capacitors are used with other components to accomplish a wide variety of jobs. For example, capacitors are used

1. In power supply filters which smooth pulsating direct current into pure direct current
2. In oscillators which convert direct current into either sinusoidal or nonsinusoidal alternating current
3. In filters which separate low-frequency alternating current from high-frequency alternating current
4. For coupling amplifiers together (to separate alternating current from direct current)
5. For power-factor correction (their leading current compensates for the lagging current of an inductive load)

The circuit in Fig. 10-35 illustrates how a capacitor works in a complete functional system. This circuit is called a *relaxation oscillator.* Its purpose is to produce a *sawtooth voltage* from a pure dc voltage. The neon lamp in the circuit fires (ionizes) at 80 V and deionizes at 70 V. When it is fired, its resistance is extremely low; when it is deionized, its resistance is high.

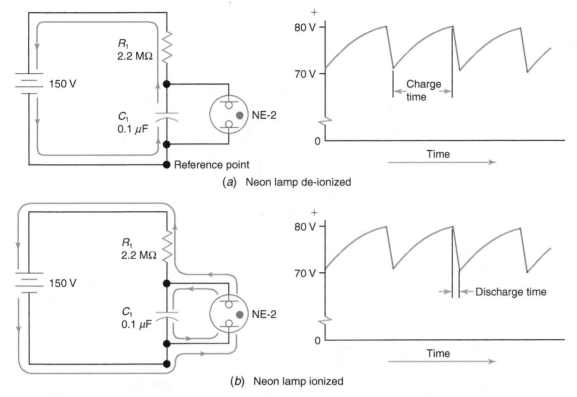

Fig. 10-35 Relaxation oscillator circuit. The capacitor discharges during the time that the lamp is ionized.

Figure 10-35(a) shows the current flow when the lamp is deionized; at this time the capacitor is charging. And, since C_1 charges through R_1, the time constant is relatively long, as indicated in the waveform graph. Once the capacitor has charged to 80 V, the lamp fires [Fig. 10-35(b)]. The internal resistance of the lamp drops to a low value when it fires. This allows the capacitor to quickly discharge through the low resistance. When the capacitor has discharged to the lamp's critical voltage (about 70 V), the lamp deionizes. The capacitor then stops discharging and again starts to recharge.

The above charging and discharging process repeats itself endlessly as long as the battery energy lasts. It produces a continuous sawtooth waveform with a 10-V peak-to-peak amplitude. Notice in Fig. 10-35 that the waveform is a fluctuating dc voltage. It varies from 70 to 80 V positive with respect to the reference point.

Suppose we wished to separate the ac saw-tooth part of the waveform from the dc part

(the +70 V). We can do this with another capacitor and resistor combination, such as C_2 and R_2 in Fig. 10-36. In this figure C_2 couples the sawtooth (ac) to R_2 while blocking the dc voltage. This is what is being referred to by the statement "A capacitor blocks direct current and passes alternating current." The broken lines in Fig. 10-36 on page 278 show the paths of current flow when the lamp is fired. When fired, the lamp carries discharge current for both C_1 and C_2 as well as current for R_1. The solid line shows the paths when the lamp is deionized. Notice that the current through R_2 is alternating current; it periodically reverses itself.

Although this circuit illustrates how a capacitor operates in a complete circuit using only the components we have studied, it is rarely (if ever) used in modern electronic systems. A sawtooth with a linear (straight line) rise can be easily produced with transistors and integrated circuits (ICs).

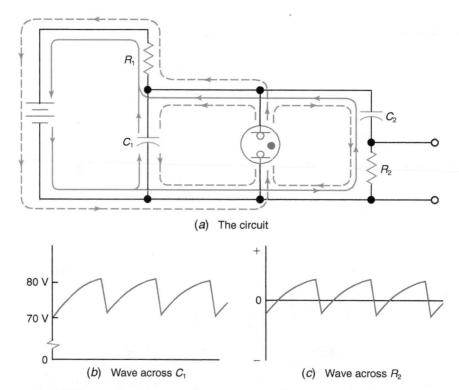

(a) The circuit

(b) Wave across C_1

(c) Wave across R_2

Fig. 10-36 Coupling capacitor. The coupling capacitor separates the alternating current from the direct current.

Self-Test

Answer the following questions.

70. How much resistance should an ohmmeter indicate across the terminals of a good ceramic capacitor?

71. When a capacitor is checked with an ohmmeter, does no deflection of the pointer always indicate an open capacitor?

72. What is another name for undesired capacitance?

73. List the minimum specifications for ordering capacitors.

74. A coupling capacitor _____ direct current and _____ alternating current.

75. Explain how a capacitor can pass the ohmmeter test and still not function correctly in a circuit.

76. What would happen to the period of the waveform in Fig. 10-35 if R_1 were reduced to 1.8 MΩ?

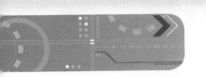
Summary

1. Capacitance is the ability to store electric energy.
2. The symbol for capacitance is C.
3. Capacitors are devices constructed to provide specific amounts of capacitance.
4. Capacitors have two plates and a dielectric.
5. A charged capacitor stores energy and creates a voltage between its plates.
6. A charged capacitor can provide energy for other parts of an electric circuit.
7. Current does not flow through the dielectric material of an ideal capacitor.
8. The farad is the base unit of capacitance.
9. The abbreviation for farad is F.
10. Capacitance is determined by plate area, distance between plates, dielectric material, and temperature.
11. Capacitors are named after their dielectric material, enclosure, construction process, or intended use.
12. Electrolytic capacitors are usually polarized. They have maximum capacitance for their size and weight.
13. Typical paper and film capacitances are less than 1 μF.
14. Ceramic and mica capacitors are limited to low values.
15. Energy-storage capacitors can provide very high power levels.
16. The curved line on a capacitor symbol identifies a specific plate on variable, polarized, paper, and film capacitors.

17. Except for an initial charging surge, capacitors block direct current.
18. The smaller of two series capacitors develops the greater voltage.
19. Capacitors control current flow in an ac circuit.
20. Current leads voltage by 90° in an ideal capacitor.
21. Capacitance uses no power or energy.
22. Energy loss in a capacitor results from dielectric and resistance losses.
23. Dissipation factor, power factor, and quality all are ways of rating a capacitor's relative energy loss.
24. Capacitive reactance is the opposition of a capacitor to sinusoidal alternating current.
25. The symbol for capacitive reactance is X_C.
26. Capacitive reactance is inversely proportional to both frequency and capacitance.
27. The total capacitance of series capacitors is less than the smallest capacitance.
28. The total reactance of series capacitors is the sum of the individual reactances.
29. The total capacitance of parallel capacitors is the sum of the individual capacitances.
30. Reactances of parallel capacitors are added reciprocally.
31. Ohm's law and Kirchhoff's laws apply to capacitor circuits.
32. A time constant defines the rate at which a capacitor charges or discharges through a resistor.

Related Formulas

Capacitance formulas:

$$C = \frac{Q}{V}$$

$$W = 0.5CV^2$$

$$X_C = \frac{1}{6.28fC}$$

$$T = RC$$

For series capacitors:

$$C_T = \frac{1}{\frac{1}{C_1} + \frac{1}{C_2} + \frac{1}{C_3} + \text{etc.}}$$

$$X_{C_T} = X_{C_1} + X_{C_2} + X_{C_3} + \text{etc.}$$

For parallel capacitors:

$$C_T = C_1 + C_2 + C_3 + \text{etc.}$$

$$X_{C_T} = \frac{1}{\frac{1}{X_{C_1}} + \frac{1}{X_{C_2}} + \frac{1}{X_{C_3}} + \text{etc.}}$$

For questions 10-1 to 10-16, determine whether each statement is true or false.

10-1. Metal oxide can be used for the dielectric in a capacitor. (10-6)

10-2. In an ac circuit, a capacitor charges every other quarter-cycle. (10-9)

10-3. For a given weight and size, a capacitor stores more energy than a battery. (10-2)

10-4. The voltage rating of a foil capacitor is a dc voltage rating. (10-6)

10-5. A farad can be defined as a coulomb per volt. (10-5)

10-6. Electrons freely travel through the dielectric material when a capacitor is being charged. (10-2)

10-7. Electrolytic capacitors are smaller and lighter than other types of capacitors that have the same capacitance and voltage rating. (10-6)

10-8. Energy-storage capacitors rarely produce a peak current of more than 0.5 A while being discharged. (10-6)

10-9. The color-band end of a paper or foil capacitor indicates the negative plate of the capacitor. (10-7)

10-10. The negative end of a variable capacitor is indicated by the curved plate of the capacitor symbol. (10-7)

10-11. Except for the charging current, a capacitor blocks direct current. (10-2)

10-12. The lowest-value capacitor develops the highest voltage in either a series ac or series dc circuit. (10-8, 10-10)

10-13. The final reading on an ohmmeter connected to a good capacitor should never exceed 100 kΩ. (10-12)

10-14. Some ceramic capacitors are self-healing. (10-6)

10-15. The plates of a metallized-film capacitor are metal oxide. (10-6)

10-16. The capacitor in a relaxation oscillator discharges through the neon lamp. (10-15)

For questions 10-17 to 10-23, supply the missing word or phrase in each statement.

10-17. The base unit of capacitance is the _____. (10-4)

10-18. The abbreviation for the base unit of capacitance is _____. (10-4)

10-19. A 5600-pF capacitor has _____ μF of capacitance. (10-4)

10-20. The symbol for capacitance is _____. (10-1)

10-21. Capacitance causes current to _____ voltage by _____ degrees. (10-9)

10-22. Reactance is _____ proportional to both _____ and _____ . (10-9)

10-23. After _____ time constants, a capacitor will be over 99 percent charged. (10-8)

Answer the following questions.

10-24. What four factors determine the capacitance of a capacitor? Also indicate how the capacitance can be increased by changing each factor. (10-5)

10-25. Where is the energy stored in a capacitor? In what form (chemical, heat, etc.) is it stored? (10-2)

10-26. What two factors determine the strength of the electric field between the plates of a capacitor? (10-3)

10-27. How are electrolytic capacitors used in ac circuits? (10-6)

10-28. List the minimum specifications needed when ordering capacitors. (10-14)

10-29. What function is performed by a coupling capacitor? (10-15)

10-30. Under what condition is the stray capacitance between two short conductors likely to be a significant factor? (10-13)

10-1. How much energy is stored in a 450-μF capacitor when it is charged to 250 V? (10-6)

10-2. How much charge is stored on the capacitor in problem 10-1? (10-4)

10-3. What is the reactance of a 0.033-μF capacitor at 10 kHz (10-9)

10-4. A 3-μF capacitor and a 2-μF capacitor are connected in parallel across a 50-V, 100-Hz source. Determine the (10-11)
   a. Total capacitance
   b. Total reactance
   c. Total current
   d. Current through the 2-μF capacitor

10-5. If the capacitors in problem 10-4 are connected in series, what is the value of (10-10)
   a. Total capacitance
   b. Total reactance
   c. Voltage across the 3-μF capacitor

10-6. What is the voltage across a capacitor after two time constants if the capacitor and a resistor are series-connected to a 200-V dc source? (10-8)

10-7. What is the time constant of a 2200-pF capacitor in series with a 2.7-MΩ resistor and a 100-V dc source? (10-8)

10-8. What is the Q of a 2-μF capacitor at 100 Hz if the ESR is 0.5 Ω? (10-9)

10-9. How many nanofarads of capacitance would 0.02-μF capacitor have? (10-4)

10-10. What is the average current when a 0.03 F capacitor is connected to a 4 $V_{peak\text{-}to\text{-}peak}$ 2-Hz source? (10-9)

10-11. A 2-μF capacitor and a 6-μF capacitor are connected in series to a 24-V battery. Determine the voltage across the 6-μF capacitor. (10-8)

10-12 What is the voltage across the 2-μF capacitor in problem 10-11? (10-8)

Critical Thinking Questions

10-1. A series combination of a 2-μF capacitor (C_1) and a 3-μF capacitor (C_2) is connected in parallel with a 1.5-μF capacitor (C_3). Determine the total capacitance and the total reactance at 100 Hz.

10-2. What value of capacitance must be in series with a 2-μF capacitor to produce a total capacitance of 1.5 μF?

10-3. What percentage of the source voltage will develop across a 1-μF capacitor that is in series with a 0.68-μF capacitor? Show how you calculated your answer.

10-4. What percentage of the source current will flow through a 0.12-μF capacitor when it is paralleled by a 0.33-μF capacitor?

10-5. Does alternating current flow in any of the conductors in Fig. 10-35? Explain.

10-6. For the formula $W = 0.5 \, CV^2$, prove that W will be in joules if C is in farads and V is in volts.

10-7. A capacitor stores 0.3 C of charge when it is charged to 500 V. How much power will it produce if it is discharged in 0.5 ms?

10-8. A 3-μF capacitor (C_1) and an 8-μF capacitor (C_2) are series-connected to a 40-V dc source. Determine the voltage on the 8-μF capacitor.

10-9. Which formula(s) would not be correct for series capacitor circuits if the capacitors were not assumed to be ideal? Why?

10-10. Using parallel circuit rules and Q, V, and C relationships, prove that $C_T = C_1 + C_2$ for parallel capacitors.

10-11. Using series circuit rules and Q, V, and C relationships, prove that

$$C_T = \frac{1}{\dfrac{1}{C_1} + \dfrac{1}{C_2}}$$

for series capacitors.

10-12. Would adding a coupling capacitor and load resistor (like C_2 and R_2 in Fig. 10-36) to the circuit in Fig. 10-35 change the period of the sawtooth waveform? Explain your answer.

1. capacitor
2. C
3. dielectric
4. dielectric
5. F
6. F
7. in the dielectric material
8. DCWV
9. Because too much electric field strength between the plates can destroy the dielectric material.
10. the distance between the plates and the voltage to which the capacitor is charged
11. farad
12. F
13. 0.0047
14. 3000
15. coulomb, volt
16. plate area, distance between plates, dielectric material, temperature
17. temperature
18. It doubles.
19. It doubles.
20. the material's ability to store energy when used as a dielectric
21. smaller
22. 0.9 mC
23. 16,667 μF
24. 400 V
25. T
26. F
27. T
28. F
29. T
30. F
31. Metallized
32. variable
33. feed-through
34. aluminum, tantalum
35. oxide
36. tantalum
37. It draws excessively high current from the source and overheats. It may explode.
38. mica, ceramic, paper, film (plastics)
39. smaller, more stable, and higher dielectric resistance
40. filter capacitor
41. in applications that require very rapid capacitor discharge

42. 6.75 J
43. negative
44. five
45. color band
46. the time required to charge or discharge a capacitor by 63.2 percent of the available voltage
47. 0.11 s
48. The 5-μF capacitor, because $V = Q/C$ and both capacitors receive the same amount of Q.
49. by producing a voltage which opposes the source voltage
50. Because current and voltage are 90° out of phase.
51. Current leads voltage by 90°.
52. Quality is the ratio of reactance to resistance.
53. dissipation factor, power factor, and quality.
54. capacitive reactance
55. X_C
56. equivalent series resistance
57. T
58. F
59. 531 Ω
60. 19.9 kHz
61. 28.3 mA
62. parallel
63. 1000 pF or 0.001 μF
64. 442 Ω
65. 1990 Ω
66. 135 V
67. a. 663.5 Ω
    b. 0.075 μF
    c. 22.5 V
    d. 11.3 mA
68. a. 53.0 Ω
    b. 188.4 mA
    c. 1500 pF
69. C_2
70. nearly infinite
71. no
72. stray capacitance
73. capacitance, tolerance, type, lead arrangement, and voltage rating
74. blocks, passes
75. The ohmmeter test does not check the value or the Q of the capacitor.
76. The period would be reduced.

CHAPTER 11

Inductance

Learning Outcomes

This chapter will help you to:

11-1 *Name* the electric quantity that opposes changes in current.

11-2 *Name and define* the base unit of inductance.

11-3 *List* four factors that are the major determiners of the inductance of an inductor.

11-4 *Explain* why: (a) some inductors resemble resistors, (b) some inductors are shielded, some inductors use laminated iron cores, and some inductors are called chokes.

11-5 *List and explain* five ratings, in addition to inductance, that are given for some inductors.

11-6 *Explain* how an inductor controls the rate at which current can increase in a dc circuit.

11-7 *Practice* using the X_L formula to determine the opposition of an ideal inductor. And, show that *V* leads *I* by 90° with an ideal inductor.

11-8 *Define* "quality" for an inductor and show how it is calculated.

11-9 *Discuss* how connecting inductors in parallel affects total inductance and total reactance.

11-10 *Discuss* how connecting inductors in series affects total inductance and total reactance.

11-11 *Calculate* the time constant for either series or parallel *RL* dc circuits.

11-12 *Reduce* mutual inductance between two inductors.

11-13 *Explain* how non-inductive wire-wound resistors are made.

Many electric and electronic devices operate on the principle of inductance. Therefore, it is important that you develop a sound understanding of inductance and inductors.

11-1 Characteristics of Inductance

Inductance is the electrical property that opposes any change in the magnitude of current in a circuit. The letter "*L*" is the symbol used to represent inductance.

Devices that are used to provide the inductance in a circuit are called *inductors*. Inductors are also known as *chokes, reactors,* and *coils.* These three names are descriptive of the way inductance behaves in a circuit. Inductance, and thus an inductor, "chokes off" and restricts sudden changes in current. Inductance reacts against (resists) changes, either increases or decreases, in current. Inductors are usually coils of wire.

Inductance is the result of a voltage being induced in a conductor. The magnetic field that induces the voltage in the conductor is produced by the conductor itself.

Inductance (*L*)

Inductors

Chokes

Reactors

Coils

💡 **You May Recall**

⚠️ . . . that previously we discussed how a magnetic field is formed around a current-carrying conductor.

When current begins to flow in a conductor, magnetic flux rings start to expand out from the conductor, as in Fig. 11-1(*a*). This expanding flux induces a small voltage in the conductor. The induced voltage has a polarity that opposes the increasing source voltage which is creating the increasing current. Thus, the inductance of the conductor opposes the rising current and tries to keep it constant. Of course, the inductance cannot completely stop the increase in current because the induced voltage is caused by the increasing flux. And the increasing flux depends on the increasing current. The inductance of the conductor, therefore, restricts only the rate at which the current can increase.

When the current in a conductor starts to decrease, as in Fig. 11-1(b), the flux starts to collapse. The collapsing flux reverses the polarity of the induced voltage from what it was when the flux was increasing. Thus, the voltage induced by a decreasing flux *aids* the source voltage and tends to keep the current from decreasing. Again, the inductance restricts the rate at which the current can change.

The amount of voltage induced in a single conductor like that in Fig. 11-1 is very small. So small, in fact, that it has no practical significance in most low-frequency electric and electronic devices. However, at high frequencies, like those used in television systems, the inductance of a single conductor can be very significant.

Self-inductance

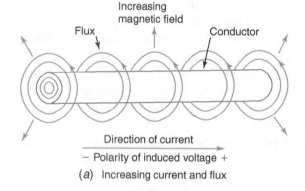

Increasing magnetic field

Flux

Conductor

Direction of current

− Polarity of induced voltage +

(*a*) Increasing current and flux

Mutual inductance

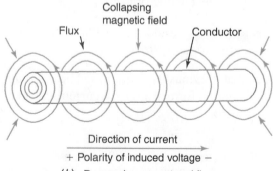

Collapsing magnetic field

Flux

Conductor

Direction of current

+ Polarity of induced voltage −

(*b*) Decreasing current and flux

Fig. 11-1 Induced voltage in a conductor. The polarity of the induced voltage is dependent on whether the current is increasing or decreasing.

Counter electromotive force (cemf)

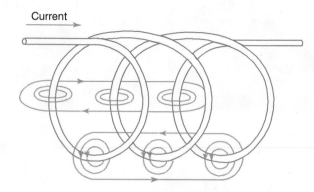

Current

Fig. 11-2 Multiturn coil. Some of the flux created by one turn links to all the other turns.

The inductance of a conductor can be greatly increased by forming the conductor into a coil as in Fig. 11-2. Now the flux produced by one turn of the coil induces voltage not only in itself but in adjacent turns as well. The long, closed flux loops in Fig. 11-2 are the result of the magnetic fields of all three turns of the coil. They are stronger than the flux created by any one of the turns. Yet, they induce a voltage into each of the three turns. The inductance of the coil is much greater than the inductance of the straight length of conductor from which it was made.

Self-Inductance

The inductance of an inductor is called *self-inductance*. It is given this name because the inductor induces voltage in itself. That is, its own changing magnetic field induces voltage in its own turns of wire. In the case of a single straight conductor, its own field induces a voltage in it.

Mutual Inductance

When the magnetic flux from one conductor induces a voltage in another, electrically isolated conductor, it is called *mutual inductance*. With mutual inductance, circuits that are electrically separated can be magnetically coupled together. A transformer uses the principle of mutual inductance. Transformers are fully discussed in Chap. 12.

Lenz's Law and CEMF

The voltage induced in a conductor or coil by its own magnetic field is called a *counter electromotive force* (cemf).

You May Recall

. . . from an earlier chapter that electromotive force (emf) is another name for voltage.

Since the induced emf (voltage) is always opposing, or countering, the action of the source voltage, it is known as a cemf. Counter electromotive force is sometimes referred to as *back electromotive force (bemf)*. The term *back* implies that the induced voltage is backward, or working against the effort of the source voltage.

The concept contained in Lenz's law is used to explain how inductance behaves. *Lenz's law* states that a cemf always has a polarity that opposes the force that created it. This idea is illustrated in Fig. 11-3(a), which shows, in schematic form, an inductor and an ac voltage source. When the voltage is increasing, as shown in the graph, the cemf opposes the source voltage. When the voltage is decreasing [Fig. 11-3(b)], the cemf aids the source voltage and tries to keep the current constant.

Energy Storage and Conversion

Another way to look at inductance is in terms of energy conversion and storage. When current flows through an inductor, the inductor builds up a magnetic field. In the process of building its magnetic field, the inductor

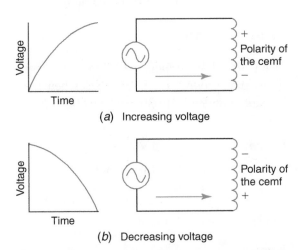

(a) Increasing voltage

(b) Decreasing voltage

Fig. 11-3 Polarity of counter electromotive force (cemf).

History of Electronics

Heinrich Lenz
In 1834 Russian physicist Heinrich Lenz stated his law concerning the polarity of induced voltages.

converts electric energy into magnetic energy. When the current increases, more electric energy is converted into magnetic energy. The inductor's magnetic field now processes more energy than it had before the current increased. When current through an inductor decreases, its magnetic field decreases. Magnetic energy from the field is converted back into electric energy in the inductor. Thus, an inductor stores energy when its current increases and returns stored energy when its current decreases. *Inductance converts no electric energy into heat energy.* Only resistance is capable of converting electric energy into heat energy. Thus, if an inductor with absolutely no resistance could be constructed, its net use of energy would be zero. For the two quarters of the ac cycle when the current is increasing (first and third quarter), it would take energy from the system. For the other two quarters of the cycle (second and fourth) when the current is decreasing, it would return the same amount of energy to the system.

Notice that inductance stores and returns energy in much the way capacitance does. During the first and third quarter-cycle, inductance converts electric energy to magnetic energy and stores the magnetic energy, whereas capacitance directly stores electric energy. During the second and fourth quarter-cycle, inductance converts its stored magnetic energy back to electric energy and returns it, whereas capacitance just returns its stored electric energy.

Back electromotive force (bemf)

Lenz's law

Answer the following questions.

1. The electric property that opposes changes in current is called _____.
2. The physical device that opposes changes in current can be called a(n) _____, _____, _____, or _____.
3. True or false. A straight wire possesses inductance.
4. True or false. A straight wire can be called an inductor.
5. True or false. Inductance converts electric energy to heat energy.
6. True or false. The cemf aids the source voltage when the current in an inductive circuit is increasing.
7. True or false. Transformers operate on the principle of self-inductance.
8. The symbol or abbreviation for inductance is _____.
9. _____ law can be used to find the polarity of the cemf in an inductor.
10. Another abbreviation for cemf is _____.
11. When current in an inductor is increasing, _____ energy is being converted to _____ energy.

11-2 Unit of Inductance— The Henry

Henry (H)

The base unit of inductance is the *henry*. This unit, named in honor of an American scientist, is abbreviated H. The henry is defined in terms of the amount of cemf produced when the current through an inductor is changing amplitude. One henry of inductance develops 1 V of cemf when the current changes at a rate of 1 A/s. This definition of a henry is shown graphically in Fig. 11-4.

A wide range of inductances are used in electric and electronic circuits. Inductances in circuits of very high frequency are often less than 1 μH. For low-frequency circuits, inductors with more than 5 H of inductance are common.

A formula for determining inductance (L) in henrys (H) is

$$L = \frac{V_{induced}}{\Delta I / \Delta t}$$

where $V_{induced}$ = the cemf in volts
ΔI = the change in coil current in amperes
Δt = the time in seconds required for the current to change

EXAMPLE 11-1

Determine the inductance of an inductor (coil) that produces 5 V of cemf when the current changes from 300 mA to 800 mA in two seconds.

Given: $V_{induced} = 5$ V
$\Delta I = 800$ mA $- 300$ mA $= 500$ mA $= 0.5$ A
$\Delta t = 2$ s

Find: Inductance (L)

Known: $L = \dfrac{V_{induced}}{\Delta I / \Delta t}$

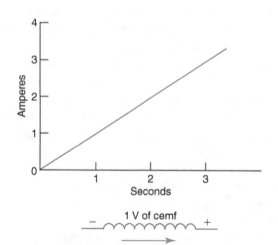

Fig. 11-4 A 1-H inductor produces 1 V of cemf when the current changes at a rate of 1 A/s.

Solution: $L = \dfrac{5\text{ V}}{0.5\text{ A}/2\text{ s}} = \dfrac{5\text{ V}}{0.25\text{ A}/\text{s}}$
$= 20\text{ H}$

Answer: The inductance is 20 henrys.

11-3 Factors Determining Inductance

The inductance of an inductor is primarily determined by four factors:

1. The type of core material
2. The number of turns of wire
3. The spacing between turns of wire
4. The diameter of the coil (or core)

The *core* of an inductor is the material that occupies the space enclosed by the turns of the inductor.

The amount of current in an *iron-core inductor* also influences its inductance. This is because the magnetic properties of the iron core change as the current changes.

Ultimately, the amount of inductance is determined by the amount of cemf produced by a specified current change. Of course, the amount of cemf depends on how much flux interacts with the conductors of the coil.

If all other factors are equal, an iron-core inductor has more inductance than an *air-core inductor*. This is because the iron has a higher permeability; that is, it is able to carry more flux. With this higher permeability, there is more flux change, and thus more cemf, for a given change in current.

Adding more turns to an inductor increases its inductance because each turn adds more

History of Electronics

Joseph Henry
American physicist Joseph Henry did extensive research on electromagnetism and discovered the principles that made the development of the telegraph possible. The fundamental unit for inductance, the henry, is named for him.

magnetic field strength to the inductor. Increasing the magnetic field strength results in more flux to cut the turns of the inductor.

When the distance between the turns of wire in a coil is increased, the inductance of the coil decreases. Figure 11-5 illustrates why this is so. With widely spaced turns [Fig. 11-5(*a*)], many of the flux lines from adjacent turns do not link together. Those lines that do not link together produce a voltage only in the turn that produced them. As the turns come closer together [Fig. 11-5(*b*)], fewer lines of flux fail to link up.

When other factors are equal, the inductor with the largest-diameter core will have the most inductance. This is because all the flux has to go through the core of an inductor. Thus a large-diameter core can handle more flux, at a specified flux density, than a small-diameter core can.

Core

Iron-core inductor

Air-core inductor

Self-Test

Answer the following questions.

12. The base unit of inductance is the _____.

13. The abbreviation for the base unit of inductance is the _____.

14. In terms of the base units of voltage, current, and time, the base unit of inductance is equal to a(n) _____.

15. Does the amount of inductance increase or decrease when more turns are added to an inductor?

16. List four ways to increase the inductance of an inductor.

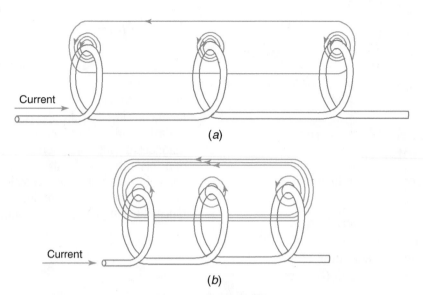

Current

(a)

Current

(b)

Fig. 11-5 Effect of turn spacing. (a) Widely spaced turns provide less inductance than (b) closely spaced turns do.

11-4 Types of Inductors

One way of classifying inductors is by the type of material used for the core of the inductor. The core may be either a magnetic material or a nonmagnetic material. The symbols for inductors with these materials are shown in Fig. 11-6.

Inductors are also classified as either fixed or variable. Figure 11-7 shows two symbols used to indicate a *variable inductor*. The most common way of varying inductance is by adjusting the position of the core material. In Fig. 11-8 the position of the ferrite core material (called a *slug*) is adjustable within the coil form. Maximum inductance occurs when the slug is positioned directly in line with the coil of wire.

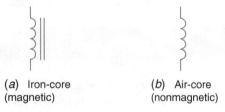

(a) Iron-core   *(b)* Air-core
(magnetic)   (nonmagnetic)

Fig. 11-6 Fixed-value inductor symbols.

Fig. 11-7 Variable inductor symbols. Either symbol can be used for magnetic or nonmagnetic cores.

Some variable inductors use a brass slug. Brass has more reluctance (opposition to flux) than air does. Therefore, the brass slug decreases inductance when it is centered in the coil.

Air-Core Inductors

An *air-core inductor,* used as part of a high-frequency circuit, is shown in Fig. 11-9. This inductor is self-supporting and requires no coil form. However, many inductors that are represented by the air-core symbol are wound on a coil form. The form may be either solid or hollow. These forms have about the same reluctance (opposition to magnetic flux) as air does. Therefore, the inductor is much like an air-core inductor; its core is nonmagnetic. These inductors may be wound on such core materials as ceramic or phenolic. They often look like the coil in Fig. 11-10(*a*). These inductors seldom have more than 5 mH of inductance.

Ferrite and Powdered-Iron Cores

The coil shown in Fig. 11-10(*a*) may also be an *iron-core inductor*. In this case the core material would be ferrite or powdered iron. The correct symbol would then be the iron-core symbol of Fig. 11-6(*a*). (On some schematic diagrams, the two solid lines in the iron-core symbol are replaced by two broken lines to represent a ferrite or powdered iron core.) Most inductors of this type have less than 200 mH

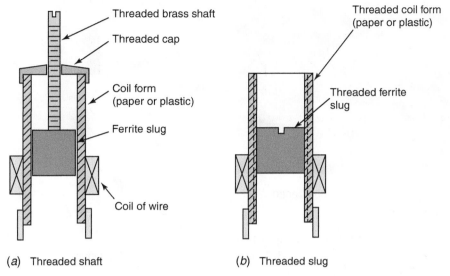

Fig. 11-9 Air-core inductor.

of inductance. They are used primarily at frequencies above the audio (sound) range. Fig. 11-10(*b*) shows another style of inductor wound on a ferrite core.

Toroid Cores

The cores of the inductors discussed so far have all been straight. The magnetic flux loops must extend through the air as well as through the core material. With a *toroid core,* the flux loops all exist within the core. Toroid cores [Fig. 11-10(*c*)] are doughnut-shaped. Each turn of wire is threaded through the center of the core, as shown in Fig. 11-10(*c*). Inductors made with toroid cores are called toroidal inductors. The toroid core is usually made from powdered iron or ferrite. Toroidal inductors can have high inductance values for their size.

Surface Mount Chip Inductor

Chip inductors are available in the nH and μH range of inductance. Fig. 11-11 shows a 68-nH chip inductor positioned inside the end

(*a*)

(*b*)

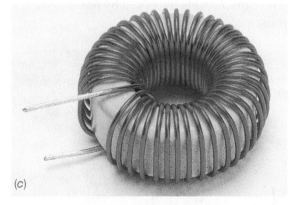

(*c*)

Toroid Core

Fig. 11-10 Miniature high-frequency inductors.
  (*a*) Either nonmagnetic or ferrite core.
  (*b*) Ferrite core.
  (*c*) Toroid core.

Fig. 11-11 Surface mount inductor positioned in the end of a small paper clip.

of a small paper clip. This inductor is approximately 1 mm thick, 1.2 mm wide, and 2 mm long. Notice in Fig. 11-11 that the connecting ends are tinned so they can be readily soldered in place on a circuit board.

Molded Inductors

Some inductors look like resistors (Fig. 11-12). These inductors are enclosed in an insulating material to protect the inductor winding. *Molded inductors* can have cores of air, ferrite, or powdered iron.

Molded inductors

Some of the resistors discussed and pictured in Chap. 4 (Sec. 4-11) are constructed similarly to the inductors wound on a nonmagnetic core. Thus, some resistors have significant inductance and some inductors have significant resistance. Resistors use high-resistance materials to minimize the number of turns needed to obtain the desired resistance. Conversely, inductors use low-resistance materials to minimize the resistance of the turns required to obtain the desired inductance. The quality (*Q*) of inductors (and capacitors) is improved by minimizing the resistance.

E and I laminations

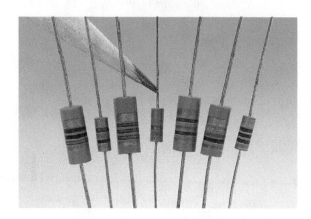

Fig. 11-12 Molded inductors.

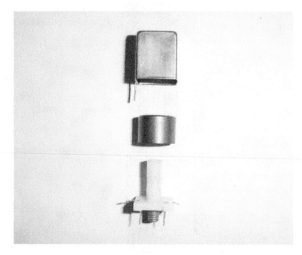

Fig. 11-13 Shielded inductor.

Special wire-wound resistors are wound so that the cemf of half of the turns cancels the cemf of the other half of the turns that are required to obtain the desired resistance. These resistors are referred to as noninductive resistors.

Shielded Inductors

Inductors are often shielded to protect them from the influence of magnetic fields other than their own. The shield is made from a magnetic material. Figure 11-13 shows an exploded view of the parts of a shielded, adjustable coil form. The coil winding is not shown. It would be wound on the cylindrical tube, or bobbin. The coil form shown in Fig. 11-13 is the type used on printed circuit boards.

Some miniature chokes (inductors), like those in Fig. 11-12, are also shielded. Their shields are encased underneath the outside molding.

Laminated Iron Core

Nearly all the large inductors used at power frequencies (60 Hz, for example) use laminated iron cores. These inductors have inductances ranging from about 0.1 to 100 H.

The typical laminated core uses laminations like those in Fig. 11-14. From this illustration it is easy to see why these laminations are called *E and I laminations*. The E laminations are stacked together to the desired thickness, as are the I laminations. The winding is put on the center leg (Fig. 11-15) of the E stack. The I stack is then positioned across the open end of the E stack.

E lamination   I lamination

Fig. 11-14 E and I laminations. These laminations are stacked in various configurations to form cores for electromagnetic devices.

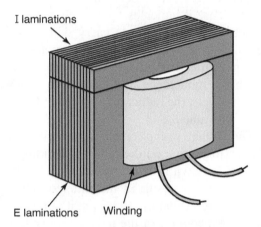

I laminations

E laminations   Winding

Fig. 11-15 Laminated-iron-core inductor. The coil fits over the center leg of the E laminations.

As seen in Fig. 11-16, the E and I laminations form two parallel paths for flux. The center leg of the E lamination is twice as wide as either of the outside legs because it has to carry twice as much flux. For a given amount and rate of current change, the laminated-iron-core inductor creates more flux than other types of inductors. This changing flux, in turn, creates cemf. This is why laminated-iron-core inductors can provide large amounts of inductance.

The inductance of an iron-core inductor is somewhat dependent on the amount of current flowing through it. The reason for this can be seen from Fig. 11-17, which illustrates the

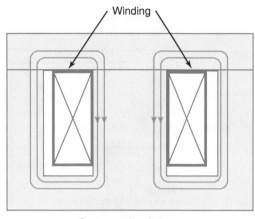

Winding

Cross-sectional view

Fig. 11-16 Flux paths in a laminated core.

magnetic characteristics of the core material. The permeability of iron and iron alloys decreases as the magnetic field strength increases. The magnetic field strength of an inductor is a function of the amount of current flowing in the winding. Refer to Fig. 11-17. Suppose the current through the inductor changes from point A to point B. The flux density would change from A' to B' and produce a certain amount of cemf. This amount of cemf would, of course, represent a certain amount of inductance. Now suppose the current in the inductor was greater and the current changed from C to D in Fig. 11-17. Although C to D is the same amount of change as A to B, it produces a much smaller change (C' to D') in flux. Thus, the inductor has less cemf and inductance at the higher current. Points E and F in Fig. 11-17 show what happens when the core of an inductor is *saturated*. Since a change in current from E to F produces almost no change in flux, there is very little inductance. Except for special applications, inductors are never operated in the saturation region of the permeability curve.

Saturated

Filter Chokes

Laminated-iron-core inductors are often referred to as *filter chokes*. These chokes are used in the filter circuits of power supplies in a wide variety of electrical and electronic equipment. The power supply is often the part of the equipment which converts alternating to direct current. The filter circuit, which includes the inductor, smooths out the fluctuating or pulsating direct current until it is nearly pure direct current.

Filter choke

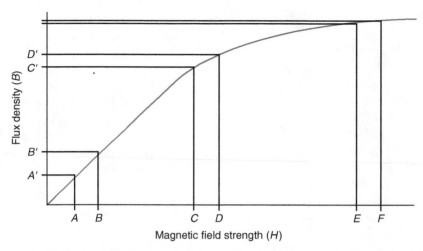

Fig. 11-17 Permeability curve. Permeability decreases as flux density and magnetic field strength increase.

There are two types of filter chokes: the *smoothing choke* and the *swinging choke*. The swinging choke is one in which the I and E laminates are butted together so that there is a minimum air gap between them. This makes the amount of inductance vary with the amount of current (Fig. 11-17). A typical swinging choke may be rated 20 H at 50 mA and 5 H at 200 mA.

The smoothing choke frequently has a small (0.1 mm) air gap between the I and E laminations. This makes the inductance less dependent on the amount of current because air does not saturate as easily as iron.

Radio-Frequency Chokes or Coils

Inductors that are used at higher frequencies are often called RF chokes or *RF coils*. Since radio was one of the early popular uses of high-frequency inductors, they became known as radio-frequency, or RF, coils. An RF coil or choke may have an air, powdered iron, or ferrite core. It may be either fixed or variable.

11-5 Ratings of Inductors

We have seen so far that one of the main ratings of an inductor is its inductance. Inductors are also rated for dc resistance, current, voltage, quality, and tolerance.

The dc resistance specifies the resistance of the wire in the winding of the inductor. This is the resistance between the terminals of the inductor that one would measure with

an ohmmeter. Therefore this dc resistance is sometimes called the *ohmic resistance*.

The current rating of an inductor is important because it indicates how much current the inductor can continuously carry without overheating. With laminated-core inductors, the current rating also indicates the current level at which the inductance was measured. At lower current levels, the inductance is greater than the specified value.

The voltage rating indicates how much voltage the insulation on the inductor winding can continuously withstand. Exceeding this voltage rating may not result in instantaneous breakdown of the insulation. However, it will shorten the life expectancy of the inductor's insulation. Voltage ratings are used mostly with laminated-core inductors. With these inductors, the core is often physically and electrically connected to the chassis of an electric device. However, the winding may be hundreds of volts positive or negative with respect to the chassis.

The *quality* of an inductor refers to the ratio of its reactance to its resistance. Generally it is desirable to have a high-quality inductor. All other factors being equal, the lower the dc resistance, the higher the quality of the inductor. Detailed information on quality is provided in Sec. 11-8.

Like all other components, inductors have manufacturer's tolerances. Precision inductors can be obtained with tolerances of less than ±1 percent. However, they are expensive. Typical inductors used in mass-produced electric and electronic devices have tolerances of ±10 percent or more.

Smoothing choke

Swinging choke

Ohmic resistance

RF coils

Quality

11-6 Inductors in DC Circuits

The behavior of an inductor in a pure dc circuit is contrasted to that of a resistor in Fig. 11-18. With a resistor [Fig. 11-18(a)], the current jumps to its maximum value almost the instant the switch is closed. When the switch is opened, it drops back to zero just as fast. An inductor in a dc circuit [Fig. 11-18(b)] forces the current to rise more slowly. This is due to the inductor's cemf. The time required for the current to reach its maximum value is dependent on the amount of inductance and resistance. With inductors of typical quality, the time is much less than 1 s. Once the current reaches its peak value, the only opposition the inductor offers is its dc resistance. When the switch in Fig. 11-18(b) is opened, the cemf of the inductor prevents the current from instantaneously dropping to zero. It does this

by ionizing the air between the switch contacts as the switch opens. As the energy stored in the inductor's magnetic field is used up, the switch contacts deionize and current stops.

When the switch in Fig. 11-18(b) is opened, the cemf of the inductor becomes much greater than the source voltage. The high voltage (cemf) generated when an inductive circuit is opened is known as an *inductive kick*. It is the voltage that ionizes the air between the switch contacts and causes the contacts to arc and burn. The inductive kick of an inductor is very high because the current drops very rapidly when the switch is opened. The difference between the source voltage and the cemf is dropped across the ionized air between the switch contacts. Kirchhoff's voltage law still applies. That is, the voltage across the switch plus the inductor's cemf still equal the source voltage. Notice the polarities in Fig. 11-19. The inductor's cemf and the voltage across the switch are series-opposing. Thus, they both can be much greater than the battery voltage. The exact value of the cemf in Fig. 11-19 when the switch opens depends upon two factors: the amount of inductance and the amount of current in the circuit before the switch is opened. The inductive kick of an inductor can be many thousands of volts. Inductive kick is the principle on which the ignition coil in an automobile operates.

The relationship between the current and voltage (cemf) in an inductor is illustrated in Fig. 11-20. The resistance R in this figure is very high relative to the ohmic resistance of the inductor. Therefore, the voltage across the inductor is almost zero once the current reaches its maximum value. Notice in Fig. 11-20 that the voltage across the inductor is maximum when the current through it is minimum and rising. Also, the voltage is minimum when the current is maximum and steady. Further, notice that the resistive current and voltage rise together.

Inductive kick

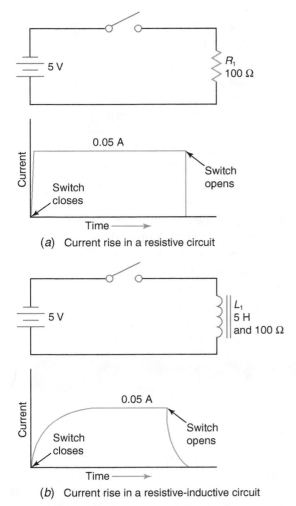

(a) Current rise in a resistive circuit

(b) Current rise in a resistive-inductive circuit

Fig. 11-18 Comparison of a resistive and a resistive-inductive dc circuit. The inductor opposes changes in current.

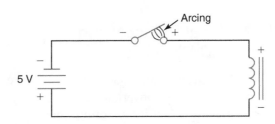

Fig. 11-19 Voltage polarities when an inductive circuit is opened.

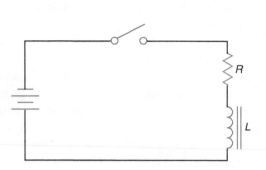
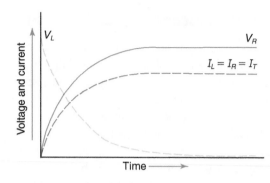

Fig. 11-20 Current and voltage relationships in an inductor. Maximum inductive voltage occurs before maximum current is reached.

Self-Test

Answer the following questions.

17. Draw the symbols for
    a. A fixed iron-core inductor
    b. A fixed air-core inductor
    c. A variable inductor
18. True or false. A slug-type core is used in an iron-core inductor.
19. True or false. When a brass core is centered inside a coil winding, the coil will have maximum inductance.
20. True or false. The air-core inductor symbol is used for all inductors that use nonmagnetic core material.
21. True or false. A ferrite-core inductor would be represented by using the symbol for an iron-core inductor.
22. True or false. A 50-mH inductor would most likely have an air core.
23. True or false. To provide maximum inductance, a ferrite core should be centered in the coil winding.
24. What type of core would be used in a 5-H inductor?

25. What is the shape of a toroidal core?
26. Why is the center leg of an E lamination wider than the outside legs?
27. Why does an inductor have a current rating?
28. What is meant by the quality of an inductor?
29. What is an RF choke?
30. What type of core does a filter choke have?
31. What is meant by inductive kick?
32. An inductive circuit and a resistive circuit have equal currents. When the switch is opened in each circuit, which circuit produces more arcing across the switch contacts?
33. Does the cemf exceed the source voltage when an inductive dc circuit is opened?
34. Does an inductor's voltage (cemf) reach its maximum value before the current reaches its maximum value?

11-7 Ideal Inductors in AC Circuits

Ideal inductor

Inductive reactance (X_L)

An *ideal inductor* is an inductor that has no resistance. It does not convert any electric energy into heat energy, and it has infinite quality. In the discussions that follow, we assume that we have ideal inductors.

Inductive Reactance

Inductance, like capacitance, controls circuit current without using power. Therefore, the opposition of an inductor to alternating current is also called reactance X. To distinguish inductive reactance from capacitive reactance, we use the symbol X_L for *inductive reactance*.

Inductive reactance is the result of the cemf of the inductor. The inductor lets just enough ac flow to produce a cemf that is equal to (and opposite to) the source voltage. This idea is illustrated in Fig. 11-21. During each half-cycle of the source (an ac generator), the cemf of the inductor produces a matching half-cycle of sinusoidal voltage. At any instant of time the two voltages (source and cemf) are equal. The reason for this can be ascertained by referring to Fig. 11-22, which shows that the voltage leads the current by 90° in an inductive circuit. Notice from the figure that the current is changing direction at the instant the source voltage is at its peak value. When the current changes direction, two things happen: (1) the polarity of the mmf and the direction of the flux change, and (2) the flux changes from a collapsing flux to an expanding flux (or vice versa). Either of these happenings would change the polarity of the cemf; but when they occur simultaneously, the polarity cannot change. Also, notice from Fig. 11-22 that the rate of current change, and therefore the rate of flux change, is greatest as the current crosses the zero reference line. Thus, the cemf is greatest as the current is changing direction.

The reactance of an inductor can be calculated with the following formula:

$$X_L - 2\pi fL = 6.28fL$$

The inductive reactance is in ohms when the frequency is in hertz and the inductance is in henrys.

From the above formula it can be seen that inductive reactance is directly proportional to both frequency and inductance. Doubling either doubles the reactance. This direct proportional relationship makes sense when one recalls two things:

1. The higher the frequency, the more rapidly the current is changing. Thus more cemf and more reactance are produced.
2. The higher the inductance, the more flux change per unit of current change. Again, more cemf and reactance are produced.

EXAMPLE 11-2

What is the reactance of a 3-H inductor when the frequency is 120 Hz?

Given:	Inductance $L = 3$ H
	Frequency $f = 120$ Hz
Find:	Inductive reactance (X_L)
Known:	$X_L = 6.28fL$
Solution:	$X_L = 6.28 \times f \times L$
	$= 6.28 \times 120$ Hz $\times 3$ H
	$= 2261$ Ω
Answer:	The 3-H inductor has 2261 Ω of reactance at 120 Hz.

$$X_L = 2\pi fL$$

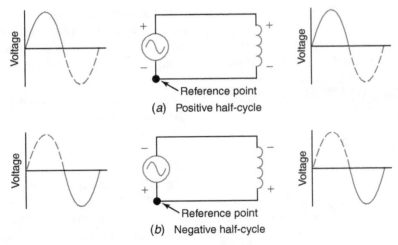

Fig. 11-21 Source and inductor ac voltages. The inductor's cemf opposes the source voltage.

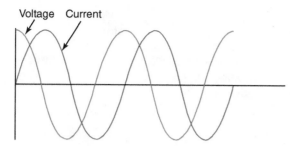

Voltage   Current

Fig. 11-22 Alternating current and voltage in an inductor. The voltage leads the current by 90°.

EXAMPLE 11-3

A 2.5-mH inductor is placed in a circuit where the frequency is 100 kHz. What is its inductive reactance?

Given: $L = 2.5$ mH
             $f = 100$ kHz
Find: X_L
Known: $X_L = 6.28fL$
Solution: 2.5 mH = 0.0025 H
             100 kHz = 100,000 Hz
             $X_L = 6.28 \times 100{,}000$ Hz
                $\times 0.0025$ H
                $= 1570\ \Omega$
Answer: The 2.5-mH inductor has 1570 Ω of reactance at 100 kHz.

In solving example 11-3, we could have left the inductance in millihenry and the frequency in kilohertz. Then, the 10^{-3} of millihenry would have canceled the 10^3 of kilohertz.

When the inductor's current and voltage are known, its reactance can be calculated by using Ohm's law.

EXAMPLE 11-4

Voltage leads the current

The ac voltage measured across an inductor is 40 V. The current measured through the inductor is 10 mA. What is its reactance?

Given: $V_L = 40$ V
             $I_L = 10$ mA
Find: X_L
Ideal inductor
Known: $X_L = \dfrac{V_L}{I_L}$

Solution: $X_L = \dfrac{40\ \text{V}}{0.01\ \text{A}} = 4000\ \Omega$
Answer: The inductor has 4000 Ω of reactance.

If the frequency of the voltage in example 11-4 is known, or measured, the inductance could then be calculated using the reactance formula.

EXAMPLE 11-5

Determine the inductance of the inductor in example 11-4 when the frequency is 500 Hz.

Given: $f = 500$ Hz
             $X_L = 4000\ \Omega$ (from example 11-4)
Find: L
Known: $X_L = 6.28fL$ rearranged, gives

             $L = \dfrac{X_L}{6.28\,f}$

Solution: $L = \dfrac{4000\ \Omega}{6.28 \times 500\ \text{Hz}} =$
             $= \dfrac{4000\ \Omega}{3140\ \text{Hz}} = 1.27$ H
Answer: The inductance is 1.27 H.

Phase Relationships of *I* and *V*

The sinusoidal cemf of the inductor in Fig. 11-21 is produced by a sinusoidal current through the inductor. This current wave, shown in Fig. 11-22, is 90° out of phase with the cemf and the source voltage. The current must be 90° out of phase because the cemf can be zero only when the current is not changing. The only instant when the current is not changing is when it is exactly at its peak value. That is, at the instant the current has just stopped rising and has not yet begun to fall, it is effectively a constant value. This is the instant at which zero cemf occurs. It can be seen from Fig. 11-22 (and Fig. 11-20) that the *voltage leads the current* in an inductor circuit. More precisely, the voltage leads the current by exactly 90° in an ideal inductor.

Power in an Inductor

The *ideal inductor* uses no power because its current and voltage are 90° out of phase. (Remember, $P = IV \cos \theta$, and $\cos 90° = 0$.) Thus, in a pure inductance (ideal inductor) both current and voltage are present but there is no net conversion of energy.

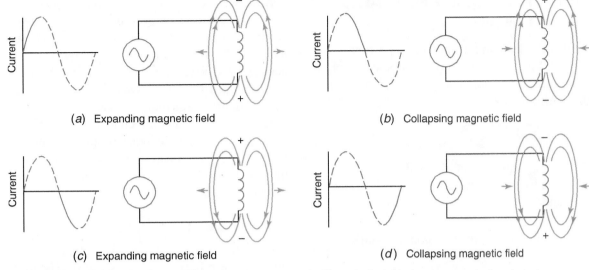

(a) Expanding magnetic field

(b) Collapsing magnetic field

(c) Expanding magnetic field

(d) Collapsing magnetic field

Fig. 11-23 Power in an inductor. Every other quarter-cycle (*b* and *d*) the inductor returns its stored energy to the source.

From an energy point of view, one can say that energy is transferred back and forth between the source and the inductor. During the quarter-cycle in which current is rising [Fig. 11-23(*a*)], energy is taken from the source (generator). The energy from the generator is converted into magnetic energy and stored in the inductor's field. During the next quarter-cycle [Fig. 11-23(*b*)], when the current is decreasing, the field of the inductor collapses. Its stored energy is converted back to electric energy and returned to the generator.

As shown in Fig. 11-23(*c*), energy is again taken from the generator during the third quarter-cycle. This energy is again returned during the fourth quarter-cycle [Fig. 11-23(*d*)]. Notice in Fig. 11-23(*b*) and (*c*) that the polarity of the cemf remains the same. Yet the magnetic field changes from a collapsing field to an expanding field. As previously explained, this is because the polarity of the magnetic field also changes [between Fig. 11-23(*b*) and (*c*)] when the direction of the current reverses.

⎍⎍ Self-Test

Answer the following questions.

35. Does an ideal inductor possess any resistance?
36. What causes inductive reactance?
37. The symbol for inductive reactance is _____.
38. The formula for calculating the reactance of an inductor is _____.
39. Doubling the frequency of an inductive circuit causes the reactance to _____.
40. Determine the reactance of the following inductances at the frequencies specified:
    a. 6 H at 60 Hz
    b. 150 mH at 10 kHz
    c. 30 μH at 250 MHz

41. What is the reactance of an inductor that drops 20 V ac when 0.5 A ac flows through it?
42. Inductance causes current to _____ voltage by _____ degrees.
43. The power consumed by an ideal inductor that draws 2 A from a 20-V ac source is _____.
44. How much inductance is needed to provide 3600 Ω of reactance to a 30-V, 400-Hz source?
45. What is the frequency when a 3-mH inductor produces 4200 Ω of reactance?

11-8 Real Inductors in AC Circuits

Real (nonideal) inductors use some power because all inductors possess resistance as well as reactance. The quality of real inductors is less than infinite.

You May Recall

. . . that quality is defined as reactance divided by resistance and that it is frequency-dependent.

For an inductor, the formula for quality Q is

$$Q = \frac{X_L}{R}$$

$Q = \dfrac{X_L}{R}$

Skin effect

Effective resistance

Impedance (Z)

EXAMPLE 11-6

What is the quality of a 10-mH coil (inductor) at 150 kHz if its resistance is 60 Ω?

Given: $L = 10$ mH
            $f = 150$ kHz
            $R = 60$ Ω

Find: Q

Known: $Q = \dfrac{X_L}{R}$ and $X_L = 6.28fL$

Solution: $X_L = 6.28 \times f \times L$
                $= 6.28 \times 150{,}000$ Hz
                $\times 0.01$ H
                $= 9420$ Ω

$$Q = \frac{9420 \ \Omega}{60 \ \Omega} = 157$$

Answer: The Q of the coil is 157.

The quality of iron-core inductors used at low frequencies is often less than 10. With air-core inductors operating at high frequencies, the quality can be more than 200. Typical RF chokes have a quality ranging from 30 to 150. The higher the quality of the coil, the less power the inductor uses. Also, the higher the quality, the farther the current and voltage are out of phase. Current and voltage are 90° out of phase only when there is no resistance.

The combined opposition offered by resistance and reactance is called *impedance*. Since an inductor has both resistance and reactance, it offers impedance to an ac current. To be technically correct, we should specify the impedance of an inductor. However, the dominant form of opposition of an inductor is reactance. Therefore, we usually talk about its opposition in terms of reactance only. When the quality of the inductor is above 5, the difference between its reactance and its impedance is less than 2 percent. Thus, the use of reactance instead of impedance when calculating the current in an inductor is quite reasonable. It is especially reasonable when you consider that the inductance may be 20 percent above or below its rated value.

Power Losses in Inductors

It has already been emphasized that real inductors use power because of their resistance. However, their resistance may actually be greater than the resistance measured by an ohmmeter. This higher-than-measured resistance is the result of the *skin effect*. The skin effect is caused by the tendency of electrons to travel close to the outer surface of a conductor (Fig. 11-24). The higher the frequency, the more pronounced the skin effect becomes. Because of the skin effect, the center of a conductor does not contribute to the current-carrying capacity of the conductor. Thus, the *effective resistance* of the conductor at high frequencies is greater than that measured by an ohmmeter.

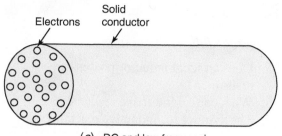

(a) DC and low frequencies

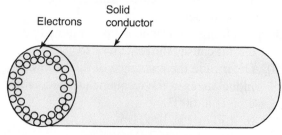

(b) Higher frequencies

Fig. 11-24 Skin effect. At high frequencies, the current concentrates near the surface of the conductor.

The skin effect can be minimized by using *litz wire,* which is a multiple-conductor cable. Each conductor in litz wire has a very thin insulation on it. The conductors are very small in diameter (about 44 gage). These small, insulated conductors are twisted together to form a very small cable. When litz wire is used, the individual conductors are soldered together at the ends of the coil. This connects all the conductors in parallel to effectively make a single wire out of the multiple-conductor cable. For a given overall diameter, litz wire provides more surface area than a single-strand conductor. Because of this greater surface area, litz wire has a lower resistance at high frequencies.

Iron-core inductors have power losses in their core material as well as in their winding. Two actions cause the core to convert electric energy into heat energy. First, the magnetic field of the winding induces a voltage in the core material. The induced voltage causes a small current to flow in the core. This current produces heat in the core. The second action that causes power loss in the core is the periodic reversal of the magnetic field. Every time the polarity of the magnetic field reverses, it creates a small amount of heat in the core. Methods of minimizing these core losses are discussed in the chapter on transformers.

Litz wire

 Self-Test

Answer the following questions.

46. Does an iron-core inductor or an air-core inductor have a higher quality rating?
47. What is the quality of a 0.3-H inductor at 20 kHz if it has an effective resistance of 100 Ω?
48. What is impedance?

49. What causes the resistance of a conductor to be greater at high frequencies than at low frequencies?
50. What is litz wire?
51. What causes the core of a laminated-iron-core inductor to heat up?

11-9 Inductors in Parallel

Parallel inductors (with no mutual inductance) can be treated just like parallel resistors. The formulas used with resistors can be used with inductors by substituting L for R. The formulas are

General method:

$$L_T = \frac{1}{\dfrac{1}{L_1} + \dfrac{1}{L_2} + \dfrac{1}{L_3} + \text{etc.}}$$

Two inductors in parallel:

$$L_T = \frac{L_1 \times L_2}{L_1 + L_2}$$

n equal inductors in parallel:

$$L_T = \frac{L}{n}$$

EXAMPLE 11-7

What is the inductance of a 0.4-H inductor and a 600-mH inductor connected in parallel?

Given: $L_1 = 0.4$ H
$L_2 = 600$ mH $= 0.6$ H

Find: L_T

Known: $L_T = \dfrac{L_1 \times L_2}{L_1 + L_2}$

Solution: $L_T = \dfrac{0.4 \text{ H} \times 0.6 \text{ H}}{0.4 \text{ H} + 0.6 \text{ H}} = \dfrac{0.24}{1.0}$
$= 0.24$ H

Answer: The total (or equivalent) inductance is 0.24 H, or 240 mH.

Parallel inductors

Notice that the total inductance is less than the smallest of the parallel inductances. It always is with parallel inductors.

Total inductive reactance

The *total inductive reactance* of parallel inductors can be found by either of two methods. The first method is to find the total inductance as in example 11-7 and then find the total reactance by using the reactance formula. The second method is to determine the reactance of the individual inductors and then, using the parallel formula, combine the individual reactances to find the total reactance. The formulas for combining *parallel inductive reactances* have the same structure as those used for parallel resistances and parallel inductances:

Parallel inductive reactances

General method:

$$X_{L_T} = \frac{1}{\dfrac{1}{X_{L_1}} + \dfrac{1}{X_{L_2}} + \dfrac{1}{X_{L_3}} + \text{etc.}}$$

Two parallel inductive reactances:

$$X_{L_T} = \frac{X_{L_1} \times X_{L_2}}{X_{L_1} + X_{L_2}}$$

n equal inductive reactances in parallel:

$$X_{L_T} = \frac{X_L}{n}$$

EXAMPLE 11-8

Using the first method, find the total inductive reactance for example 11-7 when the frequency is 20 kHz.

Given:	$L_T = 0.24$ H
	$f = 20$ kHz
Find:	X_{L_T}
Known:	$X_{L_T} = 6.28 f L_T$
Solution:	$X_{L_T} = 6.28 f L_T$
	$= 6.28 \times 20,000$ Hz
	$\times 0.24$ H
	$= 30,144 \ \Omega$
Answer:	The total inductive reactance is 30,144 Ω, or 30.144 kΩ.

EXAMPLE 11-9

Using the second method, find the total inductive reactance for example 11-7 when the frequency is 20 kHz.

Given:	$L_1 = 0.4$ H
	$L_2 = 0.6$ H
	$f = 20$ kHz
Find:	X_{L_T}
Known:	$X_L = 6.28 f L$
	$X_{L_T} = \dfrac{X_{L_1} \times X_{L_2}}{X_{L_1} + X_{L_2}}$
Solution:	$X_{L_1} = 6.28 \times 20,000 \times 0.4$
	$= 50,240 \ \Omega$
	$X_{L_2} = 6.28 \times 20,000 \times 0.6$
	$= 75,360 \ \Omega$
	$X_{L_T} = \dfrac{50,240 \times 75,360}{50,240 + 75,360}$
	$= 30,144 \ \Omega$
Answer:	The total inductive reactance is 30,144 Ω.

As you might expect, the two methods used in examples 11-8 and 11-9 give exactly the same answer.

In parallel-inductor circuits the total current splits up in inverse proportion to the inductance of the individual inductors. The lowest inductance carries the highest current (Fig. 11-25). The exact value of the current in each branch of the circuit can be found by using Ohm's law. Just replace the R in Ohm's law with X_L. The currents recorded in Fig. 11-25 were calculated by using the reactances in example 11-9:

$$I_{L_1} = \frac{V_{L_1}}{X_{L_1}} = \frac{40 \text{ V}}{50,240 \ \Omega} = 0.0008 \text{ A} = 0.8 \text{ mA}$$

$$I_{L_2} = \frac{V_{L_2}}{X_{L_2}} = \frac{40 \text{ V}}{75,360 \ \Omega} = 0.00053 \text{ A}$$
$$= 0.53 \text{ mA}$$

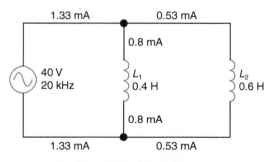

Fig. 11-25 Currents in parallel inductors. The smaller inductor carries the greater current.

$$I_T = \frac{V_T}{X_{L_T}} = \frac{40 \text{ V}}{30{,}144 \ \Omega} = 0.00133 \text{ A}$$
$$= 1.33 \text{ mA}$$

Of course, I_T can also be found by using Kirchhoff's current law. For the circuit of Fig. 11-25, we have

$$I_T = I_{L_1} + I_{L_2}$$
$$= 0.8 \text{ mA} + 0.53 \text{ mA} = 1.33 \text{ mA}$$

The current-divider formula used with parallel resistors and resistances can also be used with parallel inductors and reactances. If I_{L_1} in Fig. 11-25 was not known, it could be calculated with the current-divider formula in either of these two ways:

$$I_{L_1} = \frac{I_T \times X_{L_2}}{X_{L_1} + X_{L_2}}$$
$$= \frac{1.33 \text{ mA} \times 75.36 \text{ k}\Omega}{50.24 \text{ k}\Omega + 75.36 \text{ k}\Omega} = 0.8 \text{ mA}$$

$$I_{L_1} = \frac{I_T \times L_2}{L_1 + L_2} = \frac{1.33 \text{ mA} \times 0.6 \text{ H}}{0.4 \text{ H} + 0.6 \text{ H}} = 0.8 \text{ mA}$$

11-10 Inductors in Series

Treat *series inductances* and *series reactances* the same way you treat series resistances. The formula for series inductances is

$$L_T = L_1 + L_2 + L_3 + \text{etc.}$$

For inductive reactance in series the formula is

$$X_{L_T} = X_{L_1} + X_{L_2} + X_{L_3} + \text{etc.}$$

Again, the total reactance can also be determined by the reactance formula if the total inductance is known. That is,

$$X_{L_T} = 6.28f\,L_T$$

EXAMPLE 11-10

Using the reactance formula, find the total reactance at 60 Hz of a 3-H choke and a 5-H choke connected in series.

Given:	$L_1 = 3 \text{ H}$
	$L_2 = 5 \text{ H}$
	$f = 60 \text{ Hz}$
Find:	X_{L_T}
Known:	$X_{L_T} = 6.28f\,L_T$
	$L_T = L_1 + L_2$

Solution: $\quad L_T = 3 \text{ H} + 5 \text{ H} = 8 \text{ H}$
$X_{L_T} = 6.28 \times 60 \text{ Hz} \times 8 \text{ H}$
$= 3014 \ \Omega$

Answer: The total inductive reactance is 3014 Ω.

EXAMPLE 11-11

Find the total reactance for example 11-10 without first finding the total inductance.

Given:	$L_1 = 3 \text{ H}$
	$L_2 = 5 \text{ H}$
	$f = 60 \text{ Hz}$
Find:	X_{L_T}
Known:	$X_{L_T} = X_{L_1} + X_{L_2}$
	$X_{L_1} = 6.28\,fL_1$
	$X_{L_2} = 6.28\,fL_2$
Solution:	$X_{L_1} = 6.28 \times 60 \text{ Hz} \times 3 \text{ H}$
	$= 1130 \ \Omega$
	$X_{L_2} = 6.28 \times 60 \text{ Hz} \times 5 \text{ H}$
	$= 1884 \ \Omega$
	$X_{L_T} = 1130 \ \Omega + 1884 \ \Omega$
	$= 3014 \ \Omega$

Answer: The total reactance is 3014 Ω.

Series inductances

Series reactances

As shown in Fig. 11-26, the total voltage in a series inductor circuit splits in direct proportion to the individual inductances:

$$V_{L_1} = \frac{L_1}{L_T} \times V_T$$
$$= \frac{3 \text{ H}}{8 \text{ H}} \times 20 \text{ V} = 7.5 \text{ V}$$

$$V_{L_2} = \frac{5 \text{ H}}{8 \text{ H}} \times 20 \text{ V} = 12.5 \text{ V}$$

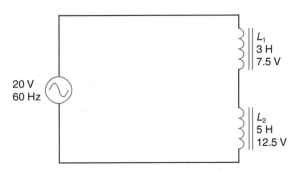

Fig. 11-26 Voltages in series inductors. The larger inductor drops the greater voltage.

The exact voltage values can also be found by using Ohm's law and the reactances of example 11-11. Again, Kirchhoff's laws apply. The total voltage equals the sum of the individual circuit voltages.

11-11 Time Constants for Inductors

You May Recall

. . . that in the previous chapter we looked at time constants for resistor-capacitor (R-C) combinations.

$$T = \frac{L}{R}$$

The concepts developed there can easily be extended to cover resistor-inductor (R-L) combinations. The only modification needed is to think in turns of current rather than voltage.

The time constant for an R-L circuit is defined as the time required for the current through the resistor-inductor to rise to 63.2 percent of its final value. For an increasing current, such as that shown in Fig. 11-27(a),

the final value is the value determined by the resistance in the circuit and the voltage applied to the circuit.

For a circuit in which the current is decreasing, the time constant is defined as the time required for the inductor's current to be reduced by 63.2 percent of its starting value. As shown in Fig. 11-27(b), it is also the time it takes for the current to decay to 36.8 percent of its former value.

The time constant of an R-L circuit can be calculated using the formula

$$T = \frac{L}{R}$$

The time constant is in seconds when L is in henrys and R is in ohms. Why the time is in seconds is shown by substituting equivalent base units into the formula, remembering that

$$\text{Henrys} = \frac{\text{volts}}{\text{amperes/seconds}}$$

and

$$\text{Ohms} = \frac{\text{volts}}{\text{amperes}}$$

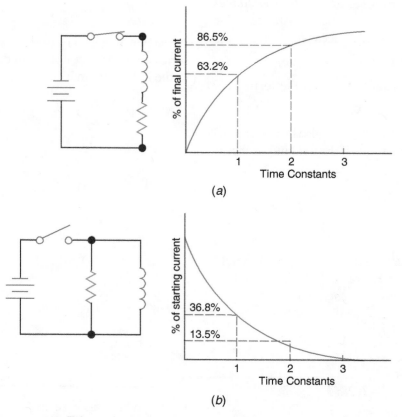

(a)

(b)

Fig. 11-27 *R-L* time constants.

Thus,

$$\text{Time constant} = \frac{\text{inductance}}{\text{resistance}} = \frac{\text{henrys}}{\text{ohms}}$$

$$= \frac{\dfrac{\text{volts}}{\text{amperes/seconds}}}{\dfrac{\text{volts}}{\text{amperes}}}$$

$$= \frac{\dfrac{\text{volts} \times \text{seconds}}{\text{amperes}}}{\dfrac{\text{volts}}{\text{amperes}}} = \text{seconds}$$

EXAMPLE 11-12

Assume that the inductor in Fig. 11-27 is an ideal inductor rated at 10 H. If the resistor is 50 Ω, what is the time constant of the circuit in Fig. 11-27?

Given:	$L = 10\text{ H}$
	$R = 50\ \Omega$
Find:	T
Known:	$T = \dfrac{L}{R}$
Solution:	$T = \dfrac{L}{R} = \dfrac{10\text{ H}}{50\ \Omega} = 0.2\text{ s}$
Answer:	The time constant is 0.2 s.

It should be noted that T for practical values of R and L is usually in fractions of a second. Even if the resistor is removed from the circuit, the ohmic (dc) resistance of multihenry inductors keeps the time constant small.

11-12 Preventing Mutual Inductance

You May Recall

... that mutual inductance occurs when magnetic flux from a component induces a voltage in an electrically isolated component.

Mutual inductance can be reduced or prevented by the following methods:

1. Axis orientation
2. Physical separation
3. Shielding

Suppose the center axes of two coils are at 90° to each other, as shown in Fig. 11-28(*a*). Under

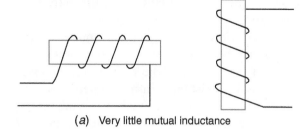

(*a*) Very little mutual inductance

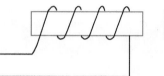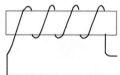

(*b*) High mutual inductance

Fig. 11-28 Axis orientation and mutual inductance.

this condition, very little of the flux from one coil cuts the other coil. We say that very little *coupling* occurs between the coils (inductors). When the axes of the coils are lined up and close together, as in Fig. 11-28(*b*), mutual inductance results.

Coupling

When inductors are physically separated, mutual inductance is reduced. The farther apart the inductors are, the less mutual inductance (coupling of flux) they have.

An inductor that is enclosed in a magnetic shield has very little mutual inductance with surrounding inductors. The flux from surrounding inductors passes through the low reluctance (high permeability) of the shield rather than through the inductor.

11-13 Undesired Inductance

As mentioned earlier, all conductors possess inductance. The inductance of a single wire, although low, is significant at very high frequencies. Often this inductance is undesirable because of its effect on the electric or electronic circuit. In high-frequency circuits, interconnecting leads are kept as short as possible to reduce inductance. Whenever possible, the inductance of the leads is used as part of the required inductance of the circuit.

As mentioned in Sec. 11-4, wire-wound and deposited-film resistors may have an appreciable amount of inductance. Their resistive elements are coils of conductive material wound on a nonmagnetic insulator form. In dc and low-frequency ac circuits, this undesired inductance has very little reactance and can be ignored. However, at higher frequencies the reactance becomes greater and the total opposition (impedance) of the resistor significantly exceeds its resistance. In many electronic circuits this is undesirable and unacceptable.

Noninductive wire-wound resistors

To minimize the above problem, special *noninductive wire-wound resistors* are produced. In these resistors, half the turns of wire are wound clockwise and half are wound counterclockwise. Thus the magnetic field of half of the turns cancels the field of the other half of the turns. These resistors are often used in high-power circuits when the load on the circuit must be independent of the frequency.

∿ Self-Test

Answer the following questions.

52. A 0.3-H inductor and a 0.6-H inductor are connected in parallel. They are connected to a 15-V, 150-Hz source.
    a. What is their total, or equivalent, inductance?
    b. What is the total reactance?
    c. What is the total current?
    d. What is the current through the 0.6-H inductor?
53. Suppose the inductors in question 52 are now connected in series rather than parallel.
    a. What is the total inductance?
    b. What is the total reactance?
    c. What is the total current?
    d. What is the voltage across the 0.6-H inductor?

54. How can you minimize the mutual inductance between two coils?
55. How is a wire-wound noninductive resistor constructed?
56. Determine the time constant of a 30-H ideal inductor connected in series with a 60-Ω resistor and a 10-V dc source.
57. A 4-mH inductor is connected in series with a 6-mH inductor to a 30-V source. Determine the voltage across the 4-mH inductor.
58. Assume the inductors in question 57 are in parallel instead of series. Determine the current through the 6-mH inductor if I_T is 10 mA.

Summary

1. Inductance opposes changes in current.
2. Inductance results from induced voltage.
3. Inductors are devices that provide inductance. Chokes, coils, and reactors are other names for inductors.
4. The symbol for inductance is L.
5. The induced voltage in an inductor is known as counter electromotive force (cemf) or back electromotive force (bemf).
6. Lenz's law is concerned with the polarity of an induced voltage (cemf).
7. A cemf opposes the change that created it.
8. Inductors convert energy back and forth between the magnetic form and the electrical form.
9. The henry is the base unit of inductance. The abbreviation for henry is H.
10. Inductance is determined by (1) core material, (2) number of turns, (3) spacing of turns, and (4) diameter of turns.
11. Inductors are rated for inductance, dc resistance, current, voltage, quality, and tolerance.
12. The dc resistance of an inductor is also called ohmic resistance.
13. The quality Q of inductors ranges from less than 10 to more than 200.
14. Current in a dc inductive circuit rises more slowly than in a dc resistive circuit.
15. In a dc inductive circuit, voltage (cemf) reaches its peak value before the current does.
16. Inductive kick causes arcing in switch contacts when an inductive circuit is opened.
17. Inductive reactance is the opposition of an inductor to alternating current.
18. The symbol for inductive reactance is X_L.
19. Inductive reactance is directly proportional to both frequency and inductance.
20. Ohm's law can be used in inductive circuits by replacing R with X_L.
21. In an inductive circuit, current lags voltage by 90°.
22. Ideal inductors use no power or energy.
23. Real inductors have resistance; therefore, they do use some power.
24. Impedance is the combined opposition of reactance and resistance.
25. The skin effect increases the effective resistance of a conductor at high frequencies.
26. Litz wire is multistrand wire designed to reduce the skin effect.
27. Core losses are caused by induced currents in the core and by periodic reversal of the magnetic field.
28. Inductors (and inductive reactances) in parallel behave like resistors in parallel. The same formulas are used except that R is replaced by L or X_L.
29. Inductors (and inductive reactances) in series behave like resistors in series.
30. The lowest series inductance drops the least voltage.
31. Mutual inductance can be reduced by axis orientation, physical separation, and shielding.
32. Undesired inductance occurs in conductors and resistors.

Related Formulas

The quality Q for an inductor:

$$Q = \frac{X_L}{R}$$

Time constant:

$$T = \frac{L}{R}$$

Inductance:

$$L = \frac{V_{induced}}{\Delta I / \Delta t}$$

Inductive reactance:

$$X_L = 6.28 \, f L$$

Related Formulas...continued

For series inductors:

$L_T = L_1 + L_2 + \text{etc.}$

$X_{L_T} = X_{L_1} + X_{L_2} + \text{etc.}$

For parallel inductors:

$$L_T = \frac{1}{\frac{1}{L_1} + \frac{1}{L_2} + \text{etc.}}$$

$$X_{L_T} = \frac{1}{\frac{1}{X_{L_1}} + \frac{1}{X_{L_2}} + \text{etc.}}$$

Chapter Review Questions

For questions 11-1 to 11-12, determine whether each statement is true or false.

11-1. A straight length of conductor has no inductance. (11-1)

11-2. A 2-H inductor would most likely have a laminated iron core. (11-4)

11-3. Magnetic shields for inductors are usually made from high-reluctance materials. (11-12)

11-4. The reactance of an inductor can be measured with an ohmmeter. (11-7)

11-5. The core material in a variable inductor is often called a toroid. (11-4)

11-6. Maximum inductance in a variable inductor occurs when the brass slug is centered within the coil core. (11-4)

11-7. The cemf exceeds the source voltage when an inductive circuit is opened. (11-6)

11-8. The iron core of an inductor converts some electric energy into heat energy. (11-8)

11-9. The quality of a coil is frequency-dependent. (11-8)

11-10. A small induced current flows in the core of an iron-core inductor. (11-8)

11-11. The lowest-value inductor drops the most voltage in a series-inductor circuit. (11-10)

11-12. The lowest-value inductor draws the most current in a parallel-inductor circuit. (11-9)

For questions 11-13 to 11-30, supply the missing word or phrase in each statement.

11-13. _____ wire is used to reduce the skin effect. (11-8)

11-14. The base unit of inductance is the _____. (11-12)

11-15. The abbreviation for the base unit of inductance is _____. (11-2)

11-16. The base unit for inductive reactance is the _____. (11-7)

11-17. The symbol for inductive reactance is _____. (11-7)

11-18. When current in an inductor is increasing, the cemf _____ the source voltage. (11-1)

11-19. Another name for cemf is _____. (11-1)

11-20. Inductive reactance is _____ proportional to frequency. (11-7)

11-21. When one coil induces a voltage in another coil, the process is called _____. (11-1)

11-22. Voltage _____ current by _____ degrees in an ideal inductor. (11-7)

11-23. Inductors are also known as _____, _____, and _____. (11-1)

11-24. The center leg of an iron-core carries _____ as much flux as an outside leg does. (11-4)

11-25. The polarity of the cemf can be determined by applying _____ law. (11-1)

11-26. The electric quantity that opposes change in current is _____. (11-1)

11-27. An inductor converts _____ energy to _____ energy while the current is increasing. (11-1)

11-28. The resistance of the turns of wire in an inductor is called _____ or _____ resistance. (11-5)

11-29. Arcing between the switch contacts when an inductive circuit is turned off is caused by _____. (11-6)

11-30. The combined opposition of reactance and resistance is called _____. (11-8)

Answer the following questions:

11-31. What are three techniques used to minimize or eliminate mutual inductance? (11-12)

11-32. What four factors determine the inductance of an inductor? How can inductance be increased using each of these factors? (11-3)

11-33. What electrical ratings are used to completely specify an iron-core inductor? (11-5)

11-34. What happens to Q when the resistance of an inductor increases? (11-5)

Chapter Review Problems

11-1. What is the time constant of a 500-mH ideal inductor connected in series with a 10-Ω resistor? (11-11)

11-2. What are the reactance and current in a circuit that consists of a 300-mH inductor connected to a 20-V, 7.5-kHz source? (11-7)

11-3. Determine the quality of a 70-mH inductor that has a resistance of 125-Ω at 35 kHz. (11-8)

11-4. Determine the total inductance and the circuit current of a 4-H inductor and a 6-H inductor series connected to a 400-Hz, 80-V supply. Also determine the voltage across the 4-H inductor. (11-10)

11-5. What is the inductance of a 5-mH inductor and a 7-mH inductor connected in parallel? (11-9)

11-6. How much inductance is needed to limit the current from a 50-V, 200-Hz source to 25 mA? (11-7)

11-7. Two inductors are connected in series to a 35-V source. Determine V_{L_1} if $L_1 = 0.36$ H and $L_2 = 0.54$ H. (11-9)

11-8. Determine the value of L needed to produce a time constant of 0.003 s when $R = 20$ Ω and $V_T = 50$ V. (11-11)

11-9. An inductor produces 4 V of cemf when its current changes at a rate of 1.2 amps in 0.5 s. What is its inductance? (11-2)

11-10. The inductor in problem 11-9 above is connected to a 24-V, 400-Hz source. Determine its reactance and current. (11-7)

11-11. How much voltage is required from an 300-Hz source to force 0.06 A through a 0.8-H inductor? (11-7)

Critical Thinking Questions

11-1. Two variable inductors are identical except that one has a brass slug, and the other a ferrite slug. Which inductor would have the larger range of inductance?

11-2. Why aren't inductor shields made from nonferrous materials such as tin or plastic?

11-3. At a given current, would a swinging choke or a smoothing choke have the larger inductance if the choke had identical coils and the same size and shape of core? Why?

11-4. Why is the total inductance of two series inductors greater than the inductance of the larger inductor?

11-5. Why are parallel inductive reactances treated like parallel resistances when figuring the total reactance?

11-6. Why does periodic reversal of the magnetic field cause a magnetic core to produce heat?

11-7. What is the frequency of the source voltage when a 0.13-A current drops 20 V across a 0.04-H inductor?

11-8. Why can we use either the values of X_L or the values of L in the current-divider formula for parallel inductor circuits?

11-9. A 200-mH (L_1) and a 0.15H (L_2) inductor are connected in parallel and draw 46 mA from the power source. Determine I_{L_2}.

11-10. Would you expect the current and voltage to be in phase in a circuit containing impedance? Why?

11-11. An inductor produces 0.5 V of cemf when the current is increased at a uniform rate from 20 mA to 60 mA in 2 ms. Determine the inductance of the inductor. Show your calculations.

1. inductance
2. choke, coil, reactor, inductor
3. T
4. F
5. F
6. F
7. F
8. L
9. Lenz's
10. bemf
11. electric, magnetic
12. henry
13. H
14. volt per ampere per second
15. increase
16. Use a core with higher permeability or less reluctance, increase the number of turns, put the turns closer together, increase the diameter of the turns.
17. a. See Fig. 11-6(*a*).
    b. See Fig. 11-6(*b*).
    c. See Fig. 11-7.
18. F
19. F
20. T
21. T
22. F
23. T
24. laminated-iron core
25. doughnut-shaped
26. Because it carries twice as much flux.
27. Because too much current will cause the winding to overheat.
28. the ratio of reactance to resistance
29. an inductor or coil designed to be used at radio frequencies
30. laminated-iron
31. the high cemf that occurs when an inductive circuit is opened
32. inductive
33. yes
34. yes
35. no
36. the cemf of the inductor
37. X_L
38. $X_L = 6.28fL$
39. double
40. a. 2262 Ω
    b. 9425 Ω
    c. 47124 Ω
41. 40 Ω
42. lag, 90°
43. zero
44. 1.43 H
45. 223 kHz
46. air-core inductor
47. 377
48. Impedance is the combined opposition of resistance and reactance.
49. skin effect
50. Litz wire is multistrand wire used to reduce the skin effect.
51. induced currents and magnetic polarity reversal
52. a. 0.2 H
    b. 188.5 Ω
    c. 79.6 mA
    d. 26.5 mA
53. a. 0.9 H
    b. 848 Ω
    c. 17.7 mA
    d. 10 V
54. Orient the axes 90° to each other, separate the coils, and shield them.
55. Half the turns are wound in one direction and the other half are wound in the opposite direction.
56. 0.5 s
57. 12 V
58. 4 mA

CHAPTER 12

Transformers

Learning Outcomes

This chapter will help you to:

12-1 *Discuss* how mutual inductance between primary and secondary windings (a) provides electrical isolation between primary and secondary windings and (b) allows the secondary voltage to be equal to, greater than, or less than, the primary voltage.

12-2 *Identify and discuss* the core and winding losses that make a transformer heat up during operation.

12-3 *Explain* what happens to the primary current and angle theta when a resistance load is connected to the secondary winding of a transformer.

12-4 *Discuss* how leakage flux is minimized in the I and E laminated core of a power transformer.

12-5 *Explain* the purpose of an isolation transformer. Also, what is the primary-to-secondary voltage ratio of an isolation transformer.

12-6 *Clarify* how a transformer can provide impedance matching between a high-impedance source and a low impedance load.

12-7 *List* the two rules that insure a transformer will not be over loaded.

12-8 *Discuss* the requirements for, and the advantages of, connecting windings in parallel or series.

12-9 *Explain* several uses of off-center-tapped windings.

12-10 *Explain* how a single laminated-iron core can accommodate the flux from all three phases of a three-phase system.

Transformers are multiple-winding inductors. They operate on the principle of mutual inductance. For a relatively simple device, they are extremely versatile. Without transformers, our present power distribution system could not exist.

12-1 Transformer Fundamentals

A transformer consists of two or more coils linked together by magnetic flux (Fig. 12-1). The changing flux from one coil (the *primary*) induces a voltage in the other coil (the *secondary*). In other words, the coils are coupled, or linked, together by mutual inductance.

Without *mutual inductance,* there would be no such thing as a transformer. The amount of mutual inductance, like the amount of self-inductance, is specified in henrys. There is 1 H of mutual induction when 1 V is induced in a coil by a current change of 1 A/s in another coil. Suppose the primary current in Fig. 12-1 changes at a rate of 1 A/s. Further, suppose the secondary voltage in Fig. 12-1 is 3 V. Then the transformer has 3 H of mutual inductance.

Symbols

The *symbol* for a *transformer* is basically two coils with their axes parallel to each other. As illustrated in Fig. 12-2, the basic symbol can be modified in many ways. These modifications are needed to more fully describe the various types of transformers.

The dashed lines in Fig. 12-2(*c*) represent the metal enclosure that houses the windings (coils). When this enclosure is made of aluminum or copper or other nonmagnetic materials, it is intended as a shield against electric fields rather than magnetic fields.

Primary

Secondary

Mutual inductance

Transformer symbols

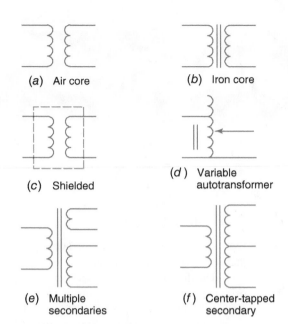

(a) Air core

(b) Iron core

(c) Shielded

(d) Variable autotransformer

(e) Multiple secondaries

(f) Center-tapped secondary

Fig. 12-2 Transformer symbols.

The symbols shown in Fig. 12-2 are not the only ones you may see for transformers. For example, iron-core transformers sometimes have shields. Thus, to indicate an iron-core transformer, the symbol of Fig. 12-2(c) would show lines indicating the iron core. Not all tapped secondaries are center-tapped. An off-center-tapped winding is represented by having more turns indicated on one side of the tap than on the other.

Primary and Secondary Windings

A transformer is a device that transfers electric energy (or power) from a primary winding to a secondary winding. Except for the autotransformer, there is no electric connection between the primary winding and the secondary winding. The primary converts the electric energy into magnetic energy. The secondary converts the magnetic energy back into electric energy. The primary and the secondary winding are said to be electrically isolated from each other but magnetically connected or coupled to one another.

A transformer receives power from a power source, such as the 23-kV output of a power company generator or the 120-V outlet in your home. The transformer winding designed to receive power from the source is called the primary winding.

A load (Fig. 12-3) takes power from the secondary winding of a transformer. Therefore, the secondary winding becomes the power source for the load.

The secondary voltage need not be the same as the primary voltage. For example, the primary of the transformer in Fig. 12-3 is connected to a 120-V supply, and the secondary delivers 20 V to the load.

Transformers are two-way devices. The winding that the manufacturer calls the secondary can be used as a primary. Of course, a primary power source of the correct voltage must be used. The winding designated as the primary then serves as a secondary. Of course, it provides power at the voltage for which it was originally designed. For example, the transformer in Fig. 12-3 can be used to provide a 120-V secondary source from a 20-V primary source, as shown in Fig. 12-4. As you can see, labeling one coil "primary" and another "secondary" is somewhat arbitrary. It is based on the intended use of the transformer, that is, the intended primary power source.

Coefficient of Coupling

The portion of the flux that links one coil to the other coil is referred to as the *coefficient of coupling* (*k*). The coefficient of coupling (*k*) can range from 0 to 1. When all the flux is coupled, the coefficient of coupling is 1. Sometimes

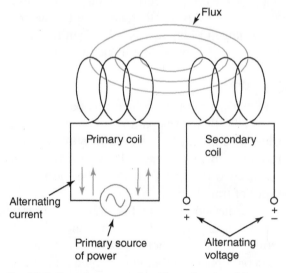

Flux

Primary coil

Secondary coil

Alternating current

Primary source of power

Alternating voltage

Coefficient of coupling

Fig. 12-1 Transformer action. The primary coil is linked to the secondary coil by magnetic flux.

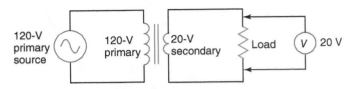

Fig. 12-3 Primary-secondary terminology. The primary winding receives power from a power source.

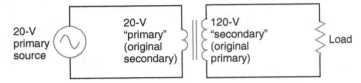

Fig. 12-4 Reversing the primary-secondary function.

the coefficient of coupling is expressed as a percentage. Thus, 100 percent coupling means a coefficient of coupling of 1.

> **You May Recall**
>
> . . . that the previous chapter discussed ways of preventing mutual inductance. When no mutual inductance exists, the coefficient of coupling is 0.

Coupling in transformers with laminated-iron cores is very close to 100 percent. This is because all the flux is concentrated in the high-permeability core on which the coils are wound. The iron core provides a complete low-reluctance path for the flux loops. Therefore, essentially none of the flux leaks into the surrounding air.

On the other hand, air-core transformers can have very low coefficients of coupling. The paths through the air surrounding the inductors offer no more reluctance than the path through the cores. Therefore, much of the flux from one coil never links up with the other coil. This flux is called *leakage flux*. It just "leaks" off into surrounding air paths. The amount of coupling in an air-core transformer can be controlled by the spacing between the coils. The farther apart the coils are, the more flux leakage occurs and the lower the percentage of coupling. The percentage of coupling can also be controlled by the axis orientation of the coils. When the axes of the coils are perpendicular to each other, the coefficient of coupling is close to zero.

The coefficient of coupling (k) can be calculated using the formula

$$k = \frac{\phi_{L_2}}{\phi_{L_1}}$$

where ϕ_{L_1} is the flux of the coil producing the flux and ϕ_{L_2} is the flux which cuts the turns of the receiving coil.

EXAMPLE 12-1

Determine k when coil L_1 is producing 6 mWb of flux and coil L_2 is being cut by 4 mWb of flux.

Given: $\phi_{L_1} = 6$ mWb,
$\phi_{L_2} = 4$ mWb

Find: k

Known: $k = \dfrac{\phi_{L_2}}{\phi_{L_1}}$

Solution: $k = \dfrac{4 \text{ mWb}}{6 \text{ mWb}} = 0.667$

Answer: Coefficient of coupling (k) is 0.667 or 66.7 percent.

Notice that k has no unit because mWb in the numerator is canceled by mWb in the denominator.

The amount of mutual inductance (L_M) between two coils is controlled by both the value of k and the inductance of each of the coils. L_M can be calculated with the formula

$$L_M = k\sqrt{L_1 L_2}$$

Coils L_1 and L_2 could be the primary and secondary coils of a transformer, or they could be any other two coils.

Leakage flux

Since the henry (H) is the unit for the inductances of L_1 and L_2, it will also be the correct unit for L_M.

EXAMPLE 12-2

Two coils, one is 0.1 H and the other is 0.3 H, are coupled together. Determine L_M when the coefficient of coupling is 0.8 or 80 percent.

Given: $L_1 = 0.1$ H

$L_2 = 0.3$ H
$k = 0.8$

Find: L_M

Known: $L_M = k\sqrt{L_1 L_2}$

Solution: $L_M = 0.8\sqrt{0.1\ H \times 0.3\ H} = 0.8$
$\sqrt{0.03\ H^2} = 0.8 \times 0.173$ H
$= 0.138$ H $= 138$ mH

Answer: The mutual inductance is 138 mH.

 Self-Test

Answer the following questions.

1. What is a transformer?
2. The two coils of a transformer are called the _____ and the _____.
3. What is the mutual inductance if a current change of 2 A/s in one coil induces 4 V in a second coil?
4. Draw the symbol for a shielded magnetic-core transformer.
5. What energy-conversion processes are involved in a transformer?
6. The _____ winding of a transformer receives power or energy from another source.
7. The _____ winding of a transformer provides power to the load.
8. Can the winding that the manufacturer of a transformer calls the secondary be used as a primary?
9. Define "coefficient of coupling."
10. Can the coefficient of coupling ever exceed 1?
11. Which have the lower coefficient of coupling, laminated-iron-core transformers or air-core transformers?
12. How can the coefficient of coupling in air-core transformers be varied?

Voltage ratio

Step up

Step down

Turns ratio

Turns per volt

Changing Voltage Values

A transformer can either *step up* or *step down* a voltage. If the primary voltage is greater than the secondary voltage, the transformer is stepping the voltage down. Thus, the transformer in Fig. 12-3 is a step-down transformer. If the voltage in the secondary exceeds the voltage in the primary (Fig. 12-4), the transformer is a step-up transformer. Some transformers with multiple secondaries have one or more step-up secondaries and one or more step-down secondaries.

Whether a secondary is a step-up or step-down winding is determined by the primary-to-secondary *turns ratio*. When the primary turns exceed the secondary turns *and the coupling is 100 percent,* the transformer steps down the voltage. In fact, with 100 percent coupling, the *turns ratio* and the *voltage ratio* are equal. Mathematically we can write

$$\frac{V_{pri}}{V_{sec}} = \frac{N_{pri}}{N_{sec}}$$

In this formula, N is the abbreviation for the number of turns. This formula can be rearranged to show that

$$\frac{N_{pri}}{V_{pri}} = \frac{N_{sec}}{V_{sec}}$$

In this new arrangement, the relationship between voltage and turns is very informative. It shows that the *turns-per-volt* ratio is the same for both the primary and the secondary.

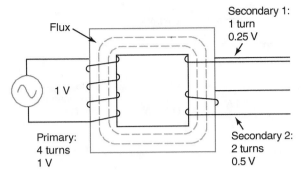

Fig. 12-5 Turns per volt. The turns-per-volt ratio is the same in all windings.

Furthermore, the turns-per-volt ratios of all secondary windings in a multiple-secondary transformer are equal. Thus, once you determine the turns-per-volt ratio of any winding, you know the ratios of all other windings.

Referring to Fig. 12-5 will help you understand why all windings have the same turns-per-volt ratio. Remember that the cemf of a coil is always equal to the ac voltage applied to the coil. Therefore, the cemf in the primary in Fig. 12-5 must be 1 V. Thus 1 V of cemf is created by the flux in the circuit. The flux, therefore, must induce 0.25 V in each turn of the primary. Notice that all the flux that produced cemf in the primary turns also goes through every secondary. Therefore, each turn in the secondary windings has 0.25 V induced in it. If it requires four turns in the primary for each volt, then four turns in a secondary will also develop 1 V.

The turns-per-volt ratio used for the windings of a laminated-iron-core transformer varies with the size of the transformer. Small transformers (less than 10-W rating) may have seven or eight turns per volt. Larger transformers (more than 500-W) may have less than one turn per volt.

The concept of turns per volt is useful in modifying the secondary voltage of a transformer.

EXAMPLE 12-3

A transformer is to be rewound to provide a voltage of 14 V. The transformer's present secondary is rated at 6.3 V and contains 20 turns. How many turns will be required for the new 14-V winding?

Given: Present secondary winding delivers 6.3 V with 20 turns

Find: Number of turns necessary to provide 14 V

Known: $\dfrac{N_{pri}}{V_{pri}} = \dfrac{N_{sec}}{V_{sec}}$

Solution: $\dfrac{N_{sec}}{V_{sec}} = \dfrac{20}{6.3} = 3.175$ turns per volt

The transformer has a 3.175 turns-per-volt ratio. Since a 14-V secondary is required

N (new secondary)
   $= 3.175$ turns per volt $\times$ 14 V
   $= 44.5$ turns or 45 turns

Answer: A 14-V winding needs 45 turns.

EXAMPLE 12-4

The designer of a transformer has calculated that the 120-V primary will have 2.6 turns per volt. Assuming an ideal, iron-core transformer ($k = 1$), how many turns will be required for a 400-V secondary?

Given: Turns per volt $= 2.6$
   $V_{pri} = 120$ V
   $V_{sec} = 400$ V
   $k = 1$

Find: N_{sec}

Known: $\dfrac{N_{pri}}{V_{pri}} = \dfrac{N_{sec}}{V_{sec}}$

Solution: $\dfrac{N_{pri}}{V_{pri}} = 2.6$ turns per volt

   $N_{sec} = 400$ V $\times$ 2.6 turns per volt
   $= 1040$ turns

Answer: The 400-V secondary will need 1040 turns.

Notice in example 12-4 that we did not need to calculate the number of turns in the primary. Also, we did not need to know the voltage of the primary. All we needed to know was the turns-per-volt ratio, and that $k = 1$.

One of the chief uses of transformers is to change voltages from one value to another. An example will show why it is so necessary to

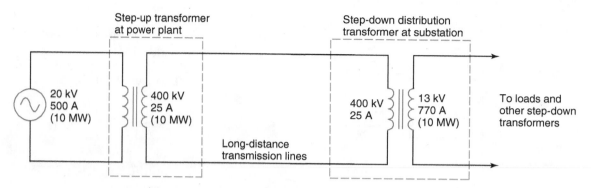

Fig. 12-6 Advantage of high-voltage power lines. The higher the voltage, the lower the current for a given amount of power.

be able to transform voltage levels. It is usually necessary to transmit electric power from the power plant where it is produced to the location where it is used. Often the distance between these two points is many hundreds of miles. Obviously, it is desirable to use as small a conductor as possible in the power lines between the two points. The size of the conductor needed is directly dependent on the amount of current it must carry. So we need to keep the current as low as possible. Assuming a 100 percent power factor, power equals current times voltage ($P = IV$). Thus, the lower the current, the higher the voltage must be for a given amount of power. However, the generators at the power plant and the loads at the point of use have a limited operating voltage. The solution to this problem is illustrated in Fig. 12-6. If the 10 MW from the generator were directly connected to the loads, the transmission lines would have to carry 500 A. With a step-up transformer at the power plant and a step-down transformer at a substation near the load, the power lines carry only 25 A. (This illustration

assumes that the power factor is 1 and that the transformer uses no power.)

Figure 12-6 also illustrates another transformer relationship. When the voltage of a transformer is stepped up, the current is stepped down, and vice versa. The exact current ratio depends upon the power factor, the power consumed by the transformer, and the voltage ratio. For the ideal transformer (one with no losses), the primary power is approximately equal to the secondary power when the transformer has a full resistance load. Stating this as a formula, we have

$$P_{pri} \simeq P_{sec}$$

This leads to the formula for the current relationship:

$$\frac{V_{pri}}{V_{sec}} \simeq \frac{I_{sec}}{I_{pri}}$$

This points out that the current varies inversely to the voltage. In other words, a step-up voltage transformer is, in effect, a step-down current transformer.

![icon] **Self-Test**

Answer the following questions.

13. True or false. With a step-down transformer, the secondary voltage is higher than the primary voltage.
14. True or false. The turns ratio equals the voltage ratio in a transformer with 100 percent coupling.
15. True or false. With 100 percent coupling, the turns-per-volt ratio of the primary

must be less than the turns-per-volt ratio of the secondary.
16. True or false. The turns-per-volt ratio of the primary is usually higher for a 2-kW transformer than for a 50-W transformer.
17. How many turns will be required for a 40-V secondary if the primary of an ideal transformer has five turns per volt and is

designed to operate from a 12-V, 60-Hz source?

18. What is one of the most common uses of a transformer?

19. Why do power systems transmit power at as high a voltage as possible?

20. When a transformer steps down voltage, does it step up or step down in current?

12-2 Efficiency of Transformers

The iron core and the copper coils of a transformer both convert some electric energy into heat energy. This, of course, is why a transformer heats up when in operation. The purpose of a transformer is not to provide heat but to transfer energy from the primary to the secondary. Therefore, any heat produced by the transformer represents inefficiency.

Since energy is equal to power times time, the efficiency of transformers is calculated in terms of power. The *efficiency* of a transformer (expressed as a percentage) is calculated by the following formula:

$$\text{Percent efficiency} = \frac{P_{sec}}{P_{pri}} \times 100$$

EXAMPLE 12-5

What is the efficiency of a transformer that requires 1880 W of primary power to provide 1730 W of secondary power?

Given: $P_{pri} = 1880\text{ W}$
$P_{sec} = 1730\text{ W}$

Find: Efficiency

Known: $\% \text{ eff.} = \frac{P_{sec}}{P_{pri}} \times 100$

Solution: $\% \text{ eff.} = \frac{1730\text{ W}}{1880\text{ W}} \times 100$
$= 92$

Answer: The transformer is 92 percent efficient.

In example 12-5, the 150-W difference between received power and delivered power is lost in the transformer. As indicated in Fig. 12-7, the power consumed by the transformer is referred to as a *power loss*. The power loss in a transformer is caused by

1. Hysteresis loss
2. Eddy current loss
3. Copper (I^2R) loss

The first two of these losses occur in the transformer core material. The last occurs in the windings. All three convert electric energy to heat energy.

Hysteresis Loss

Hysteresis loss is caused by *residual magnetism*, that is, by the magnetism that remains in a material after the magnetizing force has been removed. The core of a transformer has to reverse its magnetic polarity every time the primary current reverses direction. Every time the magnetic polarity is reversed, the residual magnetism of the previous polarity has to be overcome. This produces heat. It requires energy from the primary to produce this heat. Hysteresis loss, then, refers to the energy required to reduce the residual magnetism to zero. This loss occurs once every half-cycle just before the core is remagnetized in the opposite direction.

The *hysteresis loop* in Fig. 12-8 graphically illustrates hysteresis loss. The narrower the hysteresis loop, the lower the hysteresis loss. Therefore, the core material for transformers,

Power loss

Efficiency

Hysteresis loss

Residual magnetism

Hysteresis loop

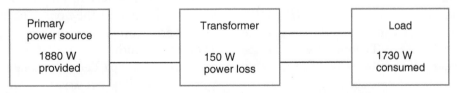

Fig. 12-7 Transformer efficiency. Power loss occurs because the transformer converts some electric energy into heat energy.

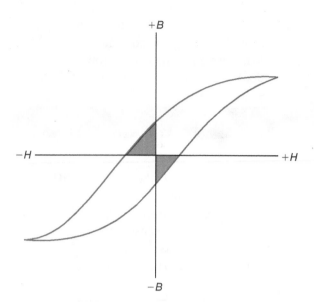

Fig. 12-8 Hysteresis loop. Magnetic energy is converted to heat energy in the shaded regions of the loop.

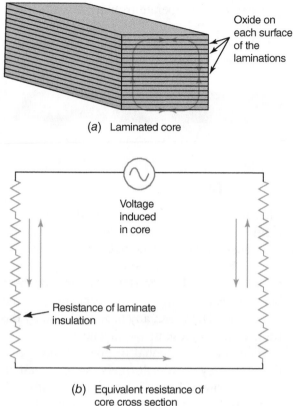

(a) Laminated core

Oxide on each surface of the laminations

Voltage induced in core

Resistance of laminate insulation

(b) Equivalent resistance of core cross section

Fig. 12-9 Reducing eddy current loss.

and other magnetic devices that operate on alternating current, should have a narrow hysteresis loop. Laminated-iron cores are made from silicon steel. Silicon steel is an alloy that has a narrow hysteresis loop and still has high permeability.

Hysteresis loss increases with an increase in the frequency of the primary current. This is one of the reasons that laminated-iron-core transformers are not used above the audio-frequency range.

Eddy Current Loss

The changing magnetic flux in the core of a transformer induces voltage into any conductors which surround it. Since the core is itself a conductor, the changing magnetic flux induces a voltage in the core as well as in the coil conductors. The voltage induced in the core causes current to circulate in the core. This current is called *eddy current*. The eddy current flowing through the resistance of the core produces heat.

The amount of heat due to eddy current is dependent on the values of both the eddy current and the induced voltage ($P = IV$). There is nothing we can do to reduce the value of the induced voltage. Therefore, we must reduce eddy current loss by reducing the value of the eddy current produced by the induced voltage. This can be done by increasing the resistance of the path through which the eddy

current must flow. (Remember, $I = V/R$.) The resistance of the core, in the plane in which eddy current flows, is increased by laminating the core. Each *lamination* of the core is insulated with a thin layer of oxide [Fig. 12-9(a)]. The oxide has a much higher resistance than the rest of the silicon-steel lamination. Notice in Fig. 12-9(a) that the eddy current would have to flow through the oxide layers in order to circulate through the core. The equivalent circuit [Fig. 12-9(b)] for the core shows that the high resistance of the oxide on each lamination effectively reduces the flow of eddy current. Thus laminating the core reduces the eddy current and its associated heat loss.

The thinner the laminations, the more series resistance the core contains and the lower the eddy current will be. However, making the laminations thinner also increases the total amount of oxide in the core. The oxide has a lower permeability than the silicon steel does. Therefore, a core with thin laminations cannot carry as high a flux density as one with thicker laminations. Flux density, of course, controls the amount of magnetic energy a given size of core can handle. This in turn controls the amount

Lamination

Eddy current

of power the transformer can handle. The transformer designer has to consider all these factors, and others, as the core is being designed.

Copper Loss

Copper loss refers to the power dissipated in the windings of a transformer. Since this loss can be calculated by $P = I^2R$, it is called the I^2R loss. The R in the formula is the ohmic, or dc, resistance of the turns in the winding.

Obviously, copper loss is minimized by using as large a conductor as possible in the windings. However, conductor size is limited by the area of the windows (openings) in the core into which the winding must fit.

EXAMPLE 12-6

Determine the power loss in a transformer that provides 480 W of secondary power at an efficiency of 87 percent.

Given: $P_{sec} = 480$ W
$\%$ eff. = 87

Find: P_{loss}

Known: $P_{loss} = P_{pri} - P_{sec}$

$\%$ eff. $= \dfrac{P_{sec}}{P_{pri}} \times 100$

rearranging:

$P_{pri} = \dfrac{P_{sec}}{\% \text{ eff.}} \times 100$

Solution: $P_{pri} = \dfrac{480 \text{ W}}{87} \times 100$

$= 551.7$ W

$P_{loss} = 551.7$ W $- 480$ W

$= 71.7$ W

Answer: The power loss in the transformer is 71.7 watts.

Primary power (W)	Secondary power (W)	Efficiency %
7	3.8	54.3
11	7.3	66.4
18	12.8	71.1
53	45.5	85.8
71	62.0	87.3

Fig. 12-10 Efficiency versus load. The transformer is most efficient when operating at its rated power.

Load and Efficiency

Maximum efficiency is obtained from a transformer when it is fully loaded. For small transformers (less than 10 W), maximum efficiency may be less than 70 percent. With transformers larger than 1000 W, it is more than 95 percent.

As the load is decreased, the efficiency of the transformer also decreases. This is because current flow in a transformer primary does not decrease in direct proportion to decreases in the load. The primary current still causes substantial core losses and copper losses even when the secondary is lightly loaded.

Data showing the changes in efficiency with changes in load are tabulated in Fig. 12-10. These data come from tests conducted on a transformer rated at 75 W.

⎍ Self-Test

Answer the following questions.

21. What causes the inefficiency in a transformer?
22. Determine the efficiency of a transformer that requires 180 W to deliver 150 W.
23. How much power does a transformer require from its source if it is 93 percent efficient and delivers 750 W?
24. List the two causes of power loss in the core of a transformer.

25. I^2R loss occurs in the _____ of a transformer.
26. Hysteresis loss is caused by _____.
27. How often must residual magnetism be overcome in the core of a transformer?
28. What is a hysteresis loop?
29. What are core laminations made of? Why?
30. Is a narrow or a wide hysteresis loop desirable for the core of a transformer?

31. What is the relationship between hysteresis loss and frequency?
32. How are eddy current losses reduced?
33. For minimum eddy current loss, should core laminations be thick or thin?
34. What is I^2R loss? How can it be minimized?
35. Which has higher efficiency, a large transformer or a small transformer?
36. When does a transformer have maximum efficiency?

12-3 Loaded and Unloaded Transformers

It has been shown that a fully loaded transformer has higher efficiency than an unloaded transformer. Many other differences are also associated with the amount of load on a transformer. These will become apparent in the discussion that follows.

An unloaded transformer acts like a simple inductor. It is a highly inductive load on the primary power source to which it is connected. The current in the primary winding is nearly 90° out of phase with the voltage [Fig. 12-11(a)]. Only the copper and core losses keep the current from being a full 90° out of phase with the voltage. The current in the primary of an unloaded transformer is called the *energizing current*. It is the current needed to set up the flux in the core. The energizing current can be rather high and still not use much power. This happens because the current and voltage are so far out of phase. The amount of energizing current is determined primarily by the inductive reactance of the primary winding.

Energizing current

When a load is connected to the secondary of a transformer, both the primary current and the angle θ change. As indicated in Fig. 12-11(b) and (c), the amount of change depends upon how heavily the transformer is loaded. For this discussion, the load on the secondary is assumed to be resistive. Therefore, the transformer is furnishing power to the load.

Figure 12-12 presents experimental data collected by testing a 75-VA transformer. Look at the columns showing values of the secondary power and the angle θ. From them you can see that the transformer appears less inductive, and more resistive, as the load increases.

The last row of numbers in Fig. 12-12 indicates that the transformer is almost purely

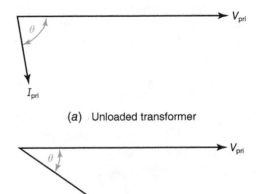

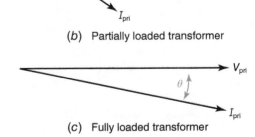

Fig. 12-11 Primary current and voltage phase relationships. When fully loaded, a transformer appears to be close to a resistive load.

resistive at 72 VA. The apparent power and the true power of the primary are nearly equal; the $\cos \theta$ (power factor) approaches 1. Thus, the primary is providing the power for the load connected to the secondary. In other words, the load on the secondary is reflected back to the primary. The primary, in turn, draws the required power from the primary power source.

EXAMPLE 12-7

Determine the efficiency of a transformer providing 550 W of secondary power when the primary voltage is 240 V, the primary current is 3 A, and the primary angle theta is 30°.

Measured values				Calculated values		
V_{pri} (volts)	I_{pri} (amperes)	P_{pri} (watts)	P_{sec} (watts)	Apparent P_{pri} (voltamperes)	Cos θ	θ (degrees)
120	0.125	3	0	15	0.20	78
120	0.135	7	3.8	16.2	0.432	64
120	0.15	11	7.3	18	0.611	52
120	0.19	18	12.8	22.8	0.789	38
120	0.46	53	45.5	55.2	0.960	16
120	0.60	71	62	72	0.986	9.6

Fig. 12-12 Primary current and angle theta (θ) versus load.

Given: $P_{sec} = 550$ W, $V_{pri} = 240$ V
$I_{pri} = 3$ A, $\theta = 30°$

Known: % eff. $= \dfrac{P_{sec}}{P_{pri}} \times 100$

$P_{pri} = V_{pri} \times I_{pri} \times \cos \theta$

Solution: From calculator or Appendix G, the cos 30° = 0.866
$P_{pri} = 240$ V $\times$ 3 A $\times$ 0.866
$= 623.5$ W

% eff. $= \dfrac{550 \text{ W}}{623.5 \text{ W}} \times 100$
$= 88.2$

Answer: The transformer with a 550-W load is 88.2 percent efficient.

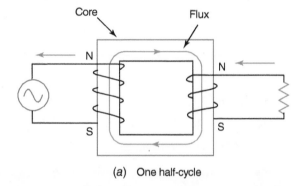

(a) One half-cycle

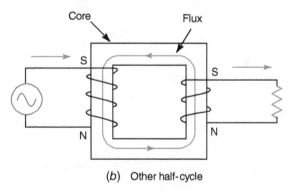

(b) Other half-cycle

Fig. 12-13 Primary and secondary magnetizing forces oppose each other. However, notice that the primary force dominates.

The secondary load is reflected back to the primary by the interaction of the primary and secondary magnetizing forces. Until now we have not mentioned a *secondary magnetizing force*. But, as soon as the secondary is loaded, current starts to flow in the secondary coil. This current in the secondary coil creates a magnetizing force that tries to produce a flux in the core material. The polarity of the magnetizing force of the secondary always opposes the magnetizing force of the primary (Fig. 12-13). Obviously, the core cannot have flux flowing in opposite directions at the same time. Either the primary or the secondary magnetizing force must dominate. The primary magnetizing force always dominates. It must dominate because the flux created by the primary is what produces the secondary voltage and current. Whenever the secondary magnetizing force increases, the primary current increases to provide a greater

primary magnetizing force. This is why the primary current increases whenever the secondary load current increases. Any primary current caused by a resistive load on the secondary is an in-phase (resistive) current (Fig. 12-14). The total primary current (Fig. 12-14 b) is composed of this reflected resistive current and the energizing current (which is mostly inductive current). When the transformer is fully loaded, the resistive current (caused by the secondary load) dominates. Thus, the fully loaded transformer appears resistive to the primary power source.

Secondary magnetizing forces

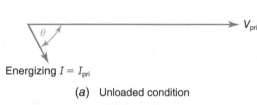

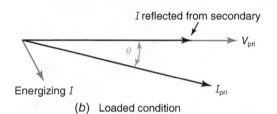

(a) Unloaded condition   (b) Loaded condition

Fig. 12-14 Reflected load current.

⌁ Self-Test

Answer the following questions.

37. Describe what happens to the following factors when a transformer's load is changed from no load to a full resistive load:
   a. Angle θ
   b. Cosine θ (power factor)
   c. Type of load the transformer presents to the primary source
38. The primary current drawn by an unloaded transformer is called the _____ current.

39. The primary current is mostly resistive when the transformer secondary is _____.
40. True or false. The resistive load on the secondary of a transformer is reflected back to the primary as an in-phase current.
41. True or false. The magnetizing force of the secondary current opposes the magnetizing force of the primary current.

12-4 Transformer Cores

Transformers can be broadly grouped by the type of core material they use. Like inductors, transformers can have either magnetic cores or air cores.

Iron-core transformers look like iron-core inductors. They use the same type of I and E laminations in their cores. However, the

Iron-core transformers

stacking of the I and E laminations in transformer cores is different from that in inductors.

You May Recall

. . . that with inductors, all the I laminations are stacked together and all the E laminations are stacked together.

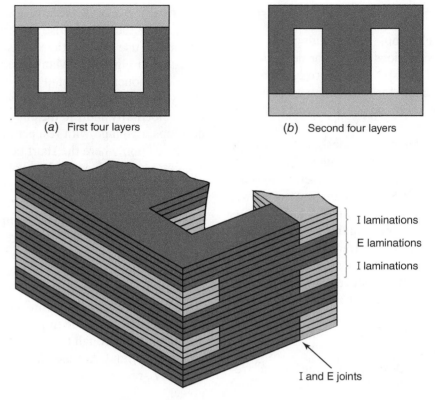

(a) First four layers   (b) Second four layers

I laminations

E laminations

I laminations

I and E joints

Fig. 12-15 Stacking of I and E laminations.

This leaves a small continuous air gap where the I and E stacks butt together. This small air gap in the inductor core aids in keeping the inductance more constant for different amounts of current. (Sometimes this air gap is increased by putting one or more layers of paper between the I and E joint.)

With transformers, the *I and E laminations* are rotated 180° every few layers (Fig. 12-15). This procedure breaks up the joint between the I and E laminations so that there is no continuous air gap. Thus flux leakage from the core is reduced to a minimum.

Laminated-iron-core transformers are used only at power and audio frequencies (frequencies up to 20 kHz). At frequencies above the audio range, their core losses become excessive.

Powdered iron and ferrite are also used as core material for magnetic-core transformers. When used in the audio range, the cores form a continuous path for the flux. *Toroidal cores* are often used at the higher audio frequencies. When used in the radio-frequency range, the core is often just a slug. Notice in Fig. 12-16

that the coils are physically separated from each other. When greater coupling is desired, the spacing between the coils is reduced. Frequently, one coil is wound on top of the other to obtain maximum coupling.

Air-core transformers are used exclusively at high (radio) frequencies. Often they are made from wire that is heavy enough to allow the individual coils to be self-supporting. Some air-core transformers are made so that the coefficient of coupling is variable. This requires an arrangement that varies either the distance between the coils or the orientation of their axes.

Air-core transformers

I and E laminations

Fig. 12-16 Radio-frequency transformer.

Toroidal cores

12-5 Types of Transformers

Electric and electronic parts catalogs list many different classifications of transformers. Usually these catalogs classify a transformer according to the application for which it was designed. Some of the more common types and their applications are discussed below.

Power transformers

Power transformers are designed to operate at power-line frequencies and voltages (usually 60 Hz and from 115 to several thousand volts). The larger transformers are for power distribution and lighting. Smaller transformers are used for rectifier or control circuits in electronic systems. Rectifier transformers are used for providing low-voltage alternating current for rectification into direct current. These transformers are also used for providing low-voltage alternating current for control circuits (relays, solenoids, etc.). Therefore, they are also called *control transformers.*

Control transformers

Transformers designed to operate at frequencies up to 20 kHz are often referred to as *audio transformers.* They are further categorized as *input, output,* and *interstage* transformers. (These terms refer to audio amplifiers.) They are transformers to (1) receive the input to the amplifier, (2) deliver the output from the amplifier, or (3) process the audio signal in the amplifier.

Audio transformers

Autotransformers

Radio-frequency transformers perform functions similar to those of audio transformers, but at radio frequencies rather than audio frequencies. Radio-frequency transformers may be either air-core or magnetic-core. They are often enclosed in a metal container (Fig. 12-16) which shields against electric fields. These transformers are used in such devices as radio and television receivers and transmitters.

Radio-frequency transformers

Some electrical and electronic equipment (such as data processing equipment and computers) is very sensitive to voltage changes. Such equipment is often powered by *constant-voltage transformers* because regular power-line voltage may vary too much. A constant-voltage transformer provides a stable secondary voltage even when the primary voltage is very unstable. Typically the primary voltage can vary from 95 to 130 V without causing more than 1 percent variation in the secondary voltage.

Constant-voltage transformers

Isolation transformers have equal primary and secondary voltages. Their purpose is to

Variable transformers

Isolation transformers

electrically isolate a piece of electrical equipment from the power distribution system. An important use of isolation transformers is illustrated in Fig. 12-17. Many electronic devices have components mounted on a metal chassis. If any of the components or the associated wiring accidentally short to the chassis, the chassis develops a voltage with respect to ground. Depending upon where the short occurs, this voltage can be as high as the line voltage powering the device. Technicians servicing this equipment can accidentally touch the chassis while power is being supplied to the equipment. If they do, they complete a circuit and receive a shock [Fig. 12-17(a)]. The shock can be fatal when the resistance through ground is low. Inserting an isolation transformer [Fig. 12-17(b)] breaks the circuit that includes the technician. Current can no longer flow from the ungrounded side of the power source through the chassis and the technician to ground. Yet the circuit containing R_1, R_2, and L_1 receives normal voltage, current, and power.

An *autotransformer* is somewhat different from the other transformers we have studied. Its primary is part of its secondary, and vice versa. With a step-up autotransformer [Fig. 12-18(a)], the secondary consists of the primary plus some additional turns. These additional turns are wound so that their induced voltage is series-aiding the cemf of the primary. A step-down autotransformer is shown in Fig. 12-18(b). Here the secondary is just a fraction of the primary. The cemf of that fraction of the primary provides the secondary voltage.

In most ways an autotransformer behaves like any other transformer. That is, a load on the secondary increases the primary current. Also, the turns-per-volt ratio of the primary and secondary is the same. The big difference between the autotransformer and other transformers is that the autotransformer does not provide electrical isolation.

Variable transformers that operate at power frequencies are usually autotransformers with an adjustable secondary. One secondary lead is connected to a carbon brush that can be adjusted up and down on the turns of the winding. The secondary voltage of most variable transformers can be adjusted from 0 to about 110 percent of the primary voltage.

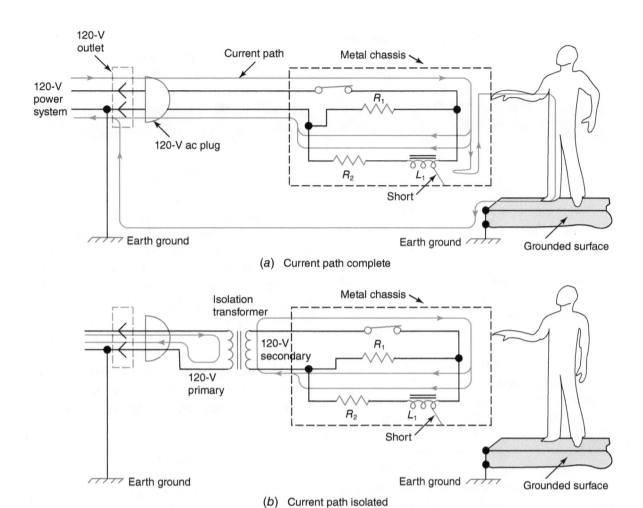

Fig. 12-17 Importance of isolation transformer.

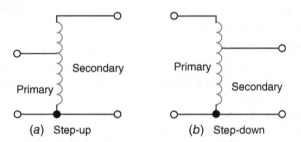

Fig. 12-18 Autotransformer. Part of the same coil is used for both the primary and the secondary.

Self-Test

Answer the following questions.

42. Why are I and E laminations rotated every few layers in transformers?

43. What types of transformers use laminated iron cores?

44. List several types of power transformers.

45. The _____ transformer provides a stable secondary voltage even though the primary voltage may vary between 95 and 130 V.

46. The _____ transformer has a secondary voltage that is equal to the primary voltage.

47. The _____ transformer does not provide electrical isolation.

48. A transformer that provides continuously variable output voltage is usually a(n) _____ transformer.

12-6 Impedance Matching

Impedance matching

We have seen that transformers are used for changing voltage levels and for isolation. The third common use of transformers is *impedance matching*.

Before we see how a transformer can match impedances, let us review why impedance matching is desirable.

Reflected impedance

You May Recall

. . . that in an earlier chapter it is shown that matching the resistance of a dc source with that of the load provides maximum transfer of power.

Impedance matching, in effect, does the same thing for ac circuits. When the internal impedance of a source is matched (equal) to the impedance of the load, maximum power is transferred from the source to the load. When the load impedance is either greater or less than the source impedance, the load power decreases. Although impedance includes both resistance and reactance, only resistance is used in the discussion which follows. Using only resistance does not change the concept; it just keeps the calculations simpler.

A transformer can make a load appear to the source to be either larger or smaller than its actual value. If the transformer is a step-up type, the load appears smaller. If it is a step-down transformer, the load appears larger. Figure 12-19 illustrates how a transformer changes the apparent value of a load. For purposes of calculation, the transformer in Fig. 12-19(*a*) is assumed to be an ideal transformer. That is, it has 100 percent efficiency, 100 percent coupling, and a 0° phase shift when loaded. If the transformer has no power losses, then its primary power equals its secondary power. To obtain 0.1 W of secondary power, the primary must provide 0.1 W. This means that the primary has to draw 0.01 A

of current from the source. Notice this is the same value of current as the 1000-Ω resistor in Fig. 12-19(*b*) draws. Thus you can see that the transformer makes the 10-Ω resistor in the secondary appear as 1000 Ω to the source. The source's output voltage, current, and power are identical in the two circuits. The transformer has matched the impedance of a 1000-Ω source to a 10-Ω load.

The 1000-Ω impedance that the primary seems to have is called *reflected impedance*. In other words, the 10 Ω in the above example is reflected back to the primary as 1000 Ω. We could say that the primary impedance of

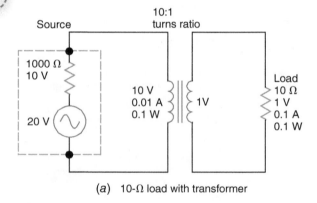

(*a*) 10-Ω load with transformer

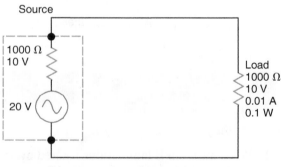

(*b*) 1000-Ω load without transformer

Fig. 12-19 Impedance-matching principle. The transformer makes the 10-Ω resistor appear to the source as a 1000-Ω resistor.

the transformer is 1000 Ω and the secondary impedance is 10 Ω. Therefore, the primary-to-secondary *impedance ratio* is 100:1. Notice in Fig. 12-19(*a*) that the turns ratio (and also the voltage ratio) is 10:1. Thus the turns ratio squared is equal to the impedance ratio. Also, the voltage ratio squared is equal to the impedance ratio. If we use the symbol Z for impedance, these relationships can be written

$$\frac{Z_{pri}}{Z_{sec}} = \left(\frac{N_{pri}}{N_{sec}}\right)^2 \quad \text{or} \quad \frac{Z_{pri}}{Z_{sec}} = \left(\frac{V_{pri}}{V_{sec}}\right)^2$$

EXAMPLE 12-8

An 8-Ω load is to be matched to a source that has 8000 Ω of internal impedance. What must the turns ratio of the transformer be?

Given: $Z_{sec} = 8\ \Omega$
 $Z_{pri} = 8000\ \Omega$

Find: $\dfrac{N_{pri}}{N_{sec}}$ (that is, the turns ratio)

Known: $\dfrac{Z_{pri}}{Z_{sec}} = \left(\dfrac{N_{pri}}{N_{sec}}\right)^2$

Solution: $\dfrac{Z_{pri}}{Z_{sec}} = \dfrac{8000\ \Omega}{8\ \Omega} = 1000$

$\left(\dfrac{N_{pri}}{N_{sec}}\right)^2 = 1000$

Turns ratio $= \dfrac{N_{pri}}{N_{sec}} = \sqrt{1000}$

$= 31.6$

Answer: The transformer must have a turns ratio of 31.6:1 (primary to secondary).

Impedance ratio

It should be remembered that while impedance matching provides maximum transfer of power from the source to the load, it only produces an efficiency of 50 percent. This can be seen in Fig. 12-19 where the load receives 0.1 W while the internal resistance of the source dissipates 0.1 W ($P = V^2/R = 10^2/1000 = 0.1$ W). As was shown in Chap. 5, increasing the load resistance or impedance increases the efficiency but reduces the amount of power delivered to the load. For example, if the load in Fig. 12-19(*b*) is increased to 9000 Ω, the efficiency increases to 90 percent and the load power decreases to 36 mW. Whether or not impedance matching is desirable depends upon specific circumstances. For example, a power company furnishing megawatts to an industry is concerned about efficiency rather than impedance matching. On the other hand, when connecting a microphone (that produces only microwatts of power) to an amplifier, the efficiency is not important but impedance matching to transfer as much as possible of the microphone output to the amplifier's input is very important.

Self-Test

Answer the following questions.

49. What does "impedance matching" mean?
50. What is the significance of impedance matching?
51. Does a load appear to the source as smaller or larger than its true value when a step-down transformer is used?

52. What does "reflected impedance" mean?
53. What is the impedance ratio of a transformer that steps down 240 V to 15 V?
54. If the transformer in question 53 is connected to a 10-Ω load, what is the value of the impedance reflected back to the source?

12-7 Transformer Ratings

To properly use a transformer, one must know its voltage and current *ratings*. Of course, from these ratings the power rating of the transformer can be calculated. Most transformers are also specified by their voltampere (apparent power) ratings.

Ratings

Voltage Rating

Manufacturers always specify the voltage rating of the primary and secondary windings. Operating the primary above rated voltage usually causes the transformer to overheat. The additional stress placed on the transformer insulation by the higher primary and secondary voltages can also be serious. Operating the primary below rated voltage does no harm, but this makes the secondary voltages lower than rated values.

The rated voltages of the secondaries are specified for full-load conditions with rated primary voltage. With no load, the secondary voltage is slightly higher than rated voltage (usually 5 to 10 percent higher).

Center-tapped

There is considerable variation in the way in which manufacturers specify *center-tapped* secondaries. For example, the secondary in Fig. 12-20 may be specified in any of the following ways:

1. 40 V C.T.
2. 20 V-0-20 V
3. 20 V each side of center

Current Rating

Manufacturers usually specify current ratings for secondary windings only. As long as the secondary current rating is not exceeded, the primary current-carrying capacity cannot be exceeded.

Exceeding the current rating of a secondary causes its voltage to fall slightly below rated value. More serious than decreased voltage, however, is the increase in I^2R loss in the secondary. The increased I^2R loss causes the winding to overheat and eventually destroys the transformer.

Power Rating

Some manufacturers specify a power rating (in watts) for their transformers. This is understood to be the power the transformer can deliver *to a resistive load*. Thus the power rating is merely the product of the current rating and the voltage rating of the secondary ($P = IV$). For multiple-secondary transformers, the power rating is the sum of the powers available from the individual secondaries ($P_T = P_1 + P_2 +$ etc.). The total power cannot be taken from a single secondary on a multiple-secondary transformer. The current rating of the individual secondaries must not be exceeded.

Voltampere Rating

The voltampere rating of a transformer is an apparent power rating. It is applicable to any type of load—resistive, reactive, or combination (impedance). The voltampere rating, like the power rating, is given for the total transformer instead of for individual secondaries. With a multiple-secondary transformer, the total voltampere rating cannot be taken from a single secondary.

A transformer can be loaded to its full voltampere rating and be delivering only a fraction of its power rating. Refer to Fig. 12-21 for an example. Here the load is a motor which has a power factor ($\cos \theta$) of 0.6. It is connected to a transformer that is rated at 750 VA and has a secondary voltage of 120 V. The motor draws 6.25 A. The transformer provides 750 VA (120 V × 6.25 A). However, the motor is drawing only 450 W ($P = IV \cos \theta = 6.25$ A × 120 V × 0.6).

As you can see, manufacturers use many ways to rate transformers. You will never overload a transformer or exceed any of its ratings if you observe two rules.

1. Never apply more than the rated voltage to the primary.
2. Never draw more than the rated current from any secondary.

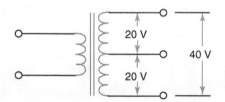

Fig. 12-20 Center-tapped secondary specifications. The secondary is classified as either 40 V center-tapped or 20 V each side of center.

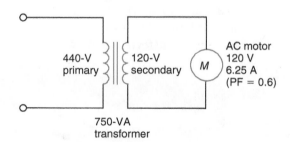

Fig. 12-21 Voltampere versus power rating. The limiting rating is the voltampere rating.

Answer the following questions.

55. True or false. It is not possible to exceed the voltampere rating of a transformer without exceeding the power rating of the transformer.

56. True or false. The power rating of a transformer can be exceeded without exceeding either the primary voltage rating or the secondary current rating.

57. True or false. The voltampere rating of a transformer cannot be exceeded without exceeding either the primary voltage rating or the secondary current rating.

58. True or false. Secondary voltages are rated at full load current and rated primary voltage.

EXAMPLE 12-9

You need to operate a 240-V device that draws 10 A and has a PF of 0.8 from a 120-V source. Determine the minimum specification for a transformer needed to do the job.

Given:	Load voltage = 240 V
	Load current = 10 A
	Load PF = 0.8
	Source voltage = 120 V
Find:	Transformer windings voltages
	Transformer power rating
Known:	$V_{pri} = V_{source}$
	$V_{sec} = V_{load}$
	$P_{apparent} = VA$
Solution:	$V_{pri} = V_{source} = 120$ V
	$V_{sec} = V_{load} = 240$ V
	$P_{apparent} = V_{load} \times I_{load} = 240$ V
	$\times$ 10 A = 2.4 kVA
Answer:	An iron-core transformer rated at 2.4 kVA (or higher) with a 120-V primary and a 240-V Secondary.

The transformer in example 12-9 could also have a 120-V primary and a 240-V secondary rated at 10 A.

12-8 Series and Parallel Windings

Some transformers are made with more than one primary as well as more than one secondary. *Multiple* transformer *windings* can be connected in series or in parallel to change the voltage or current capabilities. Either primary or secondary windings (or both) can be connected in series or in parallel.

Windings in Parallel

For parallel connections, windings should have identical ratings. They *must* have identical voltage ratings. Before windings are connected in parallel, their *phasing* must be correct. Phasing refers to the instantaneous polarity of the windings. *Correct phasing* for parallel connection is illustrated in Fig. 12-22. The polarities shown are instantaneous. Notice that the negative ends of the two secondary windings are connected together, as are the positive ends. With this phasing of the windings, no secondary current flows when there is no load. Neither winding can force current through the other because their voltages are equal and opposing. The output voltage is still 10 V, but the current capability has doubled from 3 to 6 A.

Correct phasing

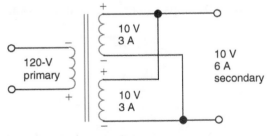

Fig. 12-22 Windings connected in parallel. The voltage remains the same, but the current capacity increases.

Multiple windings

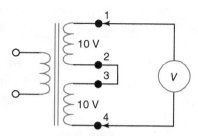

Fig. 12-23 Determining correct phasing for a parallel connection. If correctly phased, the meter will indicate 0 V.

When the two parallel windings with the same voltages have different current ratings, the current rating of the parallel combination is only double the rating of the winding with the smallest current rating.

Some transformer manufacturers indicate the phasing of windings by terminal numbers or color code. However, when phasing is not indicated, the method shown in Fig. 12-23 can be used to determine the correct phasing. If points 2 and 3 of Fig. 12-23 are of the same instantaneous polarity, the voltmeter across points 1 and 4 will indicate 0 V. This means that points 1 and 4 are also of the same polarity. If the two windings are to be operated in parallel, points 1 and 4 can be connected together. Points 2 and 3 form one terminal of the parallel winding, and points 1 and 4 form the other terminal.

If, on the other hand, the meter in Fig. 12-23 indicates 20 V, points 2 and 3 have opposite polarities. Then, for a parallel connection, points 1 and 3 can be connected together, as can points 2 and 4.

Incorrect phasing (Fig. 12-24) causes a "dead" short on the secondaries of the transformer. The voltage of each winding aids the other winding in producing secondary current. The secondary current is limited only by the resistance of the secondaries. Therefore, the current becomes very high. If not protected against overload, the transformer will soon burn out.

Some transformer primaries have two identical windings so that they can be powered from either of two voltages. The two voltages must have a 1 to 2 relationship such as 120 V and 240 V. The primaries are series connected to operate on the higher voltage. When operated on the lower voltage, the primaries are connected in parallel. Proper phasing of the primary windings can also be determined using the method of Fig. 12-23. Just use one of the secondaries as a temporary primary and apply an appropriate voltage source to it. The voltage applied to the secondary can be any value equal to *or less than* its rated value.

Windings in Series

Transformer windings can be connected in series so that they either aid or oppose each other, as in Fig. 12-25. In this figure the polarity markings indicate the instantaneous polarities for one half-cycle. In the *series-aiding* configuration [Fig. 12-25(*a*)], the output voltage is the sum of the two secondary voltages. Notice, however, that the current capability is restricted to the lower rating of the two windings. This is because all the load current flows through both secondaries.

When connected in a *series-opposing* configuration [Fig. 12-25(*b*)], the two windings produce a voltage equal to the difference between

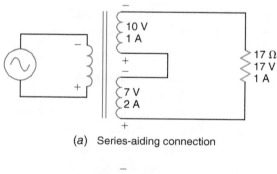

(a) Series-aiding connection

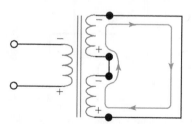

Fig. 12-24 Short circuit caused by incorrect phasing.

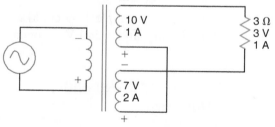

(b) Series-opposing connection

Fig. 12-25 Series-aiding and series-opposing winding connections.

the two voltages. We say the two voltages are *bucking* one another. Again note that the current is limited to the lower current rating of the two windings.

In summary, it is possible to obtain four voltages from two secondary windings:

1. The voltage of secondary 1
2. The voltage of secondary 2
3. The sum of the voltages of secondaries 1 and 2
4. The difference between the voltages of secondaries 1 and 2

Dual primary windings are connected in series for operation on the higher voltage. The connection must be series-aiding. Otherwise, the magnetic field from one winding would cancel the field from the other. The primary current would be very high. The primary would burn out immediately unless it was properly protected against overloads. Phasing the primary windings for series-aiding connections can be done as described for parallel primary windings. The connection that yields the maximum voltmeter reading is the correct connection for series-aiding primaries. Power is applied to the two terminals where the voltmeter was connected.

EXAMPLE 12-10

A transformer has a 10-V, 2-A secondary (sec 1) and a 6-V, 3-A secondary (sec 2). Determine all the voltages, and the associated current, that can be obtained from these two transformer secondary windings.

Given: Secondary 1 = 10 V at 2 A
Secondary 2 = 6 V at 3 A

Find: Voltages and currents available with series-aiding connections.
Voltages and currents available with series-opposing connections.

Known: With either aiding or opposing connections:
1. The voltage and current rating of each winding is available
2. The current rating of the series combination is that of the winding with the lowest current rating.
With aiding connections
$V_{out} = V_{sec1} + V_{sec2}$

With opposing connections
$V_{out} = V_{sec1} - V_{sec2}$

Solution: Individual secondary coils provide 10 V at 2 A and 6 V at 3 A
Series-opposing secondary coils provide 10 V − 6 V = 4 V at 2 A
Series-aiding secondary coils provide 10 V + 6 V = 16 V at 2 A

Answer: 10 V at 2 A, 6 V at 3 A, 4 V at 2 A, and 16 V at 2 A

In addition to the voltages determined in example 12-10, one could connect secondary 1 or secondary 2 in series aiding with the primary winding and use the series-aiding combination as the primary. This would increase the turns-per-volt ratio of the transformer and thus reduce the output voltage of the other secondary. It would not be advisable to use a series-opposing connection to increase the voltage of the remaining secondary because this would decrease the inductance of the primary, increase the energizing primary current, and maybe exceed the current rating of the primary winding.

12-9 Off-Center-Tapped Windings

Some transformers have an off-center-tapped primary and/or secondary. Often, the winding has two or more taps. The taps are usually near one end of the winding. Thus, changing taps produces relatively small changes in the winding's voltage rating.

The nominal voltage for homes and offices is 120 V. However, the actual voltage varies from area to area as well as according to the time of day. A transformer with primary taps for 115 V, 120 V, and 125 V can be connected to closely match the voltage typical of a given area.

Tapped secondary windings have various uses. For example, they can be used to adjust impedance-matching ratios in audio transformers. In power transformers, tapped secondaries are useful for adjusting the charging rate of a battery charger or the output of an arc welder.

Answer the following questions.

59. List two requirements for parallel-connected windings.
60. What does *phasing* refer to?
61. Two secondary windings are to be connected in series. One winding is rated for 11 V and 2 A. The other is rated for 8 V and 1.5 A.
    a. How much voltage and current are available if one instantaneous + is connected to the other instantaneous +?

b. How much voltage and current are available if + is connected to −?
62. Are incorrectly phased parallel windings likely to damage a transformer?
63. Series-connected primary windings of a dual-voltage primary are connected series _____.

64. True or False. Primary windings are never tapped for small voltage adjustments.
65. True or False. A tapped secondary winding is always center-tapped.

12-10 Three-Phase Transformers

Three-phase transformer

> 💡 **You May Recall**
>
> ... that three-phase circuits were discussed in an earlier chapter. The ideas developed in that chapter are essential in understanding three-phase transformers. Therefore, you should review that chapter section before reading this section.

Three-phase voltages can be transformed either by a single three-phase transformer or by three single-phase transformers. The end results are the same: all three of the phase voltages are changed.

The structure of a *three-phase transformer* is illustrated in Fig. 12-26. The flux in the phase 1 leg is equal to the phase 2 flux plus the phase 3 flux. Phase 2 flux equals phase 3 flux plus

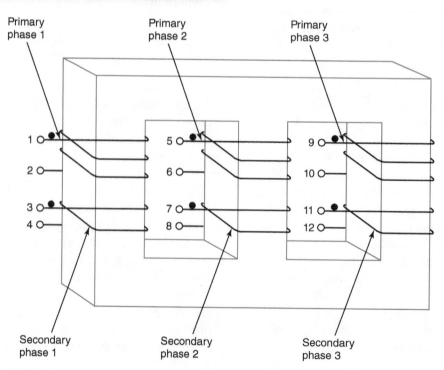

Fig. 12-26 Three-phase transformer. One phase is wound on each leg of the transformer.

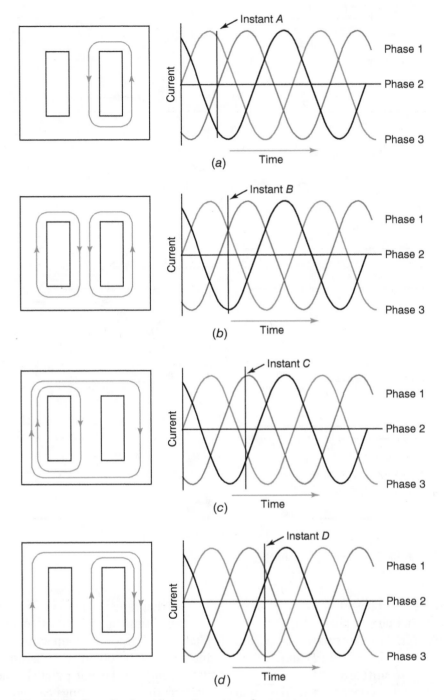

Fig. 12-27 Core flux in a three-phase transformer core.

phase 1 flux, etc. This is because the flux, like the current, of each phase is displaced by 120°.

Figure 12-27 graphically presents the idea of how the flux splits up in a three-phase transformer core. In this figure, the flux of each phase is assumed to be in step with the current in the phase. (This assumption implies that the core has no hysteresis loss.) At instant A [Fig. 12-27(a)], the phase 1 current is zero. Therefore, the phase 1 flux is also zero. At the same instant, the currents in phase 2 and phase 3 are of opposite polarities. This causes their flux to join together in legs 2 and 3. In Fig. 12-27(b) (at instant B), currents in phases 1 and 3 are both positive. This produces flux in the direction shown in both the phase 1 leg and the phase 3 leg of the core. At this same instant (instant B), phase 2 current is negative and of twice the value of either phase 1 current or phase 3 current. Thus, the phase 2 leg of the core has twice

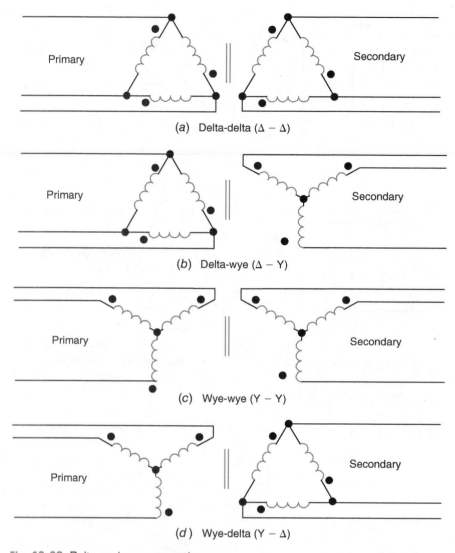

(a) Delta-delta (Δ – Δ)

(b) Delta-wye (Δ – Y)

(c) Wye-wye (Y – Y)

(d) Wye-delta (Y – Δ)

Fig. 12-28 Delta and wye connections.

as much flux as either phase 1 or phase 3. Also, the direction of the flux in phase 2 is opposite to that of the flux in the other two phases. Close inspection of Fig. 12-27(c) and (d) shows how the flux continues to shift around in the core.

The primary and secondary windings of a three-phase transformer may be either *wye-connected* or *delta-connected*. The secondary does not have to have the same configuration (wye or delta) as the primary. Figure 12-28 shows four possible ways to connect a three-phase transformer. The dots on one end of each winding indicate the beginning of each winding. Refer back to Fig. 12-26. Notice that all windings are wound in the same direction (counterclockwise) when you start at the dotted end of the winding. Identifying the start of the windings is necessary before they can be properly phased.

The diagrams in Fig. 12-28 show one way of connecting the windings to obtain correct phasing. With a wye connection, correct phasing can also be obtained by connecting all the dotted ends to the star point. In the delta connection all three windings can be reversed; just be sure that two dotted ends are not connected together. On transformers, the dotted (start) end of a winding is identified by the manufacturer. The identification may be made in several ways. Some manufacturers use a colored strip to indicate the start lead. Others use a number on a diagram mounted on the transformer.

Incorrect phasing of the primary, in either the wye or the delta configuration, causes excessively high primary current. If not protected against overload, the incorrectly phased primary can be destroyed by the excess current.

Wye-connected

Delta-connected

Improper phasing of a delta-connected secondary also causes excessive, destructive current. A quick check for correct phasing of a delta-connected secondary is shown in Fig. 12-29. If the voltmeter indicates 0 V, the windings are properly phased. The ends of the windings to which the meter is connected can be connected together to complete the delta. If the meter indicates a high voltage (twice the phase voltage), incorrect phasing exists. Reverse the lead connections on one winding at a time until the meter indicates 0 V. With some wye-delta-connected transformers, the meter in Fig. 12-29 may not indicate 0 V when properly phased. In these cases the meter will indicate the lowest reading when the phase is correct.

A wye-connected secondary provides equal line voltages when correctly phased. If the line voltages are unequal, reverse one winding at a time until the line voltages are balanced.

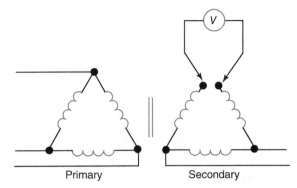

Primary   Secondary

Fig. 12-29 Checking delta phasing. When properly phased, the meter will indicate 0 V.

You May Recall

. . . that the relationships between phase and line voltages and phase and line currents were explained in an earlier chapter. These relationships apply to three-phase transformers as well as to three-phase generators.

Self-Test

Answer the following questions.

66. Three-phase transformer windings can be connected in a(n) _____ or a(n) _____ configuration.

67. When all the finish ends of the primary windings are connected together, the primary is connected in a(n) _____ configuration.

68. What is the correct phasing for a delta connection?

69. Must the primary and secondary of a three-phase transformer be connected in the same configuration?

70. What happens when a delta winding is incorrectly phased?

71. Refer to Fig. 12-26. Which terminals should be connected together to form a wye-connected primary?

Summary

1. Transformers operate on the principle of mutual inductance.
2. Mutual inductance is measured in henrys.
3. Magnetic flux links, or couples, the two coils of a transformer.
4. Primaries receive power. Secondaries deliver power.
5. Primaries and secondaries, except on autotransformers, are electrically isolated.
6. Primaries and secondaries are reversible.
7. The coefficient of coupling specifies what portion of the primary flux links with the secondary.
8. The coefficient of coupling ranges from 0 to 1. It is greatest (almost 1) with iron-core transformers.
9. Flux leakage refers to primary flux that does not couple to the secondary.
10. Turns ratio and voltage ratio are the same when the coupling is 100 percent.
11. The turns-per-volt ratio is the same in all windings of a transformer.
12. One of the major uses of transformers is to step up and step down voltages in a power transmission and distribution system.
13. When voltage is stepped up, current is stepped down.
14. When voltage is stepped down, current is stepped up.
15. Transformer losses occur in both the core and the coils.
16. Transformer loss consists of hysteresis, eddy current, and copper loss.
17. Copper loss is called I^2R loss.
18. Core loss consists of hysteresis and eddy current loss.
19. Hysteresis loss results from residual magnetism.
20. Hysteresis loss increases with increased frequency.
21. A narrow hysteresis loop means less hysteresis lost.
22. Eddy currents are currents induced in the core by primary flux.
23. Eddy currents are reduced by using laminations and oxidizing or coating the surface of the lamination with an insulating material.
24. An unloaded transformer behaves just like an inductor. The energizing current and supply voltage are nearly 90° out of phase. The input power is dissipated in the form of copper and core losses.
25. Energizing current is the current drawn by the primary when the transformer is unloaded. Its magnitude is controlled by the reactance of the primary.
26. A fully loaded transformer appears to be almost entirely resistive to the source.
27. Power factor ($\cos \theta$) approaches a value of 1 with a full resistive load on the transformer.
28. As secondary current increases, so does primary current.
29. Laminated-core transformers are used at power and audio frequencies.
30. Powdered iron and ferrite cores are used in the audio-frequency and lower radio-frequency ranges.
31. Air-core transformers are used only in the radio-frequency range. Their coupling can be controlled by the spacing and axis orientation of their coils.
32. Constant voltage transformers provide a stable secondary voltage.
33. Isolation transformers have equal primary and secondary voltages.
34. Isolation transformers help protect the service technician from receiving a shock through the chassis of electrical equipment.
35. Autotransformers use a common primary-secondary winding. They are often used as variable transformers at power frequencies.
36. Matched impedances provide maximum power transfer and 50 percent efficiency.
37. Transformers may have voltage, current, power, and voltampere ratings.
38. Transformer power ratings apply to resistive loads only.

39. Power and voltampere ratings refer to the total of all secondaries.
40. Connecting windings in parallel increases the available current but does not change the voltage rating.
41. Connecting windings in series either increases or decreases the available voltage but does not increase the current rating.

42. Parallel windings must be properly phased and have identical voltages.
43. Three single-phase transformers can be used to transform three-phase voltages.
44. Three-phase transformer windings can be connected in either delta or wye configurations.
45. The primary and secondary of a three-phase transformer need not be connected in the same configuration.

Related Formulas

$$\frac{Z_{pri}}{Z_{sec}} = \left(\frac{N_{pri}}{N_{sec}}\right)^2$$

$$\% \text{ eff.} = \frac{P_{pri}}{P_{sec}} \times 100$$

$$L_M = k\sqrt{L_1 L_2} \quad k = \frac{\phi_{L_2}}{\phi_{L_1}}$$

With 100 percent coupling:

$$\frac{V_{pri}}{V_{sec}} = \frac{N_{pri}}{N_{sec}}$$

$$\frac{N_{pri}}{V_{pri}} = \frac{N_{sec}}{V_{sec}}$$

With 100 percent efficiency:

$$\frac{V_{pri}}{V_{sec}} = \frac{I_{sec}}{I_{pri}}$$

$$\frac{I_{sec}}{I_{pri}} = \frac{N_{pri}}{N_{sec}}$$

Chapter Review Questions

For questions 12-1 to 12-13, determine whether each statement is true or false.

12-1. The same symbol is used for the air-core transformer and the powdered-iron-core transformer. (12-1)
12-2. The winding of a transformer designated as the secondary can be used as a primary. (12-1)
12-3. The secondary of a transformer converts electric energy to magnetic energy. (12-1)
12-4. In an autotransformer, part or all of a coil is used as both the primary and the secondary. (12-5)
12-5. The coefficient of coupling of an air-core transformer is greater than that of an iron-core transformer. (12-1)
12-6. When the coupling in a transformer is 100 percent, the turns ratio is numerically equal to the voltage ratio. (12-1)

12-7. In a step-up transformer, the turns-per-volt ratio in the primary is smaller than that in the secondary. (12-1)
12-8. Large transformers usually have a higher turns-per-volt ratio than small transformers do. (12-1)
12-9. A transformer is most efficient when the secondary is not loaded. (12-2)
12-10. Power factor approaches 1 when a transformer has a full resistive load. (12-3)
12-11. The primary current of a transformer increases whenever the secondary current increases. (12-3)
12-12. The primary and secondary of a three-phase transformer must be connected in the same configuration. (12-10)
12-13. Three single-phase transformers can be used to transform a three-phase voltage. (12-10)

For questions 12-14 to 12-35, supply the missing word or phrase in each statement.

12-14. Transformers operate on the principle of _____. (12-1)

12-15. The unit of mutual inductance is the _____. (12-1)

12-16. The _____ winding of a transformer receives power while the _____ winding delivers power. (12-1)

12-17. The portion of the flux of one coil that links to another coil is specified by the _____. (12-1)

12-18. Primary flux that does not couple to the secondary is known as _____ flux. (12-1)

12-19. When voltage is stepped down, current is stepped _____. (12-1)

12-20. Power loss occurs in the _____ and the _____ of a transformer. (12-2)

12-21. I^2R loss is also called _____ loss. (12-2)

12-22. In addition to I^2R loss, a transformer has _____ loss and _____ loss. (12-2)

12-23. When a transformer is fully loaded, the primary current is mostly _____. (12-3)

12-24. When a transformer has no load, the primary current is called the _____ current. (12-3)

12-25. Constant voltage transformers provide a stable _____ voltage. (12-5)

12-26. The _____ transformer has equal primary and secondary voltage. (12-5)

12-27. The _____ is a transformer that does not provide electrical isolation. (12-5)

12-28. Matched impedances will provide _____. (12-6)

12-29. Transformer power ratings apply to _____ loads. (12-7)

12-30. Connecting secondary windings in parallel increases the _____, but does not change the _____ rating of the combined windings. (12-8)

12-31. Two unequal secondary windings can provide _____ different voltages. (12-8)

12-32. Three-phase windings can be connected in either the _____ or the _____ configuration. (12-10)

12-33. Transformer cores are laminated to minimize _____ loss. (12-2)

12-34. Transformer cores are made of silicon steel to minimize _____ loss. (12-2)

12-35. A step-up transformer causes the secondary load resistance to appear to be _____ than its true value. (12-6)

Answer the following questions.

12-36. List three common uses of transformers. (12-6)

12-37. What relationship exists between the turns-pervolt ratio of the primary and secondary windings in an ideal transformer with a 2:1 voltage ratio? (12-1)

12-38. When is it desirable to have matched impedances? (12-6)

Chapter Review Problems

12-1. What is the maximum voltage and current available from a series-connected 14-V, 2-A secondary and a 7-V, 3-A secondary? (12-8)

12-2. How many turns are required for a 40-V secondary if the primary of the ideal transformer has five turns per volt? (12-1)

12-3. Determine the percent efficiency of a transformer that requires 400 W of primary power when a 60-V secondary is connected to a 12-Ω resistive load. (12-2)

12-4. A 1500-W transformer has one secondary rated at 230 V. Could this transformer power a 230-V motor which draws 7 A at a PF of 0.75? Why? (12-7)

12-5. A 10-Ω load is connected to the 6-V secondary of a transformer with a 120-V primary. What value does the load appear to be to the source? (12-6)

12-6. How much current and voltage are available from an 8-V, 2-A secondary connected in parallel with an 8-V, 3-A secondary? (12-8)

12-7. The secondary of a transformer is rated at 10 V and 5 A. How much power can it provide to a load with a PF of 0.7? (12-7)

12-8. An ideal transformer has 700 turns on its 120-V primary. How many turns are on its 24-V secondary? (12-1)

12-9. What is the apparent power rating of a transformer with a 240-V, 5-A secondary? (12-7)

12-10. A single-secondary transformer has a 2400-VA rating. Its secondary provides 120 V. What is the maximum load current it can provide when the load has a PF of 0.85? (12-7)

12-11. A 12-V, 2-A secondary is connected in series aiding to an 8-V, 3-A secondary. What is the voltage and current rating of this series connection? (12-8)

12-12. A three-phase transformer has 120-V secondary coils. What is the line voltage when these three coils are wye-connected? (12-10)

Critical Thinking Questions

12-1. The primary of a transformer has 5 turns per volt. What is the coefficient of coupling if a 40-turn secondary produces 6 V?

12-2. In the discussion of copper loss, it was stated that the size of wire in the primary winding was limited by the windows in the core. What would be the consequences of increasing the size of the window openings?

12-3. Why is it understood that the load must be pure resistance if a transformer is to be operated at its rated power output?

12-4. A transformer with 5 turns per volt has a 120-V primary, an 8-V secondary, and a 6-V secondary. If the 6-V secondary is connected in series with the primary and the series combination connected to a 100-V source, how much voltage will the 8-V secondary provide?

12-5. What might cause the voltage between the open terminals in the properly phased delta-connected secondary to be greater than 0 V?

12-6. Why doesn't an improperly phased wye-connected secondary produce destructive currents in the transformer?

12-7. A 120-V motor draws 6 A at a power factor (PF) of 0.75 from a transformer that is 90 percent efficient. The transformer draws 3 A from a 230-V source. What is the PF of the transformer primary?

12-8. How does increasing the number of primary turns affect the operation of an unloaded transformer?

12-9. Would a 10-Ω resistive load or a 10-Ω reactive load produce the larger shift of theta in the primary of a transformer? Why?

12-10. For each case below, determine the effect on system efficiency of adding an impedance-matching transformer (assume an ideal transformer). Also, determine the power of the load with and without the matching transformer.
   a. A 20-V source with 4 Ω of internal resistance driving a 16-Ω resistive load.
   b. A 15-V source with 10 Ω of internal resistance driving a 5-Ω resistive load.

12-11. Draw a schematic diagram showing how a 2-pole, 3-position rotary switch could be connected to the 4-V and 6-V secondaries of a transformer to provide 4, 6, or 10 V at the output terminals. The output voltage is to increase from 4 to 10 V as the switch is rotated clockwise.

1. two inductors coupled together by mutual inductance, or, a device that transfers power between electrically isolated circuits.
2. primary, secondary
3. 2 H
4.
5. The primary converts electric energy to magnetic energy. The secondary converts magnetic energy to electric energy.
6. primary
7. secondary
8. yes
9. It is a number that indicates the portion of the primary flux that links the secondary coil.
10. no
11. air-core transformers
12. by varying the distance between primary and secondary or by changing the axis orientation between the primary and secondary coils
13. F
14. T
15. F
16. F
17. 200
18. changing voltage levels
19. to keep current low and thus wire size small
20. steps up
21. heat losses in the core and the windings
22. 83.3 percent
23. 806.5 W
24. hysteresis loss and eddy current loss
25. windings
26. residual magnetism
27. twice each cycle
28. a curve on a graph produced by plotting magnetizing force against flux density
29. silicon steel, because it has high permeability and low hysteresis loss
30. narrow
31. As frequency increases, so does hysteresis loss.
32. by using core laminations that have oxidized or insulation-coated surfaces
33. thin
34. Power loss in the transformer windings. I^2R loss is

minimized by using large-diameter conductors in the coils.
35. large
36. when it is fully loaded
37. a. decreases from nearly 90° to nearly 0°
    b. increases from nearly 0 to nearly 1
    c. changes from inductive to resistive
38. energizing
39. fully loaded
40. T
41. T
42. to prevent a continuous air gap and thereby reduce flux leakage
43. power and audio-frequency transformers
44. distribution, lighting, rectifier, and control
45. constant voltage
46. isolation
47. auto
48. auto
49. Impedance matching means making the load impedance equal to the source impedance.
50. It provides maximum power transfer from source to load.
51. larger
52. It means the impedance of the secondary load is reflected back to the source through the transformer.
53. 256:1 (primary to secondary)
54. 2560 Ω
55. F
56. F
57. T
58. T
59. equal voltages and correct phasing
60. the momentary polarity of transformer windings
61. a. 3 V, 1.5 A
    b. 19 V, 1.5 A
62. yes
63. Aiding
64. F
65. F
66. delta, wye
67. wye
68. The start end of one winding must be connected to the finish end of the next winding.
69. no
70. The winding has excessive current flowing in it.
71. 1, 5, and 9; or 2, 6, and 10

CHAPTER 13

R, C, and L Circuits

Learning Outcomes

This chapter will help you to:

13-1 *Name* the opposition to current caused by a combination of resistance and reactance.

13-2 *Discuss* the use of the Pythagorean theorem in working with phasors.

13-3 *Solve* for values of I, V, X_C, R, or Z in either series or parallel *RC* circuits.

13-4 *Solve* for values of I, V, X_L, R, or Z in either series or parallel *RL* circuits.

13-5 *Calculate* the values for the unknown values in either series or parallel *RCL* circuits.

13-6 *Define* resonance and calculate resonant frequency, quality, bandwidth, and selectivity for both series and parallel resonant circuits.

13-7 *Discuss* four general classes of filters and specify the components used for each class.

Previous chapters have dealt with resistance, inductance, and capacitance used individually in ac circuits. This chapter deals with the results of combining two or more of these quantities in a single circuit.

13-1 Impedance

The combined opposition to current due to resistance and reactance is called *impedance*. The symbol for impedance is Z. Like resistance and reactance, impedance has base units of ohms.

Impedance

In a circuit that contains resistance and only one type of reactance, the current and voltage cannot be in phase. Nor can they be a full 90° out of phase. If the circuit contains resistance and capacitance, the current leads the voltage. If it contains resistance and inductance, the current lags the voltage. When the circuit contains all three (R, C, and L), the phase relationship depends upon the relative sizes of L and C.

Circuits containing resistance and only one type of reactance have more apparent power than true power. Their power factor is less than 1 (less than 100 percent).

Examples of common electric and electronic loads that possess impedance are motors, speakers, and earphones. These particular loads possess both inductive reactance and resistance.

1. True or false. In a circuit containing impedance, the current and voltage will be 90° out of phase.

2. Impedance is a combination of _____ and _____.
3. The symbol for impedance is _____.
4. The base unit for impedance is the _____.

13-2 Adding Phasors

💡 **You May Recall**

. . . from the chapter on power that the concepts of phasors and right triangles were introduced and used to solve power problems. Now it is time to review those concepts and expand them so that we can apply them to R, C, and L circuits.

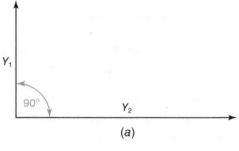

Suppose we wish to add the two phasors Y_1 and Y_2 in Fig. 13-1(a). We can add them either graphically or mathematically. First we will do it graphically, and then we will develop the mathematical method.

Graphical addition

The *graphical addition* of the phasors in Fig. 13-1(a) is illustrated in Fig. 13-1(b). The process involves the construction of a rectangle. First, line 1 is drawn from the tip of the Y_1 phasor parallel to the Y_2 phasor. Then, line 2 is drawn from the tip of the Y_2 phasor parallel to the Y_1 phasor. The point at which lines 1 and 2 cross is the tip of the resultant Y_T phasor.

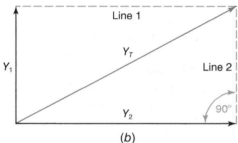

The three phasors of Fig. 13-1(b) can be rearranged into the configuration shown in Fig. 13-1(c). This rearrangement merely replaces line 2 of Fig. 13-1(b) with phasor Y_1. This is permissible because line 2 and phasor Y_1 are parallel and of the same length.

Pythagorean theorem

The rearrangement of phasors in Fig. 13-1(c) produces a right triangle. This right triangle is redrawn in Fig. 13-2 with the sides labeled. You are already familiar with the trigonometric functions listed in this figure. They are listed here merely as a review and easy reference for this chapter. Also listed in Fig. 13-2 is the formula for the relationship of the sides of a

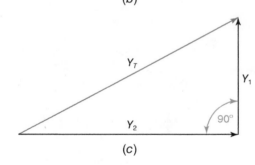

Fig. 13-1 Graphical addition of phasors. Adding Y_1 and Y_2 in (a) yields the resultant phasor (Y_T) in (b) and (c).

right triangle. This relationship, known as the *Pythagorean theorem,* applies to all right triangles regardless of the size of the triangle or of the angle θ. Thus, knowing the length of any two sides of a right triangle, you can calculate the length of the third side using the formula

$$(\text{Hypotenuse})^2 = (\text{opposite side})^2 + (\text{adjacent side})^2$$

EXAMPLE 13-1

Refer to Fig. 13-2. Suppose Y_1 represents 6 A and Y_2 represents 8 A. What is the value of the total current Y_T?

Given:	$Y_1 = 6$ A
	$Y_2 = 8$ A
Find:	Y_T
Known:	$Y_T = \sqrt{Y_1^2 + Y_2^2}$
Solution:	$Y_T = \sqrt{Y_1^2 + Y_2^2}$
	$= \sqrt{6^2 + 8^2}$
	$= \sqrt{36 + 64}$
	$= \sqrt{100} = 10$ A
Answer:	The total current is 10 A.

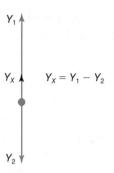

Fig. 13-3 Adding phasors separated by 180°.

Figure 13-3 illustrates how to add two phasors (Y_1 and Y_2) when the phasors are 180° apart. The resultant phasor (Y_X) is equal in length to the difference between the two phasors being added. The direction of the resultant phasor is the same as the direction of the longest phasor. As an example, assume Y_1 is 14 Ω and Y_2 is 10 Ω. Then Y_X would be 4 Ω in the direction of Y_1.

Suppose three phasors [Fig. 13-4(a)], all displaced 90° from each other, are to be added. The first step is to add the two phasors that are 180° apart. Adding Y_1 and Y_3 reduces the problem to the two phasors shown in Fig. 13-4(b). The second, and last, step involves adding the

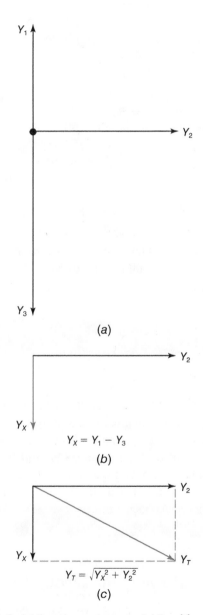

(a)

(b)

(c)

Fig. 13-4 Adding three phasors. Adding Y_1 and Y_3 in (a) results in two phasors in (b). Right triangle properties are used in (c) to find Y_T.

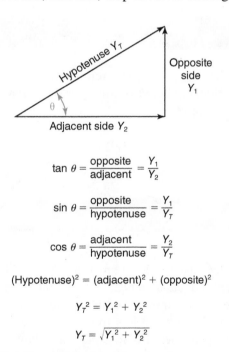

$$\tan \theta = \frac{\text{opposite}}{\text{adjacent}} = \frac{Y_1}{Y_2}$$

$$\sin \theta = \frac{\text{opposite}}{\text{hypotenuse}} = \frac{Y_1}{Y_T}$$

$$\cos \theta = \frac{\text{adjacent}}{\text{hypotenuse}} = \frac{Y_2}{Y_T}$$

$$(\text{Hypotenuse})^2 = (\text{adjacent})^2 + (\text{opposite})^2$$

$$Y_T^2 = Y_1^2 + Y_2^2$$

$$Y_T = \sqrt{Y_1^2 + Y_2^2}$$

Fig. 13-2 Properties of a right triangle.

two phasors of Fig. 13-4(*b*). This is accomplished by using the Pythagorean theorem. The result [Fig. 13-4(*c*)] is a single phasor Y_T that is equivalent to the original three phasors of Fig. 13-4(*a*). The formula given in Fig. 13-4(*c*) can be expanded so that all three phasors can be added in one step. It can be expanded by substituting $Y_1 - Y_3$ for Y_X. The result is the general formula

$$Y_T = \sqrt{(Y_1 - Y_3)^2 + Y_2^2}$$

EXAMPLE 13-2

Find the total current represented by the three current phasors in Fig. 13-5.

Given: $I_1 = 5\text{ A}$
           $I_2 = 4\text{ A}$
           $I_3 = 8\text{ A}$
Find: I_T
Known: $I_T = \sqrt{(I_1 - I_3)^2 + I_2^2}$
Solution: $I_T = \sqrt{(5 - 8)^2 + 4^2}$
             $= \sqrt{(-3)^2 + 4^2}$
             $= \sqrt{9 + 16}$
             $= \sqrt{25} = 5\text{ A}$
Answer: The three phasors represent a total current of 5 A.

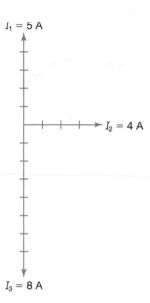

Fig. 13-5 Diagram for example 13-2.

This formula can be used to combine any three phasors that are displaced 90° from each other. The three phasors may represent voltage, current, or impedance (and the components of impedance: resistance and reactance). The only requirement is that all the phasors have the same units, that is, volts, amperes, or ohms.

Now that we know how to add phasors, we can return to problems that involve impedance. That is, we can now calculate current, voltage, and impedance in circuits containing both reactance and resistance.

 Self-Test

Answer the following questions.

5. Refer to Fig. 13-1(*a*). If $Y_1 = 7$ m and $Y_2 = 9$ m, what is the length of Y_T?
6. What is the resultant force in Fig. 13-4(*a*) when $Y_1 = 45$ N, $Y_2 = 25$ N, and $Y_3 = 30$ N?

7. The _____ theorem shows the relationship between the sides of a right triangle.

13-3 Solving *RC* Circuits

Unless stated otherwise, we will assume that the components we use are ideal devices. That is, the resistors contain only resistance, and the capacitors contain only capacitance. Actual resistors usually have small amounts of inductance and capacitance, and actual capacitors have small amounts of resistance and inductance. However, at low frequencies these undesired quantities are so small that they can be ignored without introducing significant error.

Series *RC* Circuits

Figure 13-6 shows a *series RC circuit* and the phasor diagrams that represent its voltages and oppositions. In a series circuit there is only one path for current; the capacitive current, the resistive current, and the total current are the same current. Therefore, the current phasor is often used as the reference phasor in series circuits [Fig. 13-6(b)].

In any circuit, we know that the resistive voltage and the resistive current are in phase. This means that the resistive voltage phasor [V_R in Fig. 13-6(b)] is in the same direction as I_T. We also know that the current through a capacitor leads the voltage across the capacitor by 90°. Another way of looking at this is to say that the voltage lags the current by 90°.

You May Recall

. . . from the chapter on power that phasors are assumed to rotate in a counterclockwise direction. Thus, the capacitive voltage phasor [V_C in Fig. 13-6(b)] is plotted 90° behind the current phasor.

Now that we have V_R and V_C, they can be added together to obtain V_T. Phasor V_T establishes angle θ. Angle θ shows how much the source current I_T leads the source voltage V_T. As you can see in Fig. 13-6, V_C is greater than V_R. This makes angle θ greater than 45°.

Also from Fig. 13-6(b) we can conclude that X_C must be greater than R. We make this conclusion because of the following facts:

1. $I_R = I_C$.
2. Ohm's law applies individually to both the resistor and the capacitor.
3. Ohm's law states that the voltage and the resistance or reactance are directly proportional if the current is held constant.

Thus, if V_C is greater than V_R, then X_C must also be greater than R.

Figure 13-6(c) shows that the resistance and capacitive reactance are also 90° out of phase. This is to be expected. The resistor and capacitor currents are in phase, but their voltages are out of phase. Therefore, X_C and R must also be 90° out of phase.

The phasors of Fig. 13-6(b) and (c) can be redrawn into triangles, as shown in Fig. 13-7. From Fig. 13-7 it can be seen that the

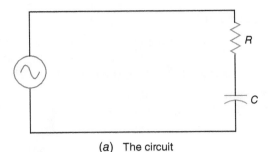

(a) The circuit

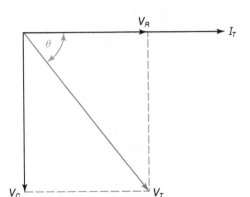

(b) Voltage phasors

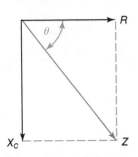

(c) *R*, X_C, and *Z* phasors

Fig. 13-6 Series *RC* circuit. The voltages and the oppositions are shown as phasors for this circuit.

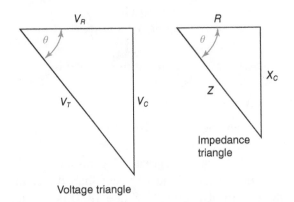

Voltage triangle

Impedance triangle

Fig. 13-7 Voltage and impedance triangles. Angle θ is the same in both triangles.

$$V_T = \sqrt{V_R^2 + V_C^2}$$

$$Z = \sqrt{R^2 + X_C^2}$$

Pythagorean theorem can be used to calculate V_T and Z:

$$V_T = \sqrt{V_R^2 + V_C^2} \quad \text{and} \quad Z = \sqrt{R^2 + X_C^2}$$

The ratio of R over X_C [in Fig. 13-6(c)] must be equal to the ratio of V_R over V_C [in Fig. 13-6(b)]. This statement is easily seen by looking at Ohm's law:

$$I_T = \frac{V_R}{R} \quad \text{and} \quad I_T = \frac{V_C}{X_C}$$

Therefore,

$$\frac{V_R}{R} = \frac{V_C}{X_C}$$

Transposing terms:

$$\frac{V_R}{V_C} = \frac{R}{X_C}$$

This means that angle θ can be figured equally well with voltages as with oppositions. In other words,

$$\cos \theta = \frac{\text{adjacent}}{\text{hypotenuse}} = \frac{V_R}{V_T} = \frac{R}{Z}$$

Now that we know the relationship between impedance, resistance, and reactance in series RC circuits, let us use it to solve an electrical problem.

EXAMPLE 13-3

Find the impedance of the circuit in Fig.13-8.

Given:	$C = 1\ \mu F$
	$R = 1000\ \Omega$
	$f = 100\ \text{Hz}$
Find:	Z
Known:	$Z = \sqrt{R^2 + X_C^2}$
	$X_C = \dfrac{1}{6.28fC}$
Solution:	$X_C = \dfrac{1}{6.28 \times 100 \times 1 \times 10^{-6}}$
	$= 1592\ \Omega$
	$Z = \sqrt{1000^2 + 1592^2}$
	$= 1880\ \Omega$
Answer:	The circuit impedance is 1880 Ω.

We can check the reasonableness of our answer very easily. Look again at Fig. 13-7 and notice that Z must be greater than either R or X_C. Also notice that Z must be less than the arithmetic sum of R and X_C. Since 1880 is greater than 1592 but less than 2592, it seems to be a reasonable answer.

Fig. 13-8 Circuit for examples 13-3, 13-4, and 13-5.

Impedance can be used in Ohm's law just like resistance and reactance. For instance, the impedance and source voltage can be used to find the total current. Then the current can be used to calculate individual voltages.

EXAMPLE 13-4

Find the current I_T, the resistive voltage V_R, and the capacitive voltage V_C for Fig. 13-8.

Given:	$V_T = 40\ \text{V}$
	$R = 1000\ \Omega$
	$X_C = 1592\ \Omega$
	(from example 13-3)
	$Z = 1880\ \Omega$
	(from example 13-3)
Find:	I_T, V_C, and V_R
Known:	Ohm's law
Solution:	$I_T = \dfrac{40\ \text{V}}{1880\ \Omega}$
	$= 0.0213\ \text{A}$
	$= 21.3\ \text{mA}$
	$V_R = 0.0213\ \text{A} \times 1000\ \Omega$
	$= 21.3\ \text{V}$
	$V_C = 0.0213\ \text{A} \times 1592\ \Omega$
	$= 33.9\ \text{V}$
Answer:	The circuit current is 21.3 mA, the resistive voltage is 21.3 V, and the capacitive voltage is 33.9 V.

We can verify our answer by checking whether the phasor sum of V_R and V_C equals the source voltage V_T:

$$V_T = \sqrt{21.3^2 + 33.9^2} = 40\ V$$

Since this calculated value of V_T agrees with the value given, we have not made any errors in our calculation.

Now we can continue our study of the circuit in Fig. 13-8 by determining angle θ and power.

EXAMPLE 13-5

For the circuit of Fig. 13-8, find angle θ and the total true power.

Given: $R = 1000\ \Omega$
$V_T = 40\ V$
$Z = 1880\ \Omega$
$I_T = 21.3\ mA$

Find: θ and P

Known: $\cos \theta = \dfrac{R}{Z}$, power formulas

Solution: $\cos \theta = \dfrac{1000\ \Omega}{1880\ \Omega} = 0.532$

$\theta = 58°$ (from Appendix G)
$P_T = IV \cos \theta$
$\quad = 0.0213 \times 40 \times 0.532$
$\quad = 0.453\ W$

Answer: The current leads the voltage by 58°, and the power in the circuit is 0.453 W.

In example 13-5, we could have found the total power without first calculating θ. Recall that only the resistance in a circuit can use power. Therefore, the power used by the resistor (P_R) must equal the total power. Thus,

$$P_T = P_R = I_R \times V_R = 0.0213\ A \times 21.3\ V$$
$$= 0.454\ W$$

This answer agrees (within round-off error) with the answer in example 13-5. Remember that $P_T = P_R$ for all circuits.

Also, in example 13-5 we could have found $\cos \theta$ using voltage ratios. This would give us

$$\cos \theta = \frac{V_R}{V_T} = \frac{21.3\ V}{40\ V} = 0.533$$

Or we could have also used the PF formula because PF = $\cos \theta$. This would result in

$$\cos \theta = PF = \frac{P}{P_{app}} = \frac{P_R}{P_{app}}$$
$$= \frac{21.3\ mA \times 21.3\ V}{21.3\ mA \times 40\ V} = 0.533$$

Both of these answers agree (within round-off error) with that obtained in example 13-5.

Suppose the frequency of the source in Fig. 13-8 were decreased. What would happen to angle θ, the impedance, and the voltage distribution? The answer to this question is illustrated

in Fig. 13-9(a) and (b). In Fig. 13-9(a) the phasor diagrams for the circuit operating at 100 Hz are shown. Figure 13-9(b) shows what happens when the frequency is decreased to 50 Hz. When the frequency decreases, the reactance of the capacitor increases. This causes an increase in the impedance and a corresponding decrease in current. Less current means less voltage across the resistor. However, the capacitive voltage increases because its reactance increased proportionately more than the current decreased. The net result is that the circuit is more capacitive than before. Angle θ has increased and the power in the circuit has decreased.

When the frequency *increases*, the results [Fig. 13-9(c)] are just the opposite: the circuit becomes more resistive, θ decreases, and the power increases.

Figure 13-10 illustrates what happens when the capacitor in Fig. 13-8 is changed. The diagrams in Fig. 13-10(a) show the situation with the original capacitor (1 μF). When the capacitance is decreased [Fig. 13-10(b)], the effects are the same as when the frequency is decreased. This is because capacitive reactance is inversely proportional to both capacitance and frequency. Decreasing capacitance causes the capacitive reactance to increase; that is, θ becomes larger.

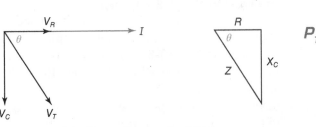

(a) Phasor diagrams for 100 Hz

(b) Phasor diagrams for 50 Hz

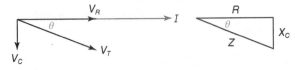

(c) Phasor diagrams for 200 Hz

Fig. 13-9 Effects of f on Z, V, and θ.

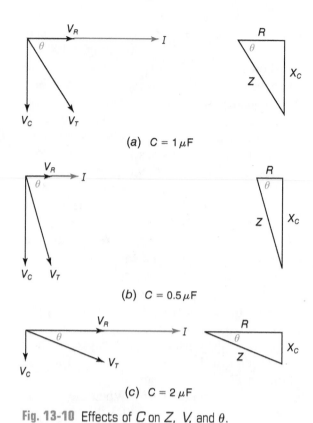

(a) $C = 1\,\mu F$

(b) $C = 0.5\,\mu F$

(c) $C = 2\,\mu F$

Fig. 13-10 Effects of C on Z, V, and θ.

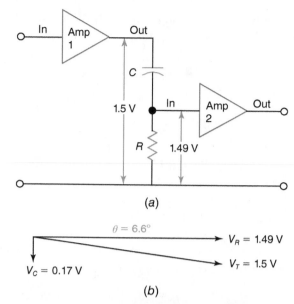

(a)

(b)

Fig. 13-11 Use of an RC circuit. The capacitor couples the amplifiers together with minimum voltage loss or phase shift.

Reference phasor

Parallel RC circuits

Increasing the capacitance, of course, decreases the capacitive reactance and causes the circuit to be more resistive [Fig. 13-10(c)].

An example of how an RC circuit is used is shown in Fig. 13-11. Amplifier 1 replaces the generator in Fig. 13-8. In other words, amplifier 1 is the source which provides the voltage across the RC circuit. Whatever voltage develops across R becomes the input voltage to the second amplifier. It is desirable to get as much as possible of the voltage from amplifier 1 to the input of amplifier 2. Therefore, the reactance of the capacitor must be low relative to the resistance of the resistor. The resistance should be about 10 times as high as the reactance. This makes the resistive voltage [Fig. 13-11(b)] nearly equal to the source voltage. Also, Fig. 13-11(b) shows that very little phase shift occurs when R is high relative to X_C. This means that the voltage input to amplifier 2 is nearly in phase with the output of amplifier 1.

Parallel RC Circuits

In a parallel circuit [Fig. 13-12(a)], the voltage is the same across all components. Obviously,

V_R, V_C, and V_T have to be in phase. Therefore, the voltage phasor is used as the *reference phasor* [Fig. 13-12(b)] in solving *parallel RC circuits*. The capacitive current still must lead the capacitive voltage by 90°. Thus, the I_C phasor is drawn 90° ahead of the voltage phasor [Fig. 13-12(b)]. As is always the case, the resistive

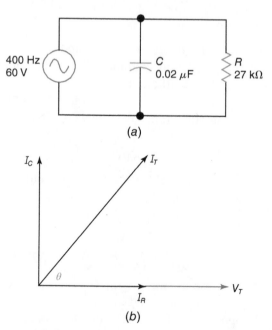

(a)

(b)

Fig. 13-12 Parallel RC circuit. The voltage phasor is the reference (0°) phasor. Notice the current phasors' relationships.

current is in phase with the resistive voltage. The I_R phasor has to be in line with the V_T phasor. Adding phasors I_C and I_R yields the I_T phasor. Since the circuit has capacitance, the source current I_T leads the source voltage V_T.

The impedance of the circuit shown in Fig. 13-12(a) can be found by using Ohm's law with V_T and I_T:

$$Z = \frac{V_T}{I_T}$$

Of course, one must calculate I_T before Z can be determined. The value of I_T can be calculated by using the Pythagorean theorem. Just express the sides of the triangle in terms of currents. The formula is

$$I_T = \sqrt{I_C^2 + I_R^2}$$

EXAMPLE 13-6

Find the impedance of the circuit in Fig. 13-12.

Given: $V_T = 60$ V
$f = 400$ Hz
$C = 0.02\ \mu F$
$R = 27$ kΩ

Find: Z

Known: Ohm's law
$X_C = \dfrac{1}{6.28fC}$
$I_T = \sqrt{I_C^2 + I_R^2}$

Solution: $X_C =$

$\dfrac{1}{6.28 \times 400 \times 0.02 \times 10^{-6}}$
$= 19,900\ \Omega = 19.9$ kΩ

$I_C = \dfrac{60\ \text{V}}{19,000\ \Omega} = 0.00301$ A
$= 3.0$ mA

$I_R = \dfrac{60\ \text{V}}{27,900\ \Omega} = 0.00222$ A
$= 2.22$ mA

$I_T = \sqrt{0.003^2 + 0.00222^2}$
$= 0.00373$ A $= 3.73$ mA

$Z = \dfrac{60\ \text{V}}{0.00373\ \text{A}} = 16,100\ \Omega$
$= 16$ kΩ

Answer: The impedance of the circuit is 16 kΩ.

Notice in this example that the impedance is less than the capacitive reactance or the resistance. This is because the reactance and the resistance are in parallel as well as 90° out of phase.

The impedance of a parallel RC circuit can also be calculated from the resistance and reactance. The formula is

$$Z = \frac{X_C \times R}{\sqrt{X_C^2 + R^2}}$$

Looking at this formula, one can see that it has the "product ÷ sum" structure needed for parallel components and the square root of the sum of squares needed for 90°-out-of-phase quantities.

This formula yields the same results as the Ohm's law method. Therefore, we can use it to check the answer arrived at in example 13-6:

$$Z = \frac{19,900 \times 27,000}{\sqrt{19,900^2 + 27,000^2}}$$
$$= 16,000\ \Omega = 16.0\ \text{kΩ}$$

Since the two methods provide the same answer (within round-off error), we can assume that no calculation errors were made.

The effects of changing f or C in a parallel RC circuit are illustrated in Fig. 13-13. Notice that the effect on angle θ is the exact opposite of the effect in a series RC circuit. Increasing C or f causes a decrease in reactance, which results in less impedance, more current, and a larger angle θ. Of course, decreasing C or f has the opposite effect.

The value of $\cos \theta$ can be calculated from either the currents or the powers. The formulas are

$$\cos \theta = \frac{P}{P_{app}} \quad \text{and} \quad \cos \theta = \frac{I_R}{I_T}$$

These are also the power-factor formulas. Once $\cos \theta$ is determined, angle θ can be determined by using a calculator or a table of trigonometric functions. We can find the power factor of the circuit in example 13-6 using the values for I_R and I_T:

$$\cos \theta = \text{PF} = \frac{I_R}{I_T} = \frac{0.00222}{0.00373}$$
$$= 0.595, \text{ or } 59.5 \text{ percent}$$

$$Z = \frac{X_C \times R}{\sqrt{X_C^2 + R^2}}$$

$$I_T = \sqrt{I_C^2 + I_R^2}$$

$$cos\ \theta = \frac{I_R}{I_T}$$

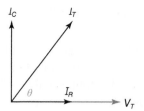
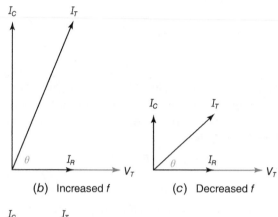

(a) Original phasor diagram

(b) Increased f

(c) Decreased f

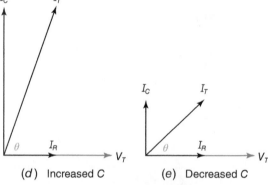

(d) Increased C

(e) Decreased C

Fig. 13-13 Effects of f and C on I and θ.

Since the total current is leading the voltage, the power factor of the circuit is 59.5 percent, leading.

The $\cos^{-1}$ function on a calculator indicates that the angle with a cosine of .595 is 53.487°.

Self-Test

Answer the following questions.

8. The current _____ the voltage in a parallel RC circuit.
9. True or false. In a series RC circuit, V_C is larger than V_T.
10. In a series RC circuit, the reference phasor is the _____ phasor.
11. True or false. If I_R is greater than I_C in a parallel RC circuit, θ will be greater than 45°.
12. In a series RC circuit, I_T leads V_T by 20°. What is the angle between R and Z for this circuit?
13. What is the phase relationship between resistance and reactance?

14. Write two formulas for finding the impedance of a series RC circuit.
15. Write the formula for combining the resistive and reactive voltages in a series RC circuit.
16. Write three formulas for finding $\cos \theta$ in a series RC circuit.
17. Which of the formulas in question 16 can also be used in a parallel RC circuit?
18. What other formula besides the one given in question 17 can be used to find $\cos \theta$ in a parallel RC circuit?
19. If the frequency of a parallel RC circuit decreases what happens to
    a. The impedance

b. The resistive current
c. The power
d. $\cos \theta$

20. If the resistance of a series *RC* circuit increases, what happens to
   a. The impedance
   b. The resistive voltage
   c. The current
   d. $\cos \theta$

21. Refer to Fig. 13-12. Change the value of C to 0.03 μF and calculate the following:

a. Z
b. P
c. θ
d. I_C

22. Refer to Fig. 13-8. Change C to 1.5 μF and calculate the following:
   a. I_T
   b. V_R
   c. P
   d. Z

13-4 Solving *RL* Circuits

Although *RL* circuits are not as common as *RC* circuits in electronics, they are important. One must know something of their characteristics to fully understand such devices as ac motors and high-frequency amplifiers.

Series *RL* Circuits

The *series RL circuit* [Fig. 13-14(a)] causes the source voltage to lead the source current. This is shown in Fig. 13-14(b), where V_T leads I_T (the reference phasor). The same general formulas used for series *RC* circuits can also be used for series *RL* circuits; merely replace the *C* with *L* and the X_C with X_L. The important formulas for series *RL* circuits are

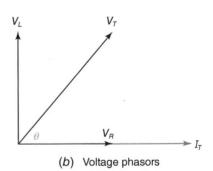

(a) The circuit

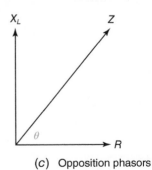

(b) Voltage phasors

(c) Opposition phasors

Fig. 13-14 Series *RL* circuit.

$$Z = \sqrt{R^2 + X_L^2}$$
$$V_T = \sqrt{V_R^2 + V_L^2}$$
$$\cos \theta = \frac{V_R}{V_T} = \frac{R}{Z} = \frac{P}{P_{app}}$$

Series *RL* circuits

$$Z = \sqrt{R^2 + X_L^2}$$
$$V_T = \sqrt{V_R^2 + V_L^2}$$

EXAMPLE 13-7

Determine the impedance, the voltage across the inductance, and the power factor for the circuit in Fig. 13-14.

Given: $V_T = 50$ V
         $f = 50$ kHz
         $L = 10$ mH
         $R = 2700 \ \Omega$

Find: Z, V_L, and PF

Known: $Z = \sqrt{R^2 + X_L^2}$
         $X_L = 6.28fL$
         PF $= \cos \theta = \dfrac{R}{Z}$

Ohm's law

Solution:

$$X_L = 6.28 \times 50 \times 10^3 \times$$
$$10 \times 10^{-3}$$
$$= 3140 \ \Omega$$

$$Z = \sqrt{2700^2 + 3140^2}$$
$$= 4141 \ \Omega$$

$$I_T = \frac{50 \ V}{4141 \ \Omega}$$
$$= 0.012 \ A = 12 \ mA$$

$$V_L = 0.012 \ A \times 3140 \ \Omega$$
$$= 37.7 \ V$$

$$PF = \frac{2700}{4141} = 0.652$$

AC motors

Answer: The impedance of the circuit is 4141 Ω, the voltage across the inductor is 37.7 V, and the power factor is 0.652.

Series *RL* and series *RC* circuits respond to changes in frequency in exactly opposite fashions. This is because inductive reactance is directly proportional to frequency whereas capacitive reactance is inversely proportional to frequency. Increasing the frequency of a series *RL* circuit causes the following to occur: X_L increases, *Z* increases, I_T decreases, V_R decreases, V_L increases, and angle θ increases. Inductive reactance is also directly proportional to inductance. Therefore, increasing the inductance has the same effect as increasing the frequency.

Power and angle θ in series *RL* and series *RC* circuits are determined in exactly the same way. The difference in the two types of circuits is that the current lags the voltage in an *RL* circuit whereas the current leads the voltage in an *RC* circuit (compare Figs. 13-6 and 13-14).

Many small, single-phase *ac motors* are started by the phase shift between two *RL* circuits. These motors have two separate windings in them. The run winding uses large-diameter wire and therefore has only a small amount of resistance. The run winding can be represented by a small resistor in series with an ideal inductor [Fig. 13-15(*a*)]. Since the run winding has a low resistance and a high reactance, its current lags far behind the voltage [Fig. 13-15(*b*)]. The start winding is wound with small-diameter wire; it

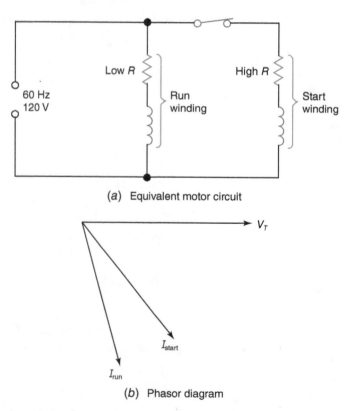

(*a*) Equivalent motor circuit

(*b*) Phasor diagram

Fig. 13-15 An ac split-phase motor circuit. (*a*) Equivalent circuit of a split-phase motor under starting conditions. (*b*) Phasor diagram showing the current in the run winding out of phase with the current in the start winding.

also has fewer turns of wire than the run winding. Therefore, the start winding has more resistance and less inductance than the run winding. This causes the start-winding current to lag the voltage less than the run-winding current does. Thus, the start-winding current is out of phase with the run-winding current [Fig. 13-15(b)]. These out-of-phase currents produce the special magnetic field needed to start the motor. Once the motor has started, a switch [Fig. 13-15(a)] disconnects the start winding from the circuit. The motor continues to run on the run winding alone. The chapter on electric motors provides the details of how such motors operate.

Parallel RL Circuits

A *parallel RL circuit* and its phasor diagram are shown in Fig. 13-16. Notice in Fig. 13-16(b) that the reference phasor is V_T and the resistive current (I_R) is in phase with V_T just as it was in the series RC circuit. Parallel RL circuits can be treated like parallel RC circuits. There are only two differences between the two types of circuits:

1. The current in the RL circuit lags the voltage, whereas the current in the RC circuit leads the voltage. [Compare Figs. 13-12(b) and 13-16(b).]
2. The two circuits respond in opposite ways to changes in frequency and inductance or capacitance. When either f or L is

increased in the parallel RL circuit, the impedance increases. This results in the inductive current and the total current decreasing and angle θ decreasing.

The following formulas are the major ones needed to work with parallel RL circuits:

$$I_T = \sqrt{I_L^2 + I_R^2}$$

$$Z = \frac{V_T}{I_T} = \frac{X_L \times R}{\sqrt{X_L^2 + R^2}}$$

$$\cos \theta = \frac{P}{P_{app}} = \frac{I_R}{I_T}$$

Notice that these formulas are the same as the formulas used for parallel RC circuits except that L is substituted for C.

$$I_T = \sqrt{I_L^2 + I_R^2}$$

$$Z = \frac{V_T}{I_T}$$

$$= \frac{X_L \times R}{\sqrt{X_L^2 + R^2}}$$

Parallel RL circuit

EXAMPLE 13-8

For the circuit of Fig. 13-16, calculate the total current, the impedance, and the power.

Given: $V_T = 50$ V
$f = 50$ kHz
$L = 10$ mH
$R = 2700 \ \Omega$

Find: I_T, Z, and P

Known: Ohm's law
$X_L = 6.28fL$
$I_T = \sqrt{I_L^2 + I_R^2}$
$P_T = P_R = I_R^2 R$

Solution: $I_R = \dfrac{50 \text{ V}}{2700 \ \Omega} = 0.0185$ A

$= 18.5$ mA
$X_L = 6.28 \times 50 \times 10^3$
$\times 10 \times 10^{-3}$
$= 3140 \ \Omega$

$I_L = \dfrac{50 \text{ V}}{3140 \ \Omega} = 0.0159$ A

$= 15.9$ mA
$I_T = \sqrt{0.0185^2 + 0.0159^2}$
$= 0.0244$ A $= 24.4$ mA

$Z = \dfrac{50 \text{ V}}{0.0244 \text{ A}} = 2049 \ \Omega$

$P = 0.0185^2 \times 2700$
$= 924$ mW

Answer: The current is 24.4 mA, the impedance is 2049 Ω, and the power is 924 mW.

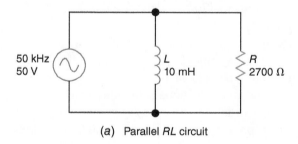

(a) Parallel RL circuit

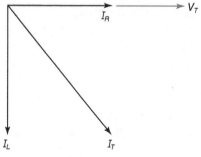

(b) Phasor diagram

Fig. 13-16 Parallel *RL* circuits. The total current lags the total voltage.

Answer the following questions.

23. In *RL* circuits, does the source current lead or lag the source voltage?
24. Suppose the inductance of a series *RL* circuit is reduced. What happens to I_T, V_L, and Z?
25. When the frequency of a parallel *RL* circuit is increased, what happens to Z and angle θ?
26. A 5-mH inductor and a 1500-Ω resistor are connected in series to an ac source. What is the impedance of this circuit at 30 kHz?

27. Determine the current and the power for the circuit in question 26 when the source voltage is 50 V.
28. Change *R* in Fig. 13-16 to 2.2 kΩ and solve for I_T and *Z*.
29. Is it necessary to know the frequency of the voltage source to determine the power in a parallel *RL* circuit? Explain.
30. List four component values that can be changed (and the direction of the change) to increase I_T in Fig. 13-16.

13-5 Solving *RCL* Circuits

The techniques used to solve *RC* and *RL* circuits can be combined to solve *RCL* circuits such as in Fig. 13-17(*a*). However, *RCL* circuits have some unusual characteristics. For instance, a reactive voltage or current can be higher than the source voltage or current.

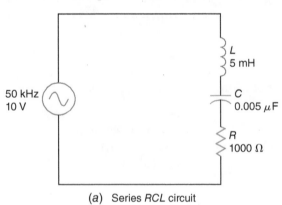

(a) Series *RCL* circuit

Net reactive voltage

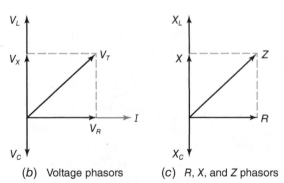

(b) Voltage phasors (c) *R*, *X*, and *Z* phasors

Fig. 13-17 Series *RCL* circuit.

Series *RCL* Circuits

Refer to the phasors in Fig. 13-17(*b*). Notice that the inductive and the capacitive voltage are 180° out of phase with each other. Adding these two voltage phasors results in V_X, which is the net reactive voltage. V_X is then added to V_R in order to determine the total, or source, voltage. The reactance and resistance phasors [Fig. 13-17(*c*)] are added in the same sequence to determine the impedance.

In Fig. 13-17(*b*) you can see that V_L is greater than V_T. At first glance this may look like an exception to Kirchhoff's voltage law; however, it is not. At any instant the sum of the instantaneous voltages of the resistor, the capacitor, and the inductor equals the source voltage. In other words, the three voltages are never of the same polarity at the same instant. This idea is illustrated in Fig. 13-18. Notice the voltage drops across each component (*L*, *C*, and *R*) in Fig. 13-18(*a*). These voltage drops are a result of the series current passing through each component. The voltage across *L*, however, is 180° out of phase with the voltage across *C*. This means that the two voltages are directly opposing one another. The result is the *net reactive voltage* shown in Fig. 13-18(*b*) across the combination of *L* and *C*.

Notice in Fig. 13-18(*a*) that both the inductive voltage and the capacitive voltage are greater than the source voltage. If the individual

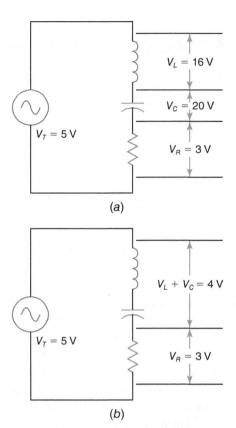

(a)

(b)

Fig. 13-18 Voltmeter readings in a series *RCL* circuit. Notice that both the inductive and the capacitive voltage in (a) exceed the source voltage.

voltages exceed the source voltage, then the individual reactances exceed the impedance. This is common in series *RCL* circuits.

Impedance and total voltage in a series circuit can be calculated with the aid of the following formulas:

$$Z = \sqrt{(X_L - X_C)^2 + R^2}$$

$$Z = \frac{V_T}{I_T}$$

$$V_T = \sqrt{(V_L - V_C)^2 + V_R^2}$$

In these formulas, $V_L - V_C$ is the same as V_X in Fig. 13-17(b), and $X_L - X_C$ is the same as X.

The power factor (cos θ) and angle θ of the series *RCL* circuits can be found by using the same formulas used for series *RL* and *RC* circuits. These formulas are PF = cos θ = R/Z and PF = cos θ = V_R/V_T. For any combination of R, C, L circuits, the general formula PF = cos θ = P/P_{app} is appropriate.

Notice from example 13-9 that X_L is greater than X_C. This causes the circuit to be inductive. The circuit produces a lagging power factor.

EXAMPLE 13-9

Find the impedance and the voltage across the resistance for the circuit in Fig. 13-17(a).

Given: $V_T = 10$ V
  $f = 50$ kHz
  $L = 5$ mH
  $C = 0.005 \ \mu$F
  $R = 1000 \ \Omega$

Find: Z and V_R

Known: Ohm's law
  $Z = \sqrt{(X_L - X_C)^2 + R^2}$
  $X_L = 6.28fL$
  $X_C = \dfrac{1}{6.28fC}$

Solution:
  $X_L = 6.28 \times 50 \times 10^3 \times 5 \times 10^{-3}$
  $= 1570 \ \Omega$

  $X_C = \dfrac{1}{6.28 \times 50 \times 10^3 \times 0.005 \times 10^{-6}}$
  $= 637 \ \Omega$

  $Z = \sqrt{(1570 - 637)^2 + 1000^2}$
  $= 1368 \ \Omega$

  $I_T = \dfrac{10 \text{ V}}{1368 \ \Omega} = 0.0073 \text{ A} = 7.3 \text{ mA}$

  $V_R = 0.0073 \text{ A} \times 1000 \ \Omega = 7.3$ V

Answer: The impedance is 1368 Ω, and the voltage across the resistance is 7.3 V.

Suppose the frequency in Fig. 13-17(a) were reduced to 25 kHz. This would cause X_L to be 50 percent of its former value, and X_C, twice its former value (785 Ω and 1274 Ω, respectively). Now the circuit would be capacitive; I_T would lead V_T, and the power factor would be leading.

ABOUT ELECTRONICS

Sharks with Powerful Memories The ocean floor is magnetized north to south except around underwater mountains, where lava flow makes magnetic spokes. Hammerhead sharks use the spokes as roads to relocate feeding grounds; they produce a current and then sense the magnetic differentials.

Parallel *RCL* Circuits

Figure 13-19 shows a *parallel RCL circuit* and
its current phasor diagram. From the phasor
diagram it is obvious that a branch current (I_C)
can exceed the total current. (This is because
the inductive current and the capacitive currents
are 180° out of phase.) In fact, both capacitive
and inductive currents can exceed the total current. For example, if *L* in Fig. 13-19(*a*) were
halved, the inductive current would double.
Both I_X and I_T would decrease; then I_L and I_C
would exceed I_T.

The formulas for working with parallel *RCL*
circuits are

$$I_T = \sqrt{(I_L - I_C)^2 + I_R^2}$$

$$\cos \theta = \frac{I_R}{I_T} = \frac{P}{P_{app}}$$

Notice that the formulas for determining $\cos \theta$
are the same as the ones used for parallel *RC*
and *RL* circuits.

$$I_T = \sqrt{(I_L - I_C)^2 + I_R^2}$$

EXAMPLE 13-10

Calculate the values of I_T and *Z* for the circuit in Fig. 13-19. (Note: X_L and X_C have the
same values as in example 13-9.)

Given: $V_T = 10\ \text{V}$
$R = 1000\ \Omega$
$X_L = 1570\ \Omega$
$X_C = 637\ \Omega$

Known: Ohm's law
$$I_T = \sqrt{(I_C - I_L)^2 + I_R^2}$$

Solution:
$$I_C = \frac{10\ \text{V}}{637\ \Omega} = 0.0157\ \text{A} = 15.7\ \text{mA}$$

$$I_L = \frac{10\ \text{V}}{1570\ \Omega} = 0.0064\ \text{A} = 6.4\ \text{mA}$$

$$I_R = \frac{10\ \text{V}}{1000\ \Omega} = 0.0010\ \text{A} = 10\ \text{mA}$$

$$I_T = \sqrt{(0.0064 - 0.0157)^2 + 0.01^2}$$

$$= 0.0137\ \text{A} = 13.7\ \text{mA}$$

$$Z = \frac{10\ \text{V}}{0.0137\ \Omega} = 730\ \Omega$$

Answer: The impedance is 730 Ω, and
the current is 13.7 mA.

The impedance of example 13-10 can also
be calculated directly from the values of *R*, *C*,
and *L* using the formula given in Appendix C,
"Formulas and Conversions."

The parallel *RCL* circuit of Fig. 13-19 and
the series *RCL* circuit of Fig. 13-17 have the
same component values. They also have the
same source voltage and frequency. As expected, the parallel circuit has less impedance
and draws more current than the series circuit
does. Inspection of the phasor diagrams for the
two circuits shows another difference between
series and parallel *RCL* circuits: one of them is
inductive and the other is capacitive. Notice in
Fig. 13-17(*b*) that the source current lags the
source voltage. Thus, this series *RCL* circuit
is inductive. In Fig. 13-19(*b*), the source current leads the source voltage. Therefore, this
parallel circuit (with the same components as
the series circuit) is capacitive. In Figs. 13-17
and 13-19, X_L is greater than X_C. If X_C were
greater than X_L, the series circuit would be capacitive and the parallel circuit would be inductive (Fig. 13-20).

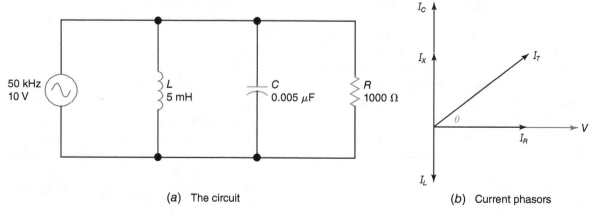

(*a*)   The circuit

(*b*)   Current phasors

Fig. 13-19 Parallel *RCL* circuit.

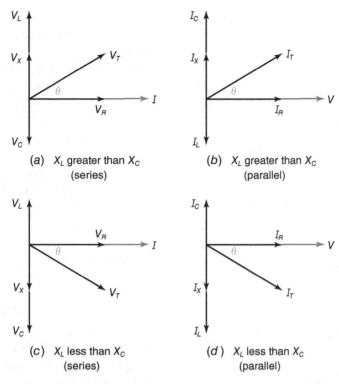

(a) X_L greater than X_C
(series)

(b) X_L greater than X_C
(parallel)

(c) X_L less than X_C
(series)

(d) X_L less than X_C
(parallel)

Fig. 13-20 Capacitive and inductive *RCL* circuits. Diagrams (*b*) and (*c*) are for capacitive circuits.

Self-Test

Answer the following questions.

31. True or false. The reactance in a series *RCL* circuit can exceed the impedance.
32. True or false. The resistance in a series *RCL* circuit can exceed the impedance.
33. True or false. The reactive current in a parallel *RCL* circuit cannot exceed the total current.
34. Write two formulas for finding the impedance of a series *RCL* circuit.
35. Write the formula for adding the branch currents in a parallel *RCL* circuit.
36. Two *RCL* circuits, one series and the other parallel, are connected to the same frequency. If X_L is greater than X_C, which circuit
    a. Is inductive
    b. Is capacitive
    c. Draws the higher source current
37. A 9-mH inductor, a 0.005-μF capacitor, and a 2000-Ω resistor are connected in series to a 30-V, 40-kHz source. Find Z, I_T, and V_L.
38. Assume that the components of question 37 are connected in parallel. Determine the impedance, the angle θ, and the current through the capacitor.
39. Is the PF in question 38 leading or lagging?

13-6 Resonance

Circuits in which $X_L = X_C$ are called *resonant circuits*. They can be either series or parallel circuits and either *RCL* or *LC* circuits. Most often, resonant circuits are *LC* circuits; the only

resistance in these circuits is that of the inductor and capacitor.

Resonant Frequency

For a given value of L and C, there is only one frequency at which X_L equals X_C. This frequency,

Resonant circuits

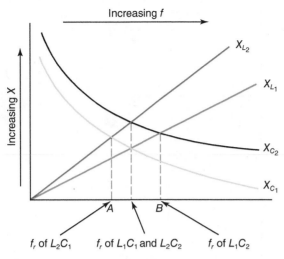

Increasing f

X_{L_2}

X_{L_1}

X_{C_2}

X_{C_1}

f_r of L_2C_1 f_r of L_1C_1 and L_2C_2 f_r of L_1C_2

Fig. 13-21 Plot of X_L and X_C versus f. At resonance, $X_L = X_C$.

Resonant frequency

$$f_r = \frac{1}{6.28\sqrt{LC}}$$

Parallel resonant circuit

called the *resonant frequency,* can be calculated with the following formula:

$$f_r = \frac{1}{6.28\sqrt{LC}}$$

In this formula, the resonant frequency f_r is in hertz if inductance is in henrys and capacitance is in farads. The resonant frequency formula is derived from the two reactance formulas. By definition, the resonant frequency is that frequency at which $X_L = X_C$. Therefore,

$$6.28fL = \frac{1}{6.28fC}$$

Solving this equation for f yields the formula for f_r given above.

Figure 13-21 further explains the concept of resonant frequency. This figure is a plot of reactance against frequency. The reactance X_{L_1} of a specific value of inductance L_1, when plotted against frequency, produces a straight line. The reactance X_{C_1} of a specific value of capacitance C_1 produces a curved line. The resonant frequency of L_1C_1 is the frequency at which the X_{L_1} line and the X_{C_1} line cross.

We have seen that L_1C_1 can be resonant at only one frequency. However, any number of other values of L and C can also be resonant at the same frequency as L_1C_1. Suppose a higher inductance and a lower capacitance, such as L_2 and C_2 of Fig. 13-21, are used. They are resonant at the same frequency as L_1C_1. One could also use a lower value of L and a higher value of C to be resonant at the same frequency as L_1C_1 and L_2C_2.

Notice two other points in Fig. 13-21. If the inductance increases to L_2 and the capacitance remains at C_1, the resonant frequency drops to point A. On the other hand, if the inductance remains at L_1 and the capacitance decreases (that is, X_C increases) to C_2, then the resonant frequency increases to point B.

The relationship between change in L or C and the resultant change in f_r is also evident from the resonant frequency formula. From the formula, it can be seen that increasing either L or C decreases f_r. Conversely, decreasing either L or C increases f_r.

EXAMPLE 13-11

What is the resonant frequency of a 10-mH inductor and a 0.005-μF capacitor?

Given:	$L = 10$ mH
	$C = 0.005\ \mu$F
Find:	f_r
Known:	$f_r = \dfrac{1}{6.28\sqrt{LC}}$

Solution:

$$f_r = \frac{1}{6.28\sqrt{10 \times 10^{-3} \times 0.005 \times 10^{-6}}}$$
$$= 22{,}500 \text{ Hz} = 22.5 \text{ kHz}$$

Answer: The resonant frequency is 22.5 kHz.

The capacitor and the inductor in the above example could be connected either in series or in parallel. The resonant frequency is the same in either case.

Parallel Resonant Circuits

The phasor diagram in Fig. 13-22 reveals the major characteristics of the *parallel resonant circuit*. In this circuit, the inductive current exactly cancels the capacitive current. Therefore, the total current and the resistive current are the same. This means that the source current and voltage are in phase, angle θ is zero, and the power factor (cos θ) is 1. Thus, the parallel resonant circuit is purely resistive.

The only current drawn from the power source in Fig. 13-22 is the current that flows through the resistor. Yet, the capacitive and inductive branches have large currents flowing in them. This is because the inductor and the capacitor are just transferring energy back and

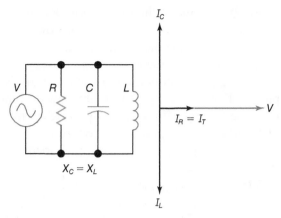

Fig. 13-22 Parallel resonant *RCL* circuit. Notice that $I_C = I_L$.

forth between themselves. This idea is graphically represented in Fig. 13-23, which shows the capacitor and inductor removed from the rest of the circuit. Suppose that the inductor and capacitor parts of the circuit are removed at the instant the capacitor is fully charged [Fig. 13-23(*a*)]. The capacitor then becomes the energy source and starts discharging through the inductor. This causes a magnetic field to

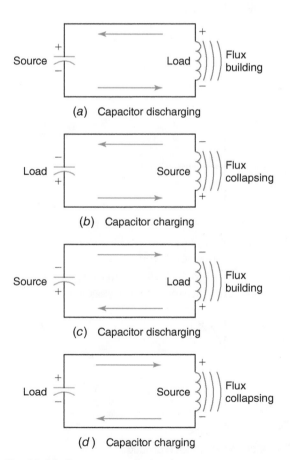

(*a*) Capacitor discharging

(*b*) Capacitor charging

(*c*) Capacitor discharging

(*d*) Capacitor charging

Fig. 13-23 Resonant tank circuit.

build up in the inductor and store the energy being transferred from the capacitor. Once the capacitor has discharged, the inductor's field starts to collapse, as in Fig. 13-23(*b*). Now the inductor is the source. It provides the energy to recharge the capacitor in a reverse polarity, as shown in Fig. 13-23(*b*). After the inductor's field has collapsed and recharged the capacitor, the capacitor again takes over as the source. It discharges [Fig. 13-23(*c*)] and again transfers energy to the inductor's magnetic field. The cycle is completed, as shown in Fig. 13-23(*d*), when the inductor's collapsing field recharges the capacitor. Then the cycle starts over again with the conditions shown in Fig. 13-23(*a*).

The parallel *LC* circuit discussed above (and shown in Fig. 13-23) is often called a *tank circuit*. "Tank circuit" is a very descriptive name because the circuit stores energy the way a tank stores liquid. A tank circuit produces a sine wave, as shown in Fig. 13-24, as the capacitor charges and discharges repeatedly. If both the capacitor and the inductor were ideal components (had no resistance), the tank circuit would produce a sine wave forever. Once the capacitor was given an initial charge and connected across the inductor, the cycling would continue indefinitely. However, all capacitors and inductors have some resistance. Therefore, some energy is converted to heat each time the capacitor charges and discharges. Thus, a real tank circuit produces a *damped waveform* like that shown in Fig. 13-24(*b*). How many cycles it takes for the waveform to completely dampen out depends on the quality of the circuit.

Tank circuit

Damped waveform

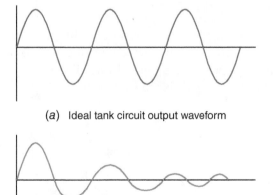

(*a*) Ideal tank circuit output waveform

(*b*) Practical tank circuit output waveform

Fig. 13-24 Tank circuit waveforms.

Suppose the LC circuit of Fig. 13-23 is connected to an ac source [Fig. 13-25(a)]. Further, let the source frequency be equal to the LC circuit's resonant frequency. If L and C were ideal components, no current would be required from the source. If no source current flowed, then the impedance of the circuit would be infinite. When real components are used, a small source current does flow. This small current, together with the source voltage, furnishes the power used by the very low resistance of the inductor and capacitor. In the phasor diagrams of Fig. 13-25, this low current is labeled I_R since it is caused by the resistance of L and C. With high-quality components, the source current is very low and the impedance is still very high.

Suppose the frequency of the source is below the resonant frequency of the LC circuit, as in Fig. 13-25(b). Then X_L is less than X_C, and I_L exceeds I_C. The difference between I_L and I_C becomes part of the current required from the source. The circuit is inductive, and its impedance has decreased from the value it had at its resonant frequency. Its impedance could be calculated by dividing the source voltage by the source current.

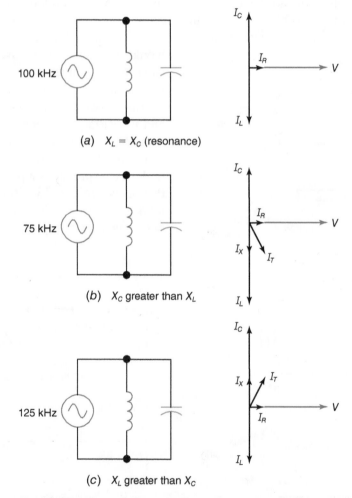

(a) $X_L = X_C$ (resonance)

(b) X_C greater than X_L

(c) X_L greater than X_C

Fig. 13-25 Parallel LC circuit. The impedance (voltage divided by total current) is greatest at resonance.

Figure 13-25(c) shows what happens when the applied frequency is greater than the resonant frequency. The circuit is capacitive, and again its impedance has decreased from the resonant frequency value.

In summary, a parallel resonant *LC* circuit has the following major characteristics:

1. Maximum impedance
2. Minimum source current
3. Phase angle almost 0°
4. High circulating inductive and capacitive currents

Series Resonant Circuits

In the series *RCL* circuit in Fig. 13-26, the frequency of the source voltage is at the resonant frequency for *L* and *C*. This means that $X_L = X_C$ and $V_L = V_C$. Since V_L and V_C are 180° apart, they cancel each other. The net voltage drop is that across the resistance. The source voltage and current are in phase, and the circuit has a power factor of 1. Since X_L and X_C cancel each other, the current in the circuit is limited only by the resistance.

Suppose the resistor in Fig. 13-26 is shorted out to produce an *LC* circuit like the one in Fig. 13-27(a). If *L* and *C* were ideal components, X_L would cancel X_C and this series resonant circuit would have zero impedance. It would be a short circuit across the source. The current would be limited only by the internal resistance of the source. The voltages across the inductor and capacitor would be extremely high. However, real inductors and capacitors have some resistance. This resistance prevents

the series resonant circuit from being a dead short. It drops an in-phase voltage (V_R in Fig. 13-27). This voltage and the line current represent the power used by the inductor and capacitor. The great majority of this power is used by the inductor.

The voltages across the capacitor and the inductor are many times greater than the source voltage in a series resonant circuit. Why this is so can be best explained by a practical example. The resistance of the typical inductor and capacitor in Fig. 13-27(a) is about 20 Ω. Therefore, the current in the circuit is

$$I_T = \frac{V_T}{Z} = \frac{5\text{ V}}{20\ \Omega} = 0.25\text{ A}$$

This 0.25 A is also the current through the inductor and the capacitor. The reactance of the inductor (which, in a resonant circuit, is the same as the reactance of the capacitor) is

$$\begin{aligned} X_L &= 6.28fL \\ &= 6.28\,(107 \times 10^3\text{ Hz})(1 \times 10^{-3}\text{ H}) \\ &= 672\ \Omega \end{aligned}$$

The voltage across the inductor can now be calculated:

$$V_L = I_L \times X_L = 0.25\text{ A} \times 672\ \Omega = 168\text{ V}$$

Thus, the voltage across the inductor (and across the capacitor) is more than 33 times (168 V versus 5 V) higher than the source voltage.

Figure 13-27(b) and (c) shows the results of operating a series *LC* circuit below and above its resonant frequency. When operating below its resonant frequency, the series circuit is capacitive because V_C is greater than V_L. Again, notice that parallel and series circuits are opposites in many ways. The parallel *LC* circuit operating below its resonant frequency [Fig. 13-25(b)] is inductive.

The major characteristics of the series resonant *LC* circuit can be summarized as follows:

1. Minimum impedance
2. Maximum source current
3. Phase angle almost 0°
4. High inductive and capacitive voltages

Notice that the only characteristic shared by the series and parallel resonant circuits is the almost 0° phase angle.

Series resonant circuit

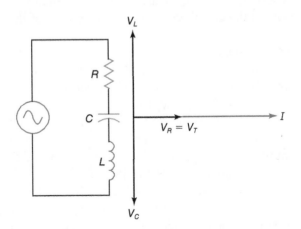

Fig. 13-26 Series resonant *RCL* circuit.

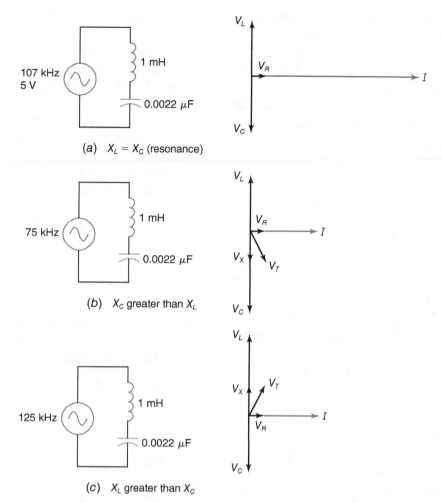

(a) $X_L = X_C$ (resonance)

(b) X_C greater than X_L

(c) X_L greater than X_C

Fig. 13-27 Series LC circuit. The impedance approaches zero at resonance.

Response Curves, Bandwidth, and Selectivity

So far, only phasor diagrams have been used to show current, voltage, and impedance of LC circuits at several frequencies. A more complete picture of the behavior of LC circuits can be shown with the *response curves* of Fig. 13-28. These curves are made by either calculating or measuring the current, voltage, or impedance at numerous frequencies above and below resonance. The values are then plotted on a graph.

Compare the curves in Fig. 13-28(a) and (b). They clearly show the major differences between series and parallel LC circuits.

The *bandwidth* (BW) of an LC circuit is expressed in hertz. It is the range of frequencies to which the circuit provides 70.7 percent or more of its maximum response. Suppose the parallel LC circuit of Fig. 13-28(a) is resonant at 500 kHz and has a resonant impedance of 100 kΩ. Further, suppose its impedance drops

to 70.7 kΩ at 495 kHz [f_{low} in Fig. 13-28(a)] and at 505 kHz [f_{high} in Fig. 13-28(a)]. Then the circuit's bandwidth is 10 kHz (495 kHz to 505 kHz). The bandwidth of a series circuit can be determined from its current-versus-frequency curve [Fig. 13-28(b)]. The edges of the bandwidth (f_{low} and f_{high}) are those frequencies at which the circuit current is 70.7 percent of the current at resonance.

The bandwidth of a circuit determines the *selectivity* of the circuit. Selectivity refers to the ability of a circuit to select one frequency (the desired frequency) out of a group of frequencies. For example, the antenna of a radio receiver receives a signal from all the local radio stations. Each station's signal is at a different frequency. It is the job of the resonant circuits in the receiver to select the frequency of one station and reject the frequencies of all the other stations. (When you tune a receiver, you are changing capacitor or inductor values to

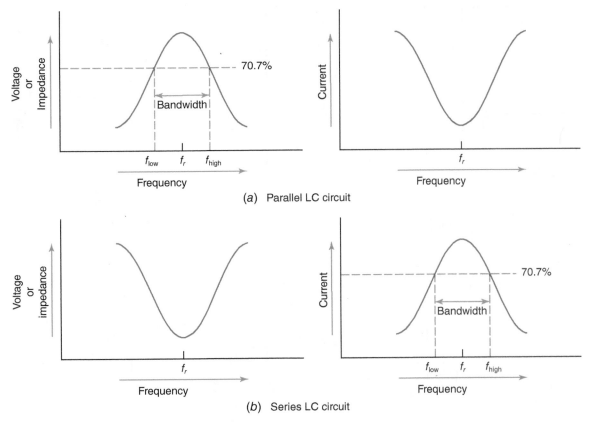

Fig. 13-28 Response curves of resonant circuits.

adjust the resonant frequency of the receiver circuits.) If the bandwidths of the receiver's circuits are too wide, two or more stations will be heard simultaneously. In this case we would say that the receiver has poor selectivity.

Quality of Resonant Circuits

The term *quality* is also used to rate resonant circuits. It is the ratio of the reactance of the inductor or capacitor to the equivalent series resistance of both components. Recall that the inductor possesses nearly all the resistance of a resonant circuit. Therefore, the quality of the inductor is, for all practical purposes, also the quality of the resonant circuit.

The quality of the resonant circuit is important for two reasons:

1. The quality determines the minimum Z of the series resonant circuit and the maximum Z of the parallel resonant circuit. These ideas are illustrated in Fig. 13-29. For the series circuit,

$$Z = \frac{V_R}{I}$$

and for the parallel circuit,

$$Z = \frac{V}{I_R}$$

From Fig. 13-29(*a*) you can see that the high-quality circuit yields the lowest Z for the series circuit. The high-quality parallel circuit yields the highest Z [Fig. 13-29(*b*)].

2. The *quality* also *determines* the *bandwidth* of the resonant circuit.

The higher the quality, the narrower the bandwidth. A narrow bandwidth (high quality) provides a larger change in impedance for a given change in frequency than a wide bandwidth does. Figure 13-30 shows why a high-quality circuit provides a greater change in impedance than a low-quality circuit does. The phasor diagrams in Fig. 13-30(*a*) are for two parallel *LC* circuits operating at the same resonant frequency. The low-quality circuit has twice as much resistive current as the high-quality circuit. Since *Z* = *V*/*I*, the low-quality circuit also has only half as much impedance as the high-quality circuit.

Quality

Quality determines bandwidth

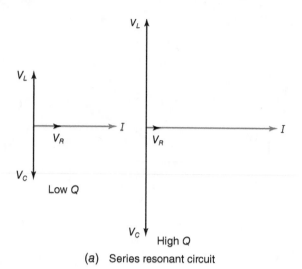

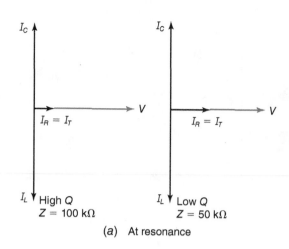

(a) At resonance

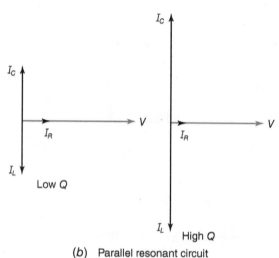

(a) Series resonant circuit

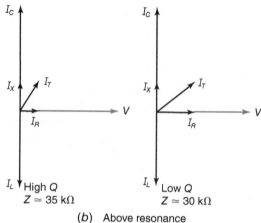

(b) Above resonance

Fig. 13-30 Effects of Q on the nonresonant Z.

(b) Parallel resonant circuit

Fig. 13-29 Effects of Q on the resonant Z.

$$BW = \frac{f_r}{Q}$$

Figure 13-30(b) shows the results of increasing the source frequency the same amount for each circuit. Notice that the impedance of the high-quality circuit dropped from 100 to 35 kΩ, a change of approximately 65 percent. For the same change in frequency, the impedance of the low-quality circuit changed only about 40 percent (50 to 30 kΩ).

The result of the greater impedance change for the high-quality circuit of Fig. 13-30 is shown in Fig. 13-31. This figure shows that the high-quality circuit yields a narrower bandwidth for either series or parallel LC circuits. From Fig. 13-31 you can see that a high-quality circuit is more selective than a low-quality circuit.

Resonant frequency, quality, and bandwidth are all interrelated. Their relationship can be expressed by the formula

$$BW = \frac{f_r}{Q}$$

Although this formula is actually only an approximation, it is accurate enough for predicting circuit behavior. If the circuit quality is 10 or more, the error will be less than 1 percent.

The quality of a series resonant circuit can also be found from the following formula:

$$Q = \frac{V_L}{V_T}$$

In this formula the voltages are the voltages at resonance. You can understand where this formula comes from by referring to Fig. 13-26 or 13-27(a). At resonance, the resistive voltage and the source voltage are equal. Also, the resistance and the reactance have the same current flowing through them. Therefore, the ratio of reactance to resistance has to be equal to the ratio of reactive voltage to resistive voltage.

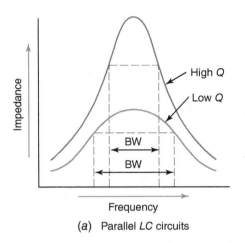

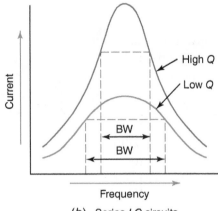

(a) Parallel *LC* circuits

(b) Series *LC* circuits

Fig. 13-31 Response curves for low- and high-*Q* circuits.

EXAMPLE 13-12

What is the quality of a circuit that is resonant at 100 kHz and has a bandwidth of 4 kHz?

Given: f_r = 100 kHz

BW = 4 kHz

Find: Q

Known: BW $= \dfrac{f_r}{Q}$

Solution: $Q = \dfrac{f_r}{\text{BW}} = \dfrac{100 \text{ kHz}}{4 \text{ kHz}} = 25$

Answer: The quality of the circuit is 25.

The preceding formula, when rearranged to $V_L = QV_T$, shows that the inductive voltage is quality times as large as the source voltage. Since $V_L = V_C$, the same can be said for the capacitive voltage.

EXAMPLE 13-13

What is the impedance (Z) and the bandwidth (BW) of a series resonant circuit that has a 0.005-μF, ideal capacitor and a 0.4-mH inductor with a Q of 50?

Given: $C = 0.005 \ \mu\text{F}$ with a Q close to infinity

$L = 0.4$ mH with a Q of 50

Circuit is series resonant

Find: Z and BW

Known: $f_r = \dfrac{1}{6.28\sqrt{LC}}$

$X_L = 6.28fL$

$Q = \dfrac{X_L}{R}$ rearranged $R = \dfrac{X_L}{Q}$

$Z = R$ in a series resonant circuit

$\text{BW} = \dfrac{f_r}{Q}$

Solution:

$f_r = \dfrac{1}{6.28\sqrt{0.005\mu\text{F} \times 0.4 \text{ mH}}} = 112.6 \text{ kHz}$

$X_L = 6.28 \times 112.6 \text{ kHz} \times 0.4 \text{ mH}$
$\quad = 282.9 \ \Omega$

$R = \dfrac{X_L}{Q} = \dfrac{282.9 \ \Omega}{50} = 5.66 \ \Omega$

$Z = R = 5.66 \ \Omega$

$\text{BW} = \dfrac{f_r}{Q} = \dfrac{112.6 \text{ kHz}}{50} = 2.25 \text{ kHz}$

Answer: The impedance is 5.66 Ω, and the bandwidth is 2.25 kHz.

 *Self-Test*

Answer the following questions.

40. At resonance, θ is approximately _____.

41. Maximum opposition is provided by a(n) _____ resonant circuit.

42. A parallel resonant circuit operating above resonance is _____.

43. A series resonant circuit operating below resonance is _____.

44. For a given resonant frequency, bandwidth is controlled by the _____ of the components.

45. The resonant frequency of a circuit can be increased by _____ the value of either L or C.

46. A parallel LC circuit is often called a(n) _____ circuit.

47. True or false. There is only one combination of L and C for each resonant frequency.

48. True or false. In a parallel resonant circuit, $X_L = Z$.

49. True or false. In any resonant circuit, $X_L = X_C$.

50. True or false. In a series resonant RCL circuit, $R = Z$.

51. True or false. High selectivity requires a wide bandwidth.

52. Determine the resonant frequency of a 5-mH inductor and a 0.03-μF capacitor.

53. What is the bandwidth of the circuit in question 52 if the resistance of the components is 5 Ω?

54. What is the bandwidth of a series LC circuit which allows 20 mA at its resonant frequency of 120 kHz and 14.14 mA at 118 kHz?

55. What is the quality of the circuit in question 54?

56. How can the selectivity of a resonant circuit be improved?

57. What is the quality of a series resonant LC circuit that produces 5 V of V_C when the source voltage is 0.1 V?

13-7 Filters

Filtering

One of the major uses of RC, RL, LC, and RCL circuits is for filtering. *Filtering* refers to separating one group of frequencies from another group of frequencies. There are four general classes of filters: band-pass, band-reject, high-pass, and low-pass.

Audio amplifiers use high-pass and low-pass filters to control the frequency response of the amplifier. Base and treble (or tone) controls adjust the resistances in the filter circuits in the amplifier.

High-frequency amplifiers use band-reject and band-pass filters to control a narrow band of frequencies. These filters select a band of frequencies to either pass on to the next amplifier or to eliminate before they get to the next amplifier.

Low-Pass Filters

Low-pass filters

Low-pass filters offer very little opposition to low-frequency signals [Fig. 13-32(a)]. For low frequencies, most of the input signal appears at the output terminals. As the signal frequency increases, the filter provides more opposition. The filter drops more of the signal voltage and leaves less signal available at the output. As the graph of the output voltage illustrates [Fig. 13-32(a)], essentially none of

High-pass filters

the high-frequency signals appears at the output terminal. Although low-pass filters can be constructed with either RL or RC circuits, RC circuits are more common.

Inspection of Fig. 13-32(b) shows why the low-pass filter behaves as described in the paragraph above. At low frequencies, the reactance of C_1 is very high relative to R_2. The impedance of $R_2 C_1$ is essentially 5 kΩ. Therefore, approximately five-sixths of the input voltage appears at the output terminals. At very high frequencies, the reactance of C_1 is very low relative to R_1. Now the impedance of $R_2 C_1$ is very low (essentially equal to X_{C_1}). Therefore, nearly all the input signal drops across R_1.

The reactance of L_1 in Fig. 13-32(c) is low (relative to R_1) at low frequencies. Thus, low frequencies pass through the filter. However, at high frequencies, the inductive reactance of L_1 is high and most of the signal drops across L_1. Very little of the high-frequency signal appears at the output terminals.

High-Pass Filters

Two simple *high-pass filters* are shown in Fig. 13-33. With the RC filter of Fig. 13-33(a), the capacitor drops nearly all the low-frequency signals. At high frequencies, the capacitive

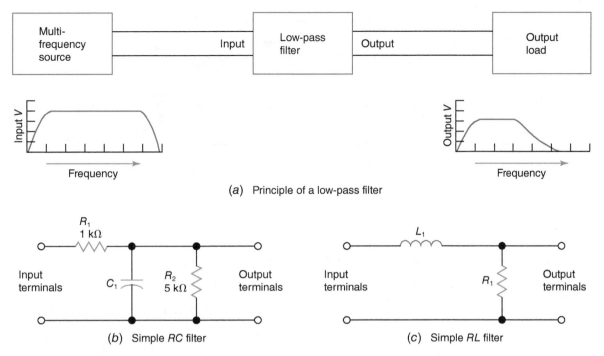

Fig. 13-32 Low-pass filters.

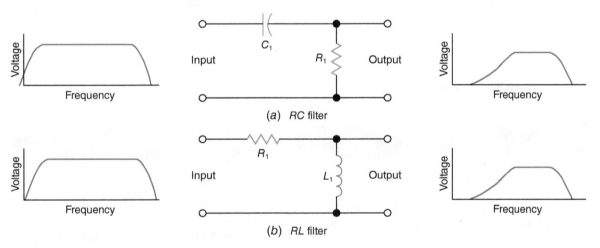

Fig. 13-33 High-pass filters.

reactance is very low. Therefore, R_1 (which provides the output) drops nearly all the high-frequency voltage. In other words the circuit passes high-frequency signals through it and blocks low-frequency signals.

You May Recall

. . . that $X_L = 6.28fL$. There-fore, L_1 in Fig. 13-33(b) drops most of the high-frequency voltage but practically none of the low-frequency voltage. Thus, a series RL circuit can serve as a high-pass filter.

Band-Pass and Band-Reject Filters

You are already familiar with *band-pass* and *band-reject filters*, sometimes called 'notch' filters. They are just series LC and parallel LC circuits. Frequencies within their response curve range are either passed or rejected (Fig. 13-34).

Notice in Fig. 13-34 that a parallel LC circuit can serve as either a band-pass or a band-reject filter. The same thing can be said for the series LC circuit. The determining factor is where the input and output terminals are connected to the circuits.

Band-pass filters

Band-reject filters

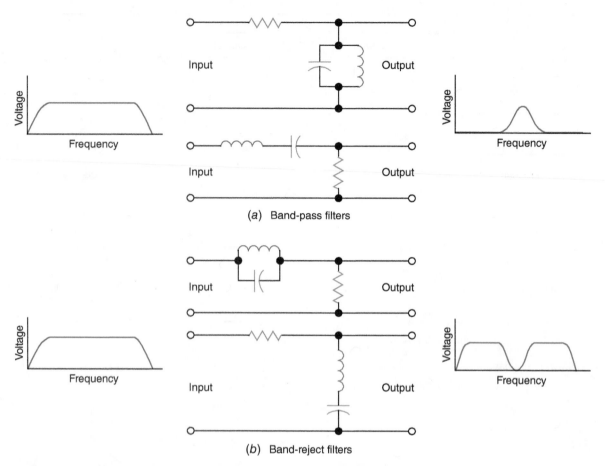

(a) Band-pass filters

(b) Band-reject filters

Fig. 13-34 Band-pass and band-reject filters.

Self-Test

Answer the following questions.

58. List four classifications of filter circuits.
59. Which of the types listed in question 58 use resonant circuits?

60. If you wanted to eliminate low-frequency noise from a stereo CD, what type of filter would you use?
61. Could the filter in question 60 be either an *RL* or an *RC* filter?

Chapter 13 Summary and Review

1. Impedance is a combination of resistance and reactance.
2. Impedance causes phase shift.
3. Impedance may cause either a leading or a lagging current.
4. The symbol for impedance is Z.
5. The ohm is the base unit of impedance.
6. In RC circuits, the current leads the voltage.
7. For series RC circuits, the following generalizations can be made:
   a. The total impedance is higher than either R or X_C.
   b. The arithmetic sum of R and X_C is greater than Z.
   c. Decreasing either f or C causes Z to increase, I to decrease, P to decrease, and θ to increase.
8. The following are characteristics of parallel RC circuits:
   a. The total impedance is less than R or X_C.
   b. Decreasing either f or C causes Z to increase, I to decrease, and θ to decrease. The power remains the same.
9. The current phasor is the reference phasor for series circuits.
10. The voltage phasor is the reference phasor for parallel circuits.
11. Resistance and reactance are 90° out of phase.
12. In all RL circuits, the current lags the voltage.
13. A series RL circuit has more Z than either R or X_L.
14. A series RL circuit has more V_T than either V_R or V_L.
15. A parallel RL circuit has less Z than either R or X_L.
16. A parallel RL circuit has more I_T than either I_R or I_L.
17. In a series RCL circuit, Z may be either less or more than X_L or X_C.
18. In a resonant RCL circuit, Z is equal to R.
19. In RCL circuits, reactive voltages and currents can be greater than the total current or voltage.
20. A series RCL circuit having X_L greater than X_C is inductive.
21. A parallel RCL circuit having X_L greater than X_C is capacitive.
22. Resonance occurs when $X_L = X_C$.
23. Increasing either L or C decreases the resonant frequency of an LC circuit.
24. A given L and C can produce only one resonant frequency.
25. For practical purposes, the quality of a resonant LC circuit is determined by the quality of the inductor.
26. All resonant circuits have a power factor (cos θ) of 1.
27. The impedance of a series resonant circuit approaches zero.
28. The impedance of a parallel resonant circuit approaches infinity.
29. A parallel LC circuit operating above its resonant frequency is capacitive.
30. A series LC circuit operating above its resonant frequency is inductive.
31. The higher the quality of a circuit, the narrower its bandwidth.
32. The narrower the bandwidth of a circuit, the more selective the circuit is.
33. The bandwidth of a circuit is equal to f_r/Q.
34. Either RC or RL circuits can be used for high-pass and low-pass filters.
35. Either series resonant or parallel resonant circuits can be used for band-pass or band-reject filters.

Related Formulas

All circuits involving combinations of R, C, and L:

$$Z = \frac{V_T}{I_T}$$

$$f_r = \frac{1}{6.28\sqrt{LC}}$$

$$\cos \theta = \frac{P}{P_{app}} = \frac{P}{I_T V_T}$$

All series circuits using combinations of R, C, and L:

$$\cos \theta = \frac{R}{Z} = \frac{V_R}{V_T}$$

All resonant circuits:
angle theta is nearly zero

$$f_r = \frac{1}{6.28\,fC} \qquad BW = \frac{f_r}{Q}$$

$$I_c = IL$$

Series resonant circuits:
  Z approaches zero

$$I_t = I_c = I_L$$

$$Q = \frac{V_L}{V_T}$$

Parallel resonant circuit:
  Z approaches infinity

All parallel circuits using combinations of R, C, and L:

$$\cos \theta = \frac{I_R}{I_T}$$

Only for series RCL circuits:

$$Z = \sqrt{(X_L - X_C)^2 + R^2}$$

$$V_T = \sqrt{(V_L - V_C)^2 + V_R^2}$$

Only for parallel RCL circuits:

$$I_T = \sqrt{(I_L - I_C)^2 + I_R^2}$$

Chapter Review Questions

For questions 13-1 to 13-13, determine whether each statement is true or false.

13-1. The Pythagorean theorem is used to determine the angles in a right triangle. (13-2)

13-2. If the capacitive reactance in a series RC circuit is greater than the resistance, angle θ is less than 45°. (13-3)

13-3. Resistance is 90° out of phase with reactance. (13-3)

13-4. Reducing the inductance in a series RL circuit causes the total power to increase. (13-4)

13-5. In a series RL circuit, the inductive voltage can be greater than the source voltage. (13-4)

13-6. The resistance in a series RCL circuit can exceed the impedance. (13-5)

13-7. In a parallel RCL circuit operating above its resonant frequency, the source current leads the source voltage. (13-5)

13-8. There is only one value of inductance that resonates with a 0.001-μF capacitor at 10,000 Hz. (13-6)

13-9. A high-quality circuit is more selective than a low-quality circuit. (13-6)

13-10. The resonant frequency of a tank circuit can be increased by increasing the capacitance. (13-6)

13-11. At resonance, the phase angle of a circuit is approximately 90°. (13-6)

13-12. A series LC circuit can be used in a band-reject filter. (13-7)

13-13. Either a series or a parallel LC circuit can be used in a band-pass filter. (13-7)

For questions 13-14 to 13-27, supply the missing word or phrase in each statement.

13-14. Impedance is a combination of _____ and _____. (13-1)

13-15. The base unit of impedance is the _____. (13-1)

13-16. The bandwidth of a circuit is determined by the _____ and the _____ of the circuit. (13-6)

13-17. The symbol for impedance is _____. (13-1)

13-18. In any RC circuit, the current _____ the voltage. (13-3)

13-19. In any RL circuit, the current _____ the voltage. (13-4)

13-20. Decreasing the frequency of a series *RL* circuit causes θ to _____, *P* to _____, *Z* to _____, and V_R to _____. (13-4)

13-21. Increasing the frequency of a parallel *RC* circuit causes *Z* to _____, θ to _____, I_C to _____, and *P* to _____. (13-3)

13-22. The _____ phasor is the reference phasor for parallel circuits. (13-3)

13-23. In a resonant *RCL* circuit, resistance and _____ have the same value. (13-5)

13-24. The quality of a resonant *LC* circuit is primarily determined by the _____. (13-6)

13-25. At resonance, the PF of a circuit is _____. (13-6)

13-26. The impedance of a series resonant *LC* circuit approaches _____. (13-6)

13-27. A series *LC* circuit operating above resonance will be _____. (13-6)

Answer the following questions.

13-28. List four types (classifications) of filter circuits. (13-7)

13-29. Which of the filter circuits in question 13-28 use resonant circuits? (13-7)

Chapter Review Problems

13-1. Calculate the bandwidth of a circuit that is resonant at 430 kHz and has a quality rating of 80. (13-6)

13-2. An 1800-Ω resistor and a 0.68-μF capacitor are connected in series to a 40-V, 200-Hz source. Calculate the following: (13-3)
   a. Impedance
   b. Voltage across the resistor
   c. Power
   d. $\cos \theta$

13-3. Assume that the resistor and capacitor in the circuit of problem 13-2 are connected in parallel instead of in series. Calculate the following: (13-3)
   a. Impedance   c. Power
   b. Total current   d. θ

13-4. What is the impedance of a 1500-Ω resistor connected in series with a 4-mH inductor and a 40-kHz source? (13-4)

13-5. Determine the resonant frequency and the bandwidth of a 6-mH inductor and a 1200-pF capacitor connected in series if the quality of the circuit is 56. (13-6)

13-6. A 0.001-μF capacitor, a 6-mH inductor, and a 1200-Ω resistor are connected in series to a 30-V, 60-kHz source. Determine the following: (13-5)
   a. Impedance
   b. Total current

   c. Voltage across the capacitor
   d. Resonant frequency

13-7. Assume the components in problem 13-6 are connected in parallel instead of in series. Calculate the following: (13-5)
   a. Impedance
   b. Total current
   c. Inductive current
   d. θ

13-8. Determine the impedance and the power factor for a series *RC* circuit when $R = 1200\ \Omega$, $C = 0.1\ \mu$F, and $f = 600$ Hz. (13-3)

13-9. Determine the *Q* of a circuit resonant at 1000 kHz with a BW of 20 kHz. (13-6)

13-10. Determine the impedance of a parallel *RC* circuit when $R = 2.2$ kΩ, $C = 0.5\ \mu$F, and $f = 100$ Hz. (13-3)

13-11. Determine the output voltage (V_{out}) of the filter in Fig. 13-33 (*a*) when $f_{in} = 200$ Hz and also when $f_{in} = 10$ kHz. For both values of f_{in}, the capacitor (C_1), the resistor (R_1), and the input voltage (V_{in}) remain 0.005 μF, 5 kΩ, and 1.0 V, respectively. (13-7)

13-12. Determine the output voltage (V_{out}) of the filter in Fig. 13-33(*b*) when $f_{in} = 200$ Hz and also when $f_{in} = 10$ kHz. For both values of f_{in}, the inductor (L_1), the resistor (R_1), and the input voltage (V_{in}) remain 50 mH, 1 kΩ, and 1.0 V, respectively. (13-7)

13-1. Two amplifiers are coupled together by an *RC* circuit. What factors will determine the amount of phase shift from the output of one amplifier to the input of the other amplifier? How would you control these factors to minimize the phase shift?

13-2. A series *RC* circuit has a 10-V, 200-Hz source and a 0.68-μF capacitor. How much resistance is required to cause the current to lead the voltage by 25°?

13-3. A parallel *RL* circuit has a 15-V, 500-Hz source and a 1200-Ω resistor. How much inductance is required to cause a 65° phase shift?

13-4. A series *RCL* circuit has an inductance of 0.1 H, a resistance of 1000 Ω, and a 20-V, 500-Hz source. Determine the capacitance required to produce a leading power factor of 0.82.

13-5. A series *RL* circuit has a 25-V, 250-Hz source and a 1000-Ω resistor. What is the inductance if the power is 95.4 mW?

13-6. Determine the value of capacitance needed to resonate with a 10-mH inductor at 450 kHz.

13-7. Determine the quality of a series *LC* circuit when *L* = 0.1 mH, *C* = 0.001 μF, and the equivalent series resistance of *L* and *C* is 20 Ω.

13-8. Refer to Fig. 13-4. Determine the value of Y_1 when Y_3 = 18 m, Y_2 = 9 m, and Y_T = 12 m at an angle between 270° and 360°.

13-9. Repeat question 13-8 when the angle of Y_T is between 0° and 90°.

13-10. Why can't we calculate *Z* in a parallel *RC* circuit using $Z = (R \times X_C)/(R + X_C)$, which is the form of the formula used for parallel resistors?

13-11. Why are *RC* filters used more often than *RL* filters at audio frequencies?

Answers to Self-Tests

1. F
2. reactance, resistance
3. Z
4. ohm
5. 11.4 m
6. 29.15 N
7. Pythagorean
8. leads
9. F
10. current
11. F
12. 20°
13. They are separated by 90°.
14. $Z = \sqrt{R^2 + X_C^2}$ and

$$Z = \frac{V_T}{I_T}$$

15. $V_T = \sqrt{V_R^2 + V_C^2}$

16. $\cos \theta = \dfrac{R}{Z} = \dfrac{V_R}{V_T} = \dfrac{P}{P_{app}}$

17. $\cos \theta = \dfrac{P}{P_{app}}$

18. $\cos \theta = \dfrac{I_R}{I_T}$

19. a. increases
    b. stays the same
    c. stays the same
    d. increases

20. a. increases
    b. increases
    c. decreases
    d. increases

21. a. 11.9 kΩ
    b. 0.133 W
    c. 63.8°
    d. 4.52 mA

22. a. 27.44 mA
    b. 27.44 V
    c. 753 mW
    d. 1458 Ω

23. lag
24. I_T increases, V_L decreases, and *Z* decreases.
25. *Z* increases and θ decreases.

26. 1771.5 Ω

27. 28.2 mA, 1.2 W

28. 27.7 mA, 1803 Ω

29. No, because $I_R = V_T/R$ and $P_T = P_R = I_R^2R$

30. decrease R, decrease L, decrease f, increase V_T

31. T

32. F

33. F

34. $Z = \dfrac{V_T}{I_T}$

$Z = \sqrt{(X_L - X_C)^2 + R^2}$

35. $I_T = \sqrt{(I_L - I_C)^2 + I_R^2}$

36. a. series circuit

b. parallel circuit

c. parallel circuit

37. 2480 Ω, 12.1 mA, 27.4 V

38. 1047 Ω 58.4°, 37.7 mA

39. leading

40. zero

41. parallel

42. capacitive

43. capacitive

44. quality

45. decreasing

46. tank

47. F

48. F

49. T

50. T

51. F

52. 12,995 Hz

53. 159 Hz

54. 4 kHz

55. 30

56. by reducing R, which will increase Q and narrow the BW

57. $Q = 50$

58. band-pass, band-reject, high-pass, and low-pass

59. band-pass, and band-reject

60. high-pass

61. yes

Electric Motors

Learning Outcomes

This chapter will help you to:

14-1 *List* the three major categories of motors and specify which category provides the most variety of types of motors.

14-2 *Discuss* the factors to be considered in deciding whether or not a specific motor should be connected to (and operated from) the available power source.

14-3 *Explain* the features of the openings on open motors with specific restrictions on the ventilation openings.

14-4 *Classify* (a) the type of motors that has no electrical connection between the power source and the rotor, and *describe* (b) the operation and characteristics of a capacitor-start motor.

14-5 *Identify* the induction motor that has a salient-pole rotor.

14-6 *Describe* (a) the three types of dc motors that use brushes, and (b) the advantages of three-phase motors over single-phase motors.

The electric motor is one of the more common electric devices in use today. It is used extensively in such diverse systems as household appliances, automobiles, computers, printers, and automatic cameras.

14-1 Motor Classifications

Electric motors can be broadly classified by the type of power source needed to operate them. The three major categories of motors, as shown in Fig. 14-1, are the *dc motor,* the *ac motor,* and the *universal motor.* The universal motor is designed to operate from either an ac or a dc power source.

Figure 14-1 also lists many of the types of motors available within the three major categories of motors. Some of the types of motors listed in Fig. 14-1 can be further subdivided. For example, there are several kinds of dc stepper motors and numerous kinds of polyphase (especially three-phase) ac motors.

Motors can also be classified by their intended use or special characteristics. Some examples are gearmotors, synchronous motors, multispeed motors, and torque motors.

A *gearmotor* is a motor that has a gear train built into the motor housing to reduce the output shaft speed and increase the shaft torque. The motor in a gearmotor may be almost any type of ac, dc, or universal motor depending on the intended application.

Synchronous motors are motors in which shaft rotation is in exact synchronization with the frequency of the power source. These motors are either single-phase or polyphase (usually three-phase) ac motors.

Multispeed motors are motors with two or more fixed speeds. Multispeed operation is often obtained by having either a tapped motor winding or a separate motor winding for each speed.

DC motors

AC motors

Universal motors

Gearmotors

Synchronous motors

Multispeed motors

Direct Current	Alternating Current	Universal
Series	Split-phase	Noncompensated
Shunt	Capacitor-start	Compensated
Compound	Permanent-split capacitor	
Permanent-magnet	Two-value-capacitor	
Brushless	Shaded-pole	
Stepper	Reluctance	
	Hysteresis	
	Repulsion	
	Repulsion-start	
	Repulsion-induction	
	Inductor	
	Consequent pole	
	Polyphase	

Fig. 14-1 Types of motors.

This is in contrast to *variable-speed motors,* in which the speed is continuously variable and is typically controlled by varying the voltage and/or frequency of the power source.

Torque motors are designed to provide maximum, or near maximum, torque when the motor is stalled and still be able to operate for an extended period of time in the stalled (locked-rotor) mode without overheating. AC, dc, or universal motors can be designed to operate as torque motors.

Motors can be grouped into one of three broad power ratings: *integral-horsepower* (ihp), *fractional-horsepower* (fhp), and *subfractional-horsepower* (sfhp). Motors rated at less than $\frac{1}{20}$ hp are classified as sfhp motors, and the power is usually expressed in millihorsepower (mhp) rather than in fractional horsepower. Fractional-horsepower motors include those motors rated from $\frac{1}{20}$ to 1 hp. Any motor rated above 1 hp is an ihp motor.

Variable-speed motors

Torque motors

Integral-horsepower

Fractional-horsepower

Subfractional-horsepower

 Self-Test

Answer the following questions.

1. On the basis of power-source requirements, what are the three categories of electric motors?
2. True or false. The speed of a variable-speed motor is usually changed by selecting a different winding in the motor.

3. True or false. Synchronous motors can be classified as dc motors.
4. True or false. A $\frac{1}{10}$-hp motor would be classified as a fractional-horsepower motor.

14-2 Motor Ratings

The parameters for which electric motors are commonly rated include voltage, current, power, speed, temperature, frequency, torque, duty cycle, service factor, and efficiency. Standards for these parameters have been established by the *National Electrical Manufacturers Association* (NEMA). The nameplate on a motor usually contains the information needed to

determine whether or not a motor is operating within its specified ratings.

Voltage Rating

Each motor is designed to operate at a specified voltage. This is the *voltage rating* given on a *motor nameplate* such as the one shown in Fig. 14-2. As a general rule, motors are usually designed so that they will meet all other ratings as long as the supply voltage is within

Voltage rating

Motor nameplate

National Electrical Manufacturers Association (NEMA)

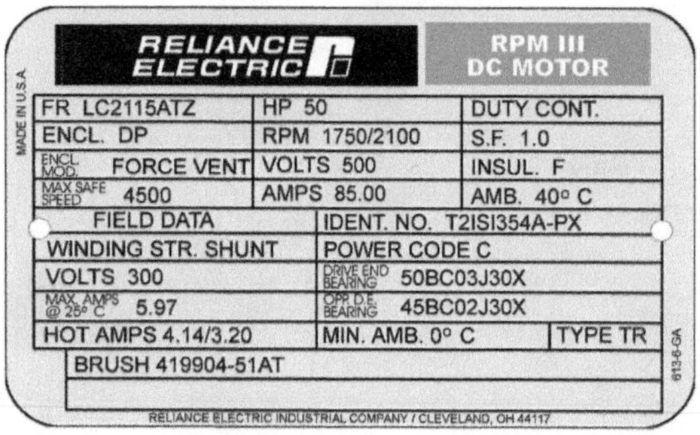

RELIANCE ELECTRIC		RPM III DC MOTOR

FR LC2115ATZ	HP 50	DUTY CONT.
ENCL. DP	RPM 1750/2100	S.F. 1.0
ENCL. MOD: FORCE VENT	VOLTS 500	INSUL. F
MAX SAFE SPEED 4500	AMPS 85.00	AMB. 40° C
FIELD DATA	IDENT. NO. T2ISI354A-PX	
WINDING STR. SHUNT	POWER CODE C	
VOLTS 300	DRIVE END BEARING 50BC03J30X	
MAX AMPS @ 25° C 5.97	OPP D.E. BEARING 45BC02J30X	
HOT AMPS 4.14/3.20	MIN. AMB. 0° C	TYPE TR
BRUSH 419904-51AT		

RELIANCE ELECTRIC INDUSTRIAL COMPANY / CLEVELAND, OH 44117

Fig. 14-2 Motor nameplate. The nameplate provides much useful information about the motor to which it is attached.

±10 percent of the rated voltage. Therefore, a 115-V motor can be operated at full load on a nominal 120-V supply and still stay below its maximum temperature rating providing that the supply voltage does not exceed 126.5 V for extended periods of time [126.5 V = 115 V + (0.10 × 115 V)].

EXAMPLE 14-1

Determine the minimum and maximum voltage that could be applied to a typical 230-V ac motor.

Given: 230-V rating and ±10 percent tolerance

Find: V_{min} and V_{max}

Known: $V_{min} = V_{rated} - (0.1 \times V_{rated})$
$V_{max} = V_{rated} + (0.1 \times V_{rated})$

Solution: $V_{min} = 230 \text{ V} - (0.1 \times 230 \text{ V})$
$= 207 \text{ V}$
$V_{max} = 230 \text{ V} + (0.1 \times 230 \text{ V})$
$= 253 \text{ V}$

Answer: The minimum and maximum voltage are 207 V and 253 V, respectively.

EXAMPLE 14-2

If the motor in example 14-1 is fully loaded, should it be operated on a nominal 220-V supply line where the voltage may have long-term variations of ±10 percent? Why?

Given: $V_{min(motor)} = 207 \text{ V}$
Find: $V_{min(supply)}$

Known: $V_{min(supply)} =$
$V_{supply} - (0.1 \times V_{supply})$

Solution: $V_{min(supply)} = 220 \text{ V}$
$- (0.1 \times 220 \text{ V})$
$= 198 \text{ V}$

Answer: No, because $V_{min(supply)}$ is less than $V_{min(motor)}$.

Notice in example 14-2 that we had to check only the minimum voltages because the supply voltage was smaller than the rated voltage of the motor. Had the nominal supply voltage been greater than the motor's rated voltage, we would have checked only the maximum voltages.

Another factor to consider when determining whether a given motor should be operated on a given supply voltage is the voltage drop caused by the lines (wires) connecting the motor to the supply. As shown in Fig. 14-3(b), the resistance of each connecting line appears in series with the motor. Therefore, if we know how much current the motor draws from the source and how much resistance the lines have, we can determine how much of the source voltage is applied to the motor. The resistance of the lines can be determined from the copper wire table in Appendix D when the gage and length of the lines are specified.

EXAMPLE 14-3

Determine the resistance of a 50-ft, two-conductor, 14-gage cable at 68°F.

Given: Ω/1000 ft of copper conductor (See Appendix D)
Cable length = 50 ft
AWG = 14

Find: Resistance of the 50-ft cable.

Known: R = length of conductor × Ω/1000 ft of conductor
Conductor length = 2 × cable length

Solution: Length = 2 × 50 = 100 ft
$R = 100 \text{ ft} \times \dfrac{2.525 \text{ } \Omega}{1000 \text{ ft}}$
$= \dfrac{2.525 \text{ } \Omega}{10} = 0.2525 \text{ } \Omega$

Answer: The resistance is 0.2525 Ω for 50 ft of 14-gage, two-conductor cable.

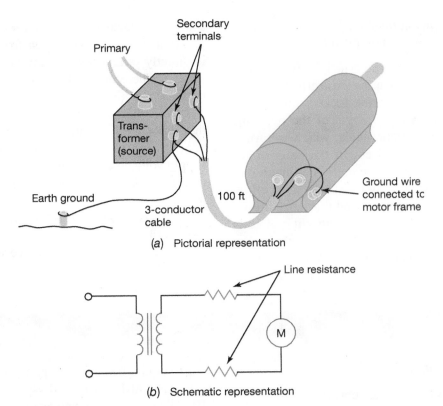

(a) Pictorial representation

(b) Schematic representation

Fig. 14-3 Voltage drop caused by the resistance of the supply lines.

EXAMPLE 14-4

Refer to Fig. 14-3(a). Assume that the minimum secondary voltage is 112 V and the maximum current required by a 120-V motor is 12 A. Will a 12-gage cable (which has a 20-A current rating) supply adequate voltage to the 120-V motor?

Given: $V_{min(supply)} = 112$ V
$V_{rated} = 120$ V
$I_{max} = 12$ A
Cable length $= 100$ ft
From Appendix D, 12-gage wire has 1.588 Ω per 1000 ft.

Find: $V_{min(motor)}$
R_{line} (resistance of the line)
V_{drop} (voltage drop of the line)
V_{motor} (voltage at the motor terminals)

Known:
$V_{min(motor)} = V_{rated} - (0.1 \times V_{rated})$
$R_{line} = (\Omega \text{ per } 1000 \text{ ft}/1000) \times \text{line length}$
$V_{drop} = R_{line} \times I_{max}$
$V_{motor} = V_{min(supply)} - V_{drop}$

Solution:
$V_{min(motor)} = 120$ V $- (0.1 \times 120$ V$)$
$= 108$ V
$R_{line} = [(1.588 \ \Omega/1000 \text{ ft})/1000] \times$
$200 \text{ ft} = 0.318 \ \Omega$
$V_{drop} = 0.318 \ \Omega \times 12$ A $= 3.82$ V
$V_{motor} = 112$ V $- 3.82$ V
$= 108.2$ V

Answer: 12-gage cable will be adequate because V_{motor} is greater than $V_{min(motor)}$.

A typical 230-V ac motor can operate on a 220-V, 230-V, or 240-V system. If a 230-V supply is not available, it would be better to operate the motor on a 240-V than on a 220-V supply. The higher voltage is preferred because, with respect to operation at the rated voltage, a higher voltage will

1. Provide greater starting or locked-rotor torque
2. Require less current at rated load

3. Run cooler at rated load
4. Be more efficient at rated load

Applying basic electrical principles will explain why higher voltage causes these effects.

For a given motor, the locked-rotor torque is dependent on the strength of the opposing or attracting magnetic fields, which in turn is dependent on the current in the coils of the motor. The current in the coils is a function of the applied voltage, coil resistance, and coil reactance. Since resistance and reactance are independent of voltage, the coil current is directly proportional to the voltage applied to the motor.

The horsepower taken from the shaft of a motor must come from the electric power source connected to the motor. For a given phase angle, the power taken from the source is directly proportional to the voltage and the current. Therefore, at full load, less current is required at a higher voltage.

If a higher voltage causes less current in the motor coils at rated power, then the copper loss (I^2R loss) will also be less at a higher supply voltage. Less copper loss means that the motor will run cooler and its efficiency will be improved.

There is one disadvantage to operating a motor at higher-than-rated voltage. At a higher voltage, a motor tends to produce slightly more noise.

Self-Test

Answer the following questions.

5. What is NEMA the abbreviation for?
6. List five parameters on which motors are rated.
7. What is the typical tolerance for the rated voltage of a motor?
8. Determine the minimum and maximum voltage at which a 240-V motor should be operated.
9. Would it be better to operate a 115-V motor at 110 V or 120 V? Why?
10. What is the disadvantage of running a motor on a voltage 8 percent greater than rated voltage?
11. A 230-V, 18-A motor is to be operated 25 ft from a 220-V ±5 percent supply. What is the minimum gage wire that can be used to avoid excess voltage drop? At 350 cmil/A can this gage wire carry 18 A?

Current Ratings

Running current

The current rating given on the motor nameplate in Fig. 14-2 is the *running current*. The running current is the current the motor draws when the motor is loaded to rated horsepower and is operated at rated voltage, frequency, and temperature. For subfractional- and fractional-horsepower motors, this is usually the only current specified on the nameplate.

Locked-rotor current

Starting current

The *locked-rotor*, or *starting*, *current* of a motor can be many times greater than the running current. For example, a ⅓-hp motor rated at 6 A may draw 40 A of locked-rotor current. This is the current the motor will draw the instant it is started. The large starting current greatly increases the instantaneous voltage drop in the supply line.

The ratio of starting current to running current can be extremely high in integral-horsepower motors. In fact, it is so high for many motors that the motors must be started on reduced voltage, or use series current-limiting resistors to start the motor. As the motor builds up speed, the voltage is increased to the rated value or the series resistors are removed. The nameplate on an ihp motor often includes the starting kilovolt-ampere (kVA) rating for the motor as well as the running current.

Power, Temperature, and Service Factor

The horsepower rating of a motor specifies the amount of power available at the motor shaft at the rated speed. Like many other electric devices, a motor is capable of delivering more than its rated power. However, unless the motor has a service factor (SF) greater than 1.0, a motor should not be operated with a load greater than its rated horsepower. To do so causes the motor's operating temperature to exceed its design limits.

Motors are designed to operate in an environment with a specified *maximum ambient temperature*. The standard ambient temperature used in most motor designs is 40°C. If a motor is operated within its other design limitations, that is, frequency, horsepower, and voltage, its operating temperature will also be within design limits when the ambient temperature is no more than 40°C.

The maximum temperature at which a motor should operate is dependent on the type of insulating material used on the motor windings. Motor insulating materials are grouped into the four classes shown in Fig. 14-4. Each of the four classes, as shown in Fig. 14-4, has a maximum operating temperature. The difference between the maximum operating temperature and the rated ambient temperature is the allowable *temperature rise* for a motor. Thus, a motor with class B insulation has a permissible temperature rise of $130° - 40°C = 90°C$, whereas a class F insulation allows a 115°C temperature rise.

Class of Insulation	Maximum Operating Temperature (°C)
A	105
B	130
F	155
H	180

Fig. 14-4 Temperature ratings of different classes of insulation.

words, they have conservative ratings that provide a small safety margin to handle unexpected increases in power required from a motor. The magnitude of this safety margin is specified by the *service factor* (SF) rating of the motor. The service factor of a motor allows one to determine the absolute maximum horsepower that a motor can provide on a continuous basis. The service factor rating is a number that can be multiplied by the horsepower rating to determine the maximum continuous horsepower the motor can provide. The additional horsepower made possible by the SF should be used only when it is ensured that frequency and voltage will be at rated values and ambient temperature will not exceed the rated value.

Maximum ambient temperature

Service factor (SF)

Temperature rise

EXAMPLE 14-5

A motor with its rating based on the standard 40°C ambient temperature uses class A insulation. What is its maximum temperature rise?

Given: $T_{amb} = 40°C$
$T_{max} = 105°C$ given in Fig. 14-4

Find: T_{rise} (maximum temperature rise)

Known: $T_{rise} = T_{max} - T_{amb}$

Solution: $T_{rise} = T_{max} - T_{amb}$
$= 105°C - 40°C = 65°C$

Answer: Maximum temperature rise is 65°C.

EXAMPLE 14-6

A certain mechanical system normally requires 0.7 hp, but under adverse conditions may require 0.9 hp. Can a ¾-hp motor with a service factor of 1.25 be used to power the system?

Given: $P_{rated} = 0.75$ hp
SF = 1.25
$P_{load} = 0.9$ hp

Find: P_{max} (maximum hp available from motor)

Known: $P_{max} = P_{rated} \times SF$

Solution: $P_{max} = 0.75$ hp $\times$ 1.25
$= 0.9375$ hp

Answer: A ¾-hp motor with SF = 1.25 will suffice because P_{max} is greater than P_{load}.

Some motors are designed and rated so that they can provide more than their rated horsepower under specified conditions. In other

The speed listed on the nameplate of a motor is the motor speed at the rated horsepower. Under light-load conditions, motor speed is

Rated speed

Starting torque

Frequency
tolerance

Stall torque

Torque (*T*)

Running torque

$$P = \frac{TS}{5252}$$

greater than the *rated speed*. When the load is greater than the rated horsepower, the speed will be less than the nameplate speed.

For ac motors, the frequency of the source voltage is also specified. In North America the specified frequency is 60 Hz for motors intended for use on commercially available power. In many European countries, power is distributed at 50 Hz, so motors designed for use in these countries are, of course, designed for 50 Hz. Most motors have a *frequency toler-ance* of ±5 percent from the design frequency providing the supply voltage is at rated values. Since the frequency of commercially distrib-uted power is closely regulated, frequency tol-erance is usually of concern only where power is locally generated.

That which produces rotation, or tends to pro-duce rotation, is known as *torque*. Common units for specifying torque are ounce-inches (oz-in.), pound-inches (lb-in.), pound-feet (lb-ft), gram-centimeters (g-cm), and newton-meter (Nm). As illustrated in Fig. 14-5, torque is a product

of the force and the distance from the center of rotation at which the force exists. The torque in Fig. 14-5(*a*) and (*b*) is the same because the force doubles when the distance is halved.

The *starting torque* (locked-rotor torque) of a motor is an important factor in applications involving high-inertia loads. Many types of motors, such as repulsion and capacitor-start motors, have starting torques that are much greater than their torques while running. With other types of motors, such as shaded-pole and reluctance motors, the starting torque is much less than the torque while running.

The torque of a motor can be specified for different running conditions. When a motor is to be used in an application that produces short-term, heavy loads, the *stall torque* (breakdown or pull-out torque) of the motor is important. Stall torque is the torque produced just before the motor stalls out or starts to stall out and reengages its starting circuit. Stall torque is considerably higher than the torque at the rated horsepower and speed.

The torque at the rated horsepower and speed, sometimes called the *running torque* (full-load torque), can be calculated from the nameplate data. The relationship between speed, power, and torque is given by the formula

$$P = \frac{TS}{5252}$$

where *P* is the power in horsepower, *T* is the torque in pound-feet, and *S* is the speed in revolutions per minute (r/min or rpm). This for-mula is often written as

$$hp = \frac{\text{lb-ft} \times \text{r/min}}{5252}$$

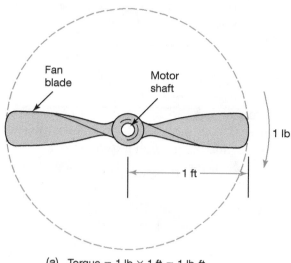

(a) Torque = 1 lb × 1 ft = 1 lb-ft

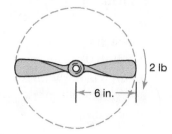

(b) Torque = 2 lb × 0.5 ft = 1 lb-ft

Fig. 14-5 Torque is the product of the distance from the point of rotation and the magnitude of the force.

EXAMPLE 14-7

What is the running torque of a 2-hp, 1740-r/min motor?

Given: *P* = 2 hp, and *S* = 1740 r/min
Find: *T* (torque)

Known: $P = \dfrac{TS}{5252}$

therefore,

$$T = \frac{5252P}{S}$$

Solution: $T = \dfrac{5252 \times 2 \text{ hp}}{1740 \text{ r/min}} = 6 \text{ lb-ft}$

Answer: The running torque is 6 lb-ft.

In some applications, such as an electric door opener, a motor operates only intermittently. Thus, some motors are rated for intermittent duty rather than continuous duty. Intermittent-duty motors are designed to operate for a specified time at rated horsepower followed by an off cycle of sufficient time to allow the motor to cool back to a specified temperature. The *duty cycle* of a motor is a ratio of the on time to the on time plus the off time. This ratio is often expressed as a percentage. For example, a motor that is on for 2 min and off for 5 min would have a duty cycle of $2/(2 + 5) = 0.29$, or 29 percent.

The efficiency of a motor is also an important characteristic of a motor. Large integral-horsepower motors are designed to be very efficient (around 95 percent), but small fractional-horsepower motors may be quite inefficient (30 to 40 percent). When a small motor is operated only intermittently, it is often cheaper to pay for the inefficiency of the motor over its useful life than it is to make the motor more efficient.

The efficiency of a given motor is a function of the load on the motor. A motor is designed to provide *maximum efficiency* at the rated horsepower. At half of the rated load, the efficiency typically decreases by about 10 percent. Of course, at no load, the efficiency is zero.

One reason the efficiency increases with load is that the power factor also increases with load. A motor with no load on it is very inductive because the motor windings are coils of wire wound on a silicon-steel core. Power is only used for the copper losses, core losses, and friction losses. Because the ratio R/X is very low, the PF is also very low. When the motor is loaded, the PF increases as the motor becomes less inductive. With a higher PF, a given amount of current delivers more power without increasing the copper loss. Thus, the efficiency increases.

EXAMPLE 14-8

Determine the efficiency of a motor that delivers 1.7 hp while drawing 12 A from a 240-V supply at a PF of 0.7.

Duty cycle

Given: $P = 1.7$ hp
$I = 12$ A
$V = 240$ V
PF $= 0.7$

Find: % eff.

Known: $\% \text{ eff.} = \dfrac{P_{\text{out}}}{P_{\text{in}}} \times 100$

$P = IV \cos \theta$
$\cos \theta = \text{PF}$
1 hp $= 746$ W

Solution: $P_{\text{in}} = 12 \text{ A} \times 240 \text{ V} \times 0.7$
$= 2016$ W
$P_{\text{out}} = 1.7 \text{ hp} \times \dfrac{746 \text{ W}}{\text{hp}}$
$= 1268.2$ W
$\% \text{ eff.} = \dfrac{1268.2 \text{ W}}{2016 \text{ W}} \times 100$
$= 62.9\%$

Maximum efficiency

Answer: The efficiency of the motor is 62.9%.

 Self-Test

Answer the following questions.

12. List two current ratings for a motor.
13. Which of the current ratings listed in question 12 is the larger value?
14. True or false. Some large motors are started on reduced-voltage or limited-current supplies.
15. True or false. The horsepower rating on the motor nameplate is the starting horsepower.

16. What is the standard ambient temperature used in rating motors?
17. What determines the maximum temperature at which a given motor should operate?
18. Define "allowable temperature rise."
19. What is the service factor of a motor?
20. It is desired to drive a 0.8-hp load with a ¾-hp motor. What would be the minimum service factor for this motor?

21. A 2.5-hp motor has an SF of 1.15. Under ideal conditions, how much continuous horsepower can it deliver?
22. True or false. The rated speed of a motor is its no-load speed.
23. True or false. The tolerance of the frequency rating of a motor is usually ±1 percent.
24. A 9-in. arm is attached to the shaft of a motor. When power is applied to the motor, the end of the arm exerts 1.3 lb of force. What is the locked-rotor torque of the motor?

25. What is the horsepower rating of a motor which produces 6.9 lb-ft of torque at its rated speed of 1140 r/min?
26. Define "stall torque."
27. Determine the efficiency of a 2-hp motor which requires 2 kW to operate at the rated horsepower.
28. True or false. For maximum efficiency, a motor should be operated at about 70 percent of its rated horsepower.
29. True or false. The PF of a motor is highest at its rated horsepower.

14-3 Motor Enclosures

The NEMA has developed standards for the dimensions of the various sizes of motor enclosures. The size (height, length, shaft diameter, etc.) is indicated by a *frame number*. There is considerable variation in the frame number for a given-horsepower motor. This variation is the result of using different insulating and core materials. As better materials have been developed, a given frame has been able to house a motor with a larger horsepower rating. For example, a 56 frame has been for many years a common size for fhp motors; but it is now possible to get multihorsepower motors in this size of frame. In general, with all other factors being equal, the larger the frame number, the larger the horsepower rating. For example, a 2-hp motor is commonly in a 145-T frame, whereas a 10-hp motor is in a 213-T frame. The 145-T frame is approximately 7 in. in diameter and 8 in. long excluding the shaft. Comparable dimensions for the 213-T frame are 10 and 11 in.

Depending on the style of enclosure, motors can be classified as either *open motors* or *totally enclosed motors*. There are many different enclosure types within each of these broad categories.

Open motors (see Fig. 14-6) have enclosures with ventilating openings which allow surrounding air to be forced through the enclosure to cool the motor windings. Various types of open motors put restrictions on the ventilating openings. These restrictions are associated with the type of environment in which the motor is intended to operate. For example, a *drip-proof motor* is an open motor in which the openings are designed so that particles or drops striking the motor enclosure at an angle no greater than 15° from the vertical will not impair the operation of the motor. *Splash-proof motors* are designed so that matter splashing within specified angles will not harm the motor. *Lint-free motors* have smooth, streamlined openings so that lint in the air will not build up and clog the openings. *Guarded motors* have the openings screened or grilled so that objects greater than a specified size and shape cannot enter the closure and make contact with electric or moving parts of the motor. Technical specifications of these and other types of open motors are published by the NEMA.

Totally enclosed motors (see Fig. 14-7) are motors that do not allow free exchange between surrounding air and air within the enclosures.

Fig. 14-6 The shaft end of a drip-proof open motor.

Fig. 14-7 A totally enclosed, fan-cooled motor.

Although they have no openings, they are not sealed or airtight. *Totally enclosed nonventilated* (TENV) *motors* have no external fans to force air over the external surface of the enclosure. *Totally enclosed fan-cooled* (TEFC)

motors have an external fan, attached to the rotor shaft, which forces air circulation around the motor. An *explosion-proof motor* is totally enclosed and designed so that an explosion of a gas inside the motor will not cause a like gas around the motor to explode also. Other types of totally enclosed motors include *waterproof motors* which can be "hosed down" and *dust-ignition-proof motors* which prevent significant amounts of explosive dust from entering the enclosure and which will not ignite explosive dust around or on the enclosure.

Many totally enclosed motors have fins on the exterior of the enclosure to aid in transferring heat from the motor windings to the external air. Larger totally enclosed motors may use circulating water or air as well as heat exchangers to aid in cooling the motor.

Explosion-proof motors

Waterproof motors

Dust-ignition-proof motors

Totally enclosed nonventilated motors

Totally enclosed fan-cooled motors

Self-Test

Answer the following questions.

30. How is the physical size of a motor specified?
31. What does the abbreviation "TEFC" stand for?

32. List two categories used to classify motor enclosures.
33. True or false. Explosionproof motors are designed to prevent gas and vapor from exploding within the motor enclosure.

14-4 Squirrel-Cage Induction Motors

As shown in Fig. 14-8, many ac motors are classified as induction motors. An *induction motor* is a motor that has no electric connections between the power sources and the rotor, yet the rotor has conductors which carry current. The current in the rotor conductors is an induced current. It is induced by the magnetic field created by the *stator windings*.

Figure 14-9 shows a stator, complete with windings, from an fhp induction motor. Notice that the start poles overlap, and are on top of, the run poles. A single silicon-steel lamination from a stator core with 36 slots to hold the coils of the stator windings is shown in Fig. 14-10. Laminations like this one are stacked together to form the stator core shown in Fig. 14-10. The stator windings are insulated from the core by *slot*

insulators and *slot wedges*. Slot insulators are installed in the core slots before the winding coils are inserted. Then the slot wedges, or slot keepers, are inserted over the winding coils to insulate and hold them in place (see Fig. 14-9). Notice in Fig 14-10 that the core slots are shaped so that they provide a *core tooth* to hold the slot keeper in place. Some stators are insulated by a thin epoxy coating rather than by slot insulators.

All parts of the stator are bonded together by an insulating varnish which is applied to the finished stator by dipping it in a varnish vat. Sometimes there are no slot keepers either; instead of slot keepers, the windings are held in place by a strong insulation varnish.

A *squirrel-cage rotor* is shown in Fig. 14-11. The rotor is called a squirrel-cage rotor because the configuration of the conductors embedded in the silicon-steel rotor core resembles the rotary cage used to provide exercise for

Slot wedges

Induction motors

Core tooth

Stator windings

Squirrel-cage rotor

Slot insulators

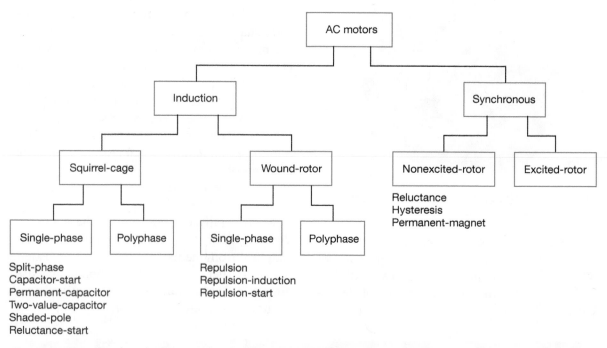

Fig. 14-8 Classification of ac motors by electrical characteristics.

Fig. 14-9 Stator with a four-pole running winding and a four-pole starting winding.

Fig. 14-10 Stator lamination and core.

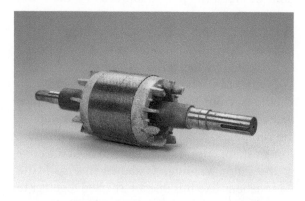

Fig. 14-11 Squirrel-cage rotor used in an induction motor.

Shorting rings

Rotor core

squirrels kept in captivity. The core of the rotor is a stack of disc laminations that have slots around the circumference. These disc laminations are pressed onto the rotor shaft. The rotor conductors occupy the slots in the core. The conductors are shorted together at each end by conducting rings. Careful inspection of Fig. 14-11 shows the *shorting rings* at each end of the *rotor core* and the conducting bars connected between the rings. If the iron core were removed, the conducting bars and rings would look like

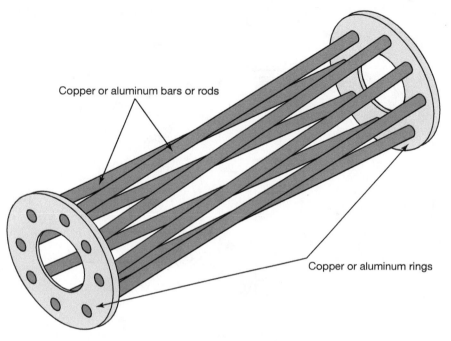

Copper or aluminum bars or rods

Copper or aluminum rings

Fig. 14-12 Conductor arrangement in a squirrel-cage rotor.

Fig. 14-12. Notice that the rotor bars are *skewed* (not parallel to the shaft). Skewing the *rotor bars* allows the bars to enter the magnetic flux of one tooth of the stator before leaving the flux of the adjacent tooth. This provides a more constant torque and reduces vibration in the motor. No separate insulating material is needed to insulate the conducting bars and rings from the rotor core because the voltage induced into these single conductors is low and the conductor resistance is extremely low compared with the resistance of the core. Core resistance is high because each lamination is oxidized to insulate it from adjacent laminations. For many rotors, the bars and rings are cast into the core after it is assembled.

 Self-Test

Answer the following questions.

34. What is the distinguishing characteristic of an induction motor?
35. What causes current in the squirrel-cage rotor?

36. True or false. Slot insulators are used in a squirrel-cage rotor to insulate the rotor bars from the rotor core.
37. True or false. The conductor bars in a squirrel-cage rotor are parallel to the rotor shaft.

Single-Phase Motors

In order to start a single-phase induction motor, some method must be found to produce two sources of magnetomotive force that are separated in both space and time. This creates a rotating magnetic field that develops a *rotational torque* on the rotor.

Why the field created by a single magnetomotive force does not produce a rotational torque to turn the rotor is illustrated in Fig. 14-13. This figure shows that when the stator coils are excited by single-phase alternating current, the pulsating magnetic field they produce induces a voltage into the rotor bars. Since the rotor

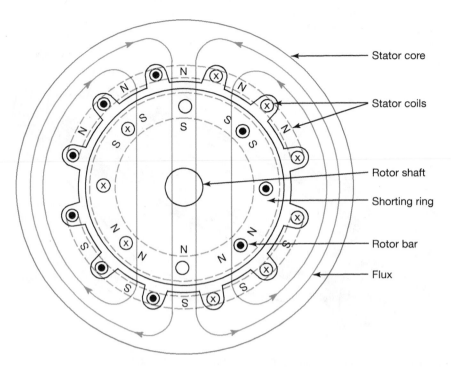

Fig. 14-13 Stator and rotor arrangement that does not provide any starting torque.

Transformer action

Run winding

Start winding

Poles

Consequent-pole motor

Split-phase internal-resistance motors

is stationary, the induced voltage is caused by *transformer action* between stator coils and the shorted rotor bars. The shorted rotor bars act like shorted turns of a transformer secondary. Notice that no voltage is induced into the top and bottom rotor bars in Fig. 14-13 because their axis is perpendicular to the axis of the stator coils. Also notice that maximum voltage is induced into the left and right rotor bars because their axis is parallel to that of the stator coils. The induced voltage causes a current to flow in the rotor bars and produces a magnetic field that is perfectly aligned with the stator field. Thus, although the rotor core is strongly attracted to the stator core, there is no rotational torque to turn the rotor.

A number of techniques are used to create a rotating magnetic field. We will look at these techniques as we discuss the various types of single-phase induction motors.

Split-Phase Motor

The *split-phase internal-resistance motor,* commonly called the split-phase motor, is the most prevalent type of fhp single-phase ac motor. Its name is descriptive of the technique it uses to create a rotating magnetic field. The rotating field is created by two out-of-phase currents which are obtained by "splitting" the single-phase ac source to provide two out-of-phase currents.

You May Recall

. . . from an earlier chapter how the phase shift caused by series *RL* circuits can produce two out-of-phase currents from a single-phase source.

A split-phase motor has two windings in the stator as illustrated in Fig. 14-14. Notice in Fig. 14-14(*a*) that all slots hold a coil for both the *run winding* and the *start winding*. In many motors, some slots will hold a coil only for the run winding or for the start winding.

Both windings in Fig. 14-14(*a*) have two *poles.* A pole is a group of coils of wire all wound in the same direction so that the magnetic field of each coil is aiding all other coils in the pole. In all motors except the *consequent-pole motor,* adjacent poles in a winding have opposite magnetic polarities. Figure 14-13 shows that the upper pole produces a north magnetic pole on the stator teeth while the bottom pole is producing a south magnetic pole. To produce opposite magnetic polarities, the coils in one pole are wound in the opposite direction from the coils in an adjacent pole.

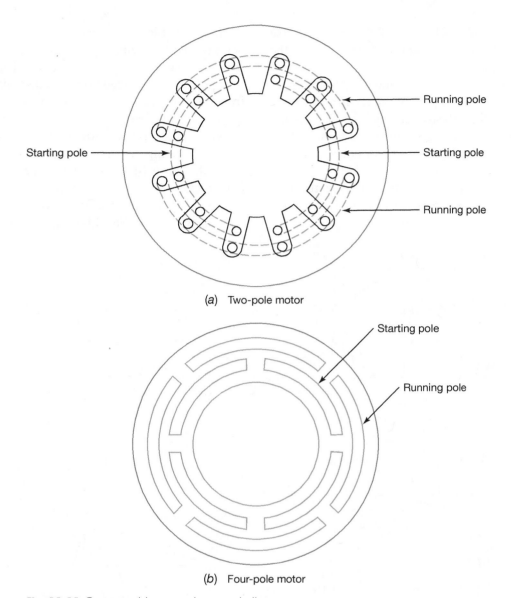

(a) Two-pole motor

Starting pole

Running pole

Starting pole

Running pole

(b) Four-pole motor

Starting pole

Running pole

Fig. 14-14 Stators with run and start windings.

Power lines or Pollution Recent studies by the United States and Finland have shown that living near power lines may increase the occurrence of leukemia in children. However, scientists are not sure whether this increase is due to high levels of pollution occurring in these power line areas. Big power lines tend to run along busy streets in older neighborhoods, so the increase could be related to conditions in older homes or to pollution from traffic.

Of course, the polarities of the poles reverse each half-cycle when the currents through the coils reverse.

Figure 14-14(b) illustrates a simplified method of showing the windings in a four-pole stator. In this illustration, individual coils are not shown; instead, a complete pole is indicated by a circular segment. The four poles of the run winding are located farthest from the center of the drawing. The wires that connect the four poles in series are omitted, as are the wires that connect to the power source.

Notice in Fig. 14-14 that the centers of the poles for the start winding are placed equidistant from the centers of adjacent run poles. Thus, the start poles are displaced 90 mechanical degrees from the run poles for a two-pole motor and 45 mechanical degrees for a four-pole motor.

When out-of-phase currents flow in the run and start windings, a rotating magnetic field is developed. Figure 14-15 shows how the magnetic field rotates 90 mechanical degrees while the start-winding current and the run-winding current change 90°. In Fig. 14-15(a), at time t_1, the

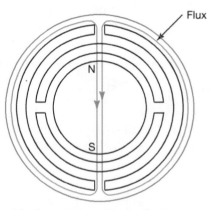

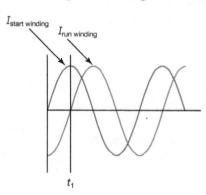

(a) $I_{run} = 0$ and $I_{start} = I_{maximum}$

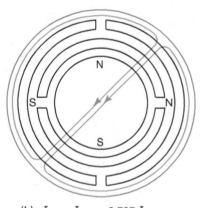

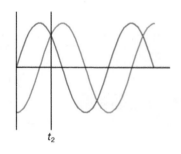

(b) $I_{run} = I_{start} = 0.707\, I_{maximum}$

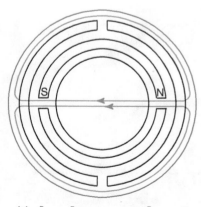

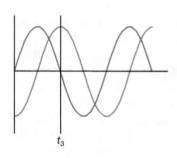

(c) $I_{run} = I_{maximum}$ and $I_{start} = 0$

Fig. 14-15 Rotating magnetic field caused by out-of-phase currents in the run and start windings.

current in the run winding is zero, so the center of the magnetic field poles is in the center of the start-winding poles. As detailed in Fig. 14-15(b), 45 electrical degrees later, the center of the field is midway between the centers of the start and run poles because each pole is contributing equally to the magnetic field. In Fig. 14-15(c), only the run winding is producing flux, so the field is now centered in the run poles. If you visualize the conditions 45° after time t_3 in Fig. 14-15(c), you can see that the start current is reversed, so the magnetic polarity of the start poles is reversed from that shown in Fig. 14-15(b). This results in the magnetic field being rotated another 45° in a clockwise direction.

The discussion based on Fig. 14-15 assumes that the run and start currents are 90° out of phase and that the run and start windings are identical, so the contribution of each winding to the magnetic field is also equal. However, neither of these assumptions has to be met to cause a rotating field. As long as the two currents are neither in phase nor 180° out of phase, the field will rotate. However, inspection of Fig. 14-16 shows that currents that are unequal in magnitude and only a few degrees out of phase produce great fluctuation in the strength of the rotating field. At time t_1 in Fig. 14-16, the field is very weak

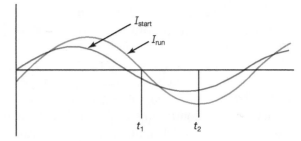

Fig. 14-16 Currents that are nearly in phase produce large pulsations in the rotating magnetic field.

because the run current is zero and the start current is very small. At time t_2, both currents are close to their maximum values, so the magnetic field is very strong compared with this strength at time t_1. The current relationships illustrated in Fig. 14-16 would provide only a very weak starting torque in a motor. When the currents are equal and 90° out of phase (Fig. 14-15), the pulsations in the strength of the rotating field are minimized because one current is maximum when the other current is zero. Current relationships like those in Fig. 14-15 provide a very strong starting torque in a motor.

How the out-of-phase currents are obtained in a split-phase motor is illustrated in Fig. 14-17. The electrical diagram for this motor is given

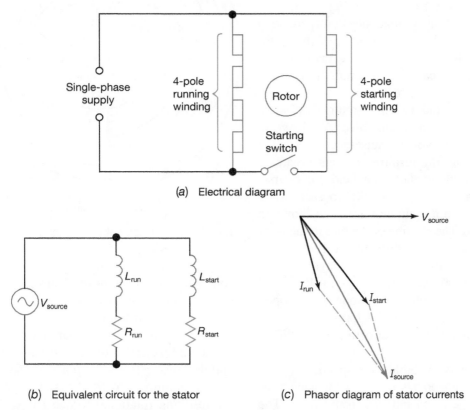

(a) Electrical diagram

(b) Equivalent circuit for the stator

(c) Phasor diagram of stator currents

Fig. 14-17 Phase relationship of the start current and run current in a split-phase motor.

in Fig. 14-17(*a*). Notice that the four-pole run winding is in parallel with the four-pole start winding while the *start switch* is closed. The equivalent circuit for the stator is shown in Fig. 14-17(*b*), in which each winding is represented by its equivalent resistance and ideal inductance. By controlling *Q*, that is, the X_L/R ratio, in each parallel branch, we can control the phase shift of the branch currents with respect to the source voltage. In the split-phase motor, the diameter of the run-winding wire is typically three or four times larger than that of the start winding. Also, the number of turns in the coils of the run winding is typically two or three times greater than in the start winding. This difference in the windings causes the run winding to have less resistance and more inductance than the start winding. The higher *Q* of the run winding results in the run current lagging behind the start current by 20° to 30°, as illustrated in Fig. 14-17(*c*). These out-of-phase

Rotating magnetic field

currents provide the *rotating magnetic field* needed to start a single-phase motor.

The rotating magnetic field produced by the stator cuts the rotor bars and induces a voltage,

Generator action

by *generator action,* into the rotor bars. Thus, the rotor has two voltages induced into it, one by generator action and one by transformer action. However, as shown in Fig. 14-13, the transformer-action voltage does not create a flux which is out of alignment with the stator flux. Thus, for our analysis, the transformer-action voltage can be ignored. Maximum generator-action voltage is induced in the bars directly under the moving stator flux when the flux is at its maximum value. As shown in Fig. 14-18(*a*), this voltage causes a current in the rotor bars, and the resulting current creates a field that is 90° to the stator field. Of course, two flux paths cannot exist 90° to each other,

Resultant flux

so the *resultant flux* in the rotor is as shown in Fig. 14-18(*b*). The shifting of the flux in the rotor by the cross fields also shifts the magnetic poles on the rotor so that they are no longer aligned with the poles in the stator. Thus, as illustrated in Fig. 14-18(*c*), a rotational torque is developed in the motor. Notice that this torque causes the rotor to turn in the same direction as the flux is rotating.

Once the rotor builds up speed, the stator field no longer has to rotate for generator-action voltage to be induced in the rotor. The

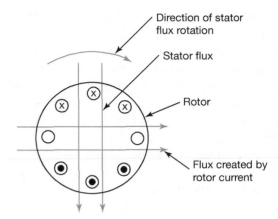

(*a*) Flux created by generator-action voltage

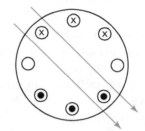

(*b*) Resultant flux in rotor

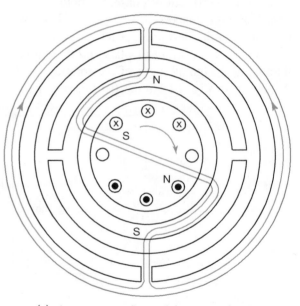

(*c*) Instantaneous flux path in rotor and stator

Fig. 14-18 Development of rotational torque in a motor.

start windings are removed from the motor circuit by opening the start switch when the motor obtains about 70 to 80 percent of its operating speed. The moving rotor bars cut the stationary stator flux as this flux sinusoidally increases and decreases. The rotor poles are still shifted out of alignment with the stator poles, and the resulting rotational torque keeps the rotor

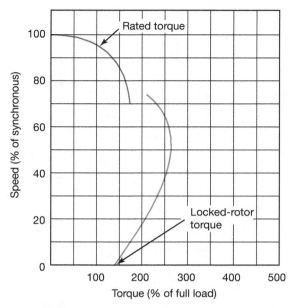

Fig. 14-19 Speed-versus-torque curve for a typical split-phase motor.

as fast as the field rotates, the synchronous speed for the two-pole motor is equal to the speed of the flux rotation, which is one revolution per cycle. For a 60-Hz supply, the synchronous speed is calculated as follows:

$$r/min = 60 \text{ cycles/s} \times 60 \text{ s/min}$$
$$\times 1 \text{ r/cycle} = 3600 \text{ r/min}$$

With a four-pole motor, the field rotates only 0.5 mechanical degrees for each electrical degree. Thus, the four-pole motor operating on a 60-Hz supply has a synchronous speed of 1800 r/min. The general formula for determining the synchronous speed of an induction motor is

$$\text{Synchronous speed} = \frac{120\,f}{\text{no. of poles}}$$

where f is the frequency in hertz and synchronous speed is in revolutions per minute.

turning as long as the load does not exceed the stall torque of the motor.

Once a motor is running, the rotor flux interacts with the stator windings. This interaction increases the effective inductance of the run winding and helps to reduce the current required by the motor. This, along with removal of the start-winding current, is why the run current is only a fraction of the start current in a split-phase motor.

The starting (locked-rotor) torque of fhp split-phase motors ranges from about 130 to 200 percent of full-load running torque. A speed-versus-torque curve for a typical split-phase motor is shown in Fig. 14-19. The motor's locked-rotor torque is 1.5 times as great as its full-load running torque. Notice from this curve that the torque increases as the motor increases speed up to about 60 percent of synchronous speed. Above 60 percent of synchronous speed, torque starts to decrease. The discontinuity in the curve is the result of the start-winding's being switched in and out of the motor circuit.

The synchronous speed of an induction motor, split-phase or any other type, is determined by the number of poles in the run winding and the frequency of the supply voltage. Figure 14-15 shows that for a two-pole motor, the field rotates 45 mechanical degrees for each 45 electrical degrees. Since the rotor can turn only

EXAMPLE 14-9

Determine the synchronous speed of a six-pole motor operating from a 220-V, 50-Hz source.

Given: No. of poles = 6
f = 50 Hz
V = 220 V
Find: Synchronous speed
Known: Synchronous speed =
$$\frac{120\,f}{\text{no. of poles}}$$
Solution: Synchronous speed =
$$\frac{120 \times 50}{6} = 1000 \text{ r/min}$$
Answer: The synchronous speed of the six-pole, 50-Hz motor is 1000 r/min.

Notice that the synchronous speed of a motor is independent of the source voltage.

Figure 14-19 shows that the split-phase motor runs at less than the synchronous speed when it is loaded. This is true of all induction motors. At the rated full-load torque, most induction motors operate at 90 to 98 percent of the synchronous speed. The difference between the synchronous speed and the rated speed of a motor is called *slip*. The slip of a motor is   **Slip** usually expressed as a percentage. The formula for calculating the percent slip is

Percent slip =
$$\frac{\text{synchronous speed} - \text{rated speed}}{\text{synchronous speed}} \times 100$$

EXAMPLE 14-10

Determine the percent slip of a four-pole, 60-Hz, split-phase motor with a rated speed of 1725 r/min.

Given: No. of poles = 4
f = 60 Hz
Rated speed = 1725 r/min

Find: % slip

Known: % slip =
$$\frac{\text{synchronous speed} - \text{rated speed}}{\text{synchronous speed}} \times 100$$
Synchronous speed = $\dfrac{120\,f}{\text{no. of poles}}$

Solution: Synchronous speed =
$$\frac{120 \times 60}{4} = 1800 \text{ r/min}$$

% slip =
$$\frac{1800 \text{ r/min} - 1725 \text{ r/min}}{1800 \text{ r/min}} \times 100 = 4.2\%$$

Answer: The slip is 4.2 percent.

Solenoid-operated switch

Centrifugal switch actuator

A disassembled split-phase motor is pictured in Fig. 14-20 to show how the start switch is operated. When assembled, the spring-loaded bobbin part of the *centrifugal switch actuator* on the rotor shaft contacts the start switch and closes the switch. When power is applied to the motor, the bobbin remains in contact with the switch as

Direction of rotation

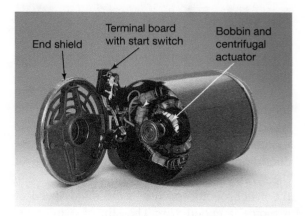

Fig. 14-20 Parts of a split-phase induction motor.

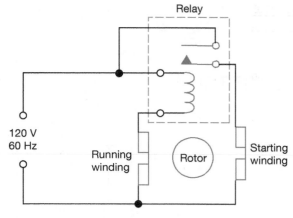

Fig. 14-21 A relay can be used to control power to the start winding in a split-phase motor.

the rotor starts to turn. When the rotor is at about 75 percent of the operating speed, the weights on the centrifugal actuator produce enough force to overcome the spring tension that holds the bobbin part of the actuator against the start switch. At this time the bobbin snaps back away from the switch and the switch opens and disconnects the start winding from the power source.

The start winding can also be disconnected by a relay or a *solenoid-operated switch*. Many of the solenoid-operated switches rely on gravity to open the switch contacts, so they must be operated in the correct physical position. The diagram for a motor using a relay starting switch is drawn in Fig. 14-21. The coil of the relay is in series with the run winding. When the motor is turned on, its run-winding current is very high, so the relay energizes and connects the start winding to the power source. When a relay is energized, the air gap between the armature and the core is eliminated, so it takes much less current to hold the armature down than it takes to pull it down. Thus, as the motor starts to increase its speed and the run-winding current starts to decrease, the relay remains closed and the start winding remains active. By the time the motor speed reaches about 75 percent of the rated speed, the run-winding current will have reduced to a value that is insufficient to hold the relay closed; the relay will open, and the start winding will be disconnected.

The *direction of rotation* of a split-phase motor can be changed by reversing the lead connections of *either* the run winding or the start winding. Reversing the leads on one winding causes the winding current to shift 180°. How this 180° shift reverses the rotation can be seen by referring back to Fig. 14-15. Notice in this figure that the flux rotates clockwise because the start current is leading the run current. Visualize inverting either waveform in this figure and you will see that the run current will lead the start current, and the flux will then rotate counterclockwise.

Self-Test

Answer the following questions.

38. True or false. A rotating magnetic field in the stator is needed to develop locked-rotor torque.
39. True or false. The voltage induced into a stationary rotor by transformer action produces a rotational torque on the rotor.
40. How are two out-of-phase currents developed in the internal-resistance split-phase motor?
41. Which winding in a split-phase motor
    a. Has the larger resistance
    b. Has the larger number of turns
    c. Uses the smaller-diameter wire
    d. Causes the larger phase shift of its current
42. Define the term "pole" as it is used in discussing motors.
43. Do adjacent poles in a motor stator have the same or opposite magnetic polarities?
44. In a six-pole motor, there are _____ mechanical degrees between adjacent start poles and _____ mechanical degrees between the center of a start pole and the center of the nearest run pole.
45. In a six-pole stator, the field rotates _____ mechanical degrees for every 90 electrical degrees.

46. For maximum starting torque, the start and run currents should be _____ degrees out of phase.
47. True or false. In an internal-resistance, split-phase motor, the start and run currents are 90° out of phase.
48. True or false. The locked-rotor torque of a split-phase motor is usually less than its full-load torque.
49. True or false. The coil of a relay used in a motor is connected in series with the start winding.
50. Determine the rated speed of a six-pole, 60-Hz motor if its slip is 4.16 percent.
51. Determine the synchronous speed of an eight-pole, 240-V, 60-Hz motor.
52. Determine the slip of a two-pole, 120-V, 60-Hz motor with a full-load speed of 3450 r/min.
53. The start switch usually opens when the motor obtains between _____ and _____ percent of the rated speed.
54. List two techniques used to connect and disconnect the start winding in a motor.
55. How can one reverse the direction of rotation in a split-phase motor?

Capacitor-Start Motor

Another popular type of single-phase motor is the *capacitor-start motor*. This motor is also a type of split-phase motor, but it does not rely on internal resistance to obtain out-of-phase currents in its run and start windings. Instead, it uses a capacitor in series with the start winding as diagrammed in Fig. 14-22(*a*). The capacitance value is selected so that its reactance is slightly greater than the reactance of the start winding

under locked-rotor conditions. As shown by the phasor diagram in Fig. 14-22(*b*), this makes the start-winding circuit capacitive and causes the start current to lead the source voltage. Thus, with the capacitor-start motor, the two winding currents can be a full 90° out of phase.

The start winding in the capacitor-start motor uses only slightly smaller wire than the run winding, and it has as many, or sometimes more, turns as the run winding. This makes the

Capacitor-start motor

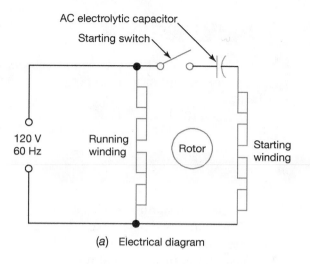

AC electrolytic capacitor

Starting switch

120 V
60 Hz

Running winding

Rotor

Starting winding

(a) Electrical diagram

I_{start}

V_{source}

I_{source}

I_{run}

(b) Phasor diagram

Fig. 14-22 Four-pole capacitor-start induction motor.

Motor-start capacitors

start-winding field about the same strength as the run-winding field.

Because the start and run currents are 90° out of phase and the fields created by these currents are about equal, the capacitor-start motor has a much stronger starting torque than the split-phase motor. The starting torque of the capacitor-start motor ranges from about 300 to 450 percent of its full-load torque.

Not only does the capacitor-start motor have more starting torque, but it requires less source current to develop a given amount of starting torque than a split-phase motor does. One of the reasons for this can be ascertained by comparing Figs. 14-17(c) and 14-22(b). Notice that the phasor

sum of two currents which are approximately 25° out of phase is much greater than that of two comparable currents that are approximately 90° out of phase. Another reason, as mentioned previously, is that equal currents that are 90° out of phase produce optimum starting torque.

Capacitor-start motors are available in standard fhp sizes. They are also available in ihp sizes up to about 7 hp.

The capacitor used in a capacitor-start motor is an ac electrolytic type. It is not designed for continuous duty. Its typical duty cycle is about 20 starts per hour. *Motor-start capacitors* are usually in the 50- to 600-μF range. A typical one is pictured in Fig. 14-23.

Except for the difference already mentioned, the capacitor-start motor and the split-phase motor are constructed in the same way. For the same size motors, the two types have the same characteristics after the start switch opens. The procedure for reversing the direction of rotation is also the same, that is, reversal of the leads to either winding.

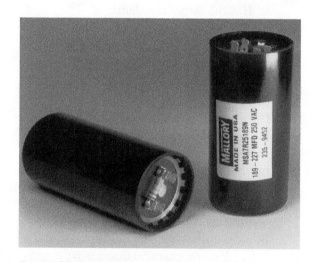

Fig. 14-23 Typical motor-start capacitor.

⚡ Self-Test

Answer the following questions.

56. In a capacitor-start motor, the run and start currents are approximately _____ degrees out of phase.

57. In terms of percentage of full-load torque, what is the range of the starting torque of capacitor-start motors?

58. What type of capacitor is used in the start circuit of a capacitor-start motor?

59. What are the advantages of the capacitor-start motor over the split-phase motor?

Permanent-Split Capacitor Motor

The *permanent-split capacitor motor* has two windings in its stator with a capacitor in series with one of the windings [see Fig. 14-24(a)]. However, it has no start switch; both windings and the capacitor are left in the circuit at all times. The rotor for this motor looks like the rotor for other split-phase motors. These motors are available in the fractional- and subfractional-horsepower sizes.

Capacitors for permanent-split capacitor motors must be rated for continuous duty. Usually an *oil-filled film capacitor* is used.

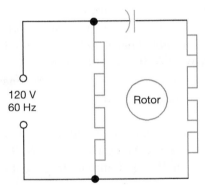

(a) Electrical diagram

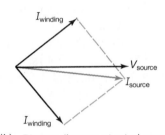

(b) Phasor diagram at rated speed

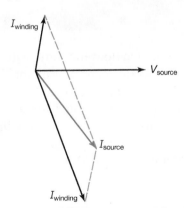

(c) Phasor diagram for locked-rotor

Fig. 14-24 Diagrams for a permanent-split capacitor motor.

The capacitance value is selected to provide a 90° phase shift between the winding currents when the motor is producing between 80 and 100 percent of the rated power [see Fig. 14-24(b)]. At this speed, the effective inductance and inductive reactance of the winding are large, so the capacitive reactance must also be large to make one of the winding circuits appear capacitive. Thus, compared with the value of a start capacitor, this capacitor is relatively small.

The permanent-split motor has a higher power factor, requires less line current, and is more efficient than other single-phase motors. The source current, as seen in Fig. 14-24(b), is small because it is the phasor sum of the two 90° winding currents. At the rated speed, the source current is shifted by the capacitive current in the winding containing the capacitor; this increases the power factor. With smaller line currents, the I^2R losses are decreased and the efficiency is increased.

Figure 14-24(c) shows that the low value of capacitance and the reduced effective inductance of the windings causes the phase difference between the winding currents to be much greater than 90° under locked-rotor conditions. This results in a very weak starting torque. The starting torque of the permanent-split capacitor is typically 50 to 80 percent of the full-load torque.

Permanent-split capacitor motors are often used in variable-speed applications. By controlling the voltage applied to one or both windings, one can control the speed of this motor over the upper 50 percent of its rated speed. As the applied voltage is decreased, the slip increases and the speed decreases.

The direction of rotation of the permanent-split capacitor motor can be changed by either of two methods. First, the leads of either winding can be reversed as was done with the split-phase and capacitor-start motor. Second, as shown in Fig. 14-25, the capacitor can be switched from one winding to the other winding. This latter method has the advantage that it is easy to do while the motor is running. If the switch in Fig. 14-25 is thrown while power is applied, the motor will rapidly decelerate until rotation stops and then immediately accelerate in the opposite direction of rotation. This method of reversal works best when the two windings

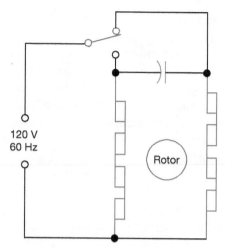

Fig. 14-25 Method of reversing rotation in a permanent-split capacitor motor.

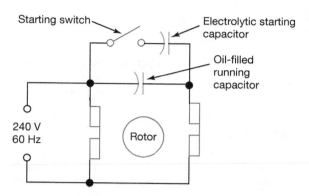

Fig. 14-26 Circuit diagram of a two-pole, two-value-capacitor motor.

are identical. With unbalanced windings, the torques are different for the different directions of rotation.

Another desirable feature of the permanent-split capacitor motor is that it is quieter than the split-phase and capacitor-start motors. This is the result of having both windings continuously energized to provide a uniform rotating field in the stator.

Two-Value Capacitor Motor

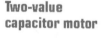

Two-value capacitor motor

An electrical diagram for a *two-value capacitor motor* is drawn in Fig. 14-26. The start switch, which mechanically looks and operates like the start switch in a split-phase motor, connects a large-value ac electrolytic capacitor in parallel with the *run capacitor* for starting the

Run capacitor

motor. Thus, the motor has the high-starting-torque and low-starting-current characteristic of a capacitor-start motor. When the motor reaches about 75 percent of its rated speed, the start switch opens and the motor runs as a permanent-split capacitor motor with its desirable characteristics of high efficiency, low source current, high power factor, and quiet operation.

Two-value capacitor motors are also called capacitor-start, capacitor-run motors. They are available in fhp and ihp motors up to about 20 hp.

The speed of the two-value capacitor motor can be varied over only a very limited range. If the speed is reduced too much, the start switch closes and puts the electrolytic capacitor back in the circuit. Of course, the electrolytic capacitor is not designed for continuous operation.

Self-Test

Answer the following questions.

60. What is the major limitation of a permanent-split capacitor motor?
61. Which type of motor is easily reversed while power is applied to the motor?
62. List five advantages of the permanent-split capacitor motor over the capacitor-start motor.

63. What function does the start switch serve in the two-value capacitor motor?
64. What type of capacitor is used in a permanent-split capacitor motor?
65. True or false. The locked-rotor winding currents in a permanent-split capacitor motor are less than 90° out of phase.

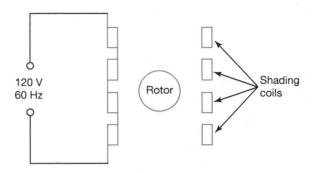

Fig. 14-27 Electrical diagram of a four-pole, shaded-pole motor.

Fig. 14-29 Stator of a two-pole shaded-pole motor. The motor uses one coil for two salient poles.

Shaded-Pole Motor

A circuit diagram for a *shaded-pole motor* is shown in Fig. 14-27. Notice that this motor has no start switch and that only one winding is connected to the power source.

Instead of a *distributed stator winding* like those used in the previously discussed motors, this motor uses a *salient-pole stator* like that pictured in Fig. 14-28. In this figure, one of the main pole coils has been removed to expose the single-turn shading coil. With the salient-pole winding, the winding for a pole is a single coil rather than a group of coils. In some sfhp motors, a single coil wrapped around the core iron may provide both poles for a two-pole motor (see Fig. 14-29).

All shaded-pole motors have a *shading coil* wrapped around part of each salient pole. The shading coils are visible in the pictures in Figs. 14-28 and 14-29. The shading coil is often

just a shorted single turn of heavy copper wire, as seen in Fig. 14-30, where the main winding has been removed.

The operation of the shaded-pole motor can be understood by applying basic principles of induced voltages and currents.

> 💡 **You May Recall**
>
> . . . from Lenz's law that induced voltages and currents always oppose the action that produced them.

Thus, when an increasing flux from the main coil induces a voltage in the shading coil, the resulting current in the shading coil creates a flux that opposes and cancels part of the main flux in the shaded part of the pole. When the flux caused by the main coil starts to decrease, the shading-coil current reverses and creates a flux that aids, and adds to, the flux in the shaded part of the pole. The net result is that the strong part of the magnetic pole shifts, or rotates, from the unshaded to the shaded part of the pole. This rotating flux is sufficient to induce, by generator action, a current in the squirrel-cage rotor and start the motor. Once started, the shaded-pole motor operates like other induction motors except that the shading coils continue to produce I^2R losses.

Shaded-pole motors are cheap to construct and easy to maintain. However, they are very inefficient, have a low PF, and have a very weak starting torque. Therefore, they are used only for subfractional and fractional horsepower to about ¼ hp.

Shaded-pole motor

Distributed stator winding

Salient-pole stator

Shading coil

Fig. 14-28 Stator for a shaded-pole motor in which there is a coil for each salient pole. One coil has been removed to show the heavy single-turn copper shading coil.

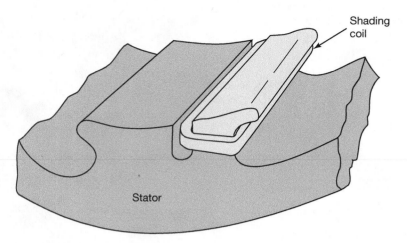

Fig. 14-30 Shading coil on a salient pole.

Reluctance-start motor

Shaded-pole motors like those we have been discussing cannot be electrically reversed. However, some small motors can be disassembled and the stator turned end-for-end to produce reverse rotation.

Reluctance-Start Motor

Occasionally a motor has a stator with salient poles shaped like the one shown in Fig. 14-31. The left side of this pole has more reluctance than the right side because of the wider air gap. More important, the flux in the left side obtains its maximum value before the flux in the right side. This is because air causes no hysteresis; that is, in air the flux is in phase with the magnetomotive force. In silicon steel, the flux lags behind the magnetomotive force. Since flux in the left part of the pole travels through more air, it leads the flux in the right part of the pole. Therefore, the field in Fig. 14-31 travels from

Dual-voltage motor

left to right and the rotor turns clockwise. Of course, the starting torque of the *reluctance-start motor* is very weak.

Some motors use a combination of shading coil and reluctance to provide increased starting torque. One side of the pole has the extra air gap and the other side has the shading coil. A stator for such a motor is pictured in Fig. 14-32.

Dual-Speed and Dual-Voltage Motors

Many motors are designed to operate on either of two source voltages that have a 1:2 ratio. For example, a two-pole, *dual-voltage motor* may be run on either 120 V or 240 V. When the motor is operated on 120 V, the coils for one pole are connected in parallel with the coils for the other pole. For 240-V operation, the coils of the two poles are series-connected. Thus, each

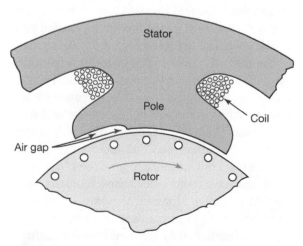

Fig. 14-31 Salient pole in a reluctance-start motor.

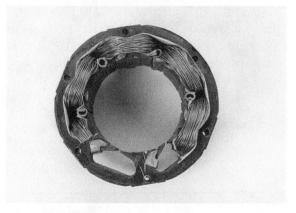

Fig 14-32 A stator that uses reluctance-start and a shading coil to produce starting torque.

coil receives the same voltage and current regardless of which source voltage is used.

A common way to make a *dual-speed motor* is to wind it with two run windings and one start winding. For example, a stator may have a four-pole run winding, a six-pole run winding, and a four-pole start winding. Regardless of the speed (1725 or 1150 r/min) at which the motor is to operate, it is always started as a four-pole motor. Then, if 1150-r/min operation is desired, the power is switched from the four-pole run winding to the six-pole run winding by an extra set of contacts on the start-switch mechanism.

 Self-Test

Answer the following questions.

66. A shaded-pole motor usually uses a(n) _____ pole stator winding.
67. List the main electrical characteristics of the shaded-pole motor.
68. Describe the shading coil in a shaded-pole motor.

69. What causes flux rotation in a reluctance-start motor?
70. Does flux shift toward or away from the shaded part of a pole?
71. True or false. Dual-voltage motors have two sets of run windings.
72. True or false. Dual-speed motors usually start on the low-speed windings.

14-5 Synchronous Motors

As an example of the many types of *synchronous motors* listed in Fig. 14-8, we will look at a *reluctance motor*. The reluctance motor is a type of induction motor that can use any of the methods discussed in Sec. 14-4 to develop a rotating stator field to get the rotor in motion. The squirrel-cage rotor in the reluctance motor is modified so that the iron core has salient poles. Figure 14-33 shows a lamination from the rotor of a typical four-pole reluctance motor.

The reluctance motor starts, like any other induction motor, by having currents induced into the rotor bars by the rotating stator flux. As the rotor gets up to about 90 to 95 percent of the synchronous speed, the salient poles on the rotor are almost lined up with magnetic poles in the stator. At this point, the rotor jumps up to the synchronous speed as the flux tries to align the rotor and stator poles to provide the shortest possible flux path that also has the minimum possible reluctance. This idea is illustrated in Fig. 14-34, where it can be seen that the flux path in Fig. 14-34(*b*) is shorter and involves no more air than in Fig. 14-34(*a*). The rotor never quite achieves perfect pole alignment as shown in Fig. 14-34(*b*) because, if it did, there would be no rotational torque due to the reluctance

effect. At a given motor load, the salient rotor poles are always the same distance from perfect alignment with the stator poles when the stator poles are at their maximum strength. As the motor load increases, the distance from perfect alignment increases. Finally, if the load becomes too great, the reluctance effect can no longer hold the rotor in approximate alignment; the motor drops out of synchronization.

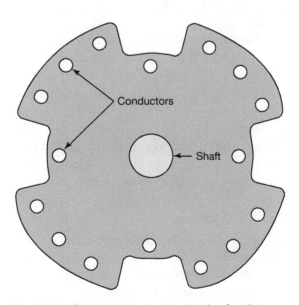

Fig. 14-33 Salient-pole rotor lamination for the rotor of a reluctance motor.

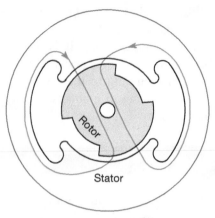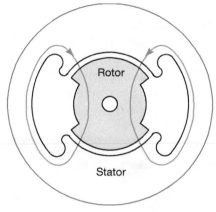

(a) Torque caused by misaligned poles   (b) Rotor and stator poles aligned

Fig. 14-34 Salient rotor poles try to align themselves with the magnetic poles in the stator.

Self-Test

Answer the following questions.

73. True or false. Salient rotor poles are needed to start a reluctance motor.

74. The reluctance motor uses a modified _____ rotor.

75. The reluctance motor is classified as a(n) _____ motor.

Internet Connection

For information on new product developments, visit the website for the Consumer Electronics Association (CEA).

DC motor

Field winding

Armature winding

Commutator

Armature

Shunt dc motor

14-6 Other Types of Motors

> **You May Recall**
>
> . . . that the operating principle of a *dc motor* was discussed in the chapter on magnetism.

In a brush-type dc motor, the stationary winding is called the *field winding* and the rotating winding is called the *armature winding*. The rotating part of the motor, which consists of the shaft, iron-core laminations, winding, and *commutator,* is referred to as the *armature.*

Shunt DC Motor

Figure 14-35(*a*) shows the diagram for a *shunt dc motor.* In this motor, the field winding can use many turns of small-diameter wire because the large source current needed under full-load conditions flows only through the armature winding. The shunt motor has a fairly

constant speed over a wide load range. However, the speed does decrease some as the load increases.

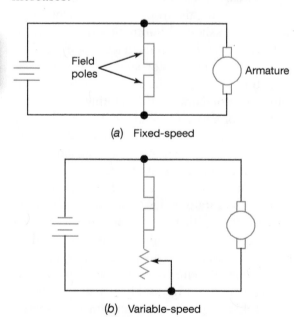

(a) Fixed-speed

(b) Variable-speed

Fig. 14-35 Shunt dc motor. The rheostat in (*b*) controls the speed of the motor.

The speed of the armature is always such that the induced voltage (cemf) in the armature winding plus the *IR* voltage (due to the resistance of armature wire) equals the source voltage. Since the armature winding uses large-diameter wire, its resistance is very small. Thus, the *IR* voltage is very small compared with the induced voltage. As the load on the motor increases, so does the armature current. This increases the *IR* voltage and decreases the cemf needed to equal the source voltage. Therefore, if the field flux remains constant, the armature must slow down or the combined *IR* voltage and cemf voltage would exceed the source voltage. Although the field flux changes slightly because of interaction between the field flux and the armature flux, the change is not significant except when the motor is operated with a very weak field flux.

Armature current in any dc motor is a function of the load on the motor. When the load is increased, armature current must increase so that a stronger armature flux will provide more torque as it reacts with the field flux.

The speed of a shunt motor can be controlled over a fairly wide range (about 1:4) by controlling the field current with a series rheostat as shown in Fig. 14-35(*b*). Decreasing the field current (and thus the field flux) forces the armature speed to increase in order to produce the required cemf. For a given load, the armature current must increase to provide the additional flux needed to increase the armature speed. Thus, the amount the speed can be increased depends on the load on the motor shaft. If the motor is operating at rated horsepower, no speed adjustment can be made.

Series DC Motor

Figure 14-36 shows a diagram of a *series dc motor*. This motor is noted for its high starting torque and its load-dependent speed.

The series motor has high starting torque because both the armature and the field are wound with large-diameter wire. When power is first applied, the current is limited only by the resistance of the windings, so a large current can flow. This large current, flowing through both windings, produces strong poles in both the armature and the field, so the resulting torque is very high. Of course, if the load is too large, the magnetic materials saturate before torque sufficient to turn the load can be developed.

With a light load on the series motor, little current is required to develop the necessary torque. However, with little current in the field winding, the field flux is weak, so the armature must rotate at a high speed to produce enough cemf to counter the source voltage. With large loads, the current is high, the field flux is strong, and the armature can, therefore, turn much more slowly and still produce the required cemf. With no load, the speed of a large series dc motor can become so high that centrifugal forces destroy the armature. Therefore, these motors are used only where there is no chance of the load being disengaged.

Compound DC Motor

The *compound motor,* shown in Fig. 14-37, has part of the field winding in series with the armature and part in parallel with the source voltage. The series part of the field winding can either aid (called *cumulative compounding*) or oppose (called *differential compounding*) the parallel part of the field winding. Since differential compounding is rarely used, we will concern ourselves only with the characteristics of the cumulative compound motor.

As expected, the cumulative compound motor has speed and starting-torque characteristics somewhere between those of the series and the

Compound motor

Cumulative compounding

Differential compounding

Series dc motor

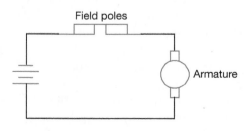

Fig. 14-36 Series dc motor.

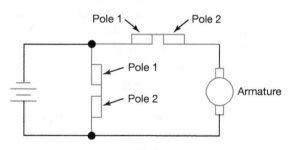

Fig. 14-37 Compound dc motor. Part of each field pole is in series with the armature, and part is in parallel with the supply.

shunt motor. Because of the series part of the field, the compound motor has more starting torque than the shunt motor. Because of the parallel part of the field, the speed of this motor is less dependent on the load. Even with no load, the parallel portion of the field provides sufficient flux to limit the speed to a nondestructive value.

Universal Motor

Universal motor

The *universal motor* has the same electrical diagram as a series dc motor. In fact, it has the same characteristics as the series dc motor, and the armature and the field are quite similar in appearance to those of the series dc motor.

The universal motor is designed to operate equally well from either an ac or a dc supply—thus the name "universal." Operating a series motor on alternating current requires a few design changes from the series dc motor. For example, the universal motor uses thin laminations in both the armature core and the field poles (core) to reduce eddy current loss when operating on alternating current.

Brushless DC Motor

Brushless dc motors are constructed in a different manner than the dc motors previously discussed. Instead of a slotted armature wound with many coils that terminate on commutator segments, the brushless dc motor has a permanent-magnet rotor. As can be seen in Fig. 14-38, the stator is wound with three sets of coils called phase A, phase B, and phase C. (Some brushless motors have only two phases; others have four phases.) Each of the three phases has two coils connected in series to produce one north pole and one south pole for each phase. In Fig. 14-38, the phase poles are labeled A and A' for phase A, B and B' for phase B, and C and C' for phase C. The three phases are connected in a wye configuration by connecting one end of the A', B', and C' coils together at an isolated (unused) star point (see Fig. 14-39). With the wye connection, two of the three phases will be energized when a dc source is connected to any two of three free ends on the wye-connected phases. The phase coils are wound so that a current flowing toward the star point creates a north pole on the phase prime pole. Thus, the current flow shown in Fig. 14-39(a) creates a magnetic north pole on pole C' and a south pole on pole C. The current flowing out of the star in Fig. 14-39(a)

creates a north pole at pole A and a south pole at pole A'. It should now be apparent that the phase currents shown in Fig. 14-39(a) create the magnetic-pole polarities shown in Fig. 14-38(a). Look carefully at the direction of current flow in Fig. 14-39(b) and verify that it produces the magnetic poles given in Fig. 14-38(e).

Fig. 14-38 shows that the permanent-magnet rotor with its fixed magnetic poles follows the rotating magnetic poles in the stator. Notice that the six magnetic-pole combinations shown in Fig. 14-38(a) through (f) produce one revolution of the rotor. Thus, each new pole combination advances the rotor another 60 mechanical degrees.

The six switches in Fig. 14-39 provide the control needed to produce any of the magnetic-polarity combinations shown in Fig. 14-38(a) through (f). Only one at a time of the (N) switches (the top row) is closed. The same applies to the (S) switches (the bottom row). Of course, the (N) and (S) switches for the same phase cannot be closed at the same time as this would create a short across the dc source. The labels on the switches in Fig. 14-39 save tracing current to determine the polarities produced by various switch-ON combinations. For example, to produce the pole pattern shown in Fig. 14-38(c), one needs a south pole on B and a north pole on C. Therefore, close switches B(S) and C(N).

The sensors in Fig. 14-38 provide a binary-digital output. That is, the output switches between either of two dc voltage levels. One output level represents a binary 0 (logic 0) and the other level represents a binary 1 (logic 1). When the south pole of a magnet passes by the sensor, it switches its output to a logic 0 if it was a logic 1. Conversely, passing a north pole by the sensor switches the output to a logic 1 if it is not at a logic 1. The output (0 or 1) of the three Hall-effect sensors is a three-digit binary code, which tells the electronic motor controller what action to take to keep the rotor turning. For example, in Fig. 14-38(f), sensor 2 is switched to 0 just before the rotor is in perfect alignment with the stator poles. This produces a new code of 001 (it had been 011), which is sent to the motor controller. This new code (001), tells the controller it is time to switch to the magnetic pole polarities shown in Fig. 14-38(a) to keep the rotor turning. This is the form of commutation used in the brushless dc motor.

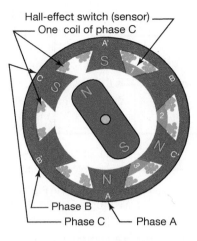

(a) Phases A and C are energized.
The output of the sensors is 000.

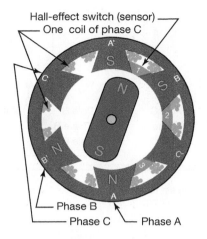

(b) Phases A and B are energized.
The output of the sensors is 100.

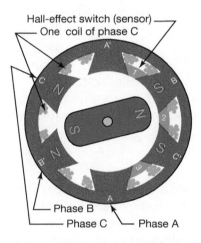

(c) Phases B and C are energized.
The output of the sensors is 110.

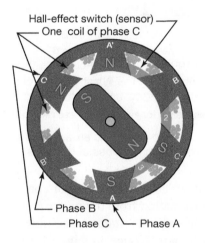

(d) Phases A and C are again energized
but magnetic polarities are reversed
from those in (a). The output of the
sensors is 111.

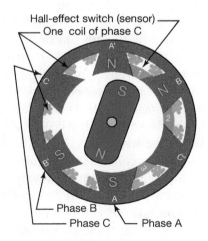

(e) Phases A and B are energized.
The output of the sensors is 011.

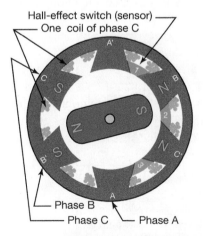

(f) Phases B and C are energized.
The output of the sensors is 001.

Fig. 14-38 Poles rotate around the stator as phase coils are energized and deenergized. Notice that the magnetic polarities of the phases change as the rotor turns clockwise.

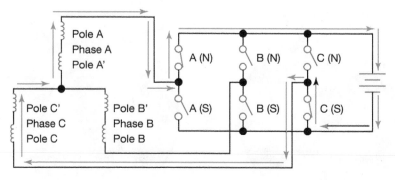

(a) Switches are set for energizing phases A and C with the magnetic polarities shown in Fig. 14-38(a).

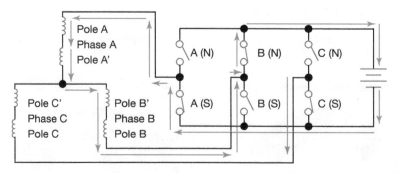

(b) Switches are set for energizing phases A and B with the magnetic polarities shown in Fig. 14-38(e).

Fig. 14-39 The settings of the six switches determine which phases are energized and the magnetic polarities of the poles.

Figure 14-40 illustrates how the motor controller receives information from the motor and responds by setting up the stator conditions requested by the motor. Notice in Fig. 14-40 that the controller sends an electronically generated dc pulse to the phase windings. The length of the pulse is determined by the time it takes for the rotor to advance far enough to

Binary code output from the Hall-effect sensors		
Sensor 1	Sensor 2	Sensor 3
Controller for a dc brushless motor		
Phase A	Phase B	Phase C
Dc voltage pulses to the appropriate phases of the motor stator		

Fig. 14-40 The output of the three sensors controls the commutation of the motor. When the controller receives a new code, it selects the appropriate phases to be energized.

generate a new binary code. (These pulses replace the battery used for conceptual purposes in Fig. 14-39.) The controller doesn't use mechanical switches to switch the dc pulses to various phase combinations. Instead, it uses solid-state (semiconductor) devices that have no moving parts and do not arc and burn contacts with each operation.

The speed of a given brushless dc motor is controlled by the amount of voltage, and thus current, the phase windings receive. More voltage means stronger magnetic poles to attract and advance the rotor poles more rapidly. Thus, more voltage increases and less voltage decreases the rpm of the motor. Rather than controlling the maximum voltage of the dc pulse, many motor controllers use pulse width modulation (PWM) to control the voltage applied to the motor phases. PWM essentially modulates the main +dc pulse with a high frequency −dc pulse. The −dc pulse periodically reduces the main pulse to zero. The net result is that the average dc applied to the motor phases is reduced and this reduces the speed of rotor rotation. Figure 14-41 illustrates this concept although the frequency of the modulating pulse

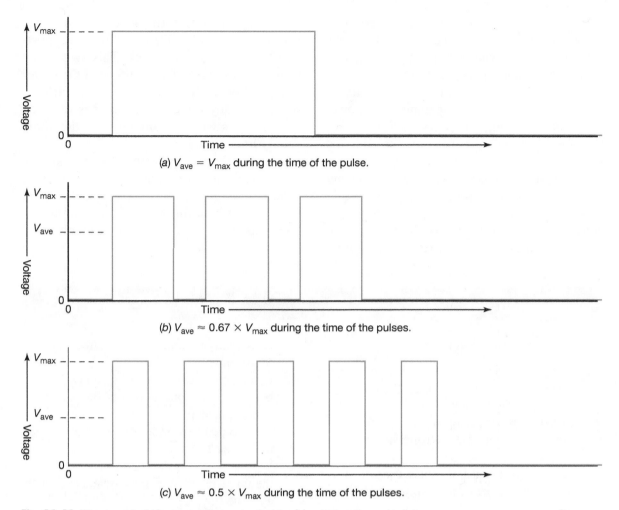

(a) $V_{ave} = V_{max}$ during the time of the pulse.

(b) $V_{ave} \approx 0.67 \times V_{max}$ during the time of the pulses.

(c) $V_{ave} \approx 0.5 \times V_{max}$ during the time of the pulses.

Fig. 14-41 The speed of the motor is controlled by V_{ave} during the pulse(s).

has been greatly reduced to simplify and clarify the figure. Notice that the total width of the main positive pulse increases as V_{ave} decreases. This is because the slower speed takes more time to produce a new binary code to send to the controller.

Not all brushless dc motors use Hall-effect sensors. The induced voltage produced in the nonenergized phase winding can be used to control commutation.

Elimination of the brushes and commutator gives the brushless dc motor many advantages over the brush-type motor. Some advantages are:

1. Can be operated in a clean environment as there is no carbon dust generated. Some of them are designed to be sterilized in an autoclave.
2. Much quieter operation—especially at higher speeds.
3. Can operate at higher rpm.

4. No electromagnetic interference produced by sparking brushes.
5. Can operate in environments that may produce explosive fumes or dust.
6. Requires less maintenance.
7. More efficient operation.

Three-Phase Motors

The stator of a *three-phase* induction *motor* has a separate winding for each phase. Therefore, the stator has a constantly rotating magnetic field and, like the permanent-split capacitor motor, needs no start switch or switch activator.

The technical details and characteristics of the many types and designs of three-phase motors are too complex to cover in this chapter. However, we can, in general terms, compare three-phase motors with single-phase motors.

Because the rotating field in three-phase motors has constant amplitude, three-phase motors produce less noise than their single-phase

Three-phase motor

counterparts. Also, they provide a constant shaft torque throughout the full cycle of the source voltage, whereas the torque of the single-phase motor pulsates during each cycle.

The efficiency of a three-phase motor is higher than that of a comparable-horsepower single-phase motor. For this and other reasons, the three-phase motor is generally smaller and lighter.

Some single-phase motors are not meant to be run in reverse rotation. Others require reversing the two leads that power either the start winding or the run winding. However, as illustrated in Fig. 14-25, reversing rotation of a single-phase, permanent-split capacitor motor requires only changing the pole position lever of a SPDT switch. This switch can be located on the external surface of the motor or the remote from the motor. The direction of rotation of any three-phase motor is reversed by merely switching any two of the three supply lines.

Finally, three-phase motors are available in a much wider range of power ratings. Whereas single-phase motors above 10 hp are not common, three-phase motors rated in hundreds of horsepower are commonly used in industry.

Self-Test

Answer the following questions.

76. The windings in brush-type dc and universal motors are the _____ and _____ windings.
77. List the three types of brush-type dc motors in order of decreasing starting torque.
78. List the three types of brush-type dc motors in order of decreasing speed regulation.
79. The universal motor has characteristics like those of the _____ dc motor.

80. How many windings does a three-phase motor have?
81. List as many differences as you can between three-phase motors and single-phase motors.
82. What type of rotor does a brushless dc motor use?
83. With a brushless dc motor, how many of the three-phase windings are energized at any one time?
84. What type of signal is produced by a Hall-effect switch sensor?

Chapter 14 Summary and Review

Summary

1. Motors can be classified by the type of power required or by the intended use.
2. Motors are available in sfhp, fhp, and ihp sizes.
3. Motors are rated for voltage, current, power, temperature, frequency, torque, duty cycle, service factor, and efficiency.
4. The National Electrical Manufacturers Association (NEMA) establishes standards for motor ratings, enclosures, and frame sizes.
5. The voltage-rating tolerance of a motor is typically ± 10 percent, and the tolerance for frequency is usually ± 5 percent.
6. Both voltage-source variation and line-voltage drops must be of concern when determining the appropriate voltage rating for a motor.
7. If possible, a motor should be operated on the positive side of its voltage tolerance rather than on the negative side.
8. The locked-rotor current is many times larger than the full-load current.
9. A motor with an SF greater than 1.0 can be operated above its rated horsepower under specified conditions.
10. Ambient temperature for motor-design purposes is usually considered to be 40°C.
11. The maximum operating temperature of a motor is determined by the class of insulation used. The classes are A (105°C), B (130°C), F (155°C), and H (180°C).
12. Three common torque ratings of a motor are locked-rotor (starting), stall (breakdown or pull-out), and running (full-load).
13. The efficiency and PF of a motor are maximum at rated horsepower.
14. Some common types of motor enclosures are dripproof, splashproof, guarded, TENV, TEFC, explosionproof, waterproof, and dust-ignitionproof.
15. Induction motors have no electric connection between the source and the rotor. Most of these motors use a squirrel-cage rotor.
16. Developing starting torque in an induction motor requires a rotating magnetic field in the stator. This field can be created by an auxiliary (start) winding, shaded poles, or differential reluctance in the poles.
17. Maximum torque results when the two-winding currents are equal in amplitude and 90° out of phase.
18. Induction motors can be designed for either dual speed or dual voltage.
19. A start switch may be either mechanically or electrically operated. This switch opens at about 75 percent of the rated speed.
20. A reluctance motor has a squirrel-cage rotor with salient poles.
21. The speed of the shunt dc motor can be varied by varying the field current.
22. The universal motor operates on either alternating or direct current and has characteristics similar to those of the series dc motor.
23. The speed of a brushless dc motor can be varied by varying the phase windings' average pulse voltage.
24. Commutation of a brushless dc motor can be controlled by the binary code produced by rotor-position sensors and sent to the motor controller.
25. The brushless dc motor uses a permanent magnet rotor.
26. Compared with single-phase motors, three-phase motors are more efficient, make less noise, provide more constant torque, are lighter and smaller, and are easier to reverse.

Related Formulas and Tables

Relationship between speed, power, and torque:

$$P = \frac{TS}{5252}$$

Synchronous speed $= \dfrac{120\,f}{\text{no. of poles}}$

Percent slip =

$$\frac{\text{synchronous speed} - \text{rated speed}}{\text{synchronous speed}} \times 100$$

Induction Motor Characteristics

| | Characteristic | | | | |
Type of Motor	Locked-Rotor Torque	Starting Current	Running Current	Efficiency	Power Factor
Split-phase	Moderate	High	Moderate	Moderate	Moderate
Capacitor-starting	Very high	Moderate	Moderate	Moderate	Moderate
Permanent-capacitor	Weak	Low	Low	High	High
Two-value-capacitor	Very high	Moderate	Low	High	High
Shaded-pole	Very weak	Very low	High	Very low	Very low
Reluctance-starting	Very weak	Very low	High	Low	Very low
Reluctance	*	*	High	Low	Low

| | Characteristic | | | | |
Type of Motor	Auxiliary Winding	Starting Switch	Common Sizes†	Usually Reversible	Variable-Speed	Synchronous
Split-phase	Yes	Yes	1 & 2	Yes	No	No
Capacitor-starting	Yes	Yes	2 & 3	Yes	No	No
Permanent-capacitor	Yes	No	1 & 2	Yes‡	Yes	No
Two-value-capacitor	Yes	Yes	2 & 3	Yes	No	No
Shaded-pole	No	No	1 & 2	No	Yes	No
Reluctance-starting	No	No	1 & 2	No	Yes	No
Reluctance	*	*	1 & 2 & 3	*	No	Yes

*Depends on technique used to start motor.
†1 = sfhp, 2 = fhp, and 3 = ihp.
‡Can be reversed while power is applied.

DC Motor Characteristics

Type of DC Motor	Characteristics		
	Speed Regulation	Starting Torque	Variable-speed
Series	Very poor	Very high	No
Shunt	Good	Moderate	Yes
Cumulative-compound	Moderate	High	No
Brushless	Very good	Moderate	Yes

Chapter Review Questions

For questions 14-1 to 14-14, determine whether each statement is true or false.

14-1. Regardless of the SF rating, a motor should not be continuously operated above its rated horsepower. (14-2)

14-2. Tolerance for the voltage rating of a motor is typically ±5 percent. (14-2)

14-3. The frequency tolerance of a motor rating is of primary concern when a motor is operated from a commercial supply. (14-2)

14-4. The run-winding current in an induction motor decreases as the motor speeds up. (14-4)

14-5. The temperature-rise rating of a motor is usually based on a 60°C ambient temperature. (14-2)

14-6. The efficiency of a motor is usually greatest at its rated power. (14-2)

14-7. The voltage drop in a line feeding a motor is greatest when the motor is at about 50 percent of its rated speed. (14-2)

14-8. An explosionproof motor prevents gas and vapors from exploding inside the motor enclosure. (14-3)

14-9. Since a squirrel-cage rotor is not connected to the power source, it does not need any conducting circuits. (14-4)

14-10. The start switch in a motor opens at about 75 percent of the rated speed. (14-4)

14-11. "Reluctance" and "reluctance-start" are two names for the same type of motor. (14-5)

14-12. The cumulative-compound dc motor has better speed regulation than the shunt dc motor. (14-6)

14-13. The compound dc motor is often operated as a variable-speed motor. (14-6)

14-14. All single-phase induction motors have a starting torque which exceeds their running torque. (14-4)

For questions 14-15 to 14-19, choose the letter that best completes each statement.

14-15. Greater starting torque is provided by a (14-6)
a. Shunt dc motor
b. Series dc motor
c. Differential compound dc motor
d. Cumulative compound dc motor

14-16. Which of these motors provides the greater starting torque? (14-4)
a. Split-phase
b. Shaded-pole
c. Permanent-split capacitor
d. Capacitor-start

14-17. Which of these motors provides the quieter operation? (14-4)
a. Split-phase
b. Capacitor-start
c. Two-value capacitor
d. Universal

14-18. Which of these motors has the greater effi-ciency? (14-4)
a. Reluctance-start
b. Shaded-pole
c. Split-phase
d. Permanent capacitor

14-19. Which of these motors would be available in a 5-hp size? (14-4)
a. Split-phase
b. Two-value capacitor
c. Permanent capacitor
d. Shaded-pole

Answer the following questions.

14-20. List three categories of motors which are based on the type of power required. (14-1)

14-21. List three categories of motors which are based on a range of horsepower. (14-1)

14-22. What is NEMA the abbreviation for? (14-2)

14-23. List three torque ratings for motors. (14-2)

14-24. Given a choice, would you operate a 230-V motor from a 220-V or a 240-V supply? Why? (14-2)

14-25. What are TEFC and TENV the abbreviations for? (14-3)

14-26. What type of action induces voltage into a rotating rotor? (14-4)

14-27. List three techniques for producing a rotating field in a stator. (14-4)

14-28. What relationships should two winding currents have to produce maximum torque? (14-4)

14-29. Differentiate between a variable-speed and a dual-speed motor. (14-4)

14-30. Why does a three-phase motor provide a non-pulsating torque? (14-6)

14-31. Is a single-phase motor or a three-phase motor of the same horsepower more efficient? (14-6)

14-32. A motor is operating at 5000 rpm in a clean-room environment. What type of motor is it likely to be? (14-3)

14-33. Are the phase windings in one type of dc motor powered by a three-phase voltage? (14-6)

Chapter Review Problems

14-1. A 240-V motor is drawing 20 A at a PF of 0.65 from a power source which is providing 225 V. The motor is producing 2.8 hp. It is connected to the power source by 50 ft of 14-gage, two-conductor copper cable. Determine the voltage at the motor terminals and the efficiency of the motor. If this is a typical motor, is it operating within voltage tolerance limits? (14-2)

14-2. Determine the running torque of a ¾-hp motor rated at 1725 r/min. (14-2)

14-3. Determine the synchronous speed of an eight-pole motor designed to operate from a 240-V, 50-Hz supply. (14-4)

14-4. A two-pole, 60-Hz motor is operating at 3400 r/min. What is the percent slip for this motor? (14-4)

14-5. Determine the maximum allowable temperature rise of a four-pole, 220-V motor with class F insulation. (14-2)

14-6. Determine the rated speed of a six-pole, 120-V, 60-Hz motor that has 5 percent slip at rated speed. (14-4)

14-7. Determine the maximum supply voltage on which a standard 115-V ac motor should be operated. (14-2)

14-8. A typical 240-V, single-phase ac motor will be operated 50 ft from a 230-V source. This motor draws 12 A when fully loaded. Determine the maximum resistance the cable should have if the 230-V source can have long-term variations of ±5 percent. (14-2)

14-9. Determine the resistance of a 75-ft two-conductor, 12-AWG cable at 68°F. (14-2)

14-10. At 350 cmil/A, how much current can a 10-AWG conductor carry? (14-2)

14-11. How much hp can a 1.5-hp motor with a SF of 1.25 produce on a continuous basis if it is operating in an environment that meets its other ratings? (14-2)

14-12. A load requires an operating speed of 3450 r/min and a torque of 0.75 lb-ft. How many pole motors are required? How much hp is required? (14-2)

14-1. A 120-V, ¾-hp motor has an SF of 1.1, an efficiency of 82 percent, and a PF of 0.77 when loaded to the value allowed by its SF. At 300 cm/A, what gage of wire is required to operate this motor when it is located 95 ft from a source that has a minimum voltage of 115 V?

14-2. What features would be required in an apparatus designed to measure the running torque of a motor?

14-3. Suppose the starting capacitor and the running capacitor in a two-value capacitor motor were accidentally interchanged. How would this affect the operation of the motor?

14-4. Why are shaded-pole motors very inefficient?

14-5. A motor is 87 percent efficient, has a PF of 0.82, and produces 2.8 hp at 220 V, which is its minimum operating voltage. What size of wire is required if the motor is to be located 100 ft from a source with minimum voltage of 225 V?

14-6. What do you think would be the results of connecting a 240-V, 60-Hz motor to a 220-V, 50-Hz source?

14-7. What are the characteristics of a motor-start capacitor, and the circuit it is used in, that limit its use to intermittent duty?

14-8. How would you expect the speed regulation and the starting torque of a differential compound motor to compare to a cumulative compound motor?

14-9. We know that $P = TS/5252$ when P is in horsepower, T is in lb-ft, and S is in revolutions per minute. We also know that one newton-meter (Nm) = 0.7376 lb-ft and horsepower = 746 W. Derive the formula for calculating P in watts, when T is in Nm, and S is in revolutions per minute.

14-10. The apparatus in Fig. 14-42 is in balance. The weight of the beam is insignificant compared to that of the ball weights. Determine the weight of the ball on the left.

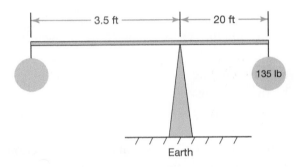

Fig. 14-42 The beam is in balance.

Answers to Self-Tests

1. DC, ac, and universal
2. F
3. F
4. T
5. National Electrical Manufacturers Association
6. any five of the following: voltage, current, power, speed, temperature, frequency, torque, duty cycle, service factor, or efficiency
7. ±10 percent
8. 216 V and 264 V
9. 120 V, because it would have more starting torque, require less current while running cooler, and be more efficient at full power output
10. The motor produces more noise at elevated voltages.
11. 13 gage. No; use the next-smaller gage number.
12. running current and starting current
13. starting current
14. T
15. F
16. 40°C
17. the class of insulation used in the motor
18. Allowable temperature rise is the temperature difference between ambient temperature and maximum permissible operating temperature.
19. Service factor specifies the factor (multiplier) by which the rated horsepower can be increased under specified conditions.
20. 1.07
21. 2.875 hp
22. F
23. F

24. 11.7 lb-in. or 0.975 lb-ft
25. 1.5 hp
26. Stall torque is the torque a motor produces just before the motor stalls out or stops rotating.
27. 74.6 percent
28. F
29. T
30. by frame number
31. totally enclosed fan-cooled
32. open and totally enclosed
33. F
34. no electrical connection between the rotor and the power source
35. voltage induced by the magnetic field created by the stator
36. F
37. F
38. T
39. F
40. by making the start-winding Q much lower than the run-winding Q
41. a. start winding
    b. run winding
    c. start winding
    d. run winding
42. a group of coils all wound in the same direction to create a magnetic field
43. opposite
44. 60, 30
45. 30
46. 90
47. F
48. F
49. F
50. 1150 r/min
51. 900 r/min
52. 4.2 percent
53. 70, 80
54. a centrifugally operated switch or a magnetically operated switch

55. by reversing the leads of either winding
56. 90
57. 300 to 450 percent
58. an intermittent-duty, ac electrolytic capacitor
59. greater starting torque and less starting current per unit of starting torque
60. very low starting torque
61. the permanent-split capacitor motor
62. quieter, higher PF, less source current, more efficient, and capability of variable-speed operation
63. It adds a large electrolytic capacitor in parallel with the run capacitor to provide excellent starting torque.
64. continuous duty
65. F
66. salient-
67. low starting torque, very low efficiency, and low PF
68. It is a shorted single turn of large copper wire around part of a salient pole.
69. the differential flux lag between the wide-air-gap and narrow-air-gap portions of the salient pole
70. toward
71. F
72. F
73. F
74. Squirrel-cage
75. synchronous
76. armature, field
77. series, compound, shunt
78. shunt, compound, series
79. series
80. three
81. Three-phase motors are quieter, smaller, lighter, more efficient, and easier to reverse. They also provide a constant torque.
82. a permanent magnet rotor
83. two
84. binary or digital or 0 and 1

Instruments and Measurements

Learning Outcomes

This chapter will help you to:

15-1 *Understand* how a digital multimeter (DMM) measures voltage, current, and resistance.

15-2 *Understand* the ratings and operation of meter movements used in analog meters.

15-3 *Calculate* (a) shunt values for analog ammeters, and *discuss* (b) current transformers, current probes and thermocouple meters.

15-4 *Calculate* (a) multiplier resistor values, and *understand* (b) voltmeter ratings.

15-5 *Understand* (a) why ac meters have frequency ratings, and *minimize* (b) meter loading and determine if meter loading has occurred.

15-6 *Understand* why the ohmmeter scale of the VOM is reverse-reading.

15-7 *Know* what instrument is used to test the condition of insulation that is to be subjected to high voltages.

15-8 *Explain* how a Wheatstone bridge is used to measure resistance.

15-9 *Discuss* (a) why an analog wattmeter measures true power, and *learn* (b) how to measure three-phase power.

15-10 *Explain* why the vibrating-reed meter has limited use.

15-11 *Discuss* two methods of measuring impedance.

15-12 *Contrast* the digital method and the ac-bridge method of measuring inductance.

In previous chapters you learned how to use meters to measure electrical quantities. In this chapter you will learn how a meter measures such things as current, voltage, resistance, power, frequency, capacitance, and inductance.

15-1 Digital Multimeter

A digital meter indirectly measures an unknown voltage by measuring the time it takes a capacitor to charge to a voltage equal to the unknown voltage. The capacitor is charged by a constant-current source, so the voltage rise on the capacitor is a linear function of time. Why the voltage rise is linear can be seen from the formulas $C = Q/V$ and $I = Q/t$, which, when rearranged and combined, yield $V = It/C$. Since I and C are constants, V is directly proportional to t.

The readout of the digital meter is connected to a circuit that counts and stores the number of cycles a reference frequency produces in the time it takes the capacitor to charge to the unknown voltage. When the unknown voltage and the capacitor voltage are equal, the stored count is displayed on the digital readout, the capacitor is discharged, and the counter is reset to zero. Then the voltage is measured again and the readout is updated.

Let us go through an example to see how the cycle count can represent the amount of unknown voltage. Suppose the capacitor is charging at a rate of 10 V/s, the reference frequency is 10 Hz, and the unknown voltage is 4 V. At 10 V/s, the capacitor voltage will reach 4 V in 0.4 s. At the end of 0.4 s, the reference frequency will have produced four cycles. The stored, and displayed, count will be 4—the value of the unknown voltage.

Although the above explanation is extremely simplified, it is representative of the technique

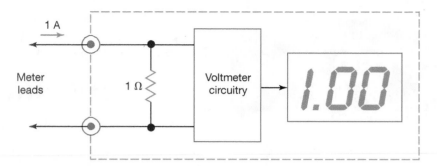

Fig. 15-1 Current function of a digital multimeter.

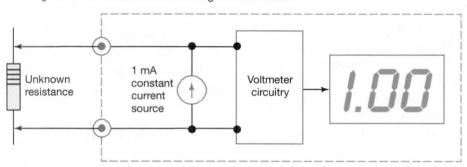

Fig. 15-2 Resistance function of a digital multimeter.

Digital multimeter (DMM)

used in many digital meters. Of course, the exact circuits are quite complex, operate at much higher frequencies, provide more than 1 V resolution, and provide multiple ranges and functions. The readout shown in Fig. 15-1 is correct if the digital multimeter (DMM) is on the 1-A current range. If it were on the 10-A range, the readout would be 01.0. Notice that on the 1-A range the readout is to the nearest 1/100 of an ampere; while on the 10-A range it is only to the nearest 1/10 of an ampere. The range-switching arrangement in the DMM selects the location of the decimal point so that the readout is in the units of the selected range, and it can indicate a reading up to the value of the range selected. Notice that the person using the DMM doesn't have to multiply or divide the readout by some factor to obtain the current reading. The same cannot be said for the user of a volt-ohm-milliammeter (VOM).

A *digital multimeter* (*DMM*) indirectly measures current with the voltmeter circuit explained above. Figure 15-1 illustrates how current is measured. As shown, 1 A flowing through a 1-Ω resistor drops 1 V, which is measured and displayed by the voltmeter circuitry. By simultaneously switching precision resistors and voltage ranges of the voltmeter circuitry, one can build various current ranges into the digital multimeter.

Figure 15-2 illustrates how the voltmeter circuitry of the DMM can be used to measure an unknown resistance. If the 1-mA constant current in Fig. 15-2 drops 1 V across the unknown resistance, the unknown resistance must be 1 kΩ. Thus, the 1-V drop across the unknown resistor, which is measured by the voltmeter circuitry, represents 1 kΩ. Various resistance ranges can be obtained by changing the value of the constant current. For example, a readout of 1.50 would represent 1.5 MΩ if the constant current were 1 μA.

Self-Test

Answer the following questions.

1. True or false. The voltmeter circuitry of the DMM is used when measuring current with the DMM.

2. True or false. The readout of a DMM displays the number of cycles produced by a reference frequency.

3. What is the measured resistance when the voltmeter circuitry of the DMM indicates 1.20 and the constant current source across the resistor is 1 μA?

4. What does a digital voltmeter measure to indirectly determine an unknown voltage?
5. What type of current sources are used in a DMM?

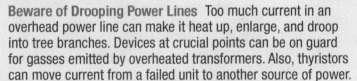

15-2 Meter Movements

Meters with moving pointers are called *analog meters*. They measure quantities by moving through an infinite number of points on a scale. The major part of any analog multimeter is the basic meter movement. Basic meter movements utilize the interaction of two magnetic fields. At least one of the fields is created by a current passing through a coil in the meter movement.

Ratings of Meter Movements

All basic meter movements have a *full-scale current rating*. This is the coil current needed to cause the meter to deflect to the maximum value of the scale. Meter movements with full-scale deflection currents as low as 5 μA are commonly available.

Another important rating of a meter movement is its *internal resistance*. Basic meter movements have appreciable resistance because of the small-diameter wire used in the coil. In general, the lower the full-scale current rating, the higher the internal resistance. A typical 1-mA meter has less than 100 Ω of resistance. A typical 50-μA meter has more than 900 Ω of resistance.

Since a meter movement has both a current and a resistance rating, it must also have a voltage rating. Usually, the manufacturer specifies only two of these three ratings. However, the third rating can easily be determined by using Ohm's law. The voltage across a meter movement (V_m) must be equal to the product of the full-scale current (I_m) and the internal resistance (R_m).

EXAMPLE 15-1

Determine the full-scale current rating of a meter with 60 Ω of internal resistance that requires 300 mV to cause full-scale deflection.

Given: $R_m = 60 \ \Omega$
$V_m = 300 \text{ mV} = 0.3 \text{ V}$
Find: I_m

Known: $V_m = I_m \times R_m$ rearranged I_m
$= V_m / R_m$

Solution: $I_m = \dfrac{0.3 \text{ V}}{60 \ \Omega} = 0.005 \text{ A} = 5 \text{ mA}$

Answer: The full-scale current rating is 5 mA.

D'Arsonval Meter Movement

The most common type of meter movement is the *d'Arsonval movement*, which was invented by the French physicist Jacques d'Arsonval (1851–1940). This meter movement is based on the interaction of the fields of a permanent magnet and an electromagnet.

The armature assembly in Fig. 15-3 contains a coil of very thin wire wound on a lightweight

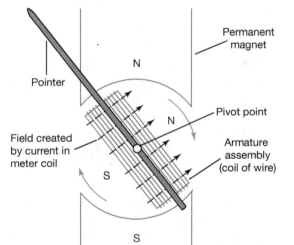

Fig. 15-3 The d'Arsonval movement principle. The electromagnetic field of the armature tries to line up with the field of the permanent magnet.

Analog meters

d'Arsonval movement

Full-scale current rating

Internal resistance

ABOUT ELECTRONICS

Beware of Drooping Power Lines Too much current in an overhead power line can make it heat up, enlarge, and droop into tree branches. Devices at crucial points can be on guard for gasses emitted by overheated transformers. Also, thyristors can move current from a failed unit to another source of power.

supporting form. The coil, coil form, and attached pointer rotate on pivot points. When current flows through the coil, a magnetic field is created, as seen in Fig. 15-3. The coil's magnetic field reacts with the permanent magnet's field to create a clockwise rotational force. This rotational force reacts against spiral springs. These springs attach to the pivot shaft on which the armature rotates. The points of the shaft rest in jewel bearings to keep friction as low as possible. This arrangement is known as a *jewel-and-pivot* suspension system.

Jewel-and-pivot

Polarized

D'Arsonval meter movements are *polarized*. If reverse current flows in the coil, a reverse torque is applied to the armature. This reverse torque can damage the meter movement and the pointer if it is excessive. Even though the d'Arsonval meter movement is polarized, you will see later how it can be used in ac as well as dc meters.

Iron-Vane Meter Movement

Iron-vane meter

In an *iron-vane meter* movement, the coil that receives the current to be measured is stationary. The field set up by the coil magnetizes two iron vanes, which then become temporary magnets (Fig. 15-4). Since the same field magnetizes both vanes, both vanes have the same magnetic polarity. Consequently, there is a force of repulsion between the two vanes. One of the vanes (the *stationary vane*) is attached to the coil form. The other vane (the *moving vane*) is mounted on the pivot shaft to which the meter pointer is attached. Thus, the magnetic force of repulsion pushes the moving vane away from the stationary vane. Of course, this force is offset by the countertorque of the spiral springs attached to the pivot shaft. The greater the current through the coil in Fig. 15-4, the stronger the magnetic repelling force; thus, the farther the moving vane rotates and the more current the pointer indicates.

Stationary vane

Moving vane

The iron-vane meter movement can operate on either alternating or direct current. When the alternating current reverses, the magnetic polarities of both vanes reverse simultaneously. Therefore, a force of repulsion is maintained throughout the cycle.

Electrodynamo-meter

Two electro-magnetic fields

The iron-vane meter movement has two shortcomings: it has an extremely *nonlinear scale,* and it requires considerable current for full-scale deflection (low sensitivity). Its most common use is as an ac ammeter.

Nonlinear scale

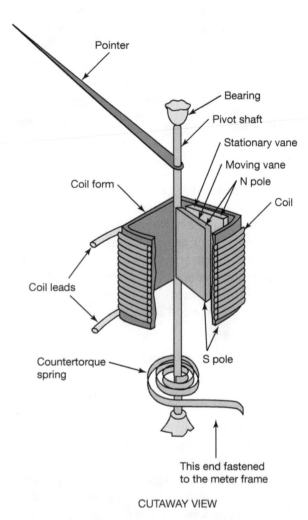

CUTAWAY VIEW

Fig. 15-4 Iron-vane meter movement. Since the movement uses no permanent magnet, it responds to alternating current as well as to direct current.

The advantages of the iron-vane meter movement are its ease of construction, ruggedness, and dependability.

Electrodynamometer Meter Movement

The *electrodynamometer* (Fig. 15-5) uses *two electromagnetic fields* in its operation. One field is created by current flowing through a pair of series-connected stationary coils. The other field is caused by current flowing through a movable coil that is attached to the pivot shaft. If the currents in the coils are in the correct directions, the pointer rotates clockwise. The rotational torque on the movable coil is caused by the opposing magnetic forces [Fig. 15-5(b)] of the three coils.

The directions of the magnetic fields shown in Fig. 15-5(b) were arbitrarily chosen; all the

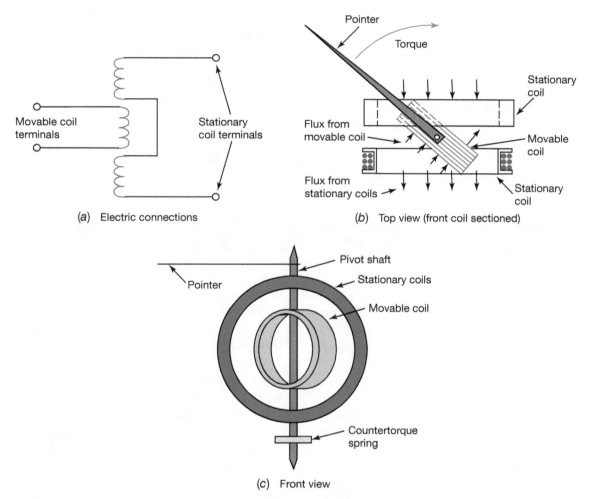

(a) Electric connections

(b) Top view (front coil sectioned)

(c) Front view

Fig. 15-5 Electrodynamometer. This meter uses two electromagnetic fields and can respond to either alternating or direct current.

arrowheads could just as well be reversed. It is only the polarity of one coil relative to the other two that is important. (The two stationary coils are permanently connected so that their fields are always aiding each other.) The electrodynamometer can operate with alternating current because the alternating current reverses direction simultaneously in all three coils. The electrodynamometer can also operate on direct current.

Electrodynamometer movements have *lower sensitivity* than d'Arsonval movements. However, electrodynamometers can be very accurate, and they are quite stable. The accuracy of the electrodynamometer movement does not depend on the strength of a permanent magnet or the permeability of the iron parts.

Although the electrodynamometer is used for measuring current and voltage, its most common use is for measuring power.

Lower sensitivity

![Self-Test icon] *Self-Test*

Answer the following questions.

6. Meter movements are rated for
_____, _____, and _____.

7. The most common type of meter movement is the _____.

8. The rotational force created by the magnetic field in a meter movement is counteracted by the reverse rotational force of a(n) _____.

9. What is the internal resistance of a meter movement that has a full-scale current

of 100 μA and a full-scale voltage of 200 mV?

10. Does a d'Arsonval meter movement respond to alternating current, direct current, or both?

11. Does an iron-vane movement respond to alternating current, direct current, or both?

12. Which of the following meter movements is the most sensitive: iron-vane, d'Arsonval, or electrodynamometer?

13. Which is more linear, a d'Arsonval movement or an iron-vane movement?

14. Does an iron-vane movement use a permanent magnet?

15. Does an electrodynamometer movement respond to alternating current, direct current, or both?

16. What causes the rotary torque in an electrodynamometer movement?

17. How many coils create the stationary magnetic field in an electrodynamometer?

15-3 Analog Ammeters

Shunt

Rectifier

Temperature coefficients

The d'Arsonval movement can be used to measure alternating current if it is connected in series with a *rectifier*, as illustrated in Fig. 15-6.

> ### You May Recall
>
> ... that a rectifier is a device that allows current to flow in only one direction.

Thus, the rectifier of Fig. 15-6 allows only one-half of an ac cycle to pass. It converts alternating current to pulsating direct current. The pulsating direct current provides the electromagnetic field of constant polarity needed to operate a d'Arsonval meter movement. The combination of a d'Arsonval movement and a rectifier produces a rectifier-type ac meter.

Shunts

External shunt

When used as an ammeter, a basic meter movement has a range equal to its full-scale deflection current. The range of an analog ammeter can be made larger by adding a shunt to the meter. A *shunt* is a resistor of very low resistance connected in parallel with the basic meter movement (Fig. 15-7). Shunts are usually made from materials with very low *temperature coefficients*. They are generally precision, low-tolerance (± 2 percent or less) resistors.

A shunt extends the range of an ammeter by diverting most of the current around the meter movement. For example, a 100-μA movement is converted to a 1-mA ammeter by shunting 900 μA around the movement. As shown in Fig. 15-7, the 1 mA (I_{circuit}) splits at the junction of the meter movement and the shunt. Then 100 μA (I_m) goes through the movement and causes full-scale deflection of the pointer. The other 900 μA (I_s) goes through the shunt.

Meter shunts are often enclosed in the same housing as the basic meter movement. However, for large dc currents (above 10 A) an external shunt is often used. *External shunts,* as shown in

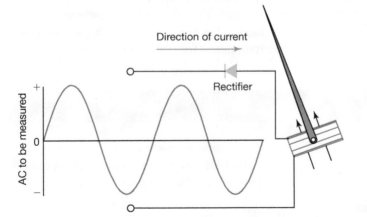
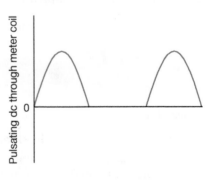

Direction of current

Rectifier

AC to be measured

Pulsating dc through meter coil

Fig. 15-6 Principle of rectification. The rectified current flows only in one direction.

EXAMPLE 15-2

It is desired to make a 1-A ammeter from a 1-mA, 80-Ω movement. Determine the resistance and power dissipation of the required shunt.

Given: $I_m = 1$ mA
$R_m = 80$ Ω

Find: R_s, P_s

Known: $R_s = \dfrac{V_s}{I_s}, V_s = V_m, V_m = I_m R_m$

$I_s = I_{circuit} - I_m, P_s = I_s V_s$

Solution: $I_s = 1\,\text{A} - 0.001\,\text{A} = 0.999\,\text{A}$
$V_m = 0.001\,\text{A} \times 80\,\Omega = 0.08\,\text{V}$

$R_s = \dfrac{0.08\,\text{V}}{0.999\,\text{A}} = 0.08\,\Omega$

$P_s = 0.999\,\text{A} \times 0.08\,\text{V}$
$= 0.0799\,\text{W}$
$= 79.9\,\text{mW}$

Answer: The shunt's resistance must be 0.08 Ω, and its power dissipation must be 79.9 mW.

Fig. 15-8, are built to handle large currents without heating up or changing resistance. Therefore, they are quite large (5 to 15 cm long).

External shunts have both a current rating and a voltage rating. Current ratings ranging from 10 to several thousand amperes are common. Voltage ratings are usually either 50 or 100 mV. The voltage rating of a shunt specifies how much voltage the shunt drops when it is carrying its rated current. The rated voltage appears between the screw terminals shown on the shunt in Fig. 15-8. When these shunts are used, leads are run from the screw terminals to a meter with a full-scale voltage range equal to the voltage rating of the shunt. The meter can be either digital or analog. Of course, the meter readout or scale will be calibrated in amperes even though it is responding to a voltage. The heavier terminals provided on some shunts are bolted into the circuit in which the current is to be measured.

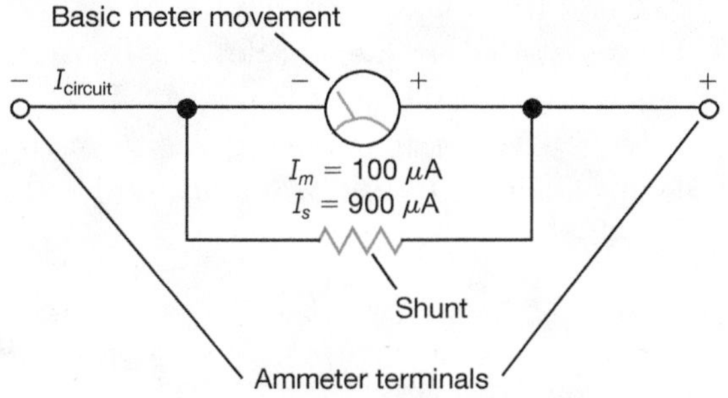

Current Transformer

When measuring large values of alternating current, a current transformer (Fig. 15-9) is used. The *current transformer* is essentially a *toroid-core transformer* that has no primary. When a current-carrying conductor is placed in the center of the current transformer, the conductor becomes the primary. An equivalent electric circuit is shown in Fig. 15-10. The current in the conductor is stepped down by transformer action.

A current transformer may have several secondaries. This allows it to be used with a variety of ammeters, either digital or analog, having different ratings.

Current transformer

Toroid-core transformer

Fig. 15-7 Ammeter with shunt. The shunt extends the range of the basic meter movement.

$I_m = 100\ \mu\text{A}$
$I_s = 900\ \mu\text{A}$

Fig. 15-8 External shunt.

Fig. 15-9 Current transformer.

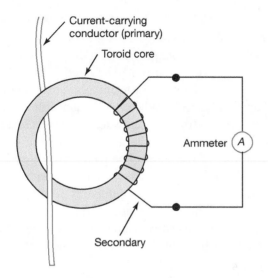

Fig. 15-10 Current transformer principle. The conductor in which current is being measured becomes a single-turn primary.

Multirange Ammeters

A multirange ammeter (or the ammeter section of a multimeter) is illustrated in Fig. 15-11. It consists of a meter movement, a number of shunts, and a rotary switch. The switch is a shorting (make-before-break) type. A *shorting switch* is necessary so that the meter movement is shunted even while it is switching ranges. With a nonshorting switch, the meter movement would carry all the current for the instant it takes to change ranges. This instant would be sufficient time to damage or burn out the meter movement.

Shorting switch

Clamp-On Ammeter

Clamp-on ammeter

A *clamp-on ammeter* (also called a clamp meter) uses the same principle as the current transformer-ammeter circuit of Fig. 15-10. However, with the clamp-on meter, the toroid

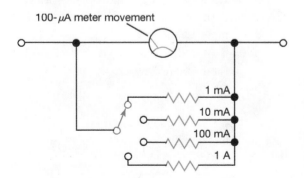

Fig. 15-11 Multirange ammeter. Switching to a smaller shunt increases the range.

core is made in two halves. The halves are hinged at one end and held together by a spring mechanism. The other end of the two halves can be separated to insert the conductor in which current is to be measured.

A clamp-on ammeter can have either a digital or an analog readout. The clamp-on mechanism of the clamp-on ammeter can be an accessory for use with a multimeter, or it can be a permanent part of a separate unit (clamp meter) that has its own readout, power source, and so on. Many clamp meters also measure other quantities (voltage, resistance, and so on) by using test leads rather than the clamp-on mechanism.

Figure 15-12 shows a clamp-on accessory being used to measure current in a large conductor without interrupting the circuit. The amount of current flow is indicated on the portable DMM to which the accessory is connected.

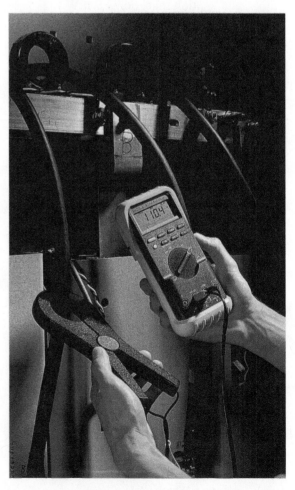

Fig. 15-12 Clamp-on accessory for a DMM.

Thermocouple Meter

Radio-frequency currents are hard to measure with the meters discussed thus far. At these high frequencies, the inductive reactance of the meter coil is high and the capacitive reactance of the capacitance between the turns of the coil is low. The combination of these two reactances make some meters useless at higher frequencies. However, the thermocouple meter avoids the reactance problem by isolating the basic meter movement from the RF currents.

The *thermocouple meter,* shown in Fig. 15-13, uses a d'Arsonval meter movement connected to a thermocouple. When current flows through the meter, the resistive element heats up and increases the temperature of the thermocouple junction.

> ### You May Recall
>
> . . . that a thermocouple is a device that converts heat energy into electric energy.

Therefore, the heated thermocouple produces a current in the meter-movement circuit. The higher the current through the resistive element, the hotter the thermocouple gets. The hotter the thermocouple, the higher the voltage it produces and the greater the current through the d'Arsonval meter movement.

Current Probe

Another technique for measuring current is to use a current probe and an oscilloscope. Current probes for oscilloscopes make it possible to measure current in a circuit without interrupting the

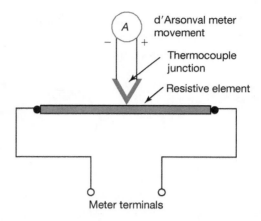

Fig. 15-13 Thermocouple meter. The current being measured does not flow through the meter movement.

circuit. The probe just clamps around a conductor in the circuit. Current probes can respond to either alternating current or direct current. Probes that respond only to alternating current are known as passive probes, and they operate on the same principle as the clamp-on accessory (Fig. 15-12) for a DMM. "Passive" means that no electronic devices are used and no external power source is needed to operate the probe. Oscilloscope probes that can measure direct current, or both alternating and direct current, are active devices. They require an external power source to operate the Hall-effect device used to detect the level of direct current in the circuit. Active probes often use external electronic circuits to enhance the capabilities of a probe. Probes are available to handle currents ranging in frequencies from direct current up to gigahertz and amplitudes from less than a milliampere up to 20 kA.

⌁ Self-Test

Answer the following questions.

18. What type of meter movement is used in a rectifier-type meter?
19. What is a rectifier?
20. List two desirable characteristics of a shunt resistor.
21. Refer to Fig. 15-11. How much current is carried by the shunt for the 10-mA range?
22. The meter movement shown in Fig. 15-11 has 1200 Ω of resistance.
    a. What is the resistance of the shunt for the 1-mA range?
    b. What is the power used by the shunt?
23. What type of rotary switch is used in a multirange ammeter?
24. Should the internal resistance of an ammeter be as low or as high as possible?
25. When are external shunts used?
26. How are external shunts rated?
27. What is a current transformer?
28. Does the use of a clamp-on meter require interruption of the circuit in which it is used?

29. What principle is used in the clamp-on ammeter?

30. What are thermocouple meters used for?

31. A thermocouple converts _____ energy into _____ energy.

15-4 Analog Voltmeters

Multiplier

Series-circuit relationships

A voltmeter consists of a basic meter movement and a *multiplier* resistor connected in series (Fig. 15-14). The resistance of the multiplier can be determined by using *series-circuit relationships* and Ohm's law because the current through the multiplier (I_{mt}) is the same as the current through the meter movement (I_m), and the multiplier voltage (V_{mt}) plus the meter-movement voltage (V_m) equals the voltmeter voltage (V_t).

Like shunts, multiplier resistors have low tolerances and temperature coefficients. Unlike shunts, multipliers have very high resistances.

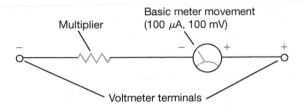

Fig. 15-14 The voltmeter multiplier extends the voltage range of the basic meter movement.

the nonshorting type. When the meter is switching ranges, the voltmeter is momentarily opened. A shorting switch would not be appropriate for a voltmeter. It would momentarily connect two multipliers in parallel. The total resistance of the paralleled multipliers would be less than that of the lowest-value multiplier. This could cause the meter movement to be overloaded.

Voltmeter Rating

The total internal resistance of a voltmeter is an important rating of the voltmeter. The internal resistance of a voltmeter is called its *input resistance*. For a multirange meter like the one in Fig. 15-15, the input resistance is different for each range. Rather than an input resistance for each range being specified, one general rating for all ranges is used. This rating is specified

EXAMPLE 15-3

Input resistance

What value multiplier is needed to make a 5-V voltmeter from a 1-mA, 100-Ω meter movement?

Given: $I_m = 1$ mA, $R_m = 100\ \Omega$

Find: R_{mt}, P_{mt}

Known: $P_{mt} = I_{mt}V_{mt},\ R_{mt} = \dfrac{V_{mt}}{I_{mt}},$

$I_{mt} = I_m,\ V_{mt} = V_t - V_m,$

$V_m = I_m R_m$

Solution: $V_m = 0.001\ \text{A} \times 100\ \Omega$

$= 0.1$ V

$V_{mt} = 5\ \text{V} - 0.1\ \text{V} = 4.9\ \text{V}$

$R_{mt} = \dfrac{4.9\ \text{V}}{0.001\ \text{A}} = 4900\ \Omega$

$P_{mt} = 0.001\ \text{A} \times 4.9\ \text{V}$

$= 0.0049\ \text{W} = 4.9\ \text{mW}$

Answer: The multiplier's resistance must be 4900 Ω, and its power rating must be greater than 4.9 mW.

A multirange voltmeter, typical of those used in the voltmeter section of a VOM, is shown in Fig. 15-15. The selector switch in Fig. 15-15 is

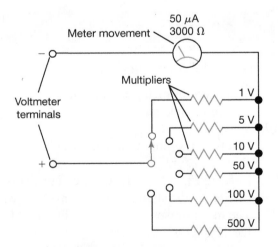

Fig. 15-15 Multirange voltmeter. Switching to a larger multiplier increases the range.

in *ohms per volt*. The *ohms-per-volt rating* indicates the input resistance for each volt of that range. For example, a 1000-Ω/V voltmeter would have 1000 Ω of input resistance on the 1-V range. On the 10-V range, it would have 10,000 Ω of input resistance. The input resistance of a voltmeter on a given range is found by multiplying the range by the ohms-per-volt rating. The ohms-per-volt rating of a voltmeter is often referred to as its *sensitivity*.

EXAMPLE 15-4

What is the input resistance of a 10,000-Ω/V voltmeter on the 20-V range?

Given: 10,000-Ω/V voltmeter
Find: Input resistance on 20-V range
Known: Input resistance
              = sensitivity × range
Solution:
Input resistance = (10,000 Ω/V) × (20 V)
              = 200,000 Ω
              = 200 kΩ
Answer: The input resistance on the 20-V range is 200 kΩ.

The sensitivity (ohms-per-volt rating) of a voltmeter is determined by the full-scale current of the meter movement. Mathematically, the relationship is

$$\text{Sensitivity} = \frac{1}{\text{full-scale current}}$$

Therefore, the sensitivity of the voltmeter of Fig. 15-15 is

$$\text{Sensitivity} = \frac{1}{0.00005 \text{ A}} = 20,000 \ \Omega/\text{V}$$

Notice that voltmeters have very *high internal resistances* and ammeters have very low internal resistances. A voltmeter should not significantly change the voltage distribution (or the load) of the circuit in which it is used. Therefore, its internal resistance must be high relative to the resistance of the load.

DMMs have the same input resistance for all voltage ranges. On the lower ranges, they usually have higher input resistance than the VOM. However, on the higher ranges, the opposite is often true. For example, a 100-kΩ/V VOM on the 500-V range has 50 MΩ of input resistance. This exceeds the input resistance of most DMMs.

Often, ac voltmeters have much lower sensitivity than dc voltmeters do. For example, a multimeter may have a 20-kΩ/V rating on the dc ranges and only a 5-kΩ/V rating on the ac ranges.

This disparity between dc and ac sensitivity in VOM ratings is caused by the rectification (Fig. 15-6) required for the d'Arsonval to respond to ac. As shown in Fig. 15-6, only one-half of each ac cycle is used to produce the magnetic field that provides rotational torque on the moving coil.

 ## Self-Test

Answer the following questions.

32. True or false. The resistor used to extend the range of a voltmeter is called a multiplier.
33. True or false. Ammeters have higher internal resistance than voltmeters do.
34. True or false. The internal resistance of the DMM is the same on all dc voltage ranges.
35. True or false. The ac voltage ranges of a multimeter usually have more internal resistance than the comparable dc voltage ranges.

36. What is the sensitivity of a voltmeter which uses a 25-μA meter movement?
37. A 150-μA, 200-Ω meter movement is used in a 50-V voltmeter. What is the power dissipation and the resistance of the multiplier?
38. What is the internal resistance of a 250-V voltmeter which uses a 50-μA meter movement?
39. Multirange voltmeters use a(n) _____ range switch.

15-5 Meter Loading

Whenever a meter is connected to a circuit, the circuit currents (and often the circuit voltages) change. This is what is meant by *meter loading*. *Significant meter loading* occurs when the current or voltage changes exceed those that might be caused by component tolerances.

In general, ac meters tend to be more prone to load a circuit than dc meters are. This is because ac voltmeters have less input resistance than dc voltmeters when both use the same meter-movement rating.

Meter loading in ac circuits is also caused by the *internal reactances* of the meter movement, rectifier, leads, and so forth. Of course, these reactances vary with frequency. Therefore, ac meters have *frequency ratings* in addition to all the ratings that dc meters have. The frequency rating tells the range of frequency over which the meter is accurate. It is important that ac meters be used only within their frequency ratings. In general DMMs have a much higher frequency rating than VOMs do.

Ammeter Loading

Ammeter loading in a dc circuit is illustrated in Fig. 15-16. In Fig. 15-16(*a*), the nominal current in the circuit is

$$I = \frac{1.5 \text{ V}}{1500 \text{ }\Omega} = 0.001 \text{ A} = 1 \text{ mA}$$

If the resistor is on either end of its tolerance, the current will be only 1 percent more or less

(0.99 to 1.01 mA). In Fig. 15-16(*b*), an ammeter has been inserted into the circuit. The measured current, if both meter and resistor were completely accurate, would be 0.95 mA. The meter has loaded the circuit.

Significant ammeter loading does not occur very often. If the resistor in Fig. 15-16 had 10 percent tolerance, the loading would not be noticed. Also, many 1-mA ammeters have less than 85 Ω of internal resistance. When ammeter loading does occur, it is usually in a *low-voltage, low-resistance circuit*.

Voltmeter Loading

Voltmeter loading is much more common than ammeter loading. It can occur in series and series-parallel circuits. It does not occur in parallel circuits because the voltmeter is connected across the power source. That is, it is in parallel with all other parts of the circuit. In general, voltmeter loading occurs in *high-resistance circuits,* such as those found in many electronic devices.

The circuits in Fig. 15-17 illustrate voltmeter loading. In Fig. 15-17(*a*), the source voltage

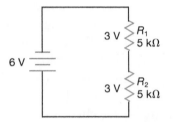

(a)   Voltages in original circuit

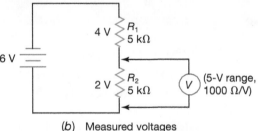

(b)   Measured voltages

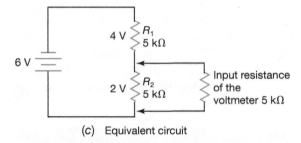

(c)   Equivalent circuit

Fig. 15-17 Voltmeter loading.

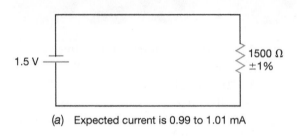

(a)   Expected current is 0.99 to 1.01 mA

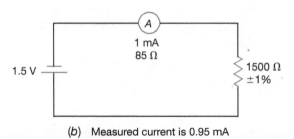

(b)   Measured current is 0.95 mA

Fig. 15-16 Ammeter loading.

Significant meter loading

Low-voltage, low-resistance circuit

Internal reactances

Voltmeter loading

Frequency ratings

High-resistance circuits

Ammeter loading

is split into equal parts by the equal resistors. However, the voltmeter in Fig. 15-17(*b*) indicates only 2 V across R_2. This is an obvious case of voltmeter loading. The input resistance of the voltmeter in Fig. 15-17(*b*) (on the 5-V range) can be calculated as follows:

$$\text{Input resistance} = \text{sensitivity} \times \text{range}$$
$$= 1000\ \Omega/\text{V} \times 5\ \text{V}$$
$$= 5000\ \Omega = 5\ \text{k}\Omega$$

As shown in the equivalent circuit of Fig. 15-17(*c*), this 5 kΩ is in parallel with the 5 kΩ of R_2. These two 5-kΩ parallel resistances have a combined resistance of 2.5 kΩ. Therefore, R_1 drops twice as much voltage as R_2 with the voltmeter across it.

If the voltmeter shown in Fig. 15-17 were a 20-kΩ/V voltmeter, the loading would be minimal. The input resistance would be 100 kΩ, and the parallel resistance of R_2 and the meter would be 4.76 kΩ. Even a 5 percent resistor at R_2 could have a lower resistance than this.

Manufacturers of electric and electronic equipment often specify the voltages at various points in their circuits. Usually, they specify either the sensitivity or the input resistance of the meter used in measuring these voltages. When using a meter different from the one specified, technicians must be aware of possible meter loading and must be sure that any *observed low voltages* are not due to meter loading rather than component failures. This can be done by analyzing the circuit diagram to determine the value of the resistances in the circuit. The voltmeter's input resistance should be 20 times greater than the resistance across which the voltage is measured. Under these conditions, the loading will change the resistance of the circuit less than 5 percent.

EXAMPLE 15-5

Assume that the voltage across R_2 in Fig. 15-17 was measured with a 2000-Ω/V voltmeter on the 5-V range. How much voltage would the meter indicate?

Given: Sensitivity = 2000 Ω/V
Range = 5 V

Find: Measured voltage across R_2
Known: Input resistance (R_1)
= sensitivity $\times$ range
Voltmeter loading
Series-parallel circuits
Solution: $R_1 = 2000\ \Omega/\text{V} \times 5\ \text{V} = 10\ \text{k}\Omega$
$$R_{1,2} = \frac{5\,\text{k}\Omega \times 10\ \text{k}\Omega}{5\ \text{k}\Omega + 10\ \text{k}\Omega} = 3.33\ \text{k}\Omega$$
$$V_{R1,2} = \frac{6\ \text{V} \times 3.33\ \text{k}\Omega}{5\ \text{k}\Omega + 3.33\ \text{k}\Omega} = 2.4\ \text{V}$$
Answer: The measured voltage is 2.4 V.

Observed low voltages

ABOUT ELECTRONICS

Reef Relief
Coral reefs take thousands of years to grow and are home to fish and other aquatic life. They also protect islands from rising sea levels due to global warming. Recognizing these benefits, Dr. Thomas Goreau and the Global Reef Alliance are attempting to grow new reefs to replace damaged ones. To accomplish this, they place steel frames on the ocean floor and connect them by wires to solar panels floating on the surface of the water. A mild electric current from the panels runs inside the frame, causing limestone to form on the frame. Coral grows well on limestone. In fact, in this environment, the coral can grow faster than it can be ruined by sewage. This reef, photographed by architect Wolf Hilbertz, is being grown to help protect the Maldives, the earth's lowest country, from the encroaching ocean.

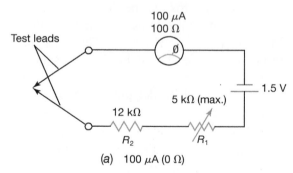
Answer the following questions.

40. What is meter loading?
41. Which is more common, ammeter loading or voltmeter loading?
42. Is meter loading as much of a problem with ac meters as it is with dc meters?
43. Does meter loading in ac circuits depend on frequency?
44. A 27-kΩ resistor in a series circuit drops 8 V. Will significant voltmeter loading occur if the voltage is measured with
    a. A 20-kΩ/V voltmeter on the 10-V range
    b. A 50-kΩ/V voltmeter on the 10-V range
    c. A DMM with 10-MΩ input resistance
45. A 4.7-MΩ resistor in a complex electronic circuit has 450 V across it. Which will give a more accurate voltage reading, a 20-kΩ/V voltmeter on the 1000-V range or a DMM with 10-MΩ input resistance?

15-6 Analog Ohmmeters

Ohmmeter

A simple *ohmmeter* (Fig. 15-18) consists of a cell, meter movement, and two resistors. The rheostat (R_1 in Fig. 15-18) is the ohms-adjust control. Adjustment of this control compensates for changes in the voltage of the cell. When the ohmmeter terminals are short-circuited together as in Fig. 15-19(*a*), current flows through the meter movement. The ohmmeter is then measuring zero resistance. If R_1 is adjusted until the meter movement indicates full scale, then full scale indicates 0 Ω. This is why the ohmmeter scale of a VOM is *reverse-reading*.

Reverse-reading

When the ohmmeter of Fig. 15-19(*a*) is ohms-adjusted, the internal resistance is

$$R_{int} = \frac{V}{I_m} = \frac{1.5 \text{ V}}{0.0001 \text{ A}}$$
$$= 15,000 \text{ Ω}$$
$$= 15 \text{ kΩ}$$

This internal resistance includes the resistance of the meter movement and both resistors (R_1 and R_2). Suppose a 15-kΩ resistor is placed between the ohmmeter terminals, as in Fig. 15-19(*b*). Now the total resistance of the circuit is 30 kΩ ($R_T = R_{int} + R_x$). The current through the meter movement is cut in half because the resistance is doubled:

$$I_m = \frac{V}{R_T} = \frac{1.5 \text{ V}}{30,000 \text{ Ω}}$$
$$= 0.00005 \text{ A}$$
$$= 50 \text{ μA}$$

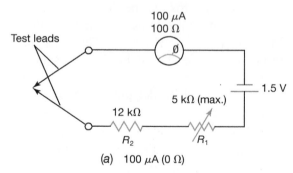

(*a*) 100 μA (0 Ω)

(*b*) 50 μA (15 kΩ)

Fig. 15-19 Measuring resistance.

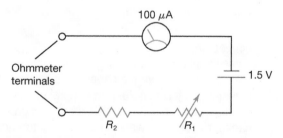

Fig. 15-18 Basic ohmmeter circuit. R_1 is adjusted to provide full-scale current when the terminals are shorted together.

When R_x in Fig. 15-19 is replaced by a 45-kΩ resistor, the total resistance is 60 kΩ ($R_T = 15$ kΩ $+ 45$ kΩ). Now the meter current is

$$I_m = \frac{1.5 \text{ V}}{60,000 \text{ Ω}}$$

$$= 0.000025 \text{ A} = 25 \text{ μA}$$

This is one-fourth of the full-scale current of the meter. Therefore, the quarter-scale point on the ohmmeter is 45 kΩ.

The above calculations illustrate how *nonlinear* an *ohmmeter scale* is. The first half of the scale covers 0 to 15 kΩ (Fig. 15-20). The next quarter of the scale covers 15 to 45 kΩ. And the last quarter has to cover the range from 45 kΩ to infinity.

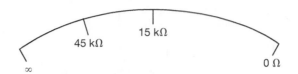

Fig. 15-20 Nonlinear ohmmeter scale.

The range of the ohmmeter circuit in Fig. 15-18 can be changed by either of two methods. First, the voltage of the cell *and* the resistance of R_2 can be increased. Second, the full-scale current of the movement can be increased by shunting, *and* the resistance of R_2 can be decreased. The first method increases the range of the ohmmeter, and the second method decreases the range.

Nonlinear ohmmeter scale

 ## Self-Test

Answer the following questions.

46. What components are used in a simple ohmmeter?
47. What is the function of the rheostat in an ohmmeter?
48. What is the center-scale resistance of an ohmmeter that has 40 Ω of internal resistance?

49. What is the internal resistance of a properly ohms-adjusted ohmmeter that has a 3-mA meter movement and a 9-V battery?
50. What are two methods of changing the range of an ohmmeter? Which method increases the range?

15-7 Insulation Testers

Insulating materials, which must withstand high voltages, have extremely high resistances. They must also be able to stand the electric stress created by the high voltage. An ordinary ohmmeter cannot do an adequate job of checking the condition of insulation that is subjected to high voltages. Therefore, the insulation in some equipment, such as transformers and motors, is checked with *insulation testers,* or *Meggers.* These instruments test the insulation's resistance by applying a high voltage across the insulation and checking for a minute leakage current. Very often the test is made between ground (the frame or chassis of the equipment) and the wiring within the equipment. The insulation is tested at a voltage greater than the highest voltage to which the insulation is normally subjected. Engineering

test procedures often specify a method for determining the test voltage. For example, certain equipment must be tested at twice the operating voltage plus 5000 V.

Because of the high voltage, insulation testers must be used with great care. Only the test leads supplied (or recommended) by the manufacturer should be used with an insulation tester. The leads should be checked regularly for signs of deterioration (checking and cracking). The manufacturer's directions and safety rules should be followed when using an insulation tester.

Insulation testers

Meggers

15-8 Wheatstone Bridge

One common use of the Wheatstone bridge (Fig. 15-21) is to measure resistance with great precision. Let us see how the bridge works for this application.

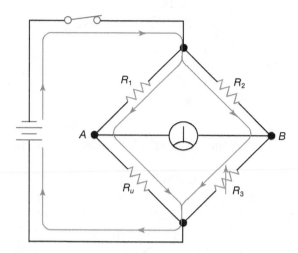

Fig. 15-21 Wheatstone bridge circuit (balanced). There is no voltage between points A and B when the bridge is balanced.

Rearranging the formula to solve for R_u yields

$$R_u = \frac{R_1}{R_2} \times R_3$$

This formula can be used to measure an unknown resistance R_u. All we need to know is the ratio of R_1/R_2 and the value of R_3.

When a Wheatstone bridge is used, the resistance to be measured is connected into the circuit and power is applied. Then R_3 is adjusted to achieve a balance (zero current through the meter). Resistor R_3 is a precision, calibrated rheostat. Its value can be read from a dial attached to its shaft. Resistors R_1 and R_2 are precision resistors of known value. The value of R_u is then found by multiplying the dial reading of R_3 by the ratio R_1/R_2. The precision with which R_u can be measured is determined by the precision (tolerances) of the other three resistors.

When a bridge is not balanced, none of the voltage, current, and resistance relationships discussed above are valid. Unbalanced bridges can be analyzed using the techniques covered in the chapter on complex circuit analysis.

Zero-center-scale meter

Galvanometer

The meter movement used in the bridge is a *zero-center-scale meter.* When the current or voltage is zero, the meter pointer is at the center of the scale. The type of meter used is called a *galvanometer.* It can measure very low current flow in either direction.

The bridge circuit in Fig. 15-21 is balanced. When the bridge is balanced, no current flows through the meter. Therefore, there is no potential difference (voltage) between points A and B. The voltage across R_1 must equal the voltage across R_2, and, of course, V_{R_u} must equal V_{R_3}. Also notice that, when the bridge is balanced,

$$I_{R_1} = I_{R_u} \qquad \text{and} \qquad I_{R_2} = I_{R_3}$$

where I_{R_u} is the current through a resistor of unknown resistance. Because of these voltage and current relationships, a definite relationship exists between the resistances of the circuit. The ratio of R_1/R_u must be equal to the ratio

Balanced bridge

of R_2/R_3. Therefore, for a *balanced bridge,* we can write

$$\frac{R_1}{R_u} = \frac{R_2}{R_3}$$

EXAMPLE 15-6

Refer to Fig. 15-21. Resistor $R_1 = 10,000\ \Omega$, $R_2 = 1000\ \Omega$, and $R_3 = 250\ \Omega$ when the bridge is balanced. What is the value of R_u?

Given: $R_1 = 10,000\ \Omega$
$R_2 = 1000\ \Omega$
$R_3 = 250\ \Omega$

Find: R_u

Known: $R_u = \dfrac{R_1}{R_2} \times R_3$

Solution: $R_u = \dfrac{10,000}{1000} \times 250$
$= 10 \times 250$
$= 2500\ \Omega$

Answer: The unknown resistance is $2500\ \Omega$.

Self-Test

Answer the following questions.

51. True or false. The Wheatstone bridge uses a zero-center-scale meter.

52. True or false. No current flows anywhere in a balanced bridge.

53. Refer to Fig. 15-21. Which resistors must drop equal voltages when the bridge is balanced?

54. What determines the accuracy of a measurement made with a bridge?

55. Refer to Fig. 15-21. Assume $R_1 = 300\ \Omega$, $R_2 = 600\ \Omega$, and $R_3 = 2000\ \Omega$ at balance. What is the value of R_u?

15-9 Wattmeters

An electrodynamometer movement (Fig. 15-5) makes a perfect *wattmeter*. The moving coil (called the *voltage coil*) is used to detect the magnitude of the circuit voltage. The stationary coils are referred to as the *current coils*. The circuit current is detected by the current coils, which are connected in series with the load (Fig. 15-22).

The stationary (current) coils are wound with large-diameter wire. This keeps the resistance that is in series with the load as low as possible. The movable (voltage) coil is wound with thin wire to keep it as light as possible. Since the movable coil responds to voltage, it has a multiplier in series with it.

Although the wattmeter in Fig. 15-22 is measuring ac power, relative polarity must still be observed. Connecting the plus-minus ($\pm$) terminals of both the voltage coil and the current coils to the power source provides correct polarity. The other end of the current coils (marked A) goes to the load. Thus, all the load current flows through the current coils. The other end of the voltage coil (marked V) returns to the power source. This puts the full source voltage across the voltage coil-multiplier circuit.

Correct polarity can also be achieved by connecting the plus-minus end of the voltage coil to the A end of the current coils. Referring to Fig. 15-22, you can see that either connection gives the same current direction in the voltage coil.

The wattmeter indicates *true power* under all conditions. It automatically compensates for phase differences between the circuit voltage and current. How this happens can be understood by referring to Fig. 15-23. In Fig. 15-23(a), the voltage and current are in phase; that is, PF = 1. Therefore, the currents through the voltage coil and current coil change at the same instant. Thus, the magnetic fields of the two coils are opposing and creating a clockwise torque during the complete cycle. However, when the current and the voltage are out of phase [Fig. 15-23(b)], the situation changes. Now for a portion of the cycle, the magnetic fields of the two coils are aiding and develop a counterclockwise torque. This reduces the net clockwise torque and thus the power indicated by the meter. When the current and voltage are 90° out of phase [Fig. 15-23(c)], the clockwise and counterclockwise torques are exactly equal. The net torque is zero, and the wattmeter indicates zero power.

Wattmeters are rated for maximum current and voltage as well as maximum power. These ratings are necessary to protect the coils in the

Wattmeter

Voltage coil

True power

Current coils

Correct polarity

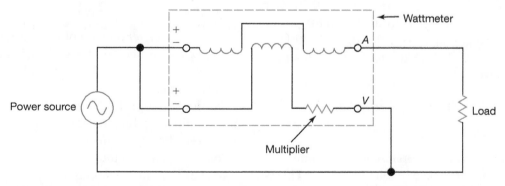

Fig. 15-22 Wattmeter connections. For clockwise torque, the instantaneous currents in the current coils and in the voltage coil must be in the same direction.

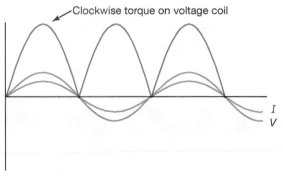

(a)   In phase (PF = 1)

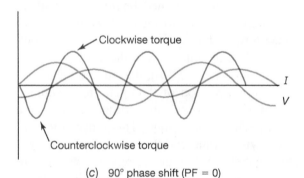

(b)   45° phase shift (PF = 0.707)

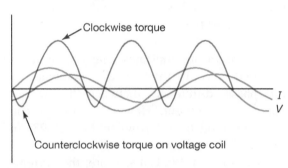

(c)   90° phase shift (PF = 0)

Fig. 15-23 How a wattmeter compensates for power factor.

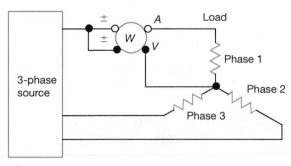

(a)   Wye-connected load

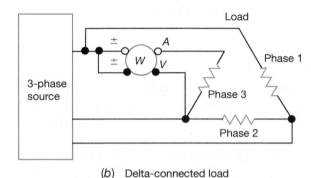

(b)   Delta-connected load

Fig. 15-24 Measuring power with balanced loads. Only one wattmeter is needed when the phase loads are equal.

electrodynamometer movement. If the *voltage rating* is exceeded, the movable coil gets too hot. Exceeding the *current rating* of the meter overloads the stationary coils. When the power factor of the load is low, the current rating is exceeded long before the meter indicates full-scale power.

Measuring Three-Phase Power

BALANCED LOAD

With a balanced three-phase load, the power in each of the three phases must be equal. Therefore, the total power of the three-phase system is equal to three times the power of any one phase. Consequently, a single wattmeter can be used to measure the power in any one of the phase loads (Fig. 15-24). The measured power is multiplied by 3 to give the total power of the system.

Notice in Fig. 15-24(*a*) that one must be able to connect the voltage coil of the wattmeter to the neutral, or star, point. Likewise, in Fig. 15-24(*b*), one has to break the junction of two of the phase loads to measure the current in a phase leg. In many three-phase loads, these junction points are not readily accessible. In such cases, the power can be measured by using two wattmeters, as described in the next section.

UNBALANCED LOAD

The *two-wattmeter method* of measuring *three-phase power* is shown in Fig. 15-25. It works equally well with wye or delta loads and with balanced or *unbalanced loads*.

Notice in Fig. 15-25 that the wattmeters are monitoring line currents and line voltages. Yet, the *algebraic sum* of the two meter readings equals the total power of the system. (Proof of this statement requires phasor analysis beyond the scope of this text.) The phrase "algebraic sum of the two meter readings"

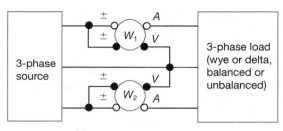

(a)   Add the meter readings

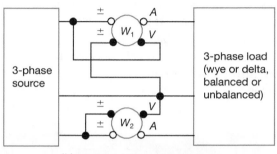

(b)   Subtract the meter readings

Fig. 15-25 Power measurement with unbalanced or balanced loads. The connections shown in (b) are needed with some unbalanced loads.

needs some explanation. Normally, the meters are connected with the polarities shown in Fig. 15-25(a). This connection usually results in clockwise deflection of both meters. If it does, the algebraic sum of the two readings is obtained by simply adding the two readings. For example, if one meter indicates 2800 W and the other indicates 1600 W, the power of the three-phase system is 4400 W (2800 W + 1600 W).

Sometimes, when the loads are extremely unbalanced, the polarities in Fig. 15-25(a) result in a *reverse deflection* of one of the meters. In this case, the polarity of the reverse-deflecting meter must be changed. This can be done by reversing the connections to the voltage coil, as shown in Fig. 15-25(b). Now the algebraic sum of the two meter readings is found by subtracting the smaller reading from the larger reading. Let us look at an example. Suppose we connect the meters as shown in Fig. 15-25(a) and W_1 deflects counterclockwise. Therefore, we reverse the voltage coil connection on W_1. This yields the connections shown in Fig. 15-25(b). Now suppose W_1 indicates 300 W and W_2 indicates 2600 W. The total power of the system is 2300 W (2600 W − 300 W).

Reverse deflection

Self-Test

Answer the following questions.

56. Why does a wattmeter use an electrodynamometer movement?
57. Which coil of a wattmeter is wound with the larger-diameter wire?
58. Why does an ac wattmeter require polarity markings?
59. Does a wattmeter indicate true power or apparent power?
60. A wattmeter is connected to a 240-V circuit that draws 10 A. The power factor is 0.5. How much power does the meter indicate?

61. A wattmeter in a balanced three-phase circuit indicates 1265 W. What is the total power of the circuit?
62. Two wattmeters are connected to a three-phase system. Both are connected for normal polarity. One meter indicates 2600 W, and the other meter indicates 2200 W. What is the power of the three-phase system?
63. If the 2200-W meter in question 62 required reverse polarity connections, what would the power be?
64. How is it possible to exceed the current rating of a wattmeter without causing the pointer to deflect beyond full scale?

15-10   Frequency Meters

Frequency can be measured with a variety of electric and electronic devices. Electronically, frequency can be measured with such devices as digital frequency counters and heterodyne *frequency meters*. These devices are capable of measuring a wide range of frequencies extending to hundreds of megahertz.

Frequency meters

Power frequency range

Vibrating-reed meter

Digital frequency meter

Electric frequency meters can measure only a narrow range of frequencies in the *power frequency range.* A *vibrating-reed meter,* a common electric frequency meter, may measure frequencies only between 58 and 62 Hz.

A *digital frequency meter* measures an unknown frequency by counting the number of cycles the frequency produces in a precisely controlled period of time. The counter circuit is incremented one count for each cycle. At the end of the time period, the final count, which represents the frequency, is displayed by the digital readout. For the next sampling of the unknown frequency, the counter is cleared, the time period is started over, and the final count in the counter is again displayed. If the measured frequency is stable, the readout does not change from sample to sample. Because the range switch selects the time period and places the decimal point in the readout, the indicated frequency is in the units specified by the range switch.

When the time period is 1 ms, the readout is in kilohertz and the range switch indicates kilohertz. For example, if the count at the end of the 1-ms period is 100, the unknown (measured) frequency must be 100 kHz because 100 counts per millisecond is equal to 100,000 counts per second.

Self-Test

Answer the following questions.

65. True or false. Digital frequency meters are limited to measuring power and audio frequencies.
66. True or false. The digital frequency meter counts the cycles produced by the unknown frequency during a precisely controlled period of time.
67. True or false. A vibrating-reed meter usually covers the entire power and audio-frequency range.

15-11   Measuring Impedance

Indirect technique

Measuring impedance

Impedance and reactance must be measured by some *indirect technique.* Of course, both impedance and reactance are functions of frequency, and so they are measured and specified at some stated frequency. In this section we will concern ourselves only with *measuring impedance* (and reactance) at power and audio frequencies.

Current-Voltage Method

Current-voltage method

One of the simplest ways to measure impedance is to measure the current and voltage and then use Ohm's law. This method is illustrated in Fig. 15-26. It works well at any low frequency, but it is especially well suited to measuring impedance at 60 Hz. The current meter must be accurate at the frequency at which the impedance is being measured. Also, the current meter must have a very low current range if high impedances are being measured.

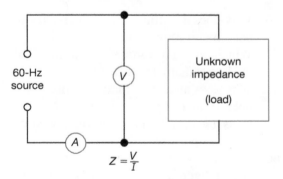

Fig. 15-26 Determining an unknown impedance. After the current and voltage are measured, Ohm's law is used to calculate impedance.

Equivalent-Resistance Method

Equivalent-resistance method

A more versatile technique of measuring impedance is shown in Fig. 15-27(*a*). When this technique is used, the source (generator) is first set to the frequency at which the impedance is to be measured. Any convenient output-voltage level (within the rating of the load) is selected.

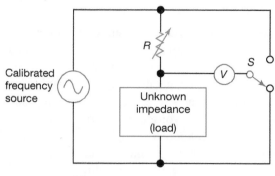

(a) With a single voltmeter

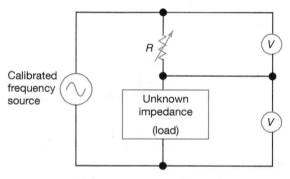

(b) With two identical voltmeters

Fig. 15-27 Impedance-measuring technique. When the voltmeter readings in (b) are equal, the resistance of R will have the same value as the unknown impedance.

Next, the voltage across the load is noted. Then the switch is moved to its upper position and the voltage across the resistor is measured. If the voltage across R is less than the voltage across the load, the value of R is increased. If the reverse is true, the value of R is decreased. The switch is then returned to its original position to measure the new value of the load voltage. Then the switch is again thrown to the up position and the resistor and load voltages are again compared. If they are not equal, R is again adjusted. This process is repeated until the measured voltages across R and the load are equal. When these voltages are equal, the impedance and resistance in this series circuit must also be equal. If R is a calibrated rheostat, the value of the impedance of the load can be read from the dial. If not, then the source voltage is removed and the value of R (which equals Z) can be measured.

When two identical voltmeters are available, the circuit in Fig. 15-27(b) is easier and faster to use. Once the generator's output voltage and frequency have been set, R is adjusted until the two meters indicate equal voltages. Then the value of R is determined.

 Self-Test

Answer the following questions.

68. What is the advantage of using two voltmeters when determining impedance by the equivalent-resistance method?

69. A speaker and a 15-Ω resistor are connected in series to a generator. They each develop 1 V at 2000 Hz. What is the impedance of the speaker at 2000 Hz?

15-12 Measuring Inductance and Capacitance

Inductance and capacitance can be measured in a variety of ways. Two common methods are presented in the following discussions.

Digital Method

Digital-voltmeter circuitry (Sec. 15-1) can be used to measure inductance indirectly by measuring the induced cemf under controlled conditions.

You May Recall

. . . that in the chapter on inductance, 1 H of inductance was shown to produce 1 V of cemf when the current changed at a rate of 1 A/s.

Therefore, in general terms, we can write

$$L = \frac{V}{I/t}$$

Now, suppose that the digital inductance meter forces a current which is rising at a rate of 1 A/s through an unknown inductor. Further, suppose that the digital voltmeter circuitry measures 2 V across the inductor. By applying the above formula, we can show that this 2 V represents 2 H of inductance. Thus,

$$L = \frac{2\text{ V}}{(1\text{ A})/(1\text{ s})} = \frac{2\text{ V}}{1\text{ A/s}} = 2\text{ H}$$

In an actual digital inductance meter, the level of current must be kept very small. Then the ohmic voltage drop (due to the resistance of the wire in the inductor) is very small compared with the cemf produced by the inductor.

A digital capacitance meter indirectly measures capacitance by using the time it takes a capacitor to charge to a specified voltage under controlled conditions.

By combining two fundamental formulas, $I = Q/t$ and $C = Q/V$, we can show the relationship between capacitance, voltage, current, and time. Rearranging $I = Q/t$ to $Q = It$ and then substituting It for Q changes $C = Q/V$ into $C = It/V$. This formula shows that C and t are directly proportional when I and V are held constant. The current I is held constant by charging the capacitor from a constant current source. The voltage V is held constant by letting the capacitor charge until the capacitor's voltage is just equal to that of a stable reference voltage. Thus, the time required to charge an unknown capacitor is a direct function of the unknown capacitance. Section 15-1 discussed how to measure time digitally by counting the cycles produced by a stable reference frequency.

When the current, voltage, and reference frequency are selected properly, the digital readout can be in microfarads, or picofarads. When the constant current is 10 mA, the reference voltage is 1 V, and the reference frequency is 10 kHz, the readout is in microfarads. For example, with these constants, the readout would be 10 if it took the capacitor 1 ms to charge (10 counts per millisecond = 10,000 counts per second = 10 kHz). Solving for C with these values of t, V, and I yields

$$C = \frac{It}{V} = \frac{(10\text{ mA}) \times (1\text{ ms})}{1\text{ V}} = 10\ \mu\text{F}$$

Thus, the readout of 10 is in microfarads.

AC-Bridge Method

The value of a capacitor or an inductor can also be determined with an *ac bridge*. The ac-bridge circuit in Fig. 15-28 is used to measure the value of the unknown capacitor C_u. To use the bridge, R_1 and R_3 are adjusted for the lowest possible indication on the voltmeter. There will be some interaction between R_1 and R_3, and so one and then the other resistor must be adjusted several times to achieve minimum voltmeter reading. When properly adjusted, the meter should indicate essentially 0 V.

The purpose of R_3 is to make the resistive losses in C_1 equal to the resistive losses in C_u. If C_u has as high a quality factor as C_1, then R_3 will be approximately 0 Ω. In fact, R_3 is needed in the bridge only when low-quality capacitors are to be measured. For our analysis of the *capacitor bridge*, we will assume that high-quality capacitors are being measured. Thus, R_3 is 0 Ω.

When R_1 and R_3 are adjusted for minimum voltmeter indication, the bridge is balanced. Under balanced conditions, the ratio of R_1 to X_{C_1} is equal to the ratio of R_2 to X_{C_u}. Mathematically, we can write

$$\frac{R_1}{X_{C_1}} = \frac{R_2}{X_{C_u}}$$

Since capacitive reactance is inversely proportional to capacitance

$$R_1 C_1 = R_2 C_u$$

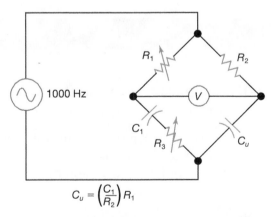

$$C_u = \left(\frac{C_1}{R_2}\right) R_1$$

Fig. 15-28 Measuring C with a bridge. The variable resistors are adjusted to obtain the lowest possible voltmeter reading.

or

$$C_u = \frac{C_1}{R_2} \times R_1$$

In the above formula, C_u will have the same unit as C_1. Also, R_1 and R_2 must have the same unit.

EXAMPLE 15-7

In Fig. 15-28, $R_2 = 2000\,\Omega$, and $C_1 = 0.002\,\mu\text{F}$. If the balance occurs when R_1 is adjusted to 1700 Ω, what is the capacitance of C_u?

Given: $R_2 = 2000\,\Omega$
             $C_1 = 0.002\,\mu\text{F}$
             $R_1 = 1700\,\Omega$

Find: C_u

Known: $C_u = \dfrac{C_1}{R_2} \times R_1$

Solution: $C_u = \dfrac{0.002 \times 10^{-6}\,\text{F}}{2000\,\Omega}$
               $\times 1700\,\Omega$
           $= 0.0017 \times 10^{-6}\,\text{F}$
           $= 0.0017\,\mu\text{F}$

Answer: The unknown capacitance is 0.0017 μF.

The *inductance bridge* in Fig. 15-29 works on the same principles as the capacitance bridge of Fig. 15-28. The formula $L_u = CR_2R_1$ can be verified by applying the rationale used to derive the capacitance-bridge formula.

Notice in Fig. 15-29 that R_3 is in parallel with C. (In some inductance bridges, it is in

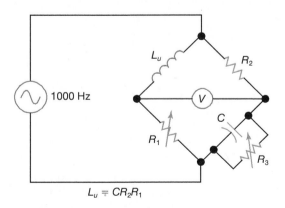

$L_u \doteq CR_2R_1$

Fig. 15-29 Measuring L with a bridge. The bridge is balanced when further adjustments of R_1 and R_2 cannot reduce the voltage indicated by the voltmeter.

series.) Again, R_3 is adjusted to provide the same amount of resistance loss in the capacitor as there is in the inductor. Since R_3 is in parallel with C, the resistive losses increase as R_3 is decreased. If the unknown inductance has a high quality factor, the value of R_3 will be very high.

The resistance of the inductor (R_{ind}) can also be determined from the bridge circuit of Fig. 15-29 by the formula

$$R_{ind} = \frac{R_1 R_2}{R_3}$$

Once the resistance and inductance of the inductor are known, it is easy to determine the quality of the inductor. (The generator frequency is known, and $Q = X_L/R = 6.28fL/R$.) Many inductance bridges have the dial attached to R_3 calibrated directly in Q values.

EXAMPLE 15-8

Suppose that the bridge in Fig. 15-29 balances when $R_1 = 2$ kΩ, $R_2 = 5$ kΩ, $R_3 = 500$ kΩ, and $C = 0.01\,\mu$F. What are the inductance and quality of the inductor?

Given: $R_1 = 2\,\text{k}\Omega$
            $R_2 = 5\,\text{k}\Omega$
            $R_3 = 500\,\text{k}\Omega$
            $C = 0.01\,\mu\text{F}$
            $f = 1000\,\text{Hz}$

Find: L and Q

Known: $L_u = CR_2R_1$

            $Q = \dfrac{X_L}{R}$

            $X_L = 6.28fL$

            $R_{ind} = \dfrac{R_2R_1}{R_3}$

Solution:
   $L_u = 0.01 \times 10^{-6} \times 5 \times 10^3 \times 2 \times 10^3$
       $= 0.1\,\text{H} = 100\,\text{mH}$
   $R_{ind} = \dfrac{5 \times 10^3 \times 2 \times 10^3}{500 \times 10^3} = 20\,\Omega$
   $X_L = 6.28 \times 1000 \times 0.1$
       $= 628\,\Omega$
   $Q = \dfrac{628}{20} = 31.4$

Answer: The inductance is 100 mH, and the quality is 31.4

Inductance bridge

Answer the following questions.

70. Name two methods of measuring capacitance and inductance.
71. What quantity is measured by a digital inductance meter to determine the value of an unknown inductor?
72. Why is a small current used when measuring inductance by the cemf method?
73. When measuring capacitance with a digital meter, which quantity (I, t, or V) is actually being measured?

74. What is the purpose of R_3 in Fig. 15-28?
75. List the three ratings of an inductor that can be determined with an inductance bridge.
76. Refer to Fig. 15-28 and find C_u when $R_2 = 2\ k\Omega$, $R_1 = 4\ k\Omega$, and $C = 1200\ pF$.
77. Refer to Fig. 15-29 and find L_u when $R_2 = 2\ k\Omega$, $R_1 = 4\ k\Omega$, and $C = 1200\ pF$.
78. Determine Q for question 77 if $R_3 = 2\ M\Omega$.

Chapter 15 Summary and Review

Summary

1. The d'Arsonval meter movement is commonly used in voltmeters, ammeters, and ohmmeters. It responds only to direct current. It is used in rectifier-type instruments to measure alternating current and voltage.

2. Electrodynamometer meter movements use stationary and moving coils to develop interacting magnetic fields. They respond to alternating or direct current and are used in wattmeters. Electrodynamometer meters have low sensitivity and high accuracy.

3. A rectifier allows current to flow in only one direction. It converts alternating current to pulsating direct current.

4. Digital voltmeters measure the time required for a capacitor to charge to the value of the unknown voltage.

5. The iron-vane meter movement has no moving coil or permanent magnet.

6. The iron-vane meter movement responds to both alternating and direct current. It has a nonlinear scale and low sensitivity.

7. Basic meter movements are rated for voltage, current, and resistance.

8. Shunts and multipliers can be used with d'Arsonval, rectifier-type, iron-vane, and electrodynamometer meters.

9. Values for shunts and multipliers can be calculated if the meter-movement ratings and the ranges are known.

10. Ammeters have very low internal resistance.

11. Voltmeters have very high internal resistance.

12. External shunts are rated in amperes and millivolts.

13. Shunts and multipliers are precision resistors having a low temperature coefficient.

14. Shunts are in parallel with the meter movement. They extend the range of ammeters.

15. Multipliers are in series with the meter movement. They extend the range of voltmeters.

16. Current transformers are used to extend the range of ac ammeters.

17. Clamp-on ammeters can measure current in an ac circuit without interrupting the circuit.

18. Thermocouple meters are used to measure high-frequency currents. They use a d'Arsonval meter movement.

19. VOMs have an ohms-per-volt rating. This rating is also called the VOM's sensitivity.

20. Input resistance = sensitivity × range

21. Ohms per volt = $\dfrac{1}{\text{full-scale current}}$

22. DMMs have an input-resistance rating that is independent of the range.

23. Meter loading causes changes in circuit currents and voltages when measurements are made.

24. Ohmmeters have a power source, rheostat, meter movement, and fixed resistor.

25. Ohmmeter scales are nonlinear and often reverse-reading.

26. Ohmmeter ranges are changed by switching voltages or shunts.

27. Insulation testers (Meggers) use high voltages to measure very high resistances (insulation).

28. Wheatstone bridges use a galvanometer that indicates zero when a bridge is balanced.

29. A bridge is balanced when the ratio of rheostat to unknown resistance equals the ratio of the two known resistors.

30. The accuracy of a bridge is determined by the tolerances of the resistors used in it.

31. In a wattmeter the moving coil responds to voltage and the stationary coil responds to current.

32. Wattmeters are polarized instruments. They indicate true power. They have current, voltage, and power ratings.

33. A single wattmeter can measure power in a balanced three-phase load.

34. Two wattmeters are required to measure power in an unbalanced three-phase load. The wattmeter readings are algebraically added.
35. A vibrating-reed meter measures only a narrow range of frequencies.
36. Impedance can be measured by the current-voltage method or by the equivalent-resistance method.

37. Inductance and capacitance can be measured by the digital method or by the ac-bridge method.
38. The digital inductance meter measures the cemf of the inductor.
39. The digital capacitor meter measures the time required to charge the capacitor.

Chapter Review Questions

For questions 15-1 to 15-16, supply the missing word or phrase in each statement.

15-1. The _____ meter movement is polarized. (15-2)
15-2. The _____ meter movement does not use a moving coil. (15-2)
15-3. The analog ohmmeter uses a(n) _____ to compensate for small changes in cell voltage. (15-6)
15-4. A(n) _____ resistor extends the range of a voltmeter. (15-4)
15-5. A(n) _____ resistor extends the range of an ammeter. (15-3)
15-6. A(n) _____ switch is used to change ranges in a multirange ammeter. (15-3)
15-7. The internal resistance of the _____ is different on each voltage range. (15-4)
15-8. A vibrating-reed meter is used to measure _____. (15-10)
15-9. A(n) _____ can measure alternating current without interrupting the circuit. (15-3)
15-10. The moving coil in a wattmeter responds to _____. (15-9)
15-11. Meter loading in an ac circuit is _____ dependent. (15-5)
15-12. The DMM measures _____ when measuring an unknown voltage. (15-1)
15-13. The DMM measures _____ when measuring an unknown resistance. (15-1)
15-14. The digital inductance meter measures _____ when measuring an unknown inductance. (15-12)
15-15. The digital capacitance meter measures _____ when measuring an unknown capacitance. (15-12)

15-16. A(n) _____ converts alternating to direct current. (15-3)

For questions 15-17 to 15-24, determine whether each statement is true or false.

15-17. A 100-μA meter movement produces a more sensitive voltmeter than a 50-μA meter movement. (15-4)
15-18. Analog ohmmeters have a nonlinear scale. (15-6)
15-19. The range of an ohmmeter can be changed only by shunting the meter movement. (15-6)
15-20. Bridge balance is obtained when the null indicator (voltmeter) indicates its lowest voltage. (15-8)
15-21. Insulation testers measure resistance under high-voltage conditions. (15-7)
15-22. When a Wheatstone bridge is in balance, no current flows in any part of the circuit. (15-8)
15-23. Ammeter loading occurs more often than voltmeter loading. (15-5)
15-24. Two wattmeters are needed to measure ac power in a balanced three-phase circuit. (15-9)

For questions 15-25 to 15-29, choose the letter that best completes each statement.

15-25. Which of these meter movements is the most sensitive? (15-2)
   a. d'Arsonval
   b. Electrodynamometer
   c. Iron-vane
15-26. Which of these meter movements cannot directly measure alternating current? (15-2)
   a. d'Arsonval
   b. Electrodynamometer
   c. Iron-vane

15-27. Which of these meter movements is used to measure power? (15-2)
  a. d'Arsonval
  b. Electrodynamometer
  c. Iron-vane

15-28. Which of these meter movements uses a permanent magnet? (15-2)
  a. d'Arsonval
  b. Electrodynamometer
  c. Iron-vane

15-29. Which of these meter movements uses a moving coil and a fixed coil? (15-2)
  a. d'Arsonval
  b. Electrodynamometer
  c. Iron-vane

Answer the following question.

15-30. List two methods of measuring impedance. (15-11)

Chapter Review Problems

15-1. What is the input resistance of a 20-kΩ/V voltmeter on the 60-V range? (15-4)

15-2. Would the meter in problem 15-1 cause significant loading if it were used to measure the voltage across a 470-kΩ resistor in a series circuit? (15-5)

15-3. What is the value of the internal resistance of a meter movement which is rated 50 μA, 200 mV? (15-2)

15-4. A 100-mA meter indicates 36 mA. Its accuracy is ± 2 percent. What are the minimum and maximum values of the current? (15-3)

15-5. You need to convert a 50-μA, 1000-Ω meter movement to a 1-mA meter. What value of resistor do you need? (15-3)

15-6. You need to convert the meter movement in problem 15-5 to a 5-V meter. What value of resistor do you need? (15-4)

15-7. What is the ohms-per-volt rating of a voltmeter which uses a 100-μA meter movement? (15-4)

15-8. A digital ohmmeter uses a 10-μA constant current source. How much resistance would be represented by a readout of 2.50? (15-1)

15-9. A 100-μA constant current source charges a capacitor to 2 V in 4 ms. Determine the capacitance. (15-12)

15-10. Refer to Fig. 15-21 and determine the value of R_u when $R_1 = 5$ kΩ, $R_2 = 10$ kΩ, and $R_3 = 9.4$ kΩ. (15-8)

15-11. A DMM is on the 5-kΩ range. A 1-mA constant current in the DMM is connected to an unknown resistance. The voltmeter circuitry displays 4.70 on the digital readout. What is the value of the unknown resistance? (15-1)

15-12. A DMM on the 500-mA range is measuring the current in a circuit. Internally, the DMM has a 100-Ω resistor connected across the input to its voltmeter circuitry. When the digital readout displays 400, what value of current is being measured? Also, when the DMM is switched from the 1-A range to the 500-mA range, how many places is the decimal point shifted? Is it shifted left or right? (15-1)

15-13. The power of a three-phase system is being checked using the two-wattmeter technique. With standard polarity connection on both wattmeters, the readings are 300 W and 900 W. What is the total power used by the system? If the voltage connections had to be reversed on one of the wattmeters to get these two readings, what would be the total power used by the system? (15-9)

Critical Thinking Questions

15-1. Two 10-kΩ, ± 10 percent series resistors are connected to a 10-V source. The voltage across one of the resistors is measured with a 3-kΩ/V voltmeter on the 5-V range. What will the meter indicate under worst-case conditions?

15-2. Why is the iron-vane meter movement so nonlinear?

15-3. Why are shunts and multipliers made from materials with very low temperature coefficients?

15-4. Why might the ac voltage ranges of a VOM have lower sensitivity than the dc voltage ranges?

15-5. An ohmmeter uses a 200-μA meter movement, a 9-V battery, and appropriate values of a series resistor and rheostat. How much resistance is represented by 15 percent deflection of the meter movement?

15-6. How long will it take a 3.3-μF capacitor to charge to 5 V when connected to a 2-mA constant-current source?

15-7. What are the characteristics of a cell or battery that requires an analog ohmmeter to have an ohms-adjust rheostat?

15-8. A precision 4.7-MΩ resistor is series connected with a precision 6.8-MΩ resistor to a 90-V source. Determine the input resistance of a DMM that indicates 46.73 V across the 6.8-MΩ resistor.

15-9. Determine the constant current source needed for a capacitance meter to readout in picofarads if the reference voltage is 10 V and the reference frequency is 500 kHz.

15-10. Two 100-kΩ, ± 10 percent resistors are series-connected to a 100-V source. The resistor voltages are measured with a 50-V meter that has a tolerance of ± 2 percent of full-scale voltage. What is the minimum voltage reading that could occur without any meter loading?

Answers to Self-Tests

1. T
2. T
3. 1.2 MΩ
4. the time required for a capacitor to charge to the unknown voltage
5. constant current source
6. full-scale current, full-scale voltage, internal resistance
7. d'Arsonval
8. spiral spring
9. 2000 Ω
10. direct current
11. both
12. the d'Arsonval
13. a d'Arsonval
14. no
15. both
16. the interaction of the magnetic fields of the stationary coils and the movable coil
17. two
18. a d'Arsonval movement
19. a device that allows current to flow only in one direction

20. low temperature coefficient and low resistance tolerance
21. 9.9 mA
22. a. 133.3 Ω
    b. 108 μW
23. a shorting type (make-before-break)
24. low
25. when measuring currents of 10 A or higher
26. by current capacity and voltage drop
27. a transformer used in conjunction with an ammeter to measure large values of alternating current
28. no
29. the current-transformer principle
30. measuring high-frequency currents
31. heat, electric
32. T
33. F
34. T
35. F
36. 40 kΩ/V
37. 7.5 mW, 333.133 kΩ
38. 5 MΩ
39. nonshorting

40. Meter loading means that the meter's internal resistance appreciably changes the circuit's current or voltage.
41. voltmeter loading
42. yes
43. yes
44. a. yes
    b. no
    c. no
45. a 20-kΩ/V voltmeter on the 1000-V range
46. a meter movement, a cell, a fixed resistor, and a variable resistor
47. The rheostat adjusts the ohmmeter for zero-resistance readings.
48. 40 Ω
49. 3000 Ω
50. Change the voltage and the series resistance, or change the series resistance and add a shunt, increasing the voltage.
51. T
52. F
53. R_1 and R_2, and R_3 and R_u
54. the tolerance of the resistors in the bridge
55. 1000 Ω
56. because the electrodynamometer movement reads true power regardless of the value of angle θ
57. the stationary, or current, coil
58. so that the current in the stationary coils will be in the correct direction relative to the current in the movable coil
59. true power
60. 1200 W
61. 3795 W
62. 4.8 kW
63. 400 W
64. The voltage may be below rated value, or the load may have a low power factor.
65. F
66. T
67. F
68. It is much faster. No switching of meter leads is required.
69. 15 Ω
70. digital method and ac-bridge method.
71. cemf
72. so that the resistance voltage will be very small compared to the cemf
73. t
74. purpose is to lower the Q of C_1 to match the Q of C_u
75. the inductance, the resistance of the inductor, and the quality of the inductor
76. 0.0024 μF
77. 9.6 mH
78. 15.1

CHAPTER 16

Residential Wiring Concepts

Learning Outcomes

This chapter will help you to:

16-1 *Realize* that new or modified residential wiring must meet the requirements of the National Electrical Code, state electrical code and local electrical code.

16-2 *Discuss* how electric energy is distributed from the place it originates to the residence where it is utilized.

16-3 *List* the components used in an overhead service entrance to a residence.

16-4 *Identify* the neutral, ground, and hot conductors in a 12-2/G cable.

16-5 *Learn* that a white conductor connected to a circuit breaker is a hot conductor and should be marked with black tape.

16-6 *Explain* (a) how the neutral side of a receptacle is identified and (b) why the neutral conductor of the cable feeding the next outlet box is not directly connected to the second screw on the neutral side of the outlet.

16-7 *Explain* why a switched-receptacle outlet uses a white conductor, marked with black tape, as a hot conductor.

16-8 *Learn* (a) that two three-way switches (SPDT) are used to control a light from two locations; and (b) controlling a light from three locations requires two three-way switches and one four-way switch.

16-9 *See* why the lighted handle of either a snap switch (SPST) or a three-way switch is lit when the light is turned off.

16-10 *Specify* which circuit interrupter or interrupters (a) protects against electric shock, (b) detects a Short between the neutral and ground conductors, and © can remove power from a branch circuit.

16-11 *Discuss* the major features of a feeder circuit.

16-12 *List* (a) the type of switches and relays used in low-voltage control circuits, and (b) the advantage of the dual-coil

16-1 Electrical Codes

Electrical codes (national, state, and local) spell out the requirements for residential, and all other, wiring systems. Although all states have adopted the *National Electrical Code* (NEC), most locales also have supplemental state and/or local codes and/or requirements. Before attempting any residential wiring, one must be aware of local and national code requirements and obtain a wiring permit when required.

This chapter explains how residential circuits and devices operate. It also explains why certain requirements and procedures exist. The rationale for a specific code requirement may be obscure. However, there is a good reason (based on electrical theory and/or practical experience) for every code requirement. A wiring system that complies with all electrical codes:

National Electrical Code (NEC)

1. Minimizes the chances of the system starting a fire
2. Minimizes the chances of occupants receiving accidental electrical shocks
3. Provides lighting that is safe and adequate, and power outlets that are safe and conveniently located.

16-2 Power Distribution

Electric energy is generated and distributed by electric power companies. It was pointed out in Chapter 1 that much of our electric energy comes from nonrenewable fossil fuels. The global population is becoming more aware of, and concern about, the limited supply and increasing cost of fossil fuel energy sources. Thus, the research on the development of technologies to utilize renewable energy is steadily increasing. The cost of developing and operating renewable energy systems is predicted to continue to decrease. For more information on renewable energy sources, see Appendix K.

As illustrated in Fig. 16-1, the output of a three-phase generator is stepped up to a very

latching relay over the single coil type, and (c) the name of the device that opens and closes the contacts that provide power to an electric heater.

16-13 *Discuss* some advantages of electronic control circuits over hard-wired relay control.

high voltage (270 kV or higher) for transfer to the point of utilization.

You May Recall

... from the discussion of transformers in Chap. 12 that it makes sense to transfer the energy at a high voltage because, for a given power, less current is required and smaller conductors can be used.

When the lines reach the outskirts of a city, the voltage is stepped down to a safer value for distribution to different areas of the city. Finally, the voltage is stepped down to a still lower value (4.7 to 13.7 kV) for single-phase distribution to residences and three-phase distribution to commercial districts.

After a residential wiring system has been inspected and approved, the local power company connects three conductors (called the *service drop*) to the system's *service conductors*. These three conductors come from the power company's step-down transformer, which is ultimately connected to one phase of a high-voltage three-phase system. As shown in Fig. 16-2, one of the three conductors is called the "neutral" and is an uninsulated (bare) conductor. It is electrically neutral (has no electrical potential with respect to earth ground) because the transformer terminal to which the neutral is connected is electrically connected to earth ground. The grounding conductor (wire) connects the *neutral conductor* (wire) to ground. Although the neutral wire is connected to ground, it is not a ground wire or a grounding wire. Because it is grounded, the neutral conductor is sometimes referred to as the *grounded conductor*. The neutral conductor can carry current (and usually does carry current) under normal (no-fault) operating conditions. The only time the neutral wire does not carry current is when the currents in the two hot wires are equal. At all other times, the neutral wire carries a current equal to the difference between the currents in the two hot wires.

Under normal operating conditions, the *grounding conductor* carries no current. The grounding conductor helps protect the residential wiring

Service drop

Service conductor

Neutral conductor

Grounded conductor

Grounding conductor

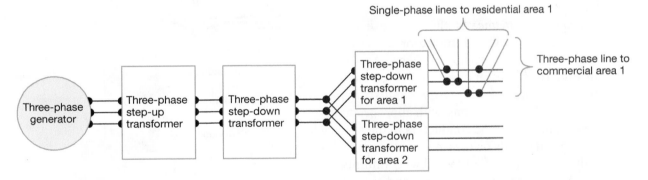

Fig. 16-1 Overview of an energy distribution system. Only single-phase power is distributed to residences.

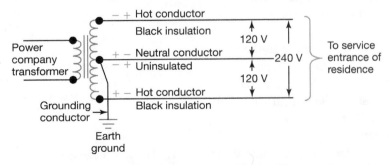

Fig. 16-2 Power company service to a residence. The transformer is either on a power pole or in an underground vault.

system from such things as lightning strikes or a primary-to-secondary short in the transformer. The other two conductors in Fig. 16-2 are insulated and are called "hot" conductors or wires. Each hot wire has a potential of 120 V with respect to the neutral and earth ground. Figure 16-2

illustrates why the two *hot conductors* can be thought of as being 180° out of phase. When one conductor is positive with respect to neutral, the other is negative with respect to neutral. Therefore, the two voltages are additive. Thus, the voltage between the two hot lines is 240 V.

⫼〰 Self-Test

Answer the following questions.

1. True or false. All states have adopted the NEC.
2. True or false. The output of most three-phase generators is 270 kV or higher.
3. Why is the power line voltage stepped down before being distributed to different areas of a city?
4. What type of power is provided to commercial and industrial areas?

5. True or false. Service drop refers to the conductors that run from the meter box to the distribution panel.
6. The neutral conductor in the _____ is a bare (uninsulated) conductor.
7. True or false. One hot conductor will be positive with respect to neutral when the other hot conductor is negative with respect to neutral.
8. The grounded conductor in the electrical system is the _____ conductor.

16-3 Service Entrance

The power company conductors connect to the *service entrance* at the *service head* for an overhead service entrance or at the meter box for an underground entrance. As shown in Fig. 16-3, the service entrance consists of the service head, *meter box, main distribution panel,* and conduit, along with appropriate fittings. The size of the conductors in the service entrance depends on the rating of the main circuit breaker and the type of wire (aluminum or copper), temperature rating, and so on.

Details of the meter box are shown in Fig. 16-4. The insulated sockets on the corners receive the power company's energy meter. The meter's internal construction completes the paths for the two hot wires from the service drop to the main circuit breaker in the main distribution panel. Notice that the neutral block in the meter box provides a continuous path for the neutral wire. This path is independent of the energy meter.

Figure 16-5 illustrates a typical main distribution panel. The distribution panel has three main elements:

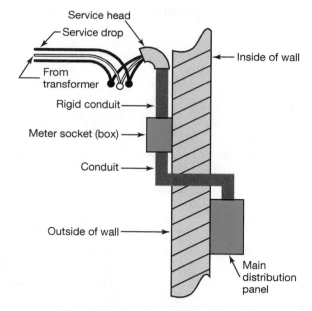

Fig. 16-3 Typical service entrance. The conduit, socket, and panel are secured to the wall. The service drop is anchored to the conduit or wall by an insulated attachment.

1. The *main circuit breaker,* which also serves as the disconnect from the power

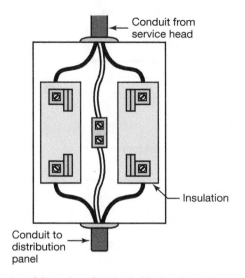

Fig. 16-4 Meter box (socket). Notice that the neutral conductor is electrically continuous.

company service. In a typical home, this breaker is rated from 100 to 200 A.

2. A parallel hot bus system that connects to the output of the main circuit breaker. These *two hot buses* provide power to the circuit breakers, which protect branch circuits and feeder circuits. The buses are constructed so that a 240-V breaker connects to each bus and 240 V is delivered to the output terminals of the 240-V breaker.

3. The *neutral/ground bus,* which provides a common connection point for all neutral conductors, grounding conductors, and the bonding jumper conductors. Sometimes a

state and/or local code requires separate buses for neutral conductors and for grounding/bonding conductors.

Notice in Fig. 16-5 that the service entrance equipment, including the neutral conductor, is connected to earth ground by one or two grounding conductors. One always goes to a grounding electrode (8-foot-long metal rod) that is driven into the ground outside the building. If the pipe from the water main is metal, a second grounding conductor connects to it. This ground system, in conjunction with all the *equipment-grounding conductors* in the branch circuits, helps protect the wiring system from such things as lightning strikes and other high-voltage surges.

Figure 16-6 shows a schematic diagram of the distribution panel in Fig. 16-5. Although the diagram clearly shows how the three branch circuits are connected, it does not emphasize some of the important details shown in Fig. 16-5. For example, Fig. 16-6 does not show that all neutrals from all branch circuits must connect directly to a neutral bus.

Equipment-grounding conductors

Two hot buses

Neutral/ground bus

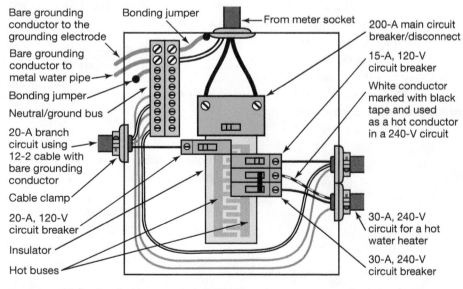

Fig. 16-5 Main distribution panel. A 240-V breaker connects to both hot buses.

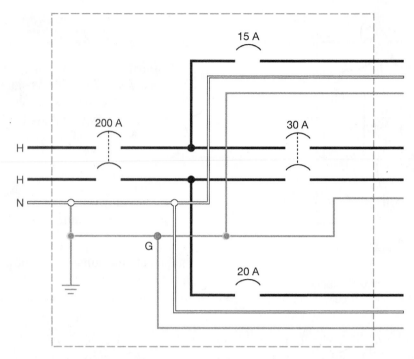

Fig. 16-6 Schematic presentation of the main distribution panel of Fig. 16-5.

16-4 Cables and Conductors

**Type NM
nonmetallic-
sheathed cable**

In residential wiring, circuits are often installed using *type NM nonmetallic-sheathed cable* (see Fig. 16-7). A cable with two 12 AWG insulated conductors and a 12 AWG bare, equipment-grounding conductor can be identified as 12-2/G or 12-2 WG. One insulated conductor is white and is used as the neutral conductor, and the other is black and used as the hot wire. When a cable has three insulated conductors (for example, 14-3/G), the third insulated conductor is red and is used as another hot conductor. A 20-A branch circuit uses 12

AWG cable, a 15-A branch uses 14 AWG, and a 30-A branch uses 10 AWG. Type NM cable with larger conductors is available for branch circuits serving ranges, ovens, and so forth.

Although the illustrations in this chapter all show nonmetallic-sheathed cable and either metal or plastic outlet boxes, many other wiring systems and materials can be used. Some other systems and materials are:

1. Armored and metal-clad cables
2. Various types of rigid metallic and nonmetallic conduit and tubing
3. Flexible metallic and nonmetallic conduit and tubing

When conduit or tubing is used, individual insulated conductors are used for hot, neutral, and grounding conductors. The hot conductor(s) can be any color except white, natural gray, or green. White and natural gray are used for neutral conductors, and green is used for grounding conductors. In some applications, the metallic conduit (with metal boxes) can serve as the grounding conductor.

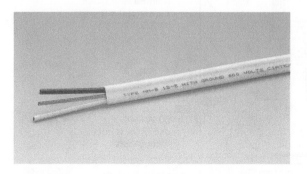

Fig. 16-7 An example of 12-2/G sheathed cable.

Answer the following questions.

9. The main components of the service entrance are the _____, _____, and _____.

10. The components provided by the main distribution panel are the _____, _____, and _____.

11. How is the service entrance equipment grounded?

12. Why are residential electrical systems grounded?

13. A 20-A branch circuit uses a cable with _____ AWG conductors.

14. Branch circuits are often wired using type _____ cable.

15. True or false. Although metal conduit is a conductor, it cannot be used as an equipment-grounding conductor.

16-5 240-V Branch Circuits

The *240-V circuit breaker* shown in Fig. 16-5 consists of two 120-V units ganged together at the factory. The two trip-reset levers are bonded together so that excess current in either hot conductor will completely disconnect the branch circuit. Excess current in only one hot conductor can be caused by a *ground fault* (that is, a short to a grounded frame) in the load device.

In a straight 240-V branch circuit like that in Fig. 16-5, the white conductor in the type NM cable must be used as a hot conductor. As shown in this figure, *black tape* has been put on this white conductor to emphasize that the

conductor is a hot conductor. Reidentifying a white or gray conductor used as a hot conductor is an NEC requirement. A white conductor is not a neutral conductor when it is connected to a circuit breaker, a switch, or a conductor that is not white or natural gray. Only white and natural gray conductors are used for neutral, and a neutral conductor must never be connected to any of the above-mentioned locations.

Some electrical devices (such as clothes dryers) require both 240 V and 120 V. The 240 V is used for the heating units, and the 120 V is used for the motor, lights, and timer. An example of a branch circuit for such a device is shown in Fig. 16-8. The only difference between this

240-V circuit breaker

Ground fault

Black tape

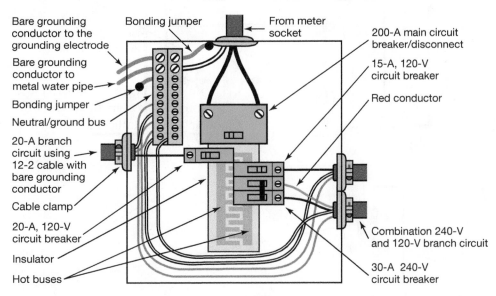

Fig. 16-8 Main distribution panel including a 240-V appliance branch with three conductors.

240-V branch and the one in Fig. 16-5 is that this one uses a cable with three insulated conductors and a bare grounding conductor (10-3 with ground). Now the white conductor does not have to be used as a hot conductor, since the

10-3 cable has a red, a black, and a white conductor. If the breaker on this circuit trips, the electrician has to determine whether the 120-V or the 240-V part of the device is faulty because a fault in either will open both hot conductors.

Self-Test

Answer the following questions.

16. True or false. A 240-V circuit breaker always disconnects both hot conductors when it is tripped.
17. True or false. Excess current in either hot conductor can trip a 240-V circuit breaker.
18. True or false. The conductor with white insulation is always a neutral conductor in a 240-V branch circuit.
19. True or false. Some 240-V branch circuits provide 120 V as well as 240 V.

16-6 120-V Branch Circuits

120-V branch circuits

Duplex receptacle

Typical *120-V branch circuits* provide power to outlet boxes for receptacles, lights, and switches. Every branch circuit is protected by a circuit breaker or a fuse at the main panel or at a subpanel.

Receptacle Outlet

Figure 16-9 illustrates how a *duplex receptacle* is wired in a 20-A branch circuit using type NM cable and a metal outlet box. Although not shown in this figure, cable clamps are provided to secure the cable to the box.

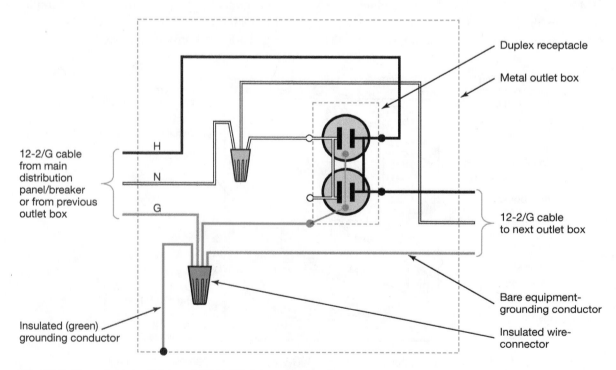

Fig. 16-9 Receptacle outlet in a 120-V, 20-A branch circuit.

The cable sheathing (again not shown) extends through the clamp to prevent the clamp from damaging the conductor insulation. Notice in Fig. 16-9 that *jumper bars* between the two neutral connections and between the two hot connections provide the duplex receptacle with current paths between the two receptacles. The wire connector on the neutral ensures that the neutral will have a continuous path even when the receptacle is removed. A complete electrical path from the neutral bus in the main distribution panel to every load (and outlet) is an absolute requirement for a residential wiring system. The equipment-grounding conductor must also have a continuous path; however, it is not electrically connected to the load. It connects only to the frame, the enclosure, and/or other metallic parts of the electrical device that provides the load.

Notice several other features of the duplex outlet. First, the neutral side of a receptacle is identified for the user as the longer of the two slots that receive the plug. For the electrician, the neutral side of the receptacle is identified by the *light-colored screws*. Second, there is only one ground screw, and it is the green screw. Third, the grounding wire from the receptacle goes to a wire connector unless the outlet box is nonmetallic and there is no outgoing cable. Then the equipment-grounding conductor terminates at the receptacle.

Often the grounding conductors are connected together with an uninsulated *crimped connector*. Also, the grounding conductor that grounds the outlet box is often a bare conductor.

Switch Outlet

The cable out of the box in Fig. 16-9 could go to a switch outlet box like the one in Fig. 16-10. The ordinary on-off wall switch is often referred to as a *snap switch*. Electrically, it is a single-pole, single-throw (SPST) switch. Again, notice that both the neutral and the ground have continuous paths to the next outlet box and to the lighting outlet box. Only hot conductors are switched or protected by circuit breakers or fuses.

If a switch fails and the hot part of the switch makes contact with the switch frame, then the metal cover plate (or the metal screws attaching a plastic cover plate) also becomes hot. That is, the cover plate or screws would be 120 V with respect to any grounded object or surface. This would be a real shock hazard—especially in kitchens, where many exposed surfaces are electrically grounded. Grounding the switch frame as shown in Fig. 16-10 (and required by the NEC) removes this shock hazard by keeping the switch frame at the same potential as all other grounded surfaces. With the switch frame grounded, the switch failure described above would trip the circuit breaker in the

Crimped connector

Jumper bars

Snap switch

Light-colored screws

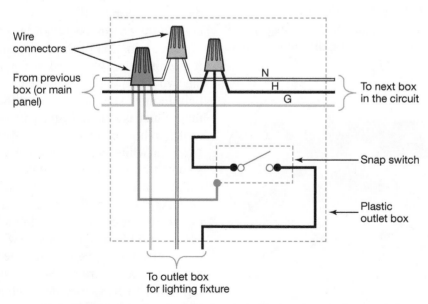

Fig. 16-10 Plastic outlet box for a wall switch. Notice that the frame of the switch is grounded.

Grounding switch frames

branch circuit. Homes wired before the NEC required *grounding switch frames* (straps) have this shock potential if plastic outlet boxes were used. If metal outlet boxes were used, and grounded, then the switch frame (mounting strap) would be grounded by the screws that attach the strap to the box.

Light Outlet

The continuation of the lighting circuit started by the switch in Fig. 16-10 is shown in Fig. 16-11. A simple light fixture with a porcelain base (like those often used in garages, shops, and unfinished basements) may not have a ground connection or flexible leads. In this case, the cable's grounding conductor connects directly to the metal outlet box. The porcelain-

Dark-colored screw

base fixture has a light-colored and a *dark-colored screw*. The neutral conductor connects to the light-colored screw. This is an important safety feature because the light-colored screw

Screw shell

connects to the *screw shell* (the part the light bulb screws into) of the fixture. When a light bulb is changed, it is relatively easy to make contact with this part of the fixture and/or light bulb. You would not want the screw shell to be hot if you accidentally touched it when your body was also touching a grounded surface! It is easier to install a porcelain fixture

Mounting strap

if you connect stranded, flexible leads to the fixture rather than connecting the cable's solid conductors directly to the fixture. Also notice that some yellow-colored wire connectors are used in Fig. 16-11. Different sizes of wire connectors, to accommodate different gages and number of conductors, have different-colored insulating covers.

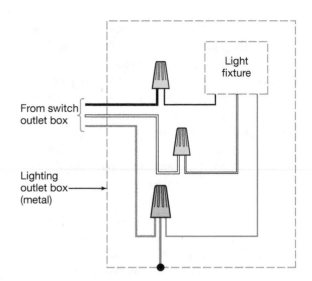

Fig. 16-11 Lighting outlet box. The light fixture has small, stranded conductors that are quite flexible.

Plastic insulated wire connectors.

Because of the physical locations of a light fixture and a wall switch, it is sometimes advantageous to have the light fixture in an outlet box with feed-through cables and the switch in the "dead-end" box. This arrangement is shown in Fig. 16-12, where plastic outlet boxes are used. If metal outlet boxes were used, the light fixture's grounding wire could be connected to the outlet box. This would be done by connecting the grounding wire to the fixture's *mounting strap*, which is screwed to the metal box. Figure 16-12 illustrates another case when a white conductor has to be used as a hot conductor. It is important that the white conductor is used in the continuously hot leg of the switching circuit instead of the switched (return) leg. If a white wire were used in the switched leg, the two white wires would be connected to the light fixture. Then it could be difficult to determine which is a neutral conductor and which is a switched hot conductor. Remember, if properly wired, a white conductor connected to either a switch or a black (or red) conductor is a hot conductor and must be reidentified with black tape. As shown in Fig. 16-12 (and Fig. 16-10), the switching device has a ground connection as required by the NEC. Snap switches used in residential wiring before the late 1990s were made without a ground screw on the mounting strap. In this case, the cable's equipment grounding conductor in a switching circuit like the one shown in Fig. 16-12 was connected

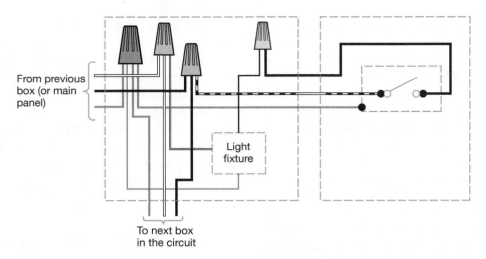

From previous box (or main panel)

Light fixture

To next box in the circuit

Fig. 16-12 Lighting circuit in plastic outlet boxes. A white wire is a hot conductor in this circuit.

directly to the box if the box was metal. If the box was plastic, the grounding conductor was not connected to anything.

The circuits in Figs. 16-9 through 16-12 may look a little confusing because of grounding wires and wire connectors in the boxes. The essence of all these circuits connected in a single branch circuit is shown in schematic form in Fig. 16-13. This figure shows that a branch circuit is just a parallel circuit with some of the branch loads controlled by switches.

Again, notice in Fig. 16-13 that the neutral (white or light gray) conductor is never grounded at any point except at the main distribution panel. Neutral conductors are insulated and meant to carry load current. In a 120-V circuit they carry the same current as the hot (black) conductor

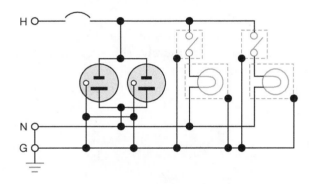

Fig. 16-13 A branch circuit in schematic form. Most branch circuits have more branches than shown here.

does. In contrast, the ground conductor (bare or green) should never carry any current unless a ground fault occurs in the system.

Answer the following questions.

20. In a typical 120-V branch circuit, an outlet box contains a(n) _____ or a(n) _____, or a(n) _____.

21. The shorter of the two slots on a receptacle is connected to the _____ conductor.

22. The neutral side of a receptacle can be identified by the _____-colored screw.

23. The ground screw on a receptacle is a(n) _____-colored screw.

24. True or false. A jumper bar connects the neutral screw pads together on a duplex receptacle.

25. When can an equipment-grounding conductor terminate on a receptacle grounding screw?

26. The screw shell of a lamp fixture should be connected to the _____ conductor.

27. True or false. A white conductor connected to a snap switch should be a continuously hot conductor.

16-7 Switched-Receptacle Outlet

The two receptacles of a duplex receptacle are connected electrically by jumper bars on both the hot and the neutral side of the duplex receptacle. When the jumper on the hot side is removed, as in Fig. 16-14(*b*), the two receptacles can be controlled independently.

Since the neutral side of each receptacle must connect to the neutral conductor, the jumper bar on the neutral side is not removed. As illustrated in Fig. 16-15, using a wire connector on the neutral ensures a continuous neutral conductor. Notice in Fig. 16-15

that the top half of the receptacle is hot all the time.

Figure 16-15 shows the connections required to control the lower half of the duplex receptacle with a wall switch. Again, notice that the white conductor in the cable to the switch outlet is the hot conductor and the black conductor is the return leg. *Switched receptacles* are a convenient way to control a table lamp. They are especially effective when a dimmer switch is used.

16-8 Multiple Switching

Sometimes it is desirable to control a lighting circuit from two or more locations. *Three-way switches* are used when control from two locations is required. When control from three or more places is desired, a combination of three-way and *four-way switches* is used.

Three-Way Switching

Figure 16-16 shows one layout for controlling a light fixture from two locations using two three-way switches. Notice that a three-way switch is the same as a single-pole, double-throw (SPDT) switch. To correctly connect a three-way switch, you must identify the *pole screw* on the switch. The dark (blackish) screw is the pole connection. The lighter-colored (brass to silver) screws are the *throw connections*.

Switched receptacles

Three-way switches

Four-way switches

Pole screw

Throw connections

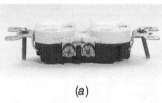

(a)

(b)

Fig. 16-14 Hot side of a duplex receptacle. Twisting off the jumper bar provides two single receptacles. (*a*) Standard configuration. (*b*) Jumper bar removed.

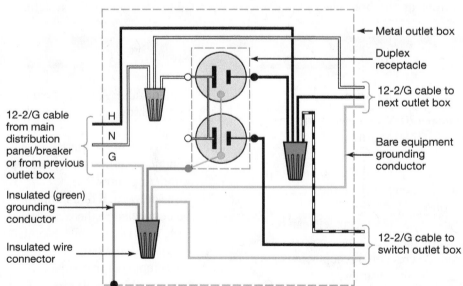

Fig. 16-15 Switched duplex receptacle. The lower half of the receptacle is controlled by a wall switch.

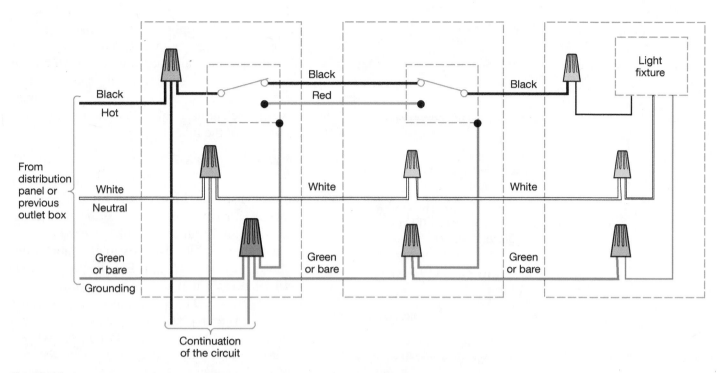

Fig. 16-16 Controlling a light from two locations using two three-way switches.

Notice in Fig. 16-16 that a cable with three insulated conductors and a grounding conductor is required between the two three-way switches. In this figure, no white conductor is used as a hot conductor because the light outlet box is the last box in the circuit. Also notice in Fig. 16-16 that the outlet boxes are not grounded—thus, they must be plastic boxes.

Because of the physical layout of the outlet boxes, it is sometimes advantageous to have a three-way switch terminate the lighting circuit. This arrangement is shown in Fig. 16-17.

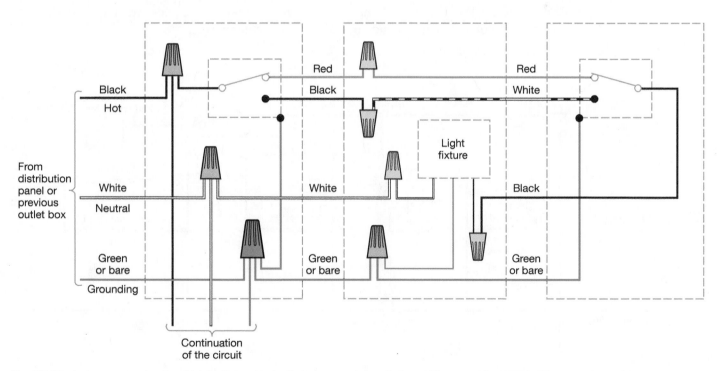

Fig. 16-17 A three-way circuit with the light physically between the switches. When a white wire is connected to a black wire, or a red wire, or a switch it is not a neutral wire.

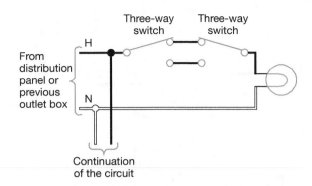

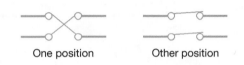

Four-way switch

Fig. 16-18 Schematic diagram of a three-way lighting circuit. The grounding conductors are not shown.

Pole connections

One position   Other position

Fig. 16-19 A four-way switch. It provides independent control from three locations when used with two three-way switches.

Three-conductor-plus-ground

Three-point control

Here, *three-conductor-plus-ground* cables are required between both switch boxes and the light fixture box. Also notice in this figure that a white conductor is used as a switched hot conductor so that the black conductor can be used for the return leg to the light fixture.

Figures 16-16 and 16-17 illustrate both physical and electrical features of the three-way lighting circuit. This may tend to make the circuit look more complex than it is. The schematic diagram in Fig. 16-18 shows that electrically it is a rather simple circuit. Notice that a three-way switch has no specific on or off position. The position in which one three-way switch turns a light on is determined by the position of the other three-way switch.

Four-Way Switching

A *four-way switch* is different from any other switch discussed in this book. Like a three-way switch, it has no on or off position. Its electrical function is rather obvious when the internal connections of its two possible positions are shown (see Fig. 16-19). Examination of Fig. 16-19 reveals that a four-way switch is like having two SPDT switches that share the same throws. The crucial restriction is that the two poles must always be on different throws. As with the three-way switch, the *pole connections* of the four-way switch are the darkest screws. Also, both pole screws are usually at one end of the oblong-shaped switch.

For *three-point control* of a light, a four-way switch is electrically connected between two three-way switches, as seen in Fig. 16-20. In this figure, the light is on. Visualize changing any one of the three switches to its other position and the light will be off. Again notice that three-conductor cables are required between the switch boxes.

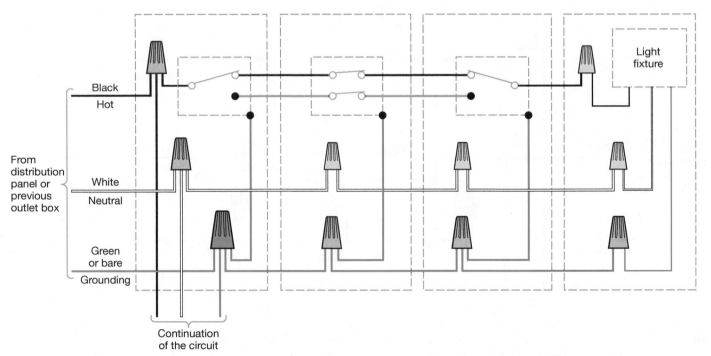

Fig. 16-20 Controlling a light from three locations using two three-way switches and one four-way switch.

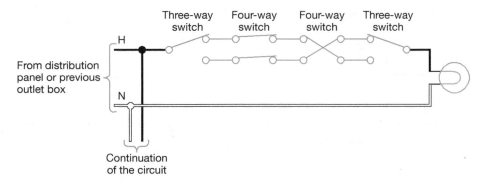

Fig. 16-21 Schematic diagram of a four-way lighting circuit. The grounding conductors are not shown.

The light outlet box does not have to be at the end of the circuit as it is in Fig. 16-20. Refer back to Fig. 16-17, and you can see that the light fixture box could be inserted between the four-way switch box and either of the three-way switch boxes. The neutral conductor would still terminate at the light fixture and a white conductor would have to be used as a hot conductor between the remaining switch box or boxes.

On very rare occasions it may be necessary to control a light from more than three locations. This can be accomplished by inserting additional four-way switches between the two three-way switches. Figure 16-21 shows, in schematic form, four-point control of a light. In this figure, the light is off. Changing the condition of any one of the four switches will turn the light on.

16-9 Lighted-Handle Switches

A lighted-handle snap (on-off) switch is shown in schematic form in Fig. 16-22. In the off position [Fig. 16-22(a)], the neon bulb is glowing (ionized). Although the lamp is off, there is still a current flowing through the lamp, the neon bulb, and the resistor. However, the neon bulb requires less than 1 mA to maintain ionization. This is a minute current compared to the lamp current when the circuit is on. For example, a 40-W lamp draws 333 mA and a 100-W lamp requires 833 mA. The cold resistance of the lamp is so small (less than 100 Ω) that it develops less than 0.1 V when the lamp is off.

Referring to Fig. 16-22(b), one can see that in the on position the resistor and neon bulb are shorted. The circuit functions like a circuit with a standard snap switch.

A three-way circuit using *lighted-handle switches* is shown in Fig. 16-23 in the off mode. Notice that the neon bulb in both switches is ionized. Careful inspection of Fig. 16-23 shows that the resistor and bulb of the S_1 switch are in parallel with the resistor and bulb of the S_2 switch. Figure 16-24 is a rearrangement of Fig.16-23 without the switches. This figure clearly shows that the two-series resistor and bulb combinations of the switches are in parallel and that the lamp current in the off position

Lighted-handle switch

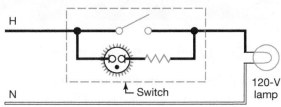

(a) Off position. Very little current is required to light the neon bulb in the switch.

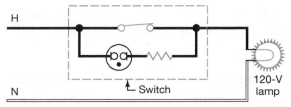

(b) On position. When the lamp is on, the neon lamp and resistor are shorted out.

Fig. 16-22 Lighted-handle switch. The handle is dimly lit when the lamp is off.

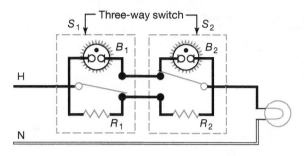

Fig. 16-23 Lighted-handle three-way switch. Both neon lamps are on when the main lamp is off.

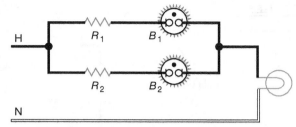

Fig. 16-24 Redrawing Fig. 16-23 without the switches shows more clearly why both neon lamps are ionized.

is equal to the sum of the currents of the two neon bulbs. Again, this is a comparatively small (less than 2 mA) current. Referring back to Fig. 16-23, you can see that toggling either switch to turn on the lamp will bypass both resistors and both neon bulbs.

Self-Test

Answer the following questions.

28. True or false. The neutral shorting bar is usually removed from a duplex receptacle when half of the duplex receptacle is switched.

29. How many insulated conductors are needed in a cable that connects outlet boxes containing three-way switches?

30. True or false. A four-way switching circuit uses three four-way switches for three-location control of a lamp.

31. A(n) _____ switch is a type of SPDT switch.

32. A four-way switch serves the same function as a DPDT switch.

33. A lighted-handle snap switch uses a(n) _____ bulb to illuminate the handle.

34. The handle of a three-way lighted-handle switch is illuminated when the lamp is turned _____.

35. True or false. Neither position of a lighted-handle snap switch will illuminate the handle when the controlled light is burned out.

16-10 Circuit Interrupters

Two main types of circuit interrupters are discussed in this section. One type removes power from a circuit when a current flows from the hot conductor to the grounding conductor. This current is referred to as leakage current. The other type removes power when unexpected, continuous arcing occurs.

Ground-Fault Circuit Interrupter

Ground-fault circuit interrupter

The NEC requires the use of a *ground-fault circuit interrupter* (GFCI) in some locations in a residence. A GFCI like the one in Fig. 16-25 is designed to be installed in an outlet box in a 120-V branch circuit. It disconnects the load plugged into it when the ground-fault current reaches about 6 mA. "Ground-fault current" refers to the current that flows from a shorted load to a grounding conductor or a ground path rather than through the normal load between the hot conductor and the neutral conductor.

Notice in Fig. 16-25 that the GFCI has the same arrangement of neutral, hot, and grounding screws that the standard duplex receptacle has. However, on the back side of the GFCI, one pair of neutral-hot screws is labeled "line" and the other pair is labeled "load." The line pair must be connected to the line (cable) that brings power to the GFCI. The load pair

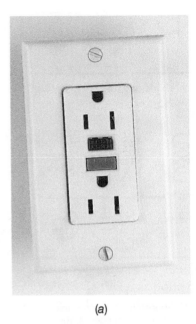

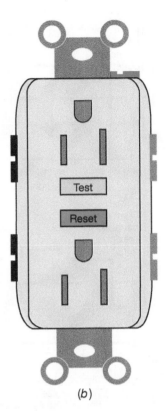

(a)   (b)

Fig. 16-25 A duplex receptacle with GFCI. Unlike a standard duplex receptacle, the incoming power line must go to specific neutral and hot screws.

connects to the cable that feeds other outlet boxes in the branch circuit.

Pressing the Test button in Fig. 16-25 causes a small ground fault current in the outlet. If the GFCI is working properly, the Reset button will pop up a small fraction of an inch from its slightly recessed position and disconnect the receptacle from the power source. To reactivate the GFCI, the Reset button is pushed back down. Manufacturers of GFCIs recommend that these units be tested periodically (about once a month) to be sure that they are working properly. Of course, if a real ground fault occurs in the branch circuit, it must be dealt with before the GFCI can be reset.

Figure 16-26 illustrates how a GFCI operates. Notice that the hot and neutral conductors run through two toroid cores. Further, notice that the currents in the two conductors are always flowing in opposite directions. Therefore, no flux is created in the toroid cores when the currents in the two conductors are of equal magnitude. However, when the currents are unequal, the larger current dominates and causes some flux in the toroid cores. This flux induces a small voltage in the multiturn

secondary of the right toroid. This secondary voltage is amplified and applied to the relay coil. Since the level of the secondary voltage of the current transformer is a function of the level of the *unbalanced current* in the primary conductors, some critical level of unbalanced current will provide enough voltage to energize the relay and disconnect the circuit. Notice that the GFCI receptacle protects all other outlets that it feeds, but it does not protect anything in the line that feeds it.

The GFCI doesn't respond to overload currents (that is the job of the circuit breaker or the fuse protecting the branch circuits). The GFCI responds only to an imbalance between the hot conductor current and the neutral conductor current in a 120-V circuit or between the two hot conductor currents in a 240-V circuit.

Any time the GFCI is connected to a power source, the coil on the left toroid is energized. This coil is the primary coil of the toroid-core transformer and it induces a small voltage in both the hot and neutral conductors, which act as single-turn secondaries of the transformer. The voltages induced in the two conductors cancel each other because at any given instant

Unbalanced current

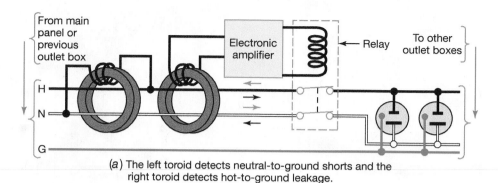

(a) The left toroid detects neutral-to-ground shorts and the right toroid detects hot-to-ground leakage.

(b) The energized coil on the left toroid includes equal, opposing voltages in the hot and the neutral conductors. When a neutral-to-ground short occurs, the neutral-conductor current exceeds the hot-conductor current.

Fig. 16-26 Principle of a GFCI. The relay energizes when the imbalance between the hot and neutral conductor currents reaches about 6 mA.

Ground current

one voltage aids the supply voltage and the other one opposes the supply voltage.

As shown in Fig. 16-26(b), any time a load-side short occurs between the neutral conductor and the grounding conductor, the induced voltage in the neutral conductor causes a current to flow in the neutral and grounding conductors. (Remember the neutral conductor and grounding conductor are bonded together at the main panel so there is a complete low-resistance circuit.) If there is no load connected to the GFCI, there will be no current in the hot conductor; so the imbalance in currents equals the neutral current and the relay opens the GFCI. If there is a load on the GFCI when this short occurs, the GFCI will open because the neutral-conductor current will equal the hot-conductor current plus the ground-conductor current.

Figure 16-27 illustrates a situation that causes an imbalance between hot and neutral currents. In this circuit, 0.05 A is shunted by the short through the frame and the ground conductor. The rest (5.95 A) of the 6 A of hot conductor current continues through the lower part of the motor winding and the neutral conductor. The

5.95 A of neutral current joins the 0.05 A of *ground current* at the main distribution panel where the neutral wire is grounded. The 0.05 A of imbalance (6 A − 5.95 A) between the neutral and hot currents would trip a GFCI and disconnect the motor and all other loads connected to other outlets protected by the GFCI.

Notice in Fig. 16-27 that the motor would continue to run well if it were not connected to a GFCI. It would not shock anyone touching the frame because the frame is grounded. However, it is very likely that the short would

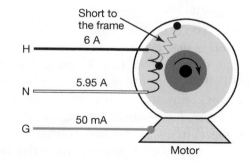

Fig. 16-27 A high *R* short to the grounded frame causes an imbalance in the hot and neutral currents.

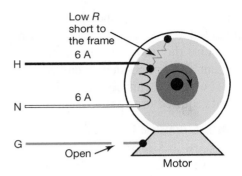

Fig. 16-28 A short to the ungrounded frame causes a severe shock hazard unless there is a GFCI.

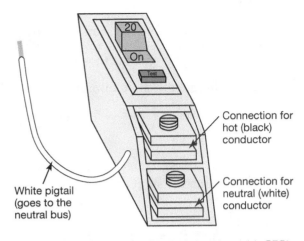

Fig. 16-29 A 120-V, 20-A circuit breaker with GFCI protection. Both the neutral conductor and the hot conductor connect to the breaker.

worsen (decrease its resistance) and eventually increase the current through the top part of the winding to a level that would ruin the winding and trip the branch-circuit circuit breaker.

The GFCI can prevent an individual from receiving a severe or fatal shock from a faulty device that is not properly grounded. Figure 16-28 shows the same circuit as in Fig. 16-27 except that the ground conductor is open and the short is a low-resistance (dead) short. As long as the motor is mounted on an ungrounded surface, it operates normally and the hot and neutral currents are equal. But what happens when an individual simultaneously touches the motor frame and a grounded surface? Then a current flows through the individual's body, the grounded surface, and the ground path back to the grounded main distribution system. Without a GFCI, the resistance of the human body determines the magnitude of this current. With a GFCI, the circuit would open when the current reached no more than 6 mA.

Ground-fault-circuit interrupters are available in configurations other than the 120-V duplex receptacles we have discussed so far. Examples of other configurations are:

1. GFCI plug for an extension cord
2. Combination single GFCI receptacle outlet and switch
3. GFCI controls for 240-V circuits
4. 15-A and 20-A circuit breakers with GFCI protection

A circuit breaker with GFCI protection is illustrated in Fig. 16-29. Notice that the neutral conductor of the branch circuit cable also connects to the circuit breaker. This is necessary for the GFCI part of the breaker to respond to unbalanced currents between the hot and neutral

conductors. The pigtail conductor coming out of the breaker connects the neutral conductor to the neutral bus after the neutral current has gone through the GFCI detection circuitry. The advantage of the GFCI circuit breaker is that it protects everything in the branch circuit including the cable that feeds the first outlet box. Its disadvantages are the extra space it requires in the distribution panel and its cost.

EXAMPLE 16-1

A 500-W resistive device, PF = 1, is operating on a 120-V branch circuit. The device develops a 40-Ω short between the hot conductor and the device's conductive surface which is connected to the grounding conductor. Determine the current in each of the three branch-circuit conductors before and after the short develops.

Given: P_{load} = 500 W
                R_{short} = 40 Ω
                V_{load} = 120 V

Find: I_H before and after the short develops

              I_N before and after the short develops

              I_G before and after the short develops

Known: $I_{load} = \dfrac{P_{load}}{V_{load}}$ when PF = 1

              $I_{load} = I_H = I_N$ before the short develops

$$I_G = 0 \text{ before the short develops}$$

$$I_G = \frac{V_{load}}{R_{short}} \text{ after the short develops}$$

Solution: Before the short develops:

$$I_H = I_N = I_{load} = \frac{P_{load}}{V_{load}} = \frac{500 \text{ W}}{120 \text{ V}}$$

$$= 4.17 \text{ A}$$

$$I_G = 0$$

After the short develops:

$$I_G = \frac{120 \text{ V}}{40 \text{ }\Omega} = 3 \text{ A}$$

$$I_H = I_{load} + I_G = 4.17 \text{ A} + 3 \text{ A}$$

$$= 7.17 \text{ A}$$

$$I_N = I_{load} = 4.17 \text{ A}$$

Answer: Before the short develops:

$$I_H = 4.17 \text{ A}, I_N = 4.17 \text{ A}, I_G = 0$$

After the short develops:

$$I_H = 7.17 \text{ A}, I_N = 4.17 \text{ A}, I_G = 3 \text{ A}$$

Arc-fault circuit interrupter

Notice in example 16-1 that the neutral-conductor current did not change when the short occurred. This is because the short occurred in the wiring that connected directly to the hot conductor. If this branch circuit was powered through a circuit breaker with GFCI protection, the breaker would have tripped long before the hot-conductor current increased to 7.17 A.

Appliance Leakage Current Interrupter (ALCI)

Like the GFCI, the ALCI protects the appliance user from electric shock when a leakage path develops from the hot conductor to ground. When the leakage current reaches 6 mA, power is removed from the appliance. Unlike the GFCI, the ALCI does not detect a short between the neutral conductor and the grounding conductor. Thus, the ALCI circuit would be like that shown in Fig. 16-26(a) except the left toroid transformer is not needed.

The ALCI device is usually built into the appliance cord plug. It is used for indoor appliances.

Equipment Leakage Current Interrupter (ELCI)

The ELCI is used to protect electric equipment from excess leakage current that could ultimately destroy the equipment. Although it

operates on the same principle as ALCI, it is not intended to provide personnel shock protection because the imbalance in neutral-conductor current and hot-conductor current required to trip the device can be many times greater than that required to trip the ALCI.

Leakage Current Detection Interrupter (LCDI)

The LCDI is a special cord set (cord and plug) designed to prevent a damaged cord from starting a fire. Many portable room air conditioners have a factory-installed LCDI cord set. The electrical-electronic interrupter part of the LCDI is built into the plug of the cord set. The cord has a metal shield (mesh or braid) around each of the insulated conductors in the cord. When a small (typically 2.5 to 3 mA) leakage current develops between any conductor and its metal shield, the LCDI trips and removes power from the equipment it is protecting.

Arc-Fault Circuit Interrupter (AFCI)

The AFCI uses sophisticated integrated electronic circuits to detect certain types of abnormalities in the shape of the 60-Hz sinusoidal waveform. When these abnormalities are detected, the AFCI is tripped and the protected circuit is disconnected from the power source. The use of AFCIs is predicted to greatly reduce the number of fires started by wiring systems.

AFCI protection in branch circuits can be provided by any one of three types of UL-listed AFCI devices. The three types (and their major features) are:

1. Branch/feeder AFCI
   This type of AFCI is part of an arc-fault-circuit-interrupter circuit breaker. The AFCI circuit breaker looks just like the GFCI circuit breaker shown in Fig. 16-29. The major difference in their physical appearance is the color of the test button, which is often a blue color for the AFCI. These breakers have been in use since about 1999. They are designed to detect arcing anywhere in the branch circuit. They can detect arcing between the conductors in a frayed extension cord, arcing between a hot conductor and ground, or arcing between the hot and the neutral conductors in the cables. Any

of these *types of arcing* are capable of starting a fire in surrounding combustible materials.

2. Combination AFCI
   The combination AFCI provides protection against all of the types of arcing listed for the branch/feeder AFCI plus it will protect against arcing caused by loose connections. This type of AFCI is usually provided as part of a duplex receptacle. It protects the receptacle it is part of and all of the rest of the branch circuit fed from it. However, it does not protect any of the branch circuit that feeds it.

3. Outlet circuit AFCI
   The outlet circuit AFCI protects only the duplex receptacle it is part of and the devices plugged into that receptacle. Parts of the branch circuit feeding the receptacle or being fed by the receptacle are not protected. This type of AFCI provides protection for the same type of arcs listed for the combination AFCI.

Since 2002, the NEC has required AFCI protection on all branch circuits providing power to bedrooms and sleeping areas in new construction. In 2008 NEC expanded the AFCI protection requirement to include a combination-type AFCI on branch circuits providing power to living and dinning rooms, closets, hallways, sunrooms, recreation/game rooms, family rooms, and libraries in addition to bedrooms and sleeping areas. Of course, AFCI protection would be a good addition to the electrical system of previously constructed dwellings.

There are several options when both AFCI and GFCI are desired on the same branch circuit. One can use an AFCI breaker and install a GFCI duplex receptacle in the first outlet box of the branch circuit. Or one could do the reverse and use a GFCI breaker and an AFCI duplex receptacle.

A duplex receptacle that incorporates both AFCI and GFCI is now being UL-tested for UL listing. Also, some manufacturers are doing research to develop, and get certification of, a circuit breaker that contains both GFCI and AFCI protection. Thus, one could have overload, GFCI, and AFCI protection in a single package.

16-11 Feeder Circuits

A circuit (cable or conductors) that feeds power from the main distribution panel to a subpanel containing circuit breakers for branch circuits is called a *feeder circuit*. An example of a feeder circuit is shown in Fig. 16-30. A number of important features should be noted about this figure:

1. The feeder cable connects to a circuit breaker of appropriate size at the main (service) panel. Thus, power to the subpanel can be controlled at the main distribution panel.

2. At the subpanel, the feeder cable connects to a circuit breaker with a current rating appropriate for the wire gage of the cable conductors. This breaker provides power to the bus bars to which branch-circuit breakers are connected. It also serves as the disconnect for the subpanel; thus, power to all branch circuits served by the subpanel can be disconnected by throwing one circuit-breaker handle at either the main panel or the subpanel.

3. In the subpanel there is a separate bus for neutral conductors and equipment-grounding conductors. Neutral conductors can be grounded only at the main (service) panel. Therefore, the neutral bus is insulated from the metal box of the subpanel. Since the subpanel box must be grounded, the equipment-grounding bus is electrically bonded to the box.

4. Branch circuits out of the subpanel are wired in the same way as those in the main panel except for the separation of the neutral and grounding conductors.

A feeder circuit such as the one in Fig. 16-30 could be used to provide electrical service to a detached garage. The feeder cable could be buried between the house and the garage. Of course, the cable would have to be a type approved for direct burial such as underground feeder (UF) cable. Depth of burial, between 6 and 24 inches, depends on the type of protection provided (if any) and the surface coverage.

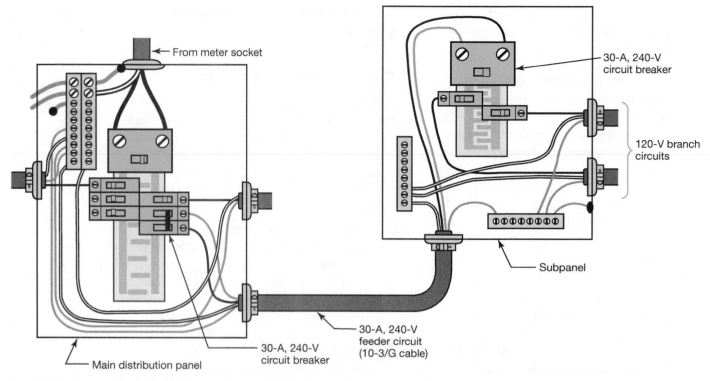

From meter socket

30-A, 240-V
circuit breaker

120-V branch
circuits

Subpanel

30-A, 240-V
feeder circuit
(10-3/G cable)

30-A, 240-V
circuit breaker

Main distribution panel

Fig. 16-30 Feeder circuit connecting a subpanel to the main panel.

Self-Test

Answer the following questions.

36. True or false. A GFCI can protect only 120-V circuits.
37. Excessive _____ current causes a GFCI to trip.
38. True or false. An AFCI circuit breaker can detect arcs in any of the loads in a branch circuit.
39. The current transformer in a GFCI uses a(n) _____ core.
40. What is a feeder circuit?
41. True or false. A subpanel contains a master circuit breaker that removes power from all its branch circuit breakers.
42. Both ends of the hot conductors of a cable used for a(n) _____ circuit terminate on circuit breakers.
43. The _____ bus in a subpanel is insulated from the panel box.
44. List three types of AFCI devices.
45. Which type of AFCI does not protect against loose connection arcing?
46. True or false. A branch circuit cannot be protected with both GFCI and AFCI.

Mechanical latching

Low-voltage control

Pulse of current

16-12 Low-Voltage Control Circuits

Low-voltage control of a lighting and/or power circuit utilizes a relay. The relay has a low-voltage (typically 20 to 28 V), low-current coil, and electrical contacts capable of switching the line voltage and current required by the load. The relay has some form of *mechanical latching* so that only a *pulse of current* is required to open or close the relay. The cutaway illustration of a relay in Fig. 16-31 is shown with its contacts closed. It will remain

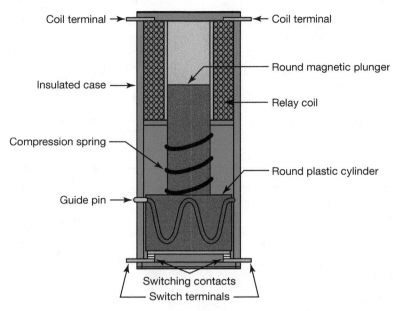

Coil terminal →

← Coil terminal

Insulated case →

Round magnetic plunger

Relay coil

Compression spring →

Round plastic cylinder

Guide pin →

Switching contacts

Switch terminals

Fig. 16-31 Low-voltage control relay with mechanical latch. Every other current pulse closes the contacts.

in that condition until another current pulse is received.

When the relay coil in Fig. 16-31 receives a current, it pulls the plunger and the top contacts up and tries to center the plunger in the coil. This opens the contacts. This action also rotates the top contacts and the plastic cylinder about 45° because of the wavy groove on the circumference of the cylinder and the stationary guide pin engaged in the groove. When the coil current stops, the compression spring forces the plunger and cylinder back to the down position. The cylinder rotates another 45° as it is forced down. Now the switch contacts are open because the top contacts and their connecting bar are 90° from (perpendicular to) the bottom contacts and their terminals. The next current pulse will cause the cylinder and the attached contacts to rotate another 90° as they are forced up and down. This, of course, aligns the upper and lower contacts and closes the relay switch.

Figure 16-32 shows how a relay such as the one in Fig. 16-31 is used in a low-voltage controlled lighting system. The relay, along with other relays, is mounted in a relay panel located close to the distribution panel. One branch circuit leading from the distribution panel to the relay panel provides power for many lights. No matter how complex the switching arrangement, the only additional power cable required is a two-conductor

ABOUT ELECTRONICS

Electric power is provided from a group of interconnected regional plants. This makes it necessary that the 60-cycle frequency of all the generators be synchronized. Your power may arrive at your home via a loop hundreds of miles long rather than by the shortest distance.

(with ground) cable from the relay panel to the light outlet box. All the cable used for switching is low-voltage, light-gage (typically No. 22 to No. 18) cable. The switch is a normally open, momentary-contact, push-button type. The *low-voltage supply* is a current-controlled type that reduces the relay coil current to a safe value if the push-button switch is held on too long or sticks in the on position.

Low-voltage supply

Multipoint control of a low-voltage controlled light is illustrated in Fig. 16-33. Notice that this figure is a repeat of Fig. 16-32 with the addition of some switches and two-conductor cable. Since *low-voltage cable* and switches are relatively cheap, switches may be located in areas quite remote from the controlled light.

Low-voltage cable

With low-voltage controlled systems, it is practical to have a master control panel, with a switch for each relay, located in the master bedroom or other convenient location. This makes it easy to turn off lights in unoccupied

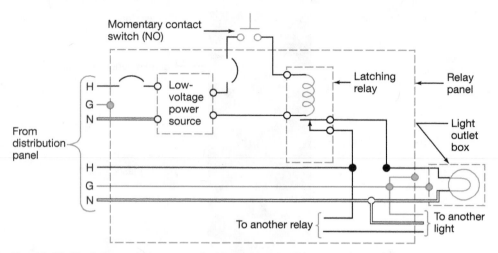

Fig. 16-32 Basic low-voltage controlled lamp circuit. The control circuitry uses a two-conductor low-voltage low-current cable.

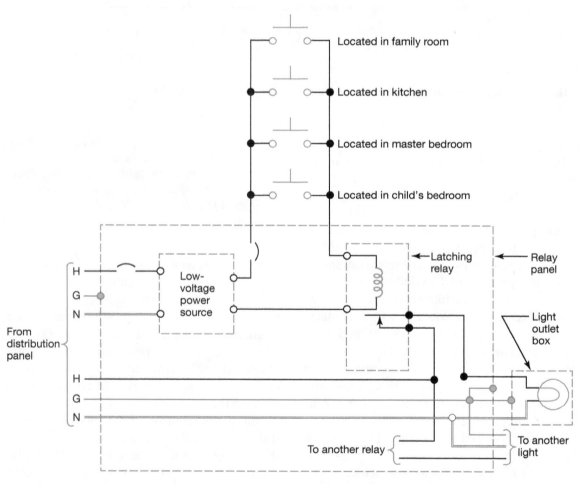

Fig. 16-33 Multipoint control of a low-voltage controlled lamp. Two-conductor, low-voltage cable is used between the switches.

rooms. Of course, there can be duplicate master control panels because any number of switches can be wired in parallel.

Remote switching of a low-voltage controlled light is more convenient when *lighted-button*

switches, which indicate when a light is on, are used. Figure 16-34 shows how indicator (pilot) lights are added to the low-voltage control circuit. For this circuit, the control relay has another set of contacts. Since the additional

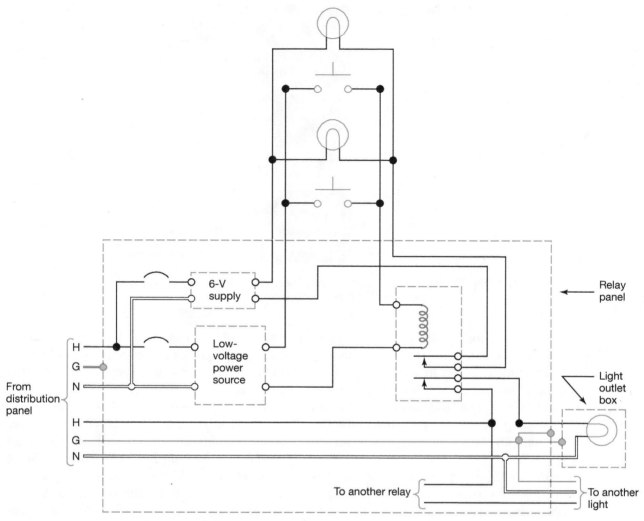

Fig. 16-34 Adding an indicator light to the control switches requires four-conductor, low-voltage cable in the control circuit.

contacts switch low currents and voltage, they can be much lighter-duty contacts than those used to switch the load (120-V lamp). The pilot (indicator) lights can be either incandescent lamps or light-emitting diodes (LEDs). The LED light requires less current, but it requires a direct current supply. With lighted switches, a four-conductor cable is required between switches. Also notice that the pilot light is on when the 120-V lamp is on. This is just the opposite of the lighted-handle switches.

Another type of *latching relay* (dual-coil) for low-voltage control is illustrated in Fig. 16-35. It is shown in the on state, with the spring holding the plunger down and the contacts closed. The upper contact is a spring contact, and when given the chance, it springs up and opens the contacts. When the upper coil receives a current pulse, the plunger is pulled

up and held up by the *compression spring*. This allows the upper contact to spring up to the off state. Operation of this relay requires a three-conductor, low-voltage cable and an SPDT *momentary-contact rocker switch*. Figure 16-36 shows the basic circuit used to control a dual-coil relay and its load.

Multipoint control of the circuit in Fig. 16-36 can be obtained by adding rocker switches in parallel using three-conductor cable. With this relay, it is not necessary to know the present state (on or off) of the light to control the light from a remote location because this relay requires dedicated pulses for on and off action.

Another feature of the dual-coil-relay system is the simplicity of the master control unit. As shown in Fig. 16-37, the unit consists of two push-button switches and one two-pole rotary switch. The rotary switch

Compression spring

Momentary-contact rocker switch

Multipoint control

Latching relay

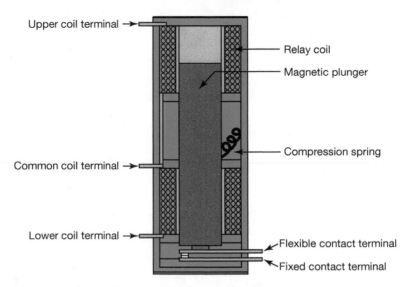

Fig. 16-35 Dual-coil latching relay. The upper coil must be pulsed to open the contacts.

Heating element

Bimetal strip

Thermostat control

can have many more positions than shown in Fig. 16-37. To operate the master unit, merely rotate the rotary switch to select the desired light and press the on or off button. To quickly turn all lights on, just hold the on switch down while rotating the rotary switch through all of its positions. Of course, all lights can be turned off just as quickly.

Figure 16-38 shows another type of low-voltage control circuit. This circuit is used to provide remote *thermostat control* of one or

more baseboard electric heaters. Notice that the transformer primary is energized all the time, so the circuit uses a little energy even when the heater is off.

When the thermostat detects the need for heat, its detection device closes a set of contacts. This allows a small (milliampere) current to flow through the small *heating element* in the 18-V secondary circuit. The heating element heats the *bimetal strip* and causes it to bend down and close a set of

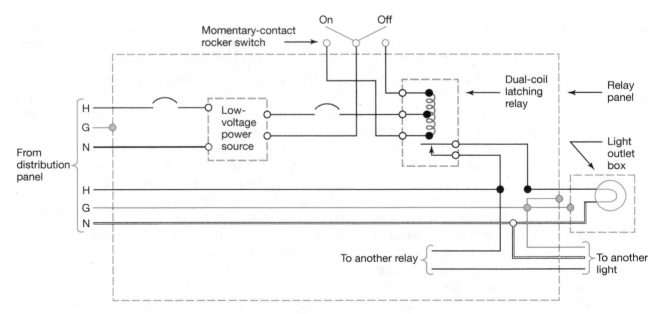

Fig. 16-36 Low-voltage controlled lamp circuit using a dual-coil relay. The control circuitry uses a three-conductor cable.

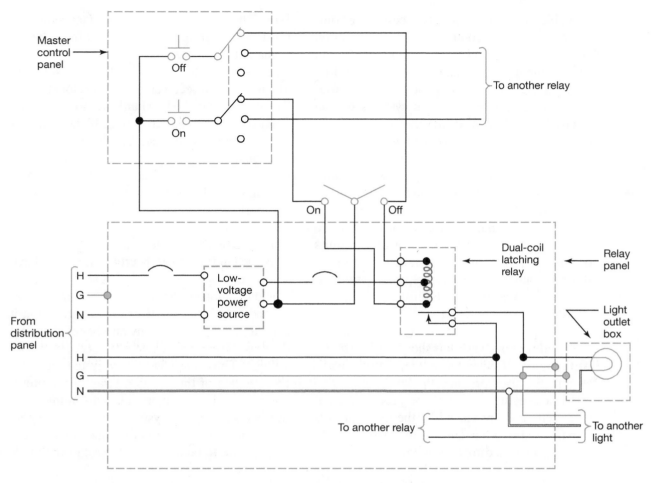

Fig. 16-37 Master control station for a dual-coil relay circuit. The master control uses two NO push-button switches and a 2P rotary switch.

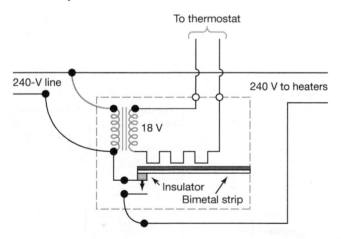

Fig. 16-38 Electric heater control circuit. Note that the required grounding conductors are not shown.

heavy-duty contacts that connect the 240-V main line to the 240-V heater load. After the room warms up, the thermostat contacts open, the bimetal strip cools and straightens out, and the 240-V line is disconnected from the heater.

16-13 Electronic Control Circuits

The systems discussed in the previous section are hard-wired systems using only switches and relays. More sophisticated systems are available. They use a microprocessor or computer interface to control a *dimming function* as well as a *switching function*. They can handle inputs from many sources including radio-frequency signals, photocells, and motion detectors. Many of these systems use an electronic equivalent of the mechanical latching relay. Some electronic systems do not require low-voltage control cables. Instead of control cables, *coded high-frequency radio signals* are sent to electronic control units. Each unit can respond to its unique code, to a universal code that all units respond to, or to a group code to which a select group of units responds. Some units are designed to fit into a standard outlet box and can replace a standard switch when a previously installed conventional system is upgraded.

A simplified block diagram illustrating an *electronic control system* is shown in Fig. 16-39. At the remote (master) unit, the controls available to the user produce the code needed to select a specific local unit or group of local units. The code produced by the user controls also instructs the local unit as to whether it is to turn off, turn on, dim, or brighten the controlled light. The *RF signal* is *modulated* (encoded) by these codes before it is transmitted from the antenna of the remote unit.

The modulated RF signal radiates to the antenna of the receivers in all the local units. The *decoders* in the local units remove the code from the modulated RF signal. If the first part of the received code matches the code settings of a local unit, the unit responds to the remainder of the code. If no match occurs, the local unit ignores the instructions contained in the last part of the code. The user controls on the local unit allow the user to override the instructions sent by the remote unit.

Notice in Fig. 16-39 that the *triac* in the local control unit operates the light. A triac is an electronic component that can switch a light off or on and hold it in the off or on condition. Or, it can turn the light on for any number of electrical degrees during each cycle of the 60-Hz supply voltage. Thus, the triac not only can perform the function of the latching relay of the previous section, but it can provide a dimming function as well. Some systems have memory so that favorite settings and the time at which the settings are to be used can be programmed.

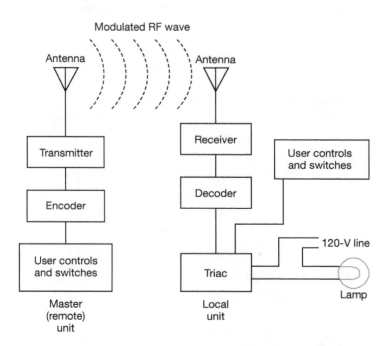

Fig. 16-39 Simplified block diagram of a radio-frequency (RF) controlled lamp.

Answer the following questions.

47. A mechanical _____ relay is used in low-voltage control circuits.
48. What is the major difference between the operation of the two-conductor relay and the three-conductor relay?
49. What types of switches are used with a two-conductor relay?
50. When is a lighted-button switch illuminated in a low-voltage control system?

51. In a low-voltage control circuit, all local and remote switches are wired in _____.
52. Does the heater control circuit use any energy when the heater is off?
53. Electronic control systems can remotely control both _____ and _____ of a light without any low-voltage cables.

Summary

1. Although the National Electrical Code is applicable to all states, most locales have supplemental requirements.
2. It is more efficient to transfer electric energy at a high voltage than at a low voltage.
3. The neutral conductor is a grounded conductor. It is grounded at the service entrance.
4. Although the neutral is grounded, it is not a ground wire.
5. Under normal operating conditions, the grounding conductor and the equipment-grounding conductors carry no current.
6. The neutral conductor carries current except in a 120-V/240-V circuit in which the currents in the hot conductors are equal.
7. A residential electrical system includes at least a service drop, a service entrance, and numerous branch circuits. Some systems include one or more feeder circuits.
8. In a branch circuit, the neutral conductor has white or natural gray insulation.
9. A conductor with white or natural gray insulation is not a neutral conductor if it is connected to a switch, a circuit breaker, or another conductor that has neither white nor natural gray insulation.
10. A straight 240-V branch circuit does not require a neutral conductor. In this circuit, a properly marked white conductor can be used as a hot conductor.
11. Neutral and equipment-grounding conductors are continuous-path conductors. They are never switched or fused.
12. The light-colored screw on a receptacle or a lamp fixture is connected to the neutral conductor.
13. The screw shell of a lamp fixture should be connected to the neutral conductor.
14. One-half of a duplex receptacle can be switched when the jumper bar on the hot side is removed.
15. Three-way switches are SPDT switches. The pole connection is the dark-colored screw.
16. Four-way switches have two poles that share two throw connections.
17. A lighted-handle switch is illuminated when the circuit is off.
18. A GFCI removes power from a circuit when (or before) the ground fault current reaches 5 mA.
19. A GFCI receptacle responds to the difference between the hot conductor and the neutral conductor currents.
20. An AFCI responds to arcing in the neutral or hot conductor connections or to arcing between the hot conductor and the neutral conductor or any ground.
21. Feeder circuits are used to power subpanels.
22. The neutral bus bar in a subpanel is insulated from ground.
23. A low-voltage control lighting system uses light-duty, low-voltage cable for all switching functions. This system provides for remote switching and master switching panels.
24. Electronic light switching systems are more versatile than hard-wired, low-voltage systems. They perform dimming as well as switching functions.

Chapter Review Questions

For questions 16-1 to 16-15, determine whether each statement is true or false.

16-1. A wiring system that is in complete compliance with the NEC is acceptable in any state or locale. (16-1)

16-2. The neutral conductor provided by the power company is a bare conductor. (16-2)

16-3. The main circuit breaker can serve as the service disconnect. (16-3)

16-4. The removal of the energy meter from the meter box disconnects only one hot conductor from the main circuit breaker. (16-3)

16-5. The ground bus and the neutral bus are usually insulated from each other in the main distribution panel. (16-4)

16-6. A 120-V, 20-A branch circuit usually uses 14-2/G cable. (16-4)

16-7. The equipment-grounding conductor in type NM cable is usually covered with green insulation. (16-4)

16-8. A 240-V circuit breaker often consists of two 120-V circuit breakers with their bodies and their trip-reset levers bonded together. (16-5)

16-9. Excess current in only one hot conductor in a 240-V branch circuit causes the circuit breaker to disconnect both hot conductors. (16-5)

16-10. All light fixtures have a ground connection. (16-6)

16-11. Usually both shorting bars are removed from a duplex receptacle when half of the receptacle is switched. (16-7)

16-12. A three-way switch has no distinct on or off positions. (16-8)

16-13. A lighted-handle three-way switch uses a low-voltage bulb or an LED to illuminate the handle. (16-9)

16-14. A three-conductor relay toggles every time it receives a current pulse. (16-12)

16-15. Electronic control systems can provide remote dimming and switching of lights without the use of low-voltage control cables. (16-13)

For questions 16-16 to 16-24, supply the missing word or phrase in each statement.

16-16. In a 12-2/G cable, the _____ conductor has white insulation. (16-4)

16-17. The conductors that connect a power company's transformer to the service entrance are called the _____ _____. (16-2)

16-18. In addition to the conduit and fittings, an overhead service entrance consists of the _____, _____, and _____. (16-3)

16-19. The conductor that connects the main distribution panel to an electrode in the earth is called a(n) _____. (16-3)

16-20. The insulation colors of the hot conductors in a 12-3/G cable in a 240-V branch circuit are _____ and _____. (16-4)

16-21. When a white-colored conductor is used in a switch circuit, the _____ colored conductor should be the continuously hot conductor. (16-6)

16-22. A four-way lighting circuit requires the use of _____ three-way switch(es) and _____ four-way switch(es). (16-8)

16-23. The cable connecting a three-way switch and a four-way switch will have _____ insulated conductors and a bare conductor. (16-8)

16-24. Subpanels are powered by a(n) _____ circuit. (16-11)

Answer the following questions.

16-25. In what form is electric energy transferred from the point of generation to the area of distribution? (16-2)

16-26. Would a 14-2/G or a 14-3/G cable be used between two outlets containing three-way switches? (16-8)

16-27. What is the inferred reference point when a conductor is called a "neutral conductor"? (16-2)

16-28. Explain how a conductor can be 120-V hot and 240-V hot at the same time. (16-2)

16-29. What colors of insulation are on the conductors in a 12-3/G cable? (16-4)

16-30. Two 12-2/G cables and a duplex receptacle outlet are in an outlet box. How many wire connectors are needed? (16-6)

16-31. How does an electrician identify the hot side of a receptacle? (16-6)

16-32. When is a white conductor connected to a snap switch? (16-6)

16-33. What causes a GFCI to disconnect a circuit? (16-10)

16-34. What causes an AFCI to disconnect a circuit? (16-10)

16-35. What type of relay is used in a low-voltage control circuit? (16-12)

16-36. Why is a wire connector used in an outlet box that accommodates two 12-2/G cables and a duplex receptacle? (16-6)

16-37. Where is the neutral conductor of a branch circuit connected when the circuit is protected by a GFCI breaker? (16-10)

16-38. Where is the pigtail on a GFCI breaker connected in the main distribution panel? (16-10)

Critical Thinking Questions

16-1. A 200-W, 120-V lighting circuit is controlled by two lighted-handle, three-way switches. Assume that the neon lamp in a switch draws 1 mA when ionized. If the circuit is on for 2 hours a day, how much energy is used by the switches in 1 year (365 days)?

16-2. At 10 cents per kilowatthour, what is the total cost of operating the circuit in question 16-1? What percentage of this cost is attributable to the lighted handles?

16-3. Discuss why having an improperly connected porcelain-base fixture might be more dangerous in a three-way circuit than in a simple on-off circuit.

16-4. Since the neutral conductor and the equipment-grounding conductor are both grounded at the service entrance, why can't the neutral also be used as the equipment ground?

16-5. Why is the neutral conductor never switched?

16-6. When three or more conductors are in a metal conduit, the total cross-sectional area of the conductors cannot exceed 40 percent of the cross-sectional area of the conduit. What might be the reason for this restriction?

16-7. Illustrate a design for a mechanical fluid-control system that is equivalent to a three-way switching circuit.

Answers to Self-Tests

1. T
2. F
3. A lower voltage is safer in building areas.
4. three-phase
5. F
6. service drop
7. T
8. neutral
9. service head, meter box, and distribution panel
10. main circuit breaker, neutral/ground bus, and hot buses
11. It is grounded by grounding conductors connected to a grounding electrode and the metal water service pipe.
12. They are grounded to help protect the system from high-voltage surges such as lightning strikes.
13. 12
14. NM
15. F
16. T

17. T
18. F
19. T
20. switch, receptacle, light fixture
21. hot or black
22. light
23. green
24. T
25. Termination on the grounding screw is allowed when a plastic outlet box is used at the end of a branch circuit.
26. neutral
27. T
28. F
29. three
30. F
31. three-way
32. F
33. neon
34. off

35. T
36. F
37. ground-fault
38. T
39. toroid
40. It is a circuit that transfers power from a main panel to a subpanel.
41. T
42. feeder
43. neutral
44. branch/feeder, combination, outlet circuit
45. branch/feeder
46. F
47. latching
48. The two-conductor relay toggles with each pulse, whereas the three-conductor relay toggles only when the appropriate coil is pulsed.
49. NO, momentary-contact, push-button switches
50. when the load it controls is turned on
51. parallel
52. yes
53. switching, dimming

Appendix A
Common Tools

Simple Testers

In addition to the meters and test equipment discussed throughout this textbook, many technicians use simple instruments to check for the presence or absence of some electrical quantity. The neon test light in Fig. A-1 responds to any voltage above about 90 V. It can be used for determining whether a 120-V outlet is providing voltage or whether a fuse or circuit breaker is open.

The continuity tester in Fig. A-2 can be used to determine whether a low-resistance path exists between two points. It cannot measure the amount of resistance, but it can determine whether a conductor is broken, a switch is defective, and so on. This tester indicates continuity by way of a light in the handle. Some testers use either a light or a buzzer to indicate continuity, and some use both.

Both of these testers are tough and rugged. They can withstand a lot of abuse, and they require little space in a tool kit.

Pliers and Cutters

The pliers and cutters shown in Fig. A-3 represent the major types used in electricity and electronics. Each type of pliers or cutter is made in a number of sizes and styles. For example, the cutters labeled 54CG, 55, 67, and 74CG are all diagonal cutters. The 74CG is a flush-cutting diagonal cutter, and the 67 is a heavy-duty diagonal cutter.

The 60CG is a side-cutting pliers. It is often called *electrician's* or *lineman's pliers.*

The 62CG is an end cutter. It is useful for trimming excess lead length after a component has been soldered onto a printed circuit board.

The pliers labeled 59CG, 72CG, 41CG, and 51CG are all long-nose or chain-nose pliers. Some long-nose pliers, like the 51CG, also have side cutters in the jaws.

The 56CG pliers is a needle-nose pliers. It is used for bending light (large-gage numbers) wire and lightly gripping small objects, especially when working space is restricted.

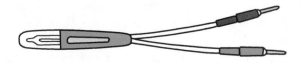

Fig. A-1 Neon voltage tester.

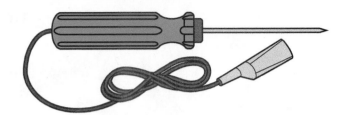

Fig. A-2 Battery-powered continuity tester.

The round-nose pliers, 71CG, is used to form a loop in the end of a conductor. The looped end of the conductor is then fastened under a screw head to make an electrical connection.

Other pliers shown in Fig. A-3 are the slip-joint (76C), the lock-joint (53CG), and the flat-nose (70CG). These are general-purpose pliers used for twisting, turning, and squeezing mechanical fasteners.

The Vise Grip pliers shown in Fig. A-4 are different from the pliers in Fig. A-3 in two ways. First, the intensity of the grip on an object is adjusted by the jaw-size adjustment screw. And second, once the operator squeezes the handles together, the pliers maintains its grip until the grip is released by the small lever on the lower handle. Vise Grip pliers are also known as *locking pliers.*

Drivers

Many drivers are used to install and remove sheet metal screws, machine screws and nuts, bolts and nuts, and the like used in electrical and electronic devices.

Screwdrivers

The most common driver tool is the screwdriver with a handle, a fixed blade (also called a *shaft* or *shank*), and one type of tip. Some common types of screw heads and screwdriver tips are shown in Fig. A-5. In this figure, the top view of the screw head is shown on the left and a pictorial view of the screwdriver tip is shown on the right. The original configuration for screw heads and screwdrivers was the slot tip. It is still in common use. Its major drawback is the ease with which the tip slips out of the slot and mars the surrounding surface.

The Phillips tip (also known as the *cross-point tip*) solves the slipping problem of the slot tip. It also makes it easier to start a screw into a hole and to apply more torque to the screw.

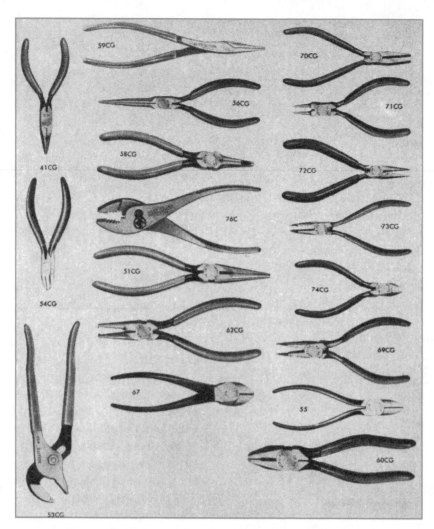

Fig. A-3 Pliers and cutters used by electrical and electronic technicians.

The square tip (sometimes called the *Robertson* or *Scrulox*) is even less likely to slip than the Phillips tip. It also can withstand more torque than either the slot tip or the Phillips tip. Both sheet metal and wood screws are available with square-drive heads.

The Torx configuration is often used with machine screws and bolts and studs. Torx heads and tips can withstand higher torque without stripping out the head or tip.

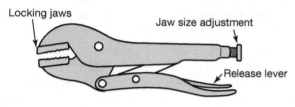

Fig. A-4 Vise Grip pliers.

Another common tip configuration, shown in Fig. A-6, is the hex (also called the *Allen*). Setscrews like those used in control knobs and belt pulleys often have hex heads. Some high-grade machine screws and bolts also have hex heads.

The clutch and spline heads shown in Fig. A-6 are not encountered very often. The spline head is very similar to the Torx head except that it has sharper corners. It is occasionally encountered on setscrews—especially on older equipment. The spline head is also known as the *Bristol* head.

One other tip, not shown here, is the Reed-Prince. It looks just like the Phillips tip. The difference is that the Reed-Prince is a blunter (wider-angle) tip. Thus, it does not fully mesh with a Phillips head screw. Nor will a Phillips tip fully engage a Reed-Prince screw head. Fortunately, the Reed-Prince configuration is rarely encountered.

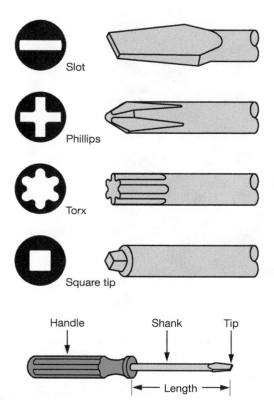

Slot

Phillips

Torx

Square tip

Handle   Shank   Tip

|← Length →|

Fig. A-5 Screwdriver parts and tip configurations.

Nut Drivers

Figure A-7 shows a nut driver, which is used to loosen and tighten hexagon nuts. These drivers are available in a wide range of both metric and inch sizes. Nut drivers are especially useful when there is not room to operate adjustable or fixed wrenches.

There are two types of nut drivers—solid shaft and hollow shaft. When the threaded end of the machine screw

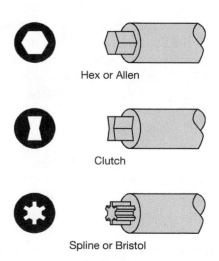

Hex or Allen

Clutch

Spline or Bristol

Fig. A-6 More screwdriver tip configurations.

Fig. A-7 Nut driver for hexagon nuts.

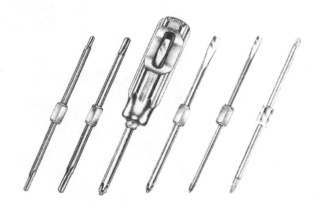

Fig. A-8 Common handle and interchangeable shafts with different tips.

or bolt extends well beyond the nut, only the hollow shaft nut driver is usable.

Interchangeable Drivers

One technique for providing driver interchangeability is shown in Fig. A-8. Here, a common handle can accommodate any one of the shafts shown. Some kits provide shafts with tips ranging from the simple slot tip to a nut driver tip.

Another technique for achieving driver interchangeability is to have a common handle with a fixed shaft. Instead of a tip, the fixed shaft has a bit holder designed to receive bits with short hex shafts. Each bit has a different tip on it. An example is shown in Fig. A-9.

When many screws or nuts are to be removed or tightened, a cordless electric screwdriver can save a lot of time and effort. These drivers have rechargeable batteries and use bits like the one in Fig. A-9.

Crimping Tools

Crimping tools (see Fig. A-10) are used to attach crimp-type terminals to the end of a conductor. With the jaws of the crimping tool open, the terminal is positioned in the

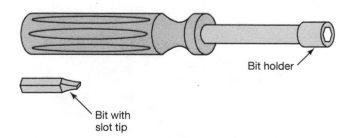

Bit holder

Bit with slot tip

Fig. A-9 Driver with interchangeable bits.

Fig. A-10 Crimp-type terminals and crimping tools.

appropriate-size groove in one of the jaws. The protrusion on the opposite jaw crimps the terminal when the jaws are closed.

Wire Strippers

Figure A-11 shows a common type of wire stripper. Each jaw has six sharpened half-circles ground into it. The matching half-circles in the opposing jaws form circles equal to the diameter of a specific wire gage. Thus, the stripper in Fig. A-9 can accommodate six different gages of wire. With this stripper, the user pulls the insulation off of the end of the wire after the jaws are closed.

Figure A-12 shows a type of wire stripper with two sets of jaws. As the handles are squeezed together, the upper jaws grip the insulated conductor and hold it in place. Then, the lower set of jaws (with the half-circles) cuts the insulation. Finally, the two sets of jaws pull apart, pulling the insulation off the end of the conductor. When the handles are released, both pairs of jaws open and return to the positions shown in Fig. A-12.

Soldering Devices

In the field, technicians often use a soldering gun (see Fig. A-13) to solder electrical and electronic components not mounted on a circuit board. Soldering guns require less than a minute to heat up.

For more delicate work, a soldering pencil, like that shown in Fig. A-13, is usually used. Some soldering pencils have built-in temperature controls that allow the technician to select the appropriate temperature.

In a shop situation, a soldering station (Fig. A-14) is often used. Soldering stations can provide temperature control, tip cleaning pads, soldering pencil holder, vacuum for desoldering, and so forth.

Wrenches and Keys

Electronic and electrical workers use a wide variety of wrenches and keys in their work.

Wrenches

Figure A-15 shows some of the wrenches required for working on electrical systems. The open-end wrench, box-end wrench, and socket are available in both metric and inch sizes. The box-end wrench makes contact with all six of the flat surfaces of a hex nut or bolt head, whereas the open-end contacts only two surfaces. However, the box-end wrench requires enough space beyond the end of the bolt to apply and remove the wrench. Some workers prefer the combination wrench (not shown) that has one box end and one open end.

The adjustable wrench is not as strong (for a given size and weight) as the other wrenches. However, it is convenient when a correct size of fixed wrench is not available, and it will do an adequate job under most conditions. This wrench is sometimes referred to as a *Crescent wrench*.

The socket on a ratchet handle (or any other handle) requires considerable clearance above the nut to be

Fig. A-11 Wire stripper used to pull insulation off a conductor.

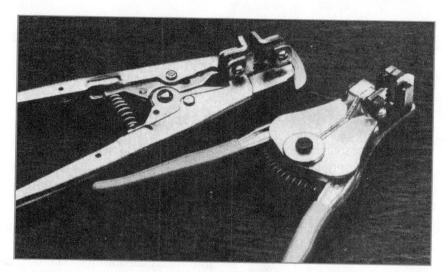

Fig. A-12 Wire stripper that automatically pulls insulation off a conductor.

removed or tightened. But a socket on a ratchet handle can remove a nut much faster than an open-end wrench or a box-end wrench can.

Keys

Figure A-16 illustrates how a variety of Allen (hex) keys can be packaged in a single tool. Not only does this tool provide a lot of size choices in a small space, but it also avoids the misplacing and/or loss of individual keys.

An individual hex key is shown in Fig. A-17. This key is occasionally called an *Allen wrench*. The long leg of the L-shaped key can be used to reach deeply recessed setscrews in pulleys and knobs. Inserting the short leg into the hex head makes it easier to apply a given amount of torque.

Hex keys with a ball end are also available (see Fig. A-18). Although the ball end does not provide as much surface contact as the regular key, it allows the key to be used at an angle to the axis of the hex head screw.

Hex key sets and hex keys are made in either metric or inch sizes. Individual keys are also available with long shafts and T handles.

Torx keys are available in the same styles as hex keys except for the ball end. Like Torx tips and Torx bits, Torx

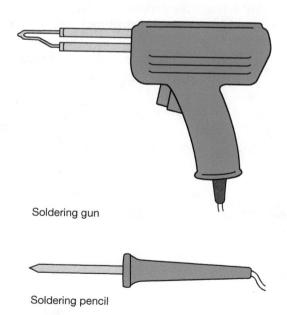

Soldering gun

Soldering pencil

Fig. A-13 Soldering devices.

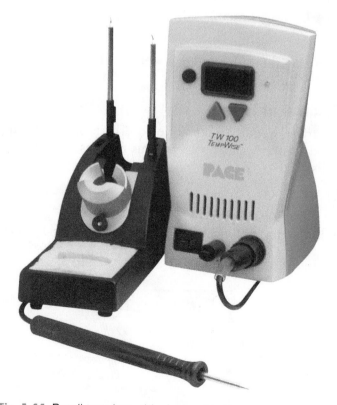

Fig. A-14 Pencil-type iron with changeable tips.

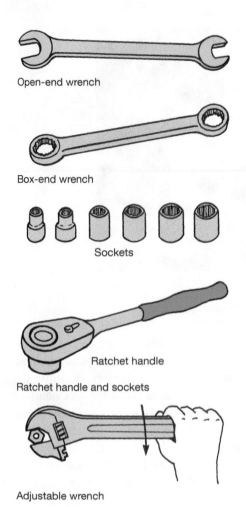

Open-end wrench

Box-end wrench

Sockets

Ratchet handle

Ratchet handle and sockets

Adjustable wrench

Fig. A-15 Common wrenches.

keys can provide high torque without stripping the Torx head of the screw or bolt.

Drills and Bits

Electrical workers often have to drill holes in metal, wood, masonry, and concrete. This requires a variety of drills and bits.

Drills

An electric hand drill like the one shown in Fig. A-19 is useful for drilling holes in wood, metal, and some

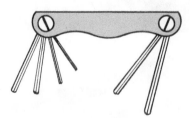

Fig. A-16 Folding Allen (hex) key set.

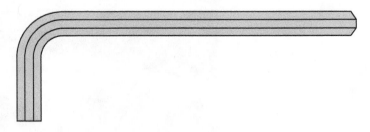

Fig. A-17 Individual hex key.

masonry. Light-duty hand drills have chucks (bit holders) that accommodate bits with up to ⅜-in. shanks. More powerful hand drills often have chucks with ½-in. or ⅝-in. chucks. Most hand drills have reverse rotation and variable speed capabilities. The desired drill speed depends on the type and size of drill bit and the type of material being drilled.

For light-duty use, a cordless (battery-powered) electric drill is very handy. Most of these drills are sold with an extra battery and a charger. Except under extreme conditions, that is, heavy-duty and continuous use, one battery will recharge before the other one is discharged.

Hammer drills are used for drilling holes in concrete and very hard masonry materials. These drills provide pounding action (thousands of strokes per minute) as well as the rotating action of a conventional drill. Electricians use these drills when attaching conduit, boxes, and other items to concrete surfaces.

Bits

The twist drill bit shown in Fig. A-19 is the most common bit used in drills. It can be used on wood, metal, plastics, and other synthetic building products. Two major grades of twist drill bits are the carbon steel types and the high-speed steel (HSS) types. For drilling holes in steel and iron, the HSS grade is recommended (if not required).

The flat wood bit, also known as the wood *spade bit,* is shown in Fig. A-20. This is a fast-cutting bit without spiral grooves to remove wood chips. When deep holes are drilled, the bit has to be periodically withdrawn to remove the chips. Electricians often use this bit to drill through wooden studs when installing electrical cables.

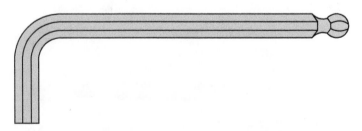

Fig. A-18 Ball-end hex key.

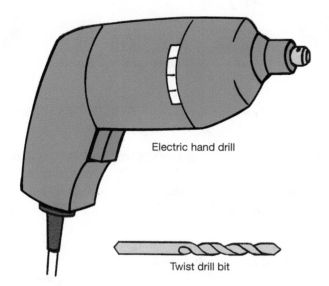

Electric hand drill

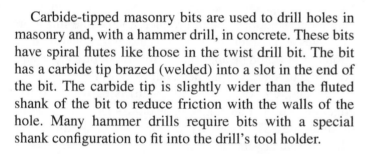

Twist drill bit

Fig. A-19 Electric hand drill and bits.

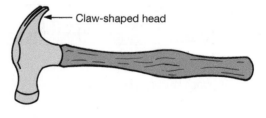

Claw-shaped head

Claw hammer

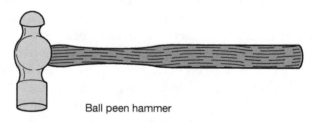

Ball peen hammer

Fig. A-21 Two common types of hammers.

Carbide-tipped masonry bits are used to drill holes in masonry and, with a hammer drill, in concrete. These bits have spiral flutes like those in the twist drill bit. The bit has a carbide tip brazed (welded) into a slot in the end of the bit. The carbide tip is slightly wider than the fluted shank of the bit to reduce friction with the walls of the hole. Many hammer drills require bits with a special shank configuration to fit into the drill's tool holder.

Hammers

The claw hammer (Fig. A-21) is a common hammer used by home owners, carpenters, and electricians. The claw on the head of the hammer can be used to pull nails and pry apart boards that are nailed together. Electricians use this hammer to drive the staples needed to hold electrical cables in place.

Ball peen hammers (Fig. A-21) are available in a wide range of sizes (weights). Since machinists make extensive use of these hammers, they are also known as machinist hammers. These hammers are used to straighten or bend metal, pound on chisels and punches, drive fasteners into holes in concrete, and so on.

Punches

Working with metal objects often requires the use of punches. Two common types of punches are the center punch and the taper punch.

The center punch (Fig. A-22) is used to mark (with an indentation) the exact center of a hole to be drilled with a twist drill bit. The punch in Fig. A-22 has to be struck with a hammer to produce a usable indentation. Automatic center punches with an internal spring-loaded "hammer" are also available. With the automatic punch, one just locates the center point and presses down on the punch until the "hammer" releases and drives the point against the metal to be marked.

When metal objects are being assembled, the holes that bolt the pieces together may be slightly out of alignment. The taper punch shown in Fig. A-23 can be used to adjust the hole alignment by working the punch side to side and up and down as it is pushed into the misaligned holes. This aligns the holes so that a bolt or machine screw can be inserted.

Saws

The hacksaw (Fig. A-24) is a metal-cutting saw. The blade can be changed very quickly to meet the requirements of the metal being cut. Cutting thin metal requires a blade with

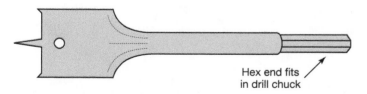

Hex end fits in drill chuck

Fig. A-20 Flat wood bit or spade bit.

Fig. A-22 Center punch with knurled body.

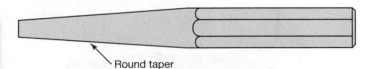

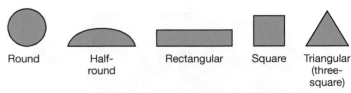

Round   Half-round   Rectangular   Square   Triangular (three-square)

Fig. A-25 Cross-sectional view of common file shapes.

Fig. A-23 Taper punch with hex body.

Round taper

more teeth per inch than the blade used to cut a thick metal. Generally, the blade should have enough teeth so that more than one tooth is in contact with the metal being cut.

The cross-cut saw (shown in Fig. A-24) is designed to cut across the grain of wood stock. For example, it can be used to cut notches in wood framing when conduit is installed.

The keyhole saw (Fig. A-24) is used to cut curved or circular cuts in wood and other soft materials such as wallboard. This saw is sometimes referred to as a *compass saw*. It has interchangeable blades so that blades of different sizes can be selected to fit the task at hand. This saw can be used to cut holes for outlet boxes in walls and ceilings when branch circuits are added or extended.

Files

Files come in a great variety of shapes, lengths, and styles. The most common shapes are shown in Fig. A-25.

Three common teeth arrangements are illustrated in Fig. A-26. The rasp tooth produces a fast-cutting, coarse-cutting file used for rapidly shaping wooden objects and

holes. With the single-cut tooth, a single tooth runs diagonally across the width of the file. This tends to provide a smoother finish than the double-cut tooth does but does not remove material as rapidly as the double-cut tooth design.

How rapidly a given tooth design removes material depends on the coarseness (spacing and depth) of the teeth. Terms used to specify the coarseness of single-cut and double-cut files are *bastard*, *second cut*, and *smooth*, with bastard being the coarsest design.

Single-cut and double-cut files are used extensively for smoothing and shaping metals. A flat (rectangular) single-cut, smooth file can produce a very fine finish on the edges and corners of a sheet of metal. A half-round or a round file would be appropriate for smoothing the inside edge of a tube after it has been cut to length.

The tang end of a file (Fig. A-26) is narrow, and the edges and point of the tang can be quite sharp—especially on small files. Therefore, a file should not be used without a handle that covers the tang.

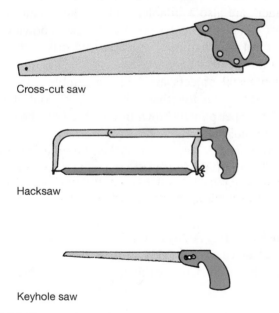

Cross-cut saw

Hacksaw

Keyhole saw

Fig. A-24 Commonly used saws.

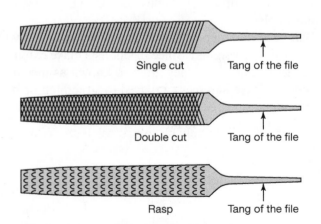

Single cut   Tang of the file

Double cut   Tang of the file

Rasp   Tang of the file

Fig. A-26 Common tooth designs for files.

Appendix B
Solder and the Soldering Process*

From Simple Task to Fine Art

Soldering is the process of joining two metals together by the use of a low-temperature melting alloy. Soldering is one of the oldest known joining techniques, first developed by the Egyptians in making weapons such as spears and swords. Since then, it has evolved into what is now used in the manufacturing of electronic assemblies. Soldering is far from the simple task it once was; it is now a fine art, one that requires care, experience, and a thorough knowledge of the fundamentals.

The importance of having high standards of workmanship cannot be overemphasized. Faulty solder joints remain a cause of equipment failure, and because of that, soldering has become a *critical skill.*

The material contained in this appendix is designed to provide the student with both the fundamental knowledge and the practical skills needed to perform many of the high-reliability soldering operations encountered in today's electronics.

Covered here are the fundamentals of the soldering process, the proper selection of solder and flux, and the use of the soldering station.

The key concept in this appendix is *high-reliability soldering.* Much of our present technology is vitally dependent on the reliability of countless, individual soldered connections. High-reliability soldering was developed in response to early failures with space equipment. Since then the concept and practice have spread into military and medical equipment. We have now come to expect it in everyday electronics as well.

The Advantage of Soldering

Soldering is the process of connecting two pieces of metal together to form a reliable electrical path. Why solder them in the first place? The two pieces of metal could be put together with nuts and bolts, or some other kind of mechanical fastening. The disadvantages of these methods are twofold. First, the reliability of the connection cannot be assured because of vibration and shock. Second, because oxidation and corrosion are continually occurring on the metal surfaces, electrical conductivity between the two surfaces would progressively decrease.

*This material is provided courtesy of PACE, Inc., Laurel, Maryland.

A soldered connection does away with both of these problems. There is no movement in the joint and no interfacing surfaces to oxidize. A continuous conductive path is formed, made possible by the characteristics of the solder itself.

The Nature of Solder

Solder used in electronics is a low-temperature melting alloy made by combining various metals in different proportions. The most common types of solder are made from tin and lead. When the proportions are equal, it is known as 50/50 solder—50 percent tin and 50 percent lead. Similarly, 60/40 solder consists of 60 percent tin and 40 percent lead. The percentages are usually marked on the various types of solder available; sometimes only the tin percentage is shown. The chemical symbol for tin is Sn; thus Sn 63 indicates a solder which contains 63 percent tin.

Pure lead (Pb) has a melting point of 327°C (621°F); pure tin, a melting point of 232°C (450°F). But when they are combined into a 60/40 solder, the melting point drops to 190°C (374°F)—lower than either of the two metals alone.

Melting generally does not take place all at once. As illustrated in Fig. B-1, 60/40 solder begins to melt at 183°C (361°F), but it has not fully melted until the temperature reaches 190°C (374°F). Between these two temperatures, the solder exists in a plastic (semiliquid) state—some, but not all, of the solder has melted.

The plastic range of solder will vary, depending on the ratio of tin to lead, as shown in Fig. B-2. Various ratios of tin to lead are shown across the top of this figure. With most ratios, melting begins at 183°C (361°F), but the full

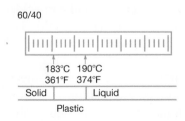

Fig. B-1 Plastic range of 60/40 solder. Melt begins at 183°C (361°F) and is complete at 190°C (374°F).

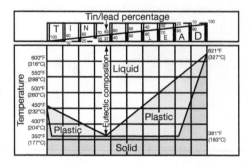

Fig. B-2 Fusion characteristics of tin/lead solders.

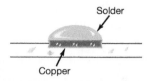

Fig. B-3 The wetting action. Molten solder dissolves and penetrates a clean copper surface, forming an intermetallic bond.

melting temperatures vary dramatically. There is one ratio of tin to lead that has no plastic state and is known as *eutectic solder*. This ratio is 63/37 (Sn 63) and it fully melts and solidifies at 183°C (361°F).

The solder most commonly used for hand soldering in electronics is the 60/40 type, but because of its plastic range, care must be taken not to move any elements of the joint during the cool-down period. Movement may cause a disturbed joint. Characteristically, this type of joint has a rough, irregular appearance and looks dull instead of bright and shiny. It is unreliable and therefore one of the types of joints that is unacceptable in high-reliability soldering.

In some situations, it is difficult to maintain a stable joint during cooling, for example, when wave soldering is used with a moving conveyor line of circuit boards during the manufacturing process. In other cases it may be necessary to use minimal heat to avoid damage to heat-sensitive components. In both of these situations, eutectic solder is the preferred choice, since it changes from a liquid to a solid during cooling with no plastic range.

The Wetting Action

To someone watching the soldering process for the first time, it looks as though the solder simply sticks the metals together like a hot-melt glue, but what actually happens is far different.

A chemical reaction takes place when the hot solder comes into contact with the copper surface. The solder dissolves and penetrates the surface. The molecules of solder and copper blend together to form a new metal alloy, one that is part copper and part solder and that has characteristics all its own. This reaction is called *wetting* and forms the intermetallic bond between the solder and copper (Fig. B-3).

Proper wetting can occur only if the surface of the copper is free of contamination and from oxide films that form when the metal is exposed to air. Also, the solder and copper surfaces need to have reached the proper temperature.

Even though the surface may look clean before soldering, there may still be a thin film of oxide covering it. When solder is applied, it acts like a drop of water on an oily surface because the oxide coating prevents the solder from coming into contact with the copper. No reaction takes place, and the solder can be easily scraped off. For a good solder bond, surface oxides must be removed during the soldering process.

The Role of Flux

Reliable solder connections can be accomplished only on clean surfaces. Some sort of cleaning process is essential in achieving successful soldered connections, but in most cases it is insufficient. This is due to the extremely rapid rate at which oxides form on the surfaces of heated metals, thus creating oxide films which prevent proper soldering. To overcome these oxide films, it is necessary to utilize materials, called *fluxes,* which consist of natural or synthetic rosins and sometimes additives called activators.

It is the function of flux to remove surface oxides and keep them removed during the soldering operation. This is accomplished because the flux action is very corrosive at or near solder melt temperatures and accounts for the flux's ability to rapidly remove metal oxides. It is the fluxing action of removing oxides and carrying them away, as well as preventing the formation of new oxides, that allows the solder to form the desired intermetallic bond.

Flux must activate at a temperature lower than solder so that it can do its job prior to the solder flowing. It volatilizes very rapidly; thus it is mandatory that the flux be activated to flow onto the work surface and not simply be volatilized by the hot iron tip if it is to provide the full benefit of the fluxing action.

There are varieties of fluxes available for many applications. For example, in soldering sheet metal, acid fluxes are used; silver brazing (which requires a much higher temperature for melting than that required by tin/lead alloys) uses a borax paste. Each of these fluxes removes oxides and, in many cases, serves additional purposes. The fluxes used in electronic hand soldering are the pure rosins, rosins combined with mild activators to accelerate the rosin's fluxing capability, low-residue/no-clean fluxes, or water-soluble fluxes. Acid fluxes or highly activated fluxes should never be used in electronic work. Various types of flux-cored solder are now in common use. They provide a convenient way to apply and control the amount of flux used at the joint (Fig. B-4).

Soldering Irons

In any kind of soldering, the primary requirement, beyond the solder itself, is heat. Heat can be applied in a number of ways—conductive (e.g., soldering iron, wave, vapor phase), convective (hot air), or radiant (IR). We are mainly concerned with the conductive method, which uses a soldering iron.

Soldering stations come in a variety of sizes and shapes, but consist basically of three main elements: a resistance heating unit; a heater block, which acts as a heat reservoir; and the tip, or bit, for transferring heat to the work. The standard production station is a variable-temperature, closed-loop system with interchangeable tips and is made with ESD-safe plastics.

Controlling Heat at the Joint

Controlling tip temperature is not the real challenge in soldering; the real challenge is to control the *heat cycle* of the work—how fast the work gets hot, how hot it gets, and how long it stays that way. This is affected by so many factors that, in reality, tip temperature is not that critical.

The first factor that needs to be considered is the *relative thermal mass* of the area to be soldered. This mass may vary over a wide range.

Consider a single land on a single-sided circuit board. There is relatively little mass, so the land heats up quickly. But on a double-sided board with plated-through holes, the mass is more than doubled. Multilayered boards may have an even greater mass, and that's before the mass of the component lead is taken into consideration. Lead mass may vary greatly, since some leads are much larger than others.

Further, there may be terminals (e.g., turret or bifurcated) mounted on the board. Again, the thermal mass is increased, and will further increase as connecting wires are added.

Each connection, then, has its particular thermal mass. How this combined mass compares with the mass of the iron tip, the "relative" thermal mass, determines the rate of the temperature rise of the work.

With a large work mass and a small iron tip, the temperature rise will be slow. With the situation reversed, using a large iron tip on a small work mass, the temperature rise of the work will be much more rapid—even though the *temperature of the tip is the same.*

Now consider the capacity of the iron itself and its ability to sustain a given flow of heat. Essentially, irons are instruments for generating and storing heat, and the reservoir is made up of both the heater block and the tip. The tip comes in various sizes and shapes; it's the *pipeline* for heat flowing into the work. For small work, a conical (pointed) tip is used, so that only a small flow of heat occurs. For large work, a large chisel tip is used, providing greater flow.

The reservoir is replenished by the heating element, but when an iron with a large tip is used to heat massive work, the reservoir may lose heat faster than it can be replenished. Thus the *size* of the reservoir becomes important: a large heating block can sustain a larger outflow longer than a small one.

An iron's capacity can be increased by using a larger heating element, thereby increasing the wattage of the iron. These two factors, block size and wattage, are what determine the iron's recovery rate.

If a great deal of heat is needed at a particular connection, the correct temperature with the right size tip is required, as is an iron with a large enough capacity and an ability to recover fast enough. *Relative thermal mass,* then, is a major consideration for controlling the heat cycle of the work.

A second factor of importance is the *surface condition* of the area to be soldered. If there are any oxides or other contaminants covering the lands or leads, there will be a barrier to the flow of heat. Then, even though the iron tip is the right size and has the correct temperature, it may not supply enough heat to the connection to melt the solder. In soldering, a cardinal rule is that a good solder connection cannot be created on a dirty surface. Before attempting to solder, the work should always be cleaned with an approved solvent to remove any grease or oil film from the surface. In some cases pretinning may be required to enhance solderability and remove heavy oxidation of the surfaces prior to soldering.

A third factor to consider is *thermal linkage*—the area of contact between the iron tip and the work.

Figure B-5 shows a cross-sectional view of an iron tip touching a round lead. The contact occurs only at the point indicated by the "X," so the linkage area is very small, not much more than a straight line along the lead.

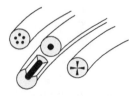

Fig. B-4 Types of cored solder, with varying solder-flux percentages.

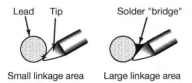

Lead　Tip　　　Solder "bridge"

Small linkage area　　Large linkage area

Fig. B-5 Cross-sectional view (left) of iron tip on a round lead. The "X" shows point of contact. Use of a solder bridge (right) increases the linkage area and speeds the transfer of heat.

The contact area can be greatly increased by applying a small amount of solder to the point of contact between the tip and workpiece. This solder heat bridge provides the thermal linkage and assures rapid heat transfer into the work.

From the aforementioned, it should now be apparent that there are many more factors than just the temperature of the iron tip that affect how quickly any particular connection is going to heat up. In reality, soldering is a very complex control problem, with a number of variables to it, each influencing the other. And what makes it so critical is *time*. The general rule for high-reliability soldering on printed circuit boards is to apply heat for no more than 2 s from the time solder starts to melt (wetting). Applying heat for longer than 2 s after wetting may cause damage to the component or board.

With all these factors to consider, the soldering process would appear to be too complex to accurately control in so short a time, but there is a simple solution—the *workpiece indicator* (WPI). This is defined as the reaction of the workpiece to the work being performed on it—a reaction that is discernible to the human senses of sight, touch, smell, sound, and taste.

Put simply, workpiece indicators are the way the work talks back to you—the way it tells you what effect you are having and how to control it so that you accomplish what you want.

In any kind of work, you become part of a closed-loop system. It begins when you take some action on the workpiece; then the workpiece reacts to what you did; you sense the change, and then modify your action to accomplish the result. It is in the sensing of the change, by sight, sound, smell, taste, or touch, that the workpiece indicators come in (Fig. B-6).

For soldering and desoldering, a primary workpiece indicator is *heat rate recognition*—observing how fast heat flows into the connection. In practice, this means observing the rate at which the solder melts, which should be within 1 to 2 s.

This indicator encompasses all the variables involved in making a satisfactory solder connection with minimum heating effects, including the capacity of the iron and its tip temperature, the surface conditions, the thermal linkage between tip and workpiece, and the relative thermal masses involved.

If the iron tip is too large for the work, the heating rate may be too fast to be controlled. If the tip is too small, it may produce a "mush" kind of melt; the heating rate will be too slow, even though the temperature at the tip is the same.

A general rule for preventing overheating is "Get in and get out as fast as you can." That means using a heated iron you can react to—one giving a 1- to 2-s dwell time on the particular connection being soldered.

Selecting the Soldering Iron and Tip

A good all-around soldering station for electronic soldering is a variable-temperature, ESD-safe station with a pencil-type iron and tips that are easily interchangeable, even when hot (Fig. B-7).

The soldering iron tip should always be fully inserted into the heating element and tightened. This will allow for maximum heat transfer from the heater to the tip.

The tip should be removed daily to prevent an oxidation scale from accumulating between the heating element and the tip. A bright, thin tinned surface must be maintained on the tip's working surface to ensure

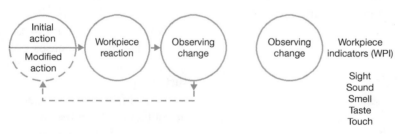

Initial action / Modified action → Workpiece reaction → Observing change

Observing change) Workpiece indicators (WPI)

Sight
Sound
Smell
Taste
Touch

Fig. B-6 Work can be viewed as a closed-loop system (left). Feedback comes from the reaction of the workpiece and is used to modify the action. Workpiece indicators (right)—changes discernible to the human senses—are the way the "work talks back to you."

Making the Solder Connection

The soldering iron tip should be applied to the area of maximum thermal mass of the connection being made. This will permit the rapid thermal elevation of the parts being soldered. Molten solder always flows toward the heat of a properly prepared connection.

When the solder connection is heated, a small amount of solder is applied to the tip to increase the thermal linkage to the area being heated. The solder is then applied to the opposite side of the connection so that the work surfaces, not the iron, melt the solder. Never melt the solder against the iron tip and allow it to flow onto a surface cooler than the solder melting temperature.

Solder, with flux, applied to a cleaned and properly heated surface will melt and flow without direct contact with the heat source and provide a smooth, even surface, feathering out to a thin edge (Fig. B-8). Improper soldering will exhibit a built-up, irregular appearance and poor filleting. The parts being soldered must be held rigidly in place until the temperature decreases to solidify the solder. This will prevent a disturbed or fractured solder joint.

Selecting cored solder of the proper diameter will aid in controlling the amount of solder being applied to the connection (e.g., a small-gage solder for a small connection; a large-gage solder for a large connection).

Removal of Flux

Cleaning may be required to remove certain types of fluxes after soldering. If cleaning is required, the flux residue should be removed as soon as possible, preferably within 1 hour after soldering.

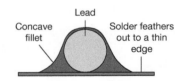

Fig. B-8 Cross-sectional view of a round lead on a flat surface.

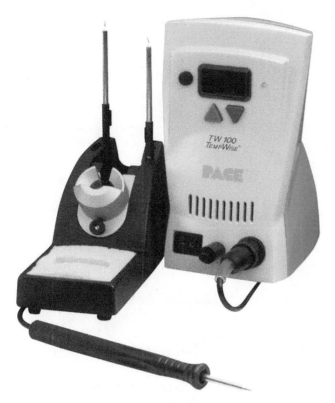

Fig. B-7 Pencil-type iron with changeable tips.

proper heat transfer and to avoid contaminating the solder connection.

The plated tip is initially prepared by holding a piece of flux-cored solder to the face so that it will tin the surface when it reaches the lowest temperature at which solder will melt. Once the tip is up to operating temperature, it will usually be too hot for good tinning, because of the rapidity of oxidation at elevated temperatures. The hot tinned tip is maintained by wiping it lightly on a damp sponge to shock off the oxides. When the iron is not being used, the tip should be coated with a layer of solder.

Appendix C
Formulas and Conversions

Amplitude Conversions

$$V_{p\text{-}p} = 2\,V_p \qquad \text{and} \qquad V_p = \frac{V_{p\text{-}p}}{2}$$

$$V_p = 1.414\,V_{rms} \qquad \text{and} \qquad V_{rms} = 0.707\,V_p$$

$$V_p = 1.57\,V_{av} \qquad \text{and} \qquad V_{av} = 0.637\,V_p$$

$$V_{av} = 0.9\,V_{rms} \qquad \text{and} \qquad V_{rms} = 1.11\,V_{av}$$

V_{rms} = effective value

These formulas can also apply to current by substituting I for V.

Bandwidth

$$\text{BW} \approx \frac{f_r}{Q}$$

Capacitance

$$C = \frac{Q}{V}$$

In parallel

$$C_T = C_1 + C_2 + C_3 + \text{etc.}$$

In series

$$C_T = \frac{1}{\dfrac{1}{C_1} + \dfrac{1}{C_2} + \dfrac{1}{C_3} + \text{etc.}}$$

$$C_T = \frac{C_1 \times C_2}{C_1 + C_2}$$

Energy storage

$$W = 0.5CV^2$$

For resonance

$$C = \frac{0.02533}{f_{r2}L}$$

Capacitive Reactance

$$X_C = \frac{1}{6.28fC}$$

In parallel

$$X_{C_T} = \frac{1}{\dfrac{1}{X_{C_1}} + \dfrac{1}{X_{C_2}} + \dfrac{1}{X_{C_3}} + \text{etc.}}$$

$$X_{C_T} = \frac{X_{C_1} \times X_{C_2}}{X_{C_1} + X_{C_2}}$$

In series

$$X_{C_T} = X_{C_1} + X_{C_2} + X_{C_3} + \text{etc.}$$

Conductance

$$G = \frac{1}{R}$$

Current (Definition)

$$I = \frac{Q}{t}$$

Current (in Impedance Circuits)

For any circuit

$$I_T = \frac{V_T}{Z}$$

For parallel circuits

$$I_T = \sqrt{I_R^2 + I_C^2}$$

$$I_T = \sqrt{I_R^2 + I_L^2}$$

$$I_T = \sqrt{I_R^2 + (I_L - I_C)^2}$$

Efficiency

$$\% \text{ eff.} = \frac{W_{out}}{W_{in}} \times 100 \qquad \% \text{ eff.} = \frac{P_{out}}{P_{in}} \times 100$$

Equivalent Circuit Conversions

$$V_{TH} = I_N R_N \qquad I_N = \frac{V_{TH}}{R_{TH}}$$

Frequency-Period Conversions

$$T = \frac{1}{f} \qquad f = \frac{1}{T}$$

Generator Frequency

$$f = \frac{r/min}{60} \times \text{pairs of poles}$$

Impedance

For any circuit

$$Z = \frac{V_T}{I_T}$$

For series circuits

$$Z = \sqrt{R^2 + X_L^2}$$
$$Z = \sqrt{R^2 + X_C^2}$$
$$Z = \sqrt{(X_L - X_C)^2 + R^2}$$

For parallel circuits

$$Z = \frac{RX_L}{\sqrt{R^2 + X_L^2}}$$

$$Z = \frac{RX_C}{\sqrt{R^2 + X_C^2}}$$

$$Z = \frac{RX_L X_C}{\sqrt{(RX_L - RX_C)^2 + (X_L^2 X_C^2)}}$$

Inductance

Quality

$$Q = \frac{X_L}{R}$$

In series

$$L_T = L_1 + L_2 + L_3 + \text{etc.}$$

For resonance

$$L = \frac{0.02533}{f_r^2 C}$$

In parallel

$$L_T = \frac{1}{\frac{1}{L_1} + \frac{1}{L_2} + \frac{1}{L_3} + \text{etc.}}$$

$$L_T = \frac{L_1 \times L_2}{L_1 + L_2}$$

Of an inductor

$$L = \frac{V_{\text{induced}}}{\Delta I / \Delta t}$$

Mutual inductance

$$L_M = k\sqrt{L_1 L_2}$$

$$k = \frac{\phi_{L_2}}{\phi_{L_1}}$$

Inductive Reactance

$$X_L = 6.28 f L$$

In parallel

$$X_{L_T} = \frac{1}{\frac{1}{X_{L_1}} + \frac{1}{X_{L_2}} + \frac{1}{X_{L_3}} + \text{etc.}}$$

$$X_{L_T} = \frac{X_{L_1} \times X_{L_2}}{X_{L_1} + X_{L_2}}$$

In series

$$X_{L_T} = X_{L_1} + X_{L_2} + X_{L_3} + \text{etc.}$$

Magnetism

$$\text{mmf} = \text{turns} \times \text{current}$$

$$H = \frac{\text{mmf}}{\text{length}} \qquad B = \frac{\phi}{\text{area}} \qquad \mu = \frac{B}{H}$$

Motor

$$\text{Synchronous speed} = \frac{120 f}{\text{no. of poles}}$$

$$\% \text{ slip} = \frac{\text{synchronous speed} - \text{rated speed}}{\text{synchronous speed}}$$

$$\text{hp} = \frac{\text{lb-ft} \times r/min}{5252}$$

$$\text{hp} = 746 \text{ W}$$

Ohm's Law

$$V = IR \qquad I = \frac{V}{R} \qquad R = \frac{V}{I}$$

Parallel Circuits

$$P_T = P_{R_1} + P_{R_2} + P_{R_3} + \text{etc.}$$

$$R_T = \frac{1}{\frac{1}{R_1} + \frac{1}{R_2} + \frac{1}{R_3} + \text{etc.}}$$

$$R_T = \frac{R_1 \times R_1}{R_1 + R_2} \text{ for two resistances}$$

$$R_T = \frac{R}{n} \text{ for } n \text{ equal resistances}$$

$$V_T = V_{R_1} = V_{R_2} = V_{R_3} = \text{etc.}$$

$$I_T = I_{R_1} + I_{R_2} + I_{R_3} + \text{etc.}$$

$$I_{R_1} = \frac{I_T R_2}{R_1 + R_2}$$

Power

$$P = IV \cos \theta \qquad P = I_R^2 R$$

$$P_{app} = IV \qquad P = V_R I_R$$

$$P = \frac{W}{t}$$

Power Factor

$$PF = \cos \theta \qquad PF = \frac{P}{P_{app}}$$

Quality

$$Q = \frac{X}{R}$$

Resistance

$$R = \frac{\text{resistivity} \times \text{length}}{\text{area}} = \frac{Kl}{A}$$

$$\text{where } K = \text{resistivity}$$

$$l = \text{length}$$

$$A = \text{area}$$

Resonant Frequency

$$f_r = \frac{1}{6.28\sqrt{LC}}$$

Series Circuits

$$P_T = P_{R_1} + P_{R_2} + P_{R_3} + \text{etc.}$$

$$R_T = R_1 + R_2 + R_3 + \text{etc.}$$

$$V_T = V_{R_1} + V_{R_2} + V_{R_3} + \text{etc.}$$

$$I_T = I_{R_1} = I_{R_2} = I_{R_3} = \text{etc.}$$

$$V_{R_1} = \frac{V_T R_1}{R_1 + R_2 + \text{etc.}}$$

Time Constant

$$T = RC \qquad T = \frac{L}{R}$$

Transformers

$$\frac{N_{pri}}{N_{sec}} = \frac{V_{pri}}{V_{sec}}$$

$$\left(\frac{N_{pri}}{N_{sec}}\right)^2 = \frac{Z_{pri}}{Z_{sec}}$$

$$\% \text{ eff.} = \frac{P_{sec}}{P_{pri}} \times 100$$

Trigonometric Functions

$$\cos \theta = \frac{\text{adjacent}}{\text{hypotenuse}}$$

$$\sin \theta = \frac{\text{opposite}}{\text{hypotenuse}}$$

$$\tan \theta = \frac{\text{opposite}}{\text{adjacent}}$$

For any circuit

$$\cos \theta = \frac{P}{P_{app}} = PF$$

For series circuits

$$\cos \theta = \frac{V_R}{V_T} = \frac{R}{Z}$$

$$\sin \theta = \frac{V_X}{V_T} = \frac{X}{Z}$$

$$\tan \theta = \frac{V_X}{V_R} = \frac{X}{R}$$

For parallel circuits

$$\cos \theta = \frac{I_R}{I_T} \qquad \sin \theta = \frac{I_X}{I_T} \qquad \tan \theta = \frac{I_X}{I_R}$$

Voltage (Definition)

$$V = \frac{W}{Q}$$

Voltage (in Impedance Circuits)

For any circuit

$$V_T = I_T Z$$

For series circuits

$$V_T = \sqrt{V_R^2 + V_C^2}$$

$$V_T = \sqrt{V_R^2 + V_L^2}$$

$$V_T = \sqrt{V_R^2 + (V_L - V_C)^2}$$

Voltage Regulation

$$\% \text{ Regulation} = \frac{V_{ML} - V_{FL}}{V_{FL}} \times 100$$

Appendix D
Copper Wire Table

AWG (B & S) gage	Standard metric size (mm)	Diameter in mils	Cross-sectional area		Ohms per 1000 ft at 20°C [68°F]	lb per 1000 ft	ft per lb
			Circular mils	Square inches			
0000	11.8	460.0	211,600	0.1662	0.04901	640.5	1.561
000	11.0	409.6	167,800	0.1318	0.06180	507.9	1.968
00	9.0	364.8	133,100	0.1045	0.07793	402.8	2.482
0	8.0	324.9	105,500	0.08289	0.09827	319.5	3.130
1	7.1	289.3	83,690	0.06573	0.1239	253.3	3.947
2	6.3	257.6	66,370	0.05213	0.1563	200.9	4.977
3	5.6	229.4	52,640	0.04134	0.1970	159.3	6.276
4	5.0	204.3	41,740	0.03278	0.2485	126.4	7.914
5	4.5	181.9	33,100	0.02600	0.3133	100.2	9.980
6	4.0	162.0	26,250	0.02062	0.3951	79.46	12.58
7	3.55	144.3	20,820	0.01635	0.4982	63.02	15.87
8	3.15	128.5	16,510	0.01297	0.6282	49.98	20.01
9	2.80	114.4	13,090	0.01028	0.7921	39.63	25.23
10	2.50	101.9	10,380	0.008155	0.9989	31.43	31.82
11	2.24	90.74	8,234	0.006467	1.260	24.92	40.12
12	2.00	80.81	6,530	0.005129	1.588	19.77	50.59
13	1.80	71.96	5,178	0.004067	2.003	15.68	63.80
14	1.60	64.08	4,107	0.003225	2.525	12.43	80.44
15	1.40	57.07	3,257	0.002558	3.184	9.858	101.4
16	1.25	50.82	2,583	0.002028	4.016	7.818	127.9
17	1.12	45.26	2,048	0.001609	5.064	6.200	161.3
18	1.00	40.30	1,624	0.001276	6.385	4.917	203.4
19	0.90	35.89	1,288	0.001012	8.051	3.899	256.5
20	0.80	31.96	1,022	0.0008023	10.15	3.092	323.4
21	0.71	28.46	810.1	0.0006363	12.80	2.452	407.8
22	0.63	25.35	642.4	0.0005046	16.14	1.945	514.2
23	0.56	22.57	509.5	0.0004002	20.36	1.542	648.4
24	0.50	20.10	404.0	0.0003173	25.67	1.223	817.7
25	0.45	17.90	320.4	0.0002517	32.37	0.9699	1,031.0
26	0.40	15.94	254.1	0.0001996	40.81	0.7692	1,300
27	0.355	14.20	201.5	0.0001583	51.47	0.6100	1,639
28	0.315	12.64	159.8	0.0001255	64.90	0.4837	2,067
29	0.280	11.26	126.7	0.00009953	81.83	0.3836	2,607
30	0.250	10.03	100.5	0.00007894	103.2	0.3042	3,287
31	0.224	8.928	79.70	0.00006260	130.1	0.2413	4,145
32	0.200	7.950	63.21	0.00004964	164.1	0.1913	5,227
33	0.180	7.080	50.13	0.00003937	206.9	0.1517	6,591
34	0.160	6.305	39.75	0.00003122	260.9	0.1203	8,310
35	0.140	5.615	31.52	0.00002476	329.0	0.09542	10,480
36	0.125	5.000	25.00	0.00001964	414.8	0.07568	13,210
37	0.112	4.453	19.83	0.00001557	523.1	0.06001	16,660
38	0.100	3.965	15.72	0.00001235	659.6	0.04759	21,010
39	0.090	3.531	12.47	0.000009793	831.8	0.03774	26,500
40	0.080	3.145	9.888	0.000007766	1049.0	0.02993	33,410

Appendix E
Resistivity of Metals and Alloys

Material	*Resistivity at 20°C ($\Omega \cdot$ cm)
Aluminum	0.00000262
Brass (66% Cu, 34% Zn)	0.0000039
Carbon (graphite form)	0.0014
Constantan (55% Cu, 45% Ni)	0.0000442
Copper (commercial annealed)	0.0000017241
German silver (18% Ni)	0.000033
Iron	0.00000971
Lead	0.0000219
Mercury	0.0000958
Monel metal (67% Ni, 30% Cu, 1.4% Fe, 1% Mn)	0.000042
Nichrome (65% Ni, 12% Cr, 23% Fe)	0.000100
Nickel	0.0000069
Phosphor bronze (4% Sn, 0.5% P, 95.5% Cu)	0.0000094
Silver	0.00000162
Steel (0.4–0.5% C)	0.000013–0.000022
Tantalum	0.0000131
Tin	0.0000114
Tungsten	0.00000548
Zinc	0.000006

Source: *Reference Data for Radio Engineers*, 6th ed., 1975, Howard W. Sams & Co., Inc.
*To convert to ohm-circular mils per foot multiply the numbers in this column by 6.02×10^6.

Appendix F

Temperature Coefficients of Resistance

Material	*Temperature Coefficient at 20°C (ohm per ohm/°C)
Aluminum	0.0039
Brass (66% Cu, 34% Zn)	0.002
Carbon (graphite form)	−0.0005
Copper (commercial annealed)	0.0039
German silver (18% Ni)	0.0004
Iron	0.0052–0.0062
Lead	0.004
Mercury	0.00089
Monel metal (67% Ni, 30% Cu, 1.4% Fe, 1% Mn)	0.002
Nichrome (65% Ni, 12% Cr, 23% Fe)	0.00017
Nickel	0.0047
Phosphor bronze (4% Sn, 0.5% P, 95.5% Cu)	0.003
Silver	0.0038
Steel (0.4–0.5% C)	0.003
Tantalum	0.003
Tin	0.0042
Tungsten	0.0045
Zinc	0.0037

Source: *Reference Data for Radio Engineers*, 6th ed., 1975, Howard W. Sams & Co., Inc.
*To convert to ppm/°C multiply the numbers in this column by 1×10^6.

Appendix G
Trigonometric Functions

Angle (degrees)	Sin	Cos	Tan	Angle (degrees)	Sin	Cos	Tan
0	0.000	1.000	0.000	46	.719	.695	1.036
1	.018	1.000	.018	47	.731	.682	1.072
2	.035	.999	.035	48	.743	.669	1.111
3	.052	.999	.052	49	.755	.656	1.150
4	.070	.998	.070	50	.766	.643	1.192
5	.087	.996	.088	51	.777	.629	1.235
6	.105	.995	.105	52	.788	.616	1.280
7	.122	.993	.123	53	.799	.602	1.327
8	.139	.990	.141	54	.809	.588	1.376
9	.156	.988	.158	55	.819	.574	1.428
10	.174	.985	.176	56	.829	.559	1.483
11	.191	.982	.194	57	.839	.545	1.540
12	.208	.978	.213	58	.848	.530	1.600
13	.225	.974	.231	59	.857	.515	1.664
14	.242	.970	.249	60	.866	.500	1.732
15	.259	.966	.268	61	.875	.485	1.804
16	.276	.961	.287	62	.883	.470	1.881
17	.292	.956	.306	63	.891	.454	1.963
18	.309	.951	.325	64	.899	.438	2.050
19	.326	.946	.344	65	.906	.423	2.145
20	.342	.940	.364	66	.914	.407	2.246
21	.358	.934	.384	67	.921	.391	2.356
22	.375	.927	.404	68	.927	.375	2.475
23	.391	.921	.425	69	.934	.358	2.605
24	.407	.914	.445	70	.940	.342	2.748
25	.423	.906	.466	71	.946	.326	2.904
26	.438	.899	.488	72	.951	.309	3.078
27	.454	.891	.510	73	.956	.292	3.271
28	.470	.883	.532	74	.961	.276	3.487
29	.485	.875	.554	75	.966	.259	3.732
30	.500	.866	.577	76	.970	.242	4.011
31	.515	.857	.601	77	.974	.225	4.332
32	.530	.848	.625	78	.978	.208	4.705
33	.545	.839	.649	79	.982	.191	5.145
34	.559	.829	.675	80	.985	.174	5.671
35	.574	.819	.700	81	.988	.156	6.314
36	.588	.809	.727	82	.990	.139	7.115
37	.602	.799	.754	83	.993	.1229	8.144
38	.616	.788	.781	84	.995	.105	9.514
39	.629	.777	.810	85	.996	.087	11.43
40	.643	.766	.839	86	.998	.070	14.30
41	.656	.755	.869	87	.999	.052	19.08
42	.669	.743	.900	88	.999	.035	28.64
43	.682	.731	.933	89	1.000	.018	57.29
44	.695	.719	.966	90	1.000	.000	—
45	.707	.707	1.000				

Some of the codes and color codes used on tantalum capacitors are shown in Fig. H-1. Notice that the colors are assigned the same values for the first and second significant figures and for the multiplier, as they are for the resistor color code. Also notice that the color code system in Fig. H-1(b) yields the value in microfarads, whereas that in Fig. H-1(c) gives the value in picofarads.

The alphanumeric coding system shown in Fig. H-2(a) is probably the most widely used system for all kinds of physically small capacitors. From the Multiplier table in Fig. H-2 you can determine that the three-number system can cover values from 0.11 pF (118) to 9,900,000 pF (995), or from 0.11 pF to 9.9 μF. Thus, the system shown in Fig. H-2(b), where the letter R is used as a decimal point, really isn't needed and is rarely used.

Very often more information than just capacitance and tolerance is provided by capacitor codes. In Fig. H-3 the temperature coefficient is given. N or P values, such as N45 or P120, could be given just as well as the NPO shown. N45 would mean a negative temperature coefficient of 45 parts per million per degree Celsius, whereas P120 would mean a positive temperature coefficient of 120 parts per million per degree Celsius.

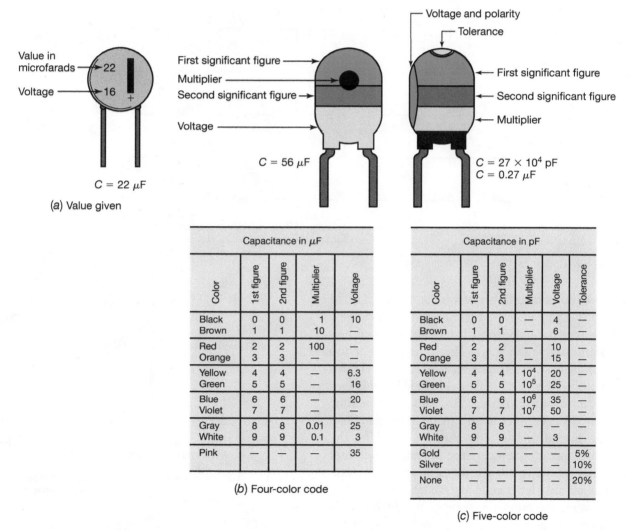

$C = 22 \ \mu\text{F}$

(a) Value given

$C = 56 \ \mu\text{F}$

$C = 27 \times 10^4 \ \text{pF}$
$C = 0.27 \ \mu\text{F}$

Capacitance in μF				
Color	1st figure	2nd figure	Multiplier	Voltage
Black	0	0	1	10
Brown	1	1	10	—
Red	2	2	100	—
Orange	3	3	—	—
Yellow	4	4	—	6.3
Green	5	5	—	16
Blue	6	6	—	20
Violet	7	7	—	—
Gray	8	8	0.01	25
White	9	9	0.1	3
Pink	—	—	—	35

(b) Four-color code

Capacitance in pF					
Color	1st figure	2nd figure	Multiplier	Voltage	Tolerance
Black	0	0	—	4	—
Brown	1	1	—	6	—
Red	2	2	—	10	—
Orange	3	3	—	15	—
Yellow	4	4	10^4	20	—
Green	5	5	10^5	25	—
Blue	6	6	10^6	35	—
Violet	7	7	10^7	50	—
Gray	8	8	—	—	—
White	9	9	—	3	—
Gold	—	—	—	—	5%
Silver	—	—	—	—	10%
None	—	—	—	—	20%

(c) Five-color code

Fig. H-1 Codes and color codes for tantalum capacitors.

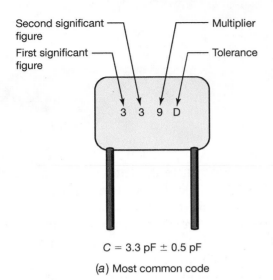

C = 3.3 pF ± 0.5 pF

(a) Most common code

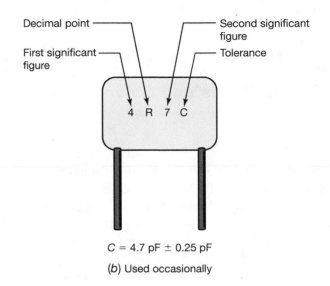

C = 4.7 pF ± 0.25 pF

(b) Used occasionally

Multiplier	
Number	Multiply by
0	1
1	10
2	100
3	10^3
4	10^4
5	10^5
8	0.01
9	0.1

Tolerance		
Letter	$C \le 10$ pF	$C > 10$ pF
B	±0.1 pF	—
C	±0.25 pF	—
D	±0.5 pF	—
F	±1.0 pF	±1%
G	±2.0 pF	±2%
H	±2.0 pF	±3%
J	—	±5%
K	—	±10%
M	—	±20%
N	—	±0.05%
P	—	+100%, −0%
Z	—	+80%, −20%

Fig. H-2 Codes for film capacitors and dipped mica capacitors.

Figure H-4 shows a coding system that differs in three aspects from that of Fig. H-3. First, and most obviously, it gives the voltage rating. Second, it does not use the three-number system to give the value. The value shown is too small to be in picofarads. Therefore, it must be in microfarads. Sometimes the value will be specified by only two numbers, 47, for example. In this case, the value would be 47 pF because disk ceramics are not made in values above 1 μF. Third, instead of the temperature coefficient, the coding system specifies temperature characteristics that are more comprehensive than a temperature coefficient. Using a letter-number-letter code and the table in Fig. H-4, one can determine the temperature range in which the capacitor can operate as well as the maximum amount the capacitance will change over this temperature range.

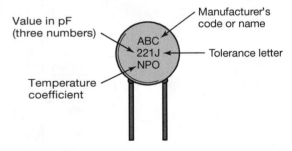

C = 220 pF ± 5% with 0 temperature coefficient

Notes: 1. Use the tolerance and multiplier charts in Fig. H-2.
2. Sometimes the voltage is also given. Otherwise, it is usually 500 V.
3. Since disk ceramics typically have a range of 1 pF to less than 1 μF, this capacitor's value is in picofarads.

Fig. H-3 One of several codes used with ceramic disk capacitors.

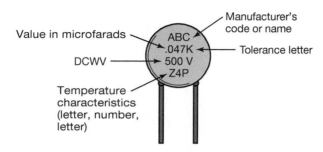

Value in microfarads

Manufacturer's code or name

DCWV

Tolerance letter

Temperature characteristics (letter, number, letter)

ABC
.047K
500 V
Z4P

$C = 0.047 \ \mu F \pm 10\%$ with no more than $\pm 10\%$ change in value over its temperature range of 10 to 65°C.

Notes: 1. Value is direct reading in microfarads because it is less than 1.
2. Use tolerance chart in Fig. H-2.

Temperature Characteristics Code					
1st letter (low temp.)		Number (high temp.)		2nd letter (*maximum* change in C over temp. range)	
X	−55°C	2	+45°C	A	+1.0%
Y	−30°C	4	+65°C	B	±1.5%
Z	+10°C	5	+85°C	C	±2.2%
		6	+105°C	D	±3.3%
		7	+125°C	E	±4.7%
				F	±7.5%
				P	±10.0%
				R	±15.0%
				S	±22.0%
				T	+22%, −33%
				U	+22%, −56%
				V	+22%, −82%

Fig. H-4 Another example of coding used on ceramic disk capacitors.

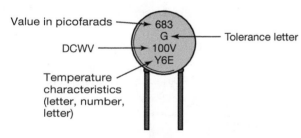

Value in picofarads

Tolerance letter

DCWV

Temperature characteristics (letter, number, letter)

683
G
100V
Y6E

$C = 68000 \ pF \ (0.068 \ \mu F) \pm 2\%$ with no more than $\pm 4.7\%$ change in value over its temperature range of −30 to +105°C.

Notes: 1. Use the multiplier and tolerance charts in Fig. H-2 to determine the value.
2. Use the temperature characteristics code in Fig. H-4.

Fig. H-5 Variation in coding used on ceramic disk capacitors.

Figure H-5 shows another variation in capacitor coding. Notice that the tolerance letter is on a separate line rather than being attached to the three-number value code.

There are many other variations in the way capacitors are coded. However, with the information given in these figures, one can usually make sense out of them.

There are many different systems used to code chip capacitors. Unfortunately, identical markings by two different manufactures can mean different things. For example, a chip capacitor coded K2 would be 240 pF for one manufacturer and 2400 pF for another manufacturer. The capacitors could look alike in color and size. So, unless one knows the manufacturer and the system used by that manufacturer, codes on chip capacitors can easily be misleading. Thus, chip codes are omitted in this appendix.

Appendix I
The Oscilloscope

Alternating currents and voltages can be measured with multimeters. They work fine for low-frequency sinusoidal waveforms. However, many alternating currents and voltages are very complex waveforms and often contain more than one frequency. These complex waveforms are referred to as *signals*, and they cannot be adequately (if at all) measured with a multimeter.

An oscilloscope (Fig. I-1) is an extremely useful instrument when one needs to troubleshoot any electrical or electronic device. With an oscilloscope, one can do such things as view the waveshape of a signal, determine the frequency, period, and voltage of a signal, and determine the phase relationship between two waveforms.

Oscilloscopes display waveforms when the phosphor coating on the inner face of a cathode-ray tube (CRT) is excited by a scanning electron beam (Fig. I-2). The electrons for the beam are provided by thermionic emission. Then they are accelerated toward the face of the CRT, formed into a beam, and deflected up and down and left and right by electric force fields applied to metal cylinders and plates inside the CRT.

Three major categories of oscilloscopes are commonly used: analog oscilloscopes, digital storage oscilloscopes (DSOs), and digital phosphor oscilloscopes (DPOs). Analog oscilloscopes use full continuous phosphor coating in the CRT, whereas digital oscilloscopes use closely spaced tiny dots of phosphor on the face of the CRT. The individual dots are arranged so that they can be scanned in the same manner as the continuously coated CRT.

Fig. I-1 A dual-channel oscilloscope.

The analog oscilloscope displays a waveform in real time. That is, when the applied signal voltage is at its peak positive voltage, the displayed waveform is also at its peak positive voltage. The displayed waveform follows the applied voltage waveform instant by instant. The only time the analog oscilloscope is not tracking and displaying the applied waveform is during the instant when the electron beam is returning from the right side to the left side of the CRT face so that the beam can start a new trace (scan). The return typically requires less than 2 percent of the scan time.

The use of a continuous phosphor coating and real-time scanning has both desirable and undesirable results. These results are caused by the fact that the amount of light emitted by phosphor materials depends on how often and how long the phosphor is excited. Thus, when the electron beam is moving slowly and follows the same path trace after trace, the displayed waveform will be very bright. If an intermittent problem causes the applied signal to occasionally have a clipped (flattened) positive half-cycle, the clipping will be much dimmer than the rest of the display. Thus, the viewer will know that the clipping is intermittent. On the negative side, if one is trying to view a sawtooth waveform with a long, slow rise time and a very fast, steep fall time, the fall time part of the waveform may be too dim to view because the electron beam is moving down so fast that the phosphor is not adequately excited.

Viewing the sawtooth waveform described above is no problem for a DSO. A digitized oscilloscope does not display a waveform in real time. Instead, it checks the value of the applied waveform at very closely spaced intervals. At each check, called a *sample*, the value is converted to a digital number and stored in the oscilloscope's memory. After the complete waveform has been sampled, the stored information is used to control the scanning electron beam in the CRT so that the beam excites only the phosphor dots needed to re-create the sampled waveform. With this system, the waveform is not re-created by a single scan of the electron beam. Instead, the beam scans the top row of dots and excites only those needed to start creating the waveform. After rapidly retracing back to the left side of the screen, the beam scans the next row of dots and excites only those required to create more of the waveform. This

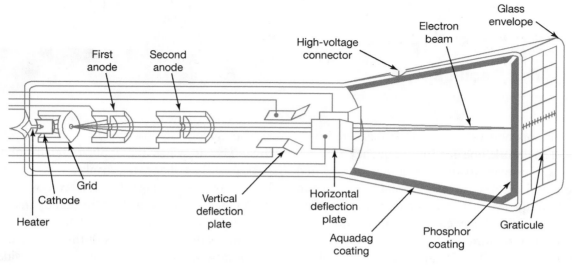

First anode
Second anode
High-voltage connector
Electron beam
Glass envelope

Grid
Cathode
Heater
Vertical deflection plate
Horizontal deflection plate
Aquadag coating
Phosphor coating
Graticule

Fig. I-2 Cutaway view of a cathode-ray tube.

process is repeated until the whole face of the CRT has been scanned (row by row) and the complete waveform has been re-created as a series of closely spaced excited dots. The dots are so close together that they look like a continuous line. With this scanning process, the electron beam does not move up or down as it scans across the CRT screen, and the stored digital numbers determine only which dots will be excited. The stored numbers do not determine how brightly the dots will glow; all excited dots will glow with equal intensity. Thus, the fall time part of the sawtooth waveform will be just as visible as the rise time part.

The DSO spends most of its operating time in processing the stored digital numbers and re-creating the waveform. Many DSOs spend less than 2 percent of the time they are receiving the input waveform in actually sampling the waveform. Therefore, the intermittent problem described above is missed unless it occurs during the short time in which the DSO is taking samples. Even if, or when, the intermittent is captured, it will be displayed just as brilliantly as the rest of the waveform.

The limitations of the DSO have been overcome by the newer DPO. By using more sophisticated digital and microprocessor circuitry, the DPO spends about as much time sampling the input waveform as the analog oscilloscope spends scanning it. Just as important, the DPO stores digital numbers that represent how often an input waveform is located on a given phosphor dot as well as digital numbers that tell which dot to illuminate when the screen is scanned. The input waveform that is to be viewed is completely sampled, and data stored, many times before it is scanned onto the CRT screen. Now the stored numbers can control not only which phosphor dot to illuminate (excite) but also how brightly the dot should

be illuminated. Thus, the DPO can view an intermittent signal or any complex signal with the same intensity grading as is available with the analog oscilloscope.

One big advantage of both the DSO and the DPO over the analog oscilloscope is the ability to "freeze" any waveform on the CRT screen. This allows one to study it for as long as necessary. Also, a waveform (or any number of waveforms) can be stored on magnetic media, such as a floppy disk, and then viewed on a computer screen.

There is tremendous variation in the number, type, and sophistication of controls available on different oscilloscopes. The manufacturer's instruction manual should be consulted, and studied, to obtain the best performance from a particular oscilloscope.

Some of the controls, and their functions, that are common to most oscilloscopes are discussed below.

- The *intensity control* determines the brightness of the waveform. Keep the brightness as low as possible when displaying an easily viewed waveform.

- The *focus control* determines the sharpness of the waveform. Adjust this control to obtain the minimum width of the trace line. The focus control and the intensity control often interact. Readjust the focus control after changing the intensity control.

- The *position controls* adjust the location of the waveform. There are two position controls. One controls the location of the waveform on the horizontal axis. The other controls the location of the waveform on the vertical axis. Adjust these controls to center the waveform both vertically and horizontally.

- The *trace rotation control* is used to align the no-signal trace parallel to the horizontal axis of the screen of the CRT. It may need adjusting when the oscilloscope is moved to a new location.

- The *volts/division control* adjusts the height (vertical size) of the waveform. (The control is sometimes called *volts/centimeter* because the divisions on the screen are often 1-cm squares.) This control has two control knobs. One is a rotary switch that makes coarse (step) adjustments. The other is a potentiometer that makes fine adjustments; it is often labeled *variable*. The position of these two knobs determines how much voltage is needed to fill the screen of the oscilloscope. The volts/division control can be used to measure the voltage applied to the input (vertical input) of the oscilloscope. When the input voltage is measured, the variable (fine adjustment) knob must be in the "calibrate" (maximum clockwise rotation) position. Then the coarse (step) adjustment indicates how much voltage is required to produce a waveform one division in height. For example, if a waveform is 4.6 divisions in height, the volts/division switch is on the 0.2 volts/division position and the variable control is in the calibrate position, then the waveform will have a peak-to-peak value of 0.92 V (V_{p-p} = 4.6 divisions × 0.2 V/division = 0.92 V).

- The *input coupling control switch* typically has three positions: AC, DC, and Ground. The Ground position removes the signal from the vertical input channel so that the no-signal trace can be vertically centered on the CRT face. This allows one to measure the value of a dc signal or the peak values of a nonsymmetrical ac signal. Alternating current coupling is used when one wants to view an ac signal that is superimposed upon a dc voltage, or, saying it another way, when one wants to view the fluctuations of a fluctuating dc voltage. Direct current coupling is used to measure the value of a dc voltage or to view both the dc component and the ac component of the input signal (Fig. I-3).

- The *math operations control(s)* allow mathematical operations to be performed on one or more signals. These controls have different names on different types of oscilloscopes. On multichannel oscilloscopes (two or four vertical input channels), the operator can choose to have two signals, applied to two different channels, added or subtracted and then displayed as a single signal. Digital oscilloscopes can do more advanced operations like division, multiplication, and Fourier transforms.

- The *time/division control* determines how many cycles of waveform are displayed on the screen. Like the volts/division control, it has two control knobs. One operates a switch that is the coarse adjustment. The other operates a potentiometer for the fine (variable) adjustment. The time/division control can be used to measure the period of a waveform. Again, the variable adjustment must be in the calibrate position in order to measure the period. For example, assume that the length of one cycle is approximately 2.8 divisions, the time/division switch is in the 1-millisecond/division (ms/div) position, and the variable knob is in the calibrate position. The period of the waveform, therefore, is:

Period (T) = 2.8 divisions × 1 ms/div = 2.8 ms

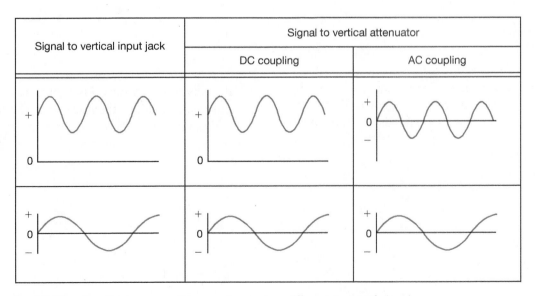

Fig. I-3 Results of using ac and dc coupling on two different types of signals.

- The *trigger controls* work together to stabilize the waveform on the CRT screen so that the waveform does not drift across the screen. This is accomplished by synchronizing the horizontal trace (called the *sweep*) with the input signal so that the input waveform is at exactly the same place on its cycle when each new sweep (trace) starts. The electron beam is forced to sweep across the screen by a sawtooth waveform that is generated by an electronic circuit. This circuit generates one cycle of sawtooth each time it is triggered (started) by a voltage of the correct polarity and value that is applied to its input. One of the trigger controls allows the operator to select where the trigger voltage will come from. Common options are (a) any one of the vertical input channels, (b) an external voltage source connected to the trigger input jack on the oscilloscope, (c) 60-Hz line voltage, and (d) special trigger voltages generated by circuits inside the oscilloscope. Two other common trigger controls are the *slope* and *level controls*. As shown in Fig. I-4, the slope control determines whether the sweep starts on the rising or falling part of the trigger signal, and the level control determines where on the rising or falling part

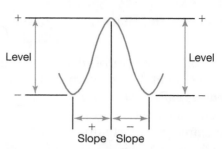

Fig. I-4 How the slope and level controls select the trigger point.

of the trigger signal the sweep will start. Another common trigger control is the *trigger mode*. A typical trigger mode control has two options: normal and auto. When set to normal mode, the horizontal sweep does not start until a trigger signal of correct polarity and magnitude is received by the sawtooth generator. In this mode, an input waveform may not be displayed if the slope and level controls are not properly adjusted. In the auto mode, the sawtooth generator is set to produce continuous sweeps when it does not receive any trigger voltage, but when it receives a trigger voltage, it is synchronized with the input waveform.

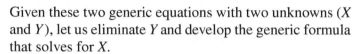

Given these two generic equations with two unknowns (X and Y), let us eliminate Y and develop the generic formula that solves for X.

$$a_1X + b_1Y = C_1 \qquad \text{Equation (1)}$$
$$a_2X + b_2Y = C_2 \qquad \text{Equation (2)}$$

To eliminate Y, multiply equation (1) by b_2, multiply equation (2) by b_1, and then subtract the resulting equation 2 from the resulting equation 1. This gives us

$$a_1b_2X + b_1b_2Y = C_1b_2 \qquad \text{Equation (1)}$$
$$\underline{a_2b_1X + b_1b_2Y = C_2b_1 \qquad \text{Equation (2)}}$$
$$a_1b_2X - a_2b_1X = C_1b_2 - C_2b_1$$

Now factor out X:

$$X(a_1b_2 - a_2b_1) = C_1b_2 - C_2b_1$$

and solve for X:

$$X = \frac{C_1b_2 - C_2b_1}{a_1b_2 - a_2b_1}$$

Now express the numerator and denominator in matrix form:

$$X = \frac{\begin{vmatrix} C_1 & b_1 \\ C_2 & b_2 \end{vmatrix}}{\begin{vmatrix} a_1 & b_1 \\ a_2 & b_2 \end{vmatrix}}$$

Notice in this formula that the numerator matrix is the matrix used to calculate the value of the determinant D_X and the denominator is the matrix used to calculate the value of the determinant D. Therefore, the determinant formula for X is

$$X = \frac{D_X}{D}$$

Appendix K
Renewable Energy Sources and Technologies

Some of the many sources of renewable energy that can be used to provide electric energy include wind, solar, water, and geothermal. The energy from these sources is converted into electrical energy in the form needed by our electrical distribution system. This form is usually alternating current at a frequency of 60 Hz and a specified voltage.

The conversion of many sources of renewable energy to 60 hertz ac is done by some type of turbine (i.e., wind, water, steam, etc.) turning the shaft of an ac generator. For a given generator with a fixed number of field poles, the output frequency is determined by the RPM of the armature. Armature RPM is controlled by the amount of energy (fuel) provided to the turbine connected to the generator shaft. The RPM must be held constant if a constant output frequency is to be maintained. If the RPM is held constant, the output voltage can be controlled by varying the current in the coils of the field poles. Before a generator is connected to an electric power grid, its frequency and output voltage must be adjusted to match that of the grid.

Once the generator is connected to the power grid, its frequency is synchronized with all the other generators connected to the grid. Now the frequency, and thus the RPM, of the armature and the turbine are locked into that of all the other generators connected to the grid. Increasing the energy (fuel) provided to the turbine can only cause a minute change in the armatures RPM and thus a very minute change in the frequency and voltage of the grid and all of the other generators connected to the grid. However, increasing energy input to the turbine does increase the power output of the generator by increasing the current output from the armature. Of course, decreasing fuel input to the turbine also produces a very minute change in the grid voltage and frequency, but the change is in the opposite direction.

Once the input energy to a turbine has been set to accommodate the requirements of the loads connected to the grid, any significant changes in the loads on the grid will cause minute changes in the frequency and voltage of the grid. This minute change is detected by electronic detectors and circuits that constantly monitor the grid frequency and voltage and make appropriate changes in the energy input to the turbines driving the generators to counteract the minute changes detected.

Renewable Energy from Water (hydroelectricity)

When the energy source is water, the amount of input energy to the turbine is controlled by the amount of water allowed to flow through the turbine. Several advantages of water as a renewable energy source are:

1. It can be stored in lakes, formed by damming a river, and used when needed either day or night. It is a very reliable source.
2. It is a green energy source that does no damage to the environment during operation.
3. In areas where water flow tends to be inadequate at times, water can be pumped into a man-made lake when water flow is high and excess electrical energy can be used to power water pumps.
4. It is one of the cheaper sources (about 4 to 11 ¢/kWh), but limited to locations (rivers) where dams can be located.
5. It is a highly developed, proven technology.

Energy can also be obtained from the tidal movement of water in and out of a larger bay or inlet that is connected to the main body of ocean water through a narrow inlet that can be gated off at high tide and allowed to flow out through a water turbine as the tide goes out (recedes). To be economically viable, this approach to using water energy requires large, regular tidal swings and the right soil conditions for building and operating the gates needed to open and close the waterway to the bay or inlet. Locations that meet all these conditions are rather limited.

Renewable Energy from Wind

The most common wind turbine in use today is the three-blade rotor (propeller) facing into the wind. The wind energy input depends on the wind speed and the pitch of the blades. Since the rotor turns at a low speed, (typically less than 30 RPM), the rotor shaft drives a gear-box that turns the ac generator at the RPM needed to produce a 60 Hz frequency. Wind speed and direction has large variations during any 24 hours, so it is continuously monitored and the blade pitch adjusted and/or the platform holding the turbine and generator is rotated so that desired turbine (rotor) speed is maintained and the rotor remains facing into the wind. Once the generator's frequency and the voltage are synchronized with the grid, the generator is connected to the grid and its frequency and voltage are essentially locked to the grid's frequency and voltage. Now, when the electronic circuitry detects any minute changes in the grid frequency and voltage, or load, the blade pitch and/or platform-to-wind orientation are automatically adjusted to correct for the detected changes.

When the rotor needs to be stopped for maintenance or because of severe wind/weather conditions, the generator is disconnected from the grid. Then the blade pitch is reduced to a minimum, the platform is rotated so that the rotor is parallel to the wind direction, and a brake system that stops the rotor shaft from turning is activated.

While wind energy systems are often said to be the cheapest form of alternative energy (4 to 6¢/kWh), their reliability is often low. Often, the wind speed is too low or too high to produce the frequency and voltage required by the grid. Also, wind energy systems require a lot of maintenance that requires down time.

Renewable Energy from Solar (Sun) Energy

Solar energy can be converted into electric energy by a PV (photovoltaic) solar panel. A single panel typically has 10 to 20 square feet of surface area to expose to sunlight, and produces 210 W to 270 W of dc power. A PV solar panel has many individual PV cells. A single cell produces about 0.5 V at about 1.5 A. Individual cells are connected in series to increase the panel output voltage and series groups are connected in parallel to increase the panel's output current and power.

The output of PV solar panels is dc. The dc output can be converted to 50 or 60 Hz ac by an electronic inverter. Inverters that can handle dc inputs to 1000 V dc and provide 400 V, 3-phase ac output are available. Some inverters can provide 2.4 mega-watts of ac power output. These inverters can be up to 98% efficient. As long as the frequency is correct and the voltage is stable, the output voltage can be transformed to the level needed by the grid before the PV solar panel system is connected to the grid.

A problem with PV cells and PV panels being connected in series is that when one cell or one panel in a series receives reduced sunlight energy, it reduces not only the current output of that one cell or panel, but the current output of the other cells or panels in series with it. This can make a very significant reduction in the power output of the PV panels and the input into the main dc-to-ac inverter. The effects of this problem can be minimized by using a low-power inverter on the output of each panel and then combining the outputs of all of these small inverters to obtain the required voltage and current.

Although the cost of operating and maintaining a PV solar farm (system) is small, the cost of building the system has been high because building solar panels has been expensive. Thus, the cost per kWh has been high. Technical developments over the past few years have greatly reduced these costs and some experts predict the cost will soon come down to, or below, 3.5¢/kWh. Building a PV solar farm requires a sizable amount of land which, in some locations, can significantly add to construction costs.

Renewable Energy from Geothermal (heat) Sources

The core of the earth is massive and hot. Thus, it is a source of geothermal energy that will be available for eons. A lot of geothermal energy is used to heat buildings. To produce electric energy, geothermal energy is used to drive a steam turbine/3-phase, 60 Hz electric generator. The output of the turbine/generator is connected to a grid after its output frequency and voltage are adjusted to the correct values. As mentioned before, adjustment is made by varying the amount of energy input into the turbine and the magnetic strength of the generator's field poles.

The ease and cost of obtaining geothermal energy suitable for driving a steam turbine varies tremendously from place to place. In some areas it is available at, or very close to, the surface of the earth. In other locations, holes over a mile deep must be drilled to find steam or water hot enough to produce steam when pumped to the earth's surface, or core material (rocks) that are hot enough to produce steam when water is injected into the hole. Of course, the cost of drilling these deep holes will vary with the type of material that must be drilled through.

Because of the variables mentioned above, the cost of converting geothermal energy to electric energy has ranged from about 4¢/kWh to 30¢/k/Wh. New advances in the use of lower temperatures to produce steam (vapor) from liquids other than water are predicted to lower the range of cost per kWh. Lower temperature geothermal energy can be obtained from shallower holes. This will significantly lower the cost/kWh.

Glossary of Terms and Symbols

Term	Definition	Symbol or Abbreviation
Alternation	One-half of a cycle	
Ammeter	Device used to measure current	
Ampere	Base unit of current (coulomb per second)	A
Ampere-hour	Unit used to show energy storage capacity of cell or battery	Ah
Ampere-turn	Base unit of magnetomotive force	$A \cdot t$
Ampere-turn per meter	Base unit of magnetic field strength	$A \cdot t/m$
Apparent power	The product of current times voltage in a circuit containing reactance and resistance	
Atom	Building block of all matter	
Bandwidth	The space, expressed in hertz, between the lowest and highest frequencies an LC circuit will respond to with at least 70.7 percent of its maximum response	BW
Battery	Two or more cells electrically connected together	
Bemf	Back electromotive force; another name for cemf	
Bimetallic strip	Sandwich of two metals with different coefficients of expansion	
Capacitance	Ability to store energy in the form of electric energy	C
Capacitive reactance	The opposition capacitance offers to alternating current	X_c
Capacitor	An electrical component that possesses capacitance	
Cell	Chemical system that produces dc voltage	
Cemf	Counter electromotive force; the voltage produced by self-inductance	
Charge	Electrical property of electrons and protons	Q
Choke	Another name for an inductor	
Circuit breaker	Device that protects a circuit from excessive currents	
Clamp-on meter	An ammeter used to measure current without physically interrupting the circuit	
Coefficient of coupling	Denotes the portion of flux from one coil that links with another coil	k
Coil	Another name for an inductor	
Compound	Matter composed of two or more elements	
Condenser	Outdated name for capacitor	
Conductance	Ability to conduct current	G
Conductors	Materials that have very low resistivity	
Constant current source	A power source that maintains a constant terminal current for all load values	
Constant voltage source	A power source that maintains a constant terminal voltage for all load values	
Continuity	Continuous path for current	
Copper loss	Refers to the conversion of electric energy to heat energy in the windings of magnetic devices	
Core loss	Refers to the conversion of electric energy to heat energy in the core material of magnetic devices	
Coulomb	Base unit of charge (6.25×10^{18} electrons)	C

Term	Definition	Symbol or Abbreviation
Current	Movement of charge in a specified direction	I
Current carrier	Charged particle (electron or ion)	
Current transformer	A transformer used to extend the range of an ac ammeter	
Cycle	That part of a periodic waveform that does not repeat itself	
Delta connection	A method of connecting a three-phase system so that the line and phase voltage are equal	
Determinant	A number calculated from the data in a matrix.	D
Dielectric	Insulation used between the plates of a capacitor	
Dielectric constant	A number that compares a material's ability to store energy to the ability of air to store energy	
Dissipation factor	A number obtained by dividing resistance by reactance. The reciprocal of Q. Used to indicate relative energy loss in a capacitor	DF
Eddy current	Current induced into the core of a magnetic device. Causes part of the core losses	
Electric field	Invisible field of force that exists between electric charges	
Electrical degree	One three-hundred-sixtieth ($\frac{1}{360}$) of an ac cycle	
Electron	Negatively charged particle of the atom	
Element	Matter composed entirely of one type of atom	
Energizing current	The primary current in an unloaded transformer	
Energy	Ability to do work	W
Farad	Base unit of capacitance; equal to one coulomb per volt	F
Filter	A circuit designed to separate one frequency, or group of frequencies, from all other frequencies	
Fluctuating direct current	Direct current that varies in amplitude but does not periodically drop to zero	
Flux	Lines of force around a magnet	ϕ
Flux density	Amount of flux per unit cross-sectional area	B
Free electrons	Electrons that are not attached (held) to any atom	
Frequency	Rapidity with which a periodic waveform repeats itself	f
Fuse	Device that protects a circuit from excessive currents	
Hall-effect voltage	Voltage produced on opposing surfaces of a current-carrying plate that is in a magnetic field.	
Henry	Base unit of inductance; one volt of cemf for a current change of one ampere per second	H
Hertz	The base unit of frequency. One cycle per second.	Hz
Horsepower	Unit of power (1 hp = 746 watts)	hp
Hydrometer	Device used to measure specific gravity	
Hysteresis	Magnetic effect caused by residual magnetism in ac-operated magnetic devices; causes part of the core losses; also causes flux to lag behind the magnetomotive force	
Impedance	The total opposition of a circuit consisting of resistance and reactance. The base unit is ohm.	Z
Induced voltage	Voltage created in a conductor when the conductor interacts with a magnetic field	
Inductance	Electrical property that opposes changes in current	L
Inductive kick	The high cemf that is generated when an inductive circuit is opened	
Inductive reactance	The opposition inductance offers to alternating current	X_L
Inductor	An electrical component that possesses inductance	
Insulators	Materials that have very high resistivity	
Internal resistance	Resistance contained within a power or energy source	

Term	Definition	Symbol or Abbreviation
Ion	An atom that has an excess or deficiency of electrons	
Iron-vane meter	A type of meter that uses a moving iron vane and a stationary coil; commonly used to measure alternating current	
Joule	Base unit of energy (newton-meter)	J
Kilowatthour	Unit of energy (1 kWh = 3,600,000 joules)	kWh
Lenz's law	States that the cemf will always oppose the force that created it	
Litz wire	A special wire used to reduce the skin effect	
Load	Device that converts electric energy to some other form of energy	
Magnetic field strength	Amount of magnetomotive force per unit length. Other terms for magnetic field strength are magnetic field intensity and magnetizing force	H
Magnetomotive force	Force that creates a magnetic field	mmf
Matrix	A group of related numbers arranged in rows and columns; a square matrix has an equal number of rows and columns	| |
Meter loading	Changes in circuit current or voltage caused by putting a meter in a circuit	
Multimeter	Electric instrument designed to measure two or more electrical quantities	
Multiplier	Precision resistor used to extend the voltage range of a meter movement	
Mutual inductance	Inductance caused by the flux from one circuit inducing a voltage into another circuit	
Neutron	Electrically neutral portion of an atom	
Node	Any point in a circuit where two or more components are joined	
Norton's theorem	A method for reducing a circuit to a two-terminal current source	
Nucleus	Center of atom, which contains protons and neutrons	
Ohm	Base unit of resistance (volt per ampere)	Ω
Ohmic resistance	The dc resistance of an inductor	
Ohm-meter	Base unit of resistivity	$\Omega \cdot m$
Ohmmeter	Device used to measure resistance	
Period	Time required to complete one cycle	T
Permeability	Ease with which flux is created in a material	μ
Phase	A time relationship between two electrical quantities	ϕ
Phase shift	The result of two waveforms being out of step with each other	
Phasing	Interconnecting transformer, generator, or motor windings so that they have the correct time (phase) relationships between them	
Phasor	A line representing alternating current or voltage at some instant of time	
Photoconductive	Changing conductance, or resistance, by changing the light energy level	
Piezoelectric	Producing voltage by applying pressure to a crystal	
Polarity	Electrical characteristic (negative or positive) of a charge	
Polarization	Accumulation of gas ions around electrode of a cell	
Potentiometer	Three-terminal variable resistor	
Power	Rate of doing work or using energy	P
Power factor	The cosine of theta; the ratio of the true power over the apparent power; also equal to resistance divided by impedance	PF
Primary cell	Cell that is not meant to be recharged	
Proton	Positively charged particle of the atom	
Pulsating direct current	A direct current that periodically returns to zero	

Term	Definition	Symbol or Abbreviation
Quality	A number that indicates the ratio of reactance to resistance	Q
Reactance	Opposition to current, which converts no energy (uses no power); a property of both inductance and capacitance	X
Reactor	Another name for an inductor	
Rectification	The process of converting ac to pulsating dc	
Relative permeability	Permeability of a material compared with permeability of air	μ_r
Reluctance	Opposition to creation of magnetic flux	$\mathcal{R}$
Residual magnetism	Magnetic flux left in temporary magnet after magnetizing force has been removed	
Resistance	Opposition to current, which converts electric energy into heat energy	R
Resistivity	Characteristic resistance of a material (resistance of a cubic meter of the material)	
Resonance	A circuit condition in which $X_L = X_C$	
Resonant frequency	That frequency at which $X_L = X_C$ for a given value of L and C	f_r
Rheostat	Two-terminal variable resistor	
Secondary cell	Cell that can be recharged	
Selectivity	The ability of a circuit to separate signals (voltages) that are at different frequencies	
Self-inductance	The process of a conductor inducing a cemf in itself	
Semiconductor	An element with four valence electrons	
Service factor	A safety factor rating for electric motors	SF
Shunt	A parallel branch or component; precision resistor used to extend current range of a meter movement	
Siemens	Base unit of conductance (ampere per volt)	S
Sine wave	A symmetrical waveform whose instantaneous value is related to the trigonometric sine function	
Skin effect	Concentration of current near the surface of a conductor, which causes the resistance of the conductor to increase with increased frequency	
Slip	Percentage of difference between synchronous and operating speed of a motor	
Specific gravity	Weight of a substance compared with weight of equal volume of water	
Superposition theorem	A method for determining unknown currents in complex multiple-source circuits	
Tank circuit	A parallel LC circuit	
Temperature coefficient	Number of units change per degree Celsius change from a specified temperature	
Tesla	Base unit of flux density	T
Theta	The phase angle between phasors	θ
Thevenin's theorem	A method for reducing a circuit to a two-terminal voltage source	
Three-phase	Three voltages or currents displaced from each other by 120 electrical degrees	$3\text{-}\phi$
Time constant	The time a capacitor requires to charge or discharge (through a resistor) 63.2 percent of the available voltage	T
Torque	Force $\times$ distance, which produces, or tends to produce, rotation	
Transformer losses	Power losses in a transformer; caused by hysteresis, eddy current, and winding resistance	
Trigonometric functions	The relationships between the angles and sides of a right triangle. The three common functions are sine, cosine, and tangent.	

Term	Definition	Symbol or Abbreviation
Turns-per-volt ratio	The number of turns for each volt in a transformer winding	
Valence electrons	Electrons in outermost shell of an atom	
Vector	A line that represents both the direction and the magnitude of a quantity. Vectors of electrical quantities are called phasors	
Vibrating-reed meter	A meter used to measure frequency—especially the low frequencies used in power systems	
Volt	Base unit of voltage (joule per coulomb)	V
Voltage	Potential energy difference (electrical pressure)	V
Voltampere	Base unit of apparent power	VA
Voltmeter	Device used to measure voltage	
Voltmeter sensitivity	Ohms-per-volt rating of voltmeter. Numerically equal to reciprocal of full-scale current of the meter movement used in voltmeter	
Watt	Base unit of power (joule per second)	W
Watthour	Unit of energy (3600 joules)	
Wattmeter	An electrical meter that measures true power	
Wattsecond	Unit of energy (1 joule)	
Weber	Base unit of flux	Wb
Wheatstone bridge	Circuit configuration used to measure electrical qualities such as resistance	
Wye connection	Connecting the phases of a three-phase system at a common point so that the line and phase currents are equal	

Photo Credits

Preface

Page **xiii** *(left)*: ©Cindy Lewis; **xiii** *(right)*: ©Lou Jones

Chapter One

Page **3** *(top)*: JGI/Getty Images; **3** *(bottom left)*: NOAA Photo Library, NOAA Central Library; OAR/ERL/ National Severe Storms Laboratory (NSSL); **3** *(bottom right)*: Adam Crowley/Getty Images; **4** *(top)*: ©Corbis; **4** *(bottom)*: ©Matt Meadows; **7** *(left)*: ©Royalty-Free/ Corbis; **7** *(right)*: ©Bettmann/Corbis; **9**: PHIL NOBLE/ Reuters/Corbis; **11**: ©Bettmann/Corbis; **12** *(left)*: ©Roger Ressmeyer/CORBIS; **12** *(top and bottom right)*: ©Royalty-Free/Corbis

Chapter Two

Page **18**: Brown Brothers; **21**: Mary Evans Picture Library; **25**: ©Mark Steinmetz/Amanita Pictures; 26: Courtesy of Westinghouse Electric Corp.; **27**: ©Siede Preis/Getty Images; **30**: ©Bettmann/Corbis; **34**: Deutsches Museum, Munich; **36**: ©Bettmann/Corbis

Chapter Three

Page **45**: Brown Brothers; **48**: ©Bettmann/Corbis; **53**: Simpson Electric Company; **54**: Simpson Electric Company; **55**: Simpson Electric Company; **56** *(all)*: Courtesy of Fluke Corporation. Reproduced with permission.

Chapter Four

Page **68** *(top)*: Exide Power Systems; **68**: *(bottom)* ©Tony Freeman/PhotoEdit; **70**: ©Richard Fowler; **76**: ©Matt Meadows; **77**: ©Cindy Schroeder; **78**: ©Mark Steinmetz/Amanita Pictures; **79**: Electric Power Research Institute; **80**: ©Andrew Lambelt Photography/Photo Researchers, Inc.; **81**: ©Mark Steinmetz/Amanita Pictures; **82**: ©Mark Steinmetz/Amanita Pictures; **84**: ©Mark Steinmetz/Amanita Pictures; **87** *(top)*: ©Mark Steinmetz/ Amanita Pictures; **87** *(bottom)*: ©Matt Meadows; **88** *(both)*: ©Richard Fowler; **89**: ©Mark Steinmetz/Amanita Pictures; **90**: ©Mark Steinmetz/Amanita Pictures; **91** *(top left)*: ©Mark Steinmetz/Amanita Pictures; **91** *(bottom left)*: ©Mark Steinmetz/Amanita Pictures; **91** *(right)*: L.S. Starrett; **95** *(all)*: Courtesy of Littelfuse, Inc.; **96**: ©Mark Steinmetz/Amanita Pictures; **97** *(left)*: Courtesy of Raychem Corp; **97** *(right both)*: ©Mark Steinmetz/Amanita Pictures; **98**: Courtesy of Eaton Corporation

Chapter Five

Page **108**: ©Doug Martin/Photo Researchers, Inc.; **115**: ©Bettmann/Corbis.

Chapter Six

Page **144**: ©Bettmann/Corbis; **172**: Courtesy of W. Atlee Burpee

Chapter Seven

Pages **178**: ©Richard Fowler; **179** *(all)*: ©Richard Fowler; **180** *(all)*: ©Richard Fowler; **182** *(all)*: ©Richard Fowler; **183**: ©Richard Fowler; **187**: ©Mark Steinmetz/ Amanita Pictures; **188** *(both)*: ©Richard Fowler; **189** *(both)*: ©Richard Fowler; **190** *(all)*: ©Richard Fowler; **192**: ©Bettmann/Corbis; **193**: ©Bettmann/Corbis; **195**: ©Bettmann/Corbis; **196** *(left)*: ©Baldwin H. Ward & Kathryn C. Ward/Corbis; **196** *(right)*: ©Spencer Grant/ PhotoEdit; **198** *(both)*: ©Richard Fowler

Chapter Eight

Page **208**: ©Bettmann/Corbis; **217**: ©Andy Sacks

Chapter Nine

Page **236**: B. Hathaway for Rod Millen Special Vehicles, www.rodmillen.com

Chapter Ten

Page **253**: ©Don Mason/Corbis; **257** *(all)*: ©Mark Steinmetz/Amanita Pictures; **259**: ©Mark Steinmetz/Amanita Pictures; **261**: High Energy Corp.

Chapter Eleven

Page **285**: ©Novisti/Sovfoto; **287**: ©Bettmann/Corbis; **289** *(all)*: ©Mark Steinmetz/Amanita Pictures; **290** *(top left)*: ©Richard Fowler; **290** *(bottom left)*: ©Matt Meadows; **290** *(top right)*: Courtesy of J. W. Miller Magnetics

Chapter Twelve

Page **320**: Electric Power Research Institute; **321**: Courtesy of J. W. Miller Magnetics

Chapter Thirteen

Page **348**: Elco, The Electric Launch Company

Chapter Fourteen

Page **380**: Courtesy of Reliance Electronic; **381**: Courtesy of Reliance Electronic; **382** *(all)*: ©Matt Meadows; **385**: ©Bob Daemmrich; **390**: ©Matt Meadows; **392**: North American Mallory Capacitor Company; **395** *(left)*: ©Richard Fowler; **386** *(right)*: ©Matt Meadows; **396**: ©Matt Meadows

Chapter Fifteen

Page **417** *(both)*: Simpson Electric Company; **418**: Courtesy of Fluke Corporation. Reproduced with permission.; **423**: Wolf H. Hilbertz

Sixteen

Pages **444**: ©Matt Meadows; **448**: ©Matt Meadows; **450**: ©Matt Meadows; **455**: ©Mark Steinmetz/Amanita Pictures

Appendix photos

Page **474**: Courtesy Xcelite, Cooper Industries; **476**: Courtesy of Ideal Industries; **477** *(top)*: Courtesy of Ideal Industries; **477** *(all)*: PACE, Inc.; **497**: Reproduced by permission of Tektronix, Inc.

Index

A

A · t (ampere-turn), 192
ac bridge, 432–433
ac circuits, 231–246
    capacitors in, 266–270
    inductors in, 294–299
    power factor, 242–243
    power in out-of-phase, 232–240
    power in resistive, 231–232
    three-phase circuits, 244
    true power and apparent power, 240–242
ac generator, 214–216
ac motors, 350, 372
Active source (superposition theorem), 156
Adjacent side, 238
Adjustable resistors, 80, 81
AFCI (arc-fault circuit interrupter), 458–459
Ah (ampere-hours), 66
Air-core inductor, 287, 288
Air-core transformers, 321
ALCI (appliance leakage current
    interrupter), 458
Algebraic addition, 156
Algebraic sum, 428
Alkaline-manganese dioxide cells, 73
Allen keys, 477
Allen tip screwdrivers, 474
Allen wrench, 477
Alloys, resistivity of, 491
Alnico, 185
Alternating current (ac), 205–227; see also
    ac circuits
    ac generator, 214–216
    advantages of, 216
    defined, 20
    history of, 195
    quantifying, 207–211
    sine wave, 206, 211–214
    terminology, 205
    three-phase, 217–226
    waveforms, 205–207
Alternating (ac) voltage, 205
Alternation (alternating current), 207
American wire gage (AWG), 91
Ammeters, 53, 416–419, 422; see also
    Volt-ohm-milliammeters (VOMs)
Ampere, 21–23
Ampère, André Marie, 21
Ampere-hours (Ah), 66
Ampere-turn (A · t), 192
Amplification, 112
Amplitude
    of ac waveform, 209–211
    conversions, 486
Analog ammeters, 416–419
Analog meters, 413
Analog multimeters, 54–56

Analog ohmmeters, 424–425
Analog panel meters, 53
Analog voltmeters, 420–421
Apparent power, 240–242
Appliance leakage current interrupter (ALCI),
    458
Approximations, 110–111
Arc-fault circuit interrupter (AFCI), 458–459
Arcing (switches), 89, 459
Armature, 198, 215
Armature coil, 196
Armature winding, 398
Assumed polarity, 144
Atoms, 6
Audio transformers, 322
Autoranging, 56
Autotransformers, 322
Average value (V_{av}), 209
AWG (American wire gage), 91

B

Back electromotive force (bemf), 285
Balanced bridge, 426
Balanced delta, 220
Balanced load, 428
Balanced wye, 222
Ball peen hammers, 479
Band-pass filters, 365–366
Band-reject filters, 365–366
Bandwidth (BW), 360–362, 486
Base 10 number system, 33
Base units, 2
Bastard files, 480
Batteries
    capacity, 66
    defined, 65
    and internal resistance, 66–67
    ratings of, 66
Bayonet base (lamps), 76
bemf (back electromotive force), 285
Bimetal strip, 77, 464
Bits, 478–479
Black tape, 445
Blackouts, 54
Blowing characteristic (fuses), 95
Boats, 348
Bohr, Niels, 11
Box-end wrench, 476
Branch (parallel circuits), 114, 117
Branch currents, 115
Bridge circuits, 142–146
Bristol head screwdrivers, 474
Brushes (generators), 190, 196, 215
Brushless dc motors, 400–403
Bucking, 329
Busbar wire, 92–93
BW (bandwidth), 360–362, 486

C

Cables, 90–93
    low-voltage, 461
    residential wiring, 444
Capacitance; see also Capacitors
    factors determining, 254–255
    formulas, 486
    measuring, 431–433
    schematic symbols, 262
    terminology, 250
    undesired, 276
    unit of, 252–254
    voltage rating, 252
Capacitive reactance, 231, 267–269, 273, 486
Capacitor bridge, 432
Capacitor motors
    capacitor-start motor, 391–392
    permanent split, 393–394
    two-value, 393–394
Capacitors, 256–278
    ceramic, 259
    as circuit components, 98
    codes, 261, 494–496
    in dc circuits, 262–266
    detecting faulty, 275–276
    electrolytic, 256–258
    energy-storage, 260–261
    film, 258–259
    mica, 259
    motor-start, 392
    paper, 258–259
    in parallel, 273–274
    in series, 271–273
    specific-use, 259–260
    specifications, 276
    super-, 261
    uses of, 276–278
Carbon-composition resistors, 83
Carbon-film resistors, 84
Carbon-zinc cells, 72–73
Cars, electric, 236
Cell capacity, 66
Cells; see also specific types
    defined, 65
    and internal resistance, 66–67
    ratings of, 66
cemf (counter electromotive force), 284–285
Center punch, 479
Center-tapped secondaries, 326
Centrifugal switch actuator, 390
Cents per kilowatthour, 51–52
Ceramic capacitors, 259
Cermet, 83–84
Changing flux, 191
Charge; see Electric charge
Charge/discharge cycles, 71
Charged capacitor, 251